Classical and Recent Aspects of Power System Optimization

AF763889

Classical and Recent Aspects of Power System Optimization

Edited by

Ahmed F. Zobaa

Shady H.E Abdel Aleem

Almoataz Youssef Abdelaziz

ELSEVIER

ACADEMIC PRESS

An imprint of Elsevier

Academic Press is an imprint of Elsevier
125 London Wall, London EC2Y 5AS, United Kingdom
525 B Street, Suite 1650, San Diego, CA 92101, United States
50 Hampshire Street, 5th Floor, Cambridge, MA 02139, United States
The Boulevard, Langford Lane, Kidlington, Oxford OX5 1GB, United Kingdom

© 2018 Elsevier Inc. All rights reserved.

No part of this publication may be reproduced or transmitted in any form or by any means, electronic or mechanical, including photocopying, recording, or any information storage and retrieval system, without permission in writing from the publisher. Details on how to seek permission, further information about the Publisher's permissions policies and our arrangements with organizations such as the Copyright Clearance Center and the Copyright Licensing Agency, can be found at our website: www.elsevier.com/permissions.

This book and the individual contributions contained in it are protected under copyright by the Publisher (other than as may be noted herein).

Notices

Knowledge and best practice in this field are constantly changing. As new research and experience broaden our understanding, changes in research methods, professional practices, or medical treatment may become necessary.

Practitioners and researchers must always rely on their own experience and knowledge in evaluating and using any information, methods, compounds, or experiments described herein. In using such information or methods they should be mindful of their own safety and the safety of others, including parties for whom they have a professional responsibility.

To the fullest extent of the law, neither the Publisher nor the authors, contributors, or editors, assume any liability for any injury and/or damage to persons or property as a matter of products liability, negligence or otherwise, or from any use or operation of any methods, products, instructions, or ideas contained in the material herein.

Library of Congress Cataloging-in-Publication Data
A catalog record for this book is available from the Library of Congress

British Library Cataloguing-in-Publication Data
A catalogue record for this book is available from the British Library

ISBN 978-0-12-812441-3

For information on all Academic Press publications visit our website at https://www.elsevier.com/books-and-journals

ELSEVIER Book Aid International Working together to grow libraries in developing countries

www.elsevier.com • www.bookaid.org

Publisher: Joe Hayton
Acquisition Editor: Lisa Reading
Editorial Project Manager: Jennifer Pierce
Production Project Manager: Sruthi Satheesh
Cover Designer: Matthew Limbert

Typeset by SPi Global, India

God's Design is Engineering's Most Wonderful Optimal Design

Contents

Contributors

Numbers in parentheses indicate the pages on which the authors' contributions begin.

Shady H.E. Abdel Aleem (185), Higher Institute of Engineering, Cairo, Egypt

Almoataz Y. Abdelaziz (185), Ain Shams University, Cairo, Egypt

Abdollah Ahmadi (33), The Australian Energy Research Institute and the School of Electrical Engineering and Telecommunications, The University of New South Wales, Sydney, NSW, Australia

Juan Manuel Alemany (1), National University of Rio Cuarto, Rio Cuarto, Argentina

Mihail Hristov Antchev (219), Section Power Electronics, Faculty of Electronic Engineering and Technology, Technical University-Sofia, Sofia, Bulgaria

Shahab Bahrami (137), Department of Electrical and Computer Engineering, University of British Columbia, Vancouver, BC, Canada

R. Balamurugan (63), Electrical and Electronics Engineering, Annamalai University, Chidambaram, India

Daniel Barbosa (255), Polytechnic School, Federal University of Bahia, Salvador, BA, Brazil

S. Baskar (99, 463), Thiagarajar College of Engineering, Madurai, India

Kianoosh Boroojeni (137), School of Computing and Information Sciences, Florida International University, Miami, FL, United States

Pierre J. Cilliers (421), University of Cape Town, Cape Town; South African National Space Agency, Hermanus, South Africa

Carlos A. Coello Coello (505), Centro de Investigacion y de Estudios Avanzados del Instituto Politecnico Nacional, Zapopan; Av. Instituto Politécnico Nacional 2508, Madero, Mexico

Michael Diekerhof (331), E.ON ERC; Institute for Automation of Complex Power Systems, Faculty of Electrical Engineering and Information Technology, RWTH Aachen University, Aachen, Germany

Mohamed Ebeed (157), Department of Electrical Engineering, Faculty of Engineering, Sohag University, Sohag, Egypt

Ricardo A.S. Fernandes (255), Department of Electrical Engineering, Federal University of São Carlos, São Carlos, SP, Brazil

Foad H. Gandoman (33), Research group MOBI—Mobility, Logistics, and Automotive Technology Research Center, Free University of Brussels, Brussels; Flanders Make, 3001 Heverlee, Belgium

M. Hadi Amini (137), Department of Electrical and Computer Engineering, Carnegie Mellon University, Pittsburgh, PA, United States

Omid Homaee (33), School of Electrical Engineering, Iran University of Science and Technology, Tehran, Iran

Sherif M. Ismael (185), Engineering for the Petroleum and Process Industries (Enppi), Cairo, Egypt

Rupamathi Jaddivada (137), Electrical Engineering and Computer Science, Massachusetts Institute of Technology, Cambridge, MA, United States

Francisco Jurado (157), Department of Electrical Engineering, University of Jaén, Linares, Spain

Salah Kamel (157), Department of Electrical Engineering, Faculty of Engineering, Aswan University, Aswan, Egypt

Farhad Kamyab (137), Department of Chemical and Process Engineering, University of Surrey, Guildford, United Kingdom

Mohammad Kiani-Moghaddam (285), Department of Electrical Engineering, Shahid Beheshti University, Tehran, Iran

Guilherme G. Lage (255), Department of Electrical Engineering, Federal University of São Carlos, São Carlos, SP, Brazil

L. Lakshminarasimman (63), Electrical and Electronics Engineering, Annamalai University, Chidambaram, India

Jeremy Lin (1), DNV-GL Energy Advisory Group, Dallas, TX, United States

Sajad Madadi (407), Faculty of Electrical and Computer Engineering, University of Tabriz, Tabriz, Iran

Fernando Magnago (1), National University of Rio Cuarto, Rio Cuarto, Argentina

Hani Mavalizadeh (33), School of Electrical Engineering, Iran University of Science and Technology, Tehran, Iran

Miguel A. Medina (505), Centro Nacional de Control de Energia, Mexico

Sakshi Mishra (137), Transmission Planning, American Electric Power, Tulsa, OK, United States

Behnam Mohammadi-Ivatloo (19, 389, 407), Faculty of Electrical and Computer Engineering, University of Tabriz, Tabriz, Iran

Antonello Monti (331), E.ON ERC; Institute for Automation of Complex Power Systems, Faculty of Electrical Engineering and Information Technology, RWTH Aachen University, Aachen, Germany

Morteza Nazari-Heris (19, 407), Faculty of Electrical and Computer Engineering, University of Tabriz, Tabriz, Iran

Hamidreza Nazaripouya (361), Electrical and Computer Engineering, Winston Chung Global Energy Center, University of California, Riverside, USA

Mário Oleskovicz (255), Department of Electrical and Computer Engineering, University of São Paulo, São Carlos, SP, Brazil

David T.O. Oyedokun (421), University of Cape Town, Cape Town, South Africa

M. Pesaran Hajiabbas (389), Faculty of Electrical and Computer Engineering, University of Tabriz, Tabriz, Iran

S. Rajasekar (99, 463), NEC Laboratories Singapore, NEC Asia Pacific Pte. Ltd, Singapore, Singapore

Juan M. Ramirez (505), Centro de Investigacion y de Estudios Avanzados del Instituto Politecnico Nacional, Zapopan, Mexico

Sebastian Schwarz (331), E.ON ERC; Institute for Automation of Complex Power Systems, Faculty of Electrical Engineering and Information Technology, RWTH Aachen University, Aachen, Germany

Adel M. Sharaf (33), Engineering and Research Company of SHARAF Energy Systems, Incorporated, Fredericton, NB, Canada

Heidar A. Shayanfar (33), School of Electrical Engineering, Iran University of Science and Technology, Tehran, Iran

Mojtaba Shivaie (285), Faculty of Electrical Engineering and Robotic, Shahrood University of Technology, Shahrood, Iran

Guilherme Spavieri (255), Department of Electrical Engineering, Federal University of São Carlos, São Carlos, SP, Brazil

S. Tamil Selvi (99, 463), SSN College of Engineering, Chennai, India

M.J. Vahid-Pakdel (389), Faculty of Electrical and Computer Engineering, University of Tabriz, Tabriz, Iran

M. Vanithasri (63), Electrical and Electronics Engineering, Annamalai University, Chidambaram, India

Yubo Wang (361), Department of Mechanical Engineering, University of California, Los Angeles, USA

Philip D. Weinsier (285), Department of Applied Electrical Engineering, Bowling Green State University, Firelands, OH, United States

Paul Weng (137), University of Michigan-Shanghai Jiaotong University Joint Institute (UM-SJTU JI), Shanghai, China

Yinliang Xu (137), Tsinghua-Berkeley Shenzhen Institute (TBSI), Lab of Smart Grid and Renewable Energy, Shenzhen, China

Ahmed F. Zobaa (185), Brunel University London, Uxbridge, Middlesex, United Kingdom

Preface

In 1855, after the first industrial revolution, the word "optimization," or making the best of anything, was originated. In mathematics, the best solution is the maximum or minimum of certain variables subject to a set of constraints. In life, we strive for the best in all things, and we always think about how we can get better results with less effort and less cost in the least time. In power systems, optimization is the art of solving problems and deciding the best alternative following a set of criteria in less time, with less human and material costs.

In the past, mathematically based optimization methods were the primary tools in the formulation and solving of numerical problems. Currently, with the accelerated evolution of computer-based technologies and availability of user-friendly software capable of performing complex tasks with a large number of decision variables and high computational speeds, meta-heuristic algorithms are widely employed for global optimization problems.

Planning, development, management, and operation of modern power systems can be rationally considered to be long-term or short-term optimization problems. In this book, conventional and recent meta-heuristic optimization methods and algorithms being used for power system studies are presented and discussed. On one hand, the classic aspects of optimization in power systems studies (such as optimal power flow, economic dispatch, and unit commitment optimization problems) are introduced and discussed, taking into consideration the recent developments in power systems that have led to new challenges in these areas. On the other hand, hot-topic optimization problems for modern power systems and smart grids are demonstrated and discussed.

This book aims to bring together several aspects of optimization for power systems which, up to now, have been scattered in different articles and books. It also introduces innovative research outcomes, programs, algorithms, and approaches that consolidate the present and future, and opportunities and challenges of the power systems from techno-economic engineering perspectives to improve readiness for the challenges of modern and future power systems.

The book focuses on classical and recent aspects of power system optimization. It is organized into eighteen chapters, each of which begins with the fundamental structure of the problem required for a fundamental understanding of the methods introduced.

Chapter 1: This chapter summarizes the state-of-the-art of the application of optimization methods and tools in the planning and operation of power systems to aid in their efficiency. The chapter describes numerous algorithms used in these optimization methods for solving different power system problems. Several areas that deserve attention and require the utilization of new optimization methods are also mentioned. A significant challenge associated with gradually increasing system sizes tests the limits of the existing algorithms, particularly in real-time applications. Furthermore, the growing penetration of renewable energy resources in the system brings uncertainties to the current problem formulations. The use of new technologies in dealing with those issues has also been highlighted in the chapter.

Chapter 2: This chapter provides an overview of uncertainty modeling methods in power system problems with a focus on the robust optimization (RO) method. The RO method is a practical way to handle the uncertainties associated with a variety of power system parameters, including electricity price, power generation of renewable energy sources, and load demands. The power system problems studied through application of the RO method are classified, and a review is prepared for researchers in this area. Finally, the RO method is applied to a case study to familiarize the reader with RO implementation on power system problems. The simulation results are reported, and the obtained solution is analyzed to ensure the performance of the RO method in modeling the uncertainty parameters.

Chapter 3: There are many optimization problems in the world, and researchers are constantly working to develop efficient algorithms for solving them. One popular approach for solving these problems is the particle swarm optimization (PSO) algorithm. This chapter presents some efficient self-sealing/self-adjusting, adaptive, error-driven PSO algorithms. The proposed algorithms are based on the modification of inertia factor in the speed and position equations at each iteration of the PSO algorithm. The error-per-iteration rate is used to modify the inertia factor to improve convergence speed and efficiency. These methods are implemented in several test cases, and the results are presented for each method. The obtained results are compared to a conventional PSO algorithm to show the efficiency of the proposed methods. Following validation, the best algorithm is selected and utilized to solve a reactive market clearing problem.

Chapter 4: The application of the radial movement optimization (RMO) to different types of nonconvex economic dispatch (ED) problems and fixed-head hydrothermal unit problems are introduced in this chapter. Four sample test systems are considered. For all the types of ED problems, RMO provides better simulation results compared to other algorithms in terms of high-quality solution and stable convergence.

Chapter 5: This chapter presents a study of the application of the metaheuristic-based GA approach to solve power system generator maintenance scheduling (GMS) problems, taking cost criterion as its objective and reliability criterion as the constraint. The application of this GA approach has been demonstrated to solve the GMS problem for small, 3-unit, medium

5-unit, and a large 21-unit test systems using an integer-based GA algorithm. The appropriate crossover and mutation probabilities of a GA approach are determined after a number of trials for each of the test systems. The test results of GA are compared with other mathematically based approaches such as decomposition approach and dynamic programming and with that of stochastic technique MDPSO. The optimal results obtained in this chapter show that the GA approach is more robust and suitable than mathematical programming techniques, irrespective of the size of the generator maintenance scheduling problems under consideration. Furthermore, upon statistical analysis, GA is found to have a faster convergence for the complex GMS problem.

Chapter 6: In this chapter, distributed/decentralized approaches to solve the optimal power flow (OPF) problem in a power system are reviewed, with a focus on the widely used DC power flow formulation, which is a simplified power flow model with an acceptable approximation for transmission networks. The main advantage of using the DCOPF formulation is to obtain a linearized model of the power system. State-of-the-art decomposition techniques to address the scalability of the OPF analysis are reviewed. Then, application of the Lagrangian Relaxation (LR) decomposition and augmented LR in designing decentralized DCOPF algorithms are discussed. The chapter also provides an overview of consensus-based distributed optimization method to deal with the DCOPF problem in a microgrid.

Chapter 7: This chapter presents a survey related to the optimal power flow (OPF) problem. The chapter describes the OPF problem formulation, including the common objective functions of power system, control variables, and operating constraints, the conventional methods that have been employed to solve the OPF problem (including the pros and cons of these methods), and the recent optimization techniques that have been applied for OPF. In addition, the recent optimization algorithms have been categorized based on inspiration methods such as evolutionary, human, nature, bio-inspired, and physics-inspired techniques.

Chapter 8: Optimal distribution system planning and proper selection of conductors can reduce electrical power losses while complying with the technical operational limits and economic aspects. In this chapter, a comprehensive overview of the problem of optimal conductor selection is presented. Further, the grasshopper optimization (GO) algorithm is proposed to solve the problem of optimal conductor selection in radial distribution networks. The proposed algorithm is applied to two test systems. The results achieved are compared with other methods available in the literature. The comparative analysis shows the effectiveness of the proposed algorithm.

Chapter 9: This chapter presents the classical and recent aspects of active power filters for power quality improvement. Types of active power filters, power circuits for realization of active power filters, and their control systems are presented and discussed in detail.

Chapter 10: This chapter presents the formulation of the capacitor model parameterization problem when the voltage across the capacitors' terminals has harmonic distortions. The optimization-based parameterization methodology determines the parameters of an electrical model by minimizing the difference between the reference current acquired in laboratory tests and the calculated current flowing through the terminals of the parameterized power capacitor model., A methodology for acquiring the necessary reference currents in laboratory tests to realize the actual parameterization is also presented. Different cases are considered to show the effectiveness of optimization-based parameterization methodology. Such an optimization-based methodology is applicable as a prior tool to support capacitor allocation studies in power distribution systems.

Chapter 11: In this chapter, the authors present the 2D-cdEMSA, a two-level computational, multidimensional, continuous/discrete, enhanced melody search algorithm designed to improve performance of the music-inspired optimization algorithms in dealing with complex, nonconvex, large-scale, nonlinear mixed-integer optimization problems. The newly proposed 2D-cdEMSA is organized by multiple music players with a two-level improvisation. Both single and group improvisation levels are used to improvise new melodies. Using multiple musicians with different tastes, ideas, and experiences in the context of the interactive relationships among musicians can cause the musicians to choose better pitches. In addition, a midterm dynamic planning problem of the medium voltage (MV), open-loop distribution networks (DyMODN) in the presence of distributed generation resources has been employed to evaluate the performance of the 2D-cdEMSA, compared with other existing optimization algorithms. The proposed DyMODN problem, with a new point of view, modeled and scrutinized the probabilistic characteristics of the power demand needed to provide a well-designed framework of distribution network expansion planning. Investment costs—operational cost along with maintenance cost—and costumer outage costs were considered in the optimization problem as three objective functions. The DyMODN problem takes into account Kirchhoff's current and voltage rules, operational constraints on the capacities of the distribution network equipment, constraints related to the financial resources, and the logical constraints. In this chapter, the DyMODN satisfied the challenges of distribution network expansion planning, and accurately modeled the probabilistic characteristics of the power demand. In addition, the authors performed a thorough analysis in which the 2D-cdEMSA was compared with state-of-the-art optimization algorithms. Additionally, the performance of the optimization algorithms, especially the 2D-cdEMSA, was thoroughly assessed through an index of cost saving. The simulation results show that the newly proposed 2D-cdEMSA outperformed the other optimization algorithms in all indices of the DyMODN problem. Hence, the 2D-cdEMSA seems to surface as an effective optimization algorithm for future large-scale engineering optimization problems with big data.

Chapter 12: This chapter provides an overview of recent challenges and applications in demand-side management for city districts. Relevant requirements for the efficient deployment of flexibility are discussed and lead to the application of a state-of-the-art architecture and distributed optimization techniques. The work links the relevant flexibility user, such as the balance responsible party or distribution system operator, to the flexibility provider in an architecture recommended by the Smart Grids Task Force of the European Commission. As major demand-side flexibility use case within demand response, the standard portfolio coordination optimization problem on the district level is introduced and discussed. Although multiple flexibility sources are introduced, the application focuses on flexibility from electrothermal heating systems equipped with thermal storage due to their growing value in residential sectors. The first design choice from the standpoint of hierarchical architecture and other requirements is distributed optimization. Therefore, an example application of the alternating direction method of multipliers proves how an aggregation service provider can successfully balance demand and supply across temporal scales. Furthermore, derived from the requirements, uncertainty must be considered within demand response applications. In this work, model predictive control and robust optimization yield a reduction in energy deviations from the offline (day-ahead) and beforehand-scheduled planning of flexible devices compared to the uncertainty affected by intraday operation. However, the application also shows that consideration of uncertainty always comes with a certain cost, such as computational intensive calculations and increasing energy costs. The latter, though, depends on the value of charged imbalance/adjustment cost. The additional adjustment costs could overwhelm the total energy costs of robust optimization and make the robust optimization with model predictive control an appealing strategy to reduce the total energy costs, including adjustments.

Chapter 13: In this chapter, two demand-side managements (DSMs) are studied. One is for microgrids, and treats all loads and generations under the same bus using lumped models, while the other is for distribution systems and considers the power flow between different buses using the networked model. In both models, to account for the stochastic nature of DERs, a SAA-based stochastic optimization framework is proposed. For the distribution grids case, a decentralized framework is studied to alleviate the centralized computational burden. Numerical results show that on one hand, decentralized framework and SAA-based stochastic optimization are complementary. The decentralized method makes the computational demanding numerical approximation much more scalable. On the other hand, the networked model does show some benefits over the lumped model in terms of better managing the network voltage and economic operations, but at the cost of more computational complexity. For DSM designers, there is no simple answer in which DSM outperforms the other; selection is highly dependent on the application scenario.

For future researchers, asynchronous decentralized methods will be studied to account for communication delay between different nodes.

Chapter 14: The increasing number of gas-fired power plants in the electrical network and the augmentation of their share in the energy provision pattern have coupled two widespread energy systems, electricity and natural gas, to each other. Furthermore, the entrance of the new technologies such as power to gas units or microturbines among the energy hubs concept has reinforced the stated linkage. This coupling brings a valuable optimization potential despite increasing complexity. In addition, considering the natural gas constraints gives us a comprehensive and realistic view of the power system optimization problems. This chapter presents an explanation about multicarrier energy systems in addition to their formulations and simulations. The topic of multicarrier energy system is introduced. The basics of integrated energy systems formation are discussed with explanations about fundamentals of an energy hub and review of some literature in this regard. Further, the different sizes of multicarrier energy systems problems are explored. Diverse scales of these problems from a mono-energy hub optimization to widespread integrated systems investigation have been mentioned. More detailed discussion about the simultaneous optimization of electrical and natural gas networks considering their technical constraints has been further provided. In addition, a comprehensive mathematical formulation of optimal energy flow of multicarrier energy systems (electricity-gas) is presented. The simulation of proposed formulation has been introduced, and its results have been demonstrated.

Chapter 15: This chapter studied the optimal robust scheduling of MGs in the day-ahead market, where the studied MG contains hydrothermal systems, wind farms, and a pump storage unit. In addition, the studied MG is interconnected with the main grid, which makes it capable of buying power during the scheduling time horizon. The uncertainty associated with power market price has been considered and studied using robust optimization method. An effective and strong optimization technique is employed for providing the minimum operating cost of the studied MG. The obtained solutions prove the practicality and high performance of the proposed model for obtaining optimal robust scheduling of hydrothermal system, WT, and pumped storage unit.

Chapter 16: The motivation for including a chapter on geomagnetically induced currents (GIC) in the context of power system optimization is to familiarize the reader with a potential threat to power system stability and availability, which requires the inclusion of particular elements in the system to optimize the resilience of the system to this threat. The flow of GIC through transmission lines and transformers across a power network could have negative consequences. These include an increase in the reactive power demanded by GIC-laden transformers, transformers operating within the region of nonlinearity due to half-wave saturation, excessive heating in transformers leading to thermal damage, malfunction of transmission line protection schemes, and voltage problems in affected sections of the network. The first part of the chapter focuses on the origins and propagation of space weather events from the sun to the Earth and provides an overview of the indices that are used to characterize the intensity of GMDs. Some models are presented which are used for prediction of the arrival time and intensity of GMDs from solar wind parameters. The first part concludes with the means available to power system operators to be notified about space weather events that may lead to GMDs. It gives an overview of the coupling between space weather and the electrical power system. Once the coupling has been established, the resulting effect of the flow on GIC in the power system can be discussed. The second part of the chapter provides a methodology for calculating GICs in electrical power systems, and discusses aspects of modeling, measurement, and mitigation of GICs in power systems.

Chapter 17: The multi-objective problem formulation is a prevailing tool in decision making, as it analyzes all the trade-offs among the conflicting criteria. Multi-objective formulation for transformer design optimization (TDO) problem presented in this chapter aims to improve reliability, design, and efficiency of the transformer and, in turn, its life expectancy. To satisfy all the needs of various types of customers, manufacturers, utilities, etc., minimization of competing objectives, purchases cost, and total life time cost is necessary. Thermal and practical manufacturing constraints have been considered.

Chapter 18: This chapter proposed a decomposition-based multi-objective teaching learning algorithm (MOTLA/D) for solving a reactive power system problem. The effectiveness and performance of MOTLA/D were compared with respect to those of MOEA/D, which represents a state-of-the-art algorithm in three cases of study. The results indicate that the proposed algorithm was able to obtain better solutions than MOEA/D in all the analyzed cases. In this chapter, an improvement of both reactive losses and voltage stability were attained.

Finally, this book aims to introduce good practice with new research outcomes, programs, and ideas that join several aspects of optimization in power systems from different perspectives. It is a tool for the planners, designers, operators, and practicing engineers of electrical power systems who are concerned with optimization problems in power systems. Likewise, it is a key resource for advanced students, postgraduates, academics, and researchers who have some background in electrical power systems to use this material to bring themselves up to date with recent research in the field.

The Editors
Ahmed Zobaa
Shady Abdel Aleem
Almoataz Abdelaziz

Chapter 1

Optimization Methods Applied to Power Systems: Current Practices and Challenges

Jeremy Lin*, Fernando Magnago[†] and Juan Manuel Alemany[†]
[*]DNV-GL Energy Advisory Group, Dallas, TX, United States, [†]National University of Rio Cuarto, Rio Cuarto, Argentina

1 INTRODUCTION

Planning and decision-making in energy, in general, involve the allocation of a large amount of resources which significantly impacts all the actors in an economy. The main objective of planning in power systems is to define an appropriate strategy with regards to generation and transmission to efficiently utilize the system. In a general framework for energy efficiency, optimization applied to the efficient use of electric resources is essential. In the case of power systems, applied optimization is related to the planning and scheduling of the resources to aid in the system operation [1].

In this chapter, we will begin by outlining the key problems in power systems that are naturally suitable for applications of classical optimization methods to solve these problems. For example, unit commitment (UC), economic dispatch (ED), and optimal power flow (OPF) are three key problems which are critical to the economically efficient operation of the power system.

UC is a large-scale, nonconvex, mixed-integer programming problem which requires effective and efficient numerical methods to obtain its solution. ED is a step that needs to be done after completing the UC process, while the OPF solves the ED by adding additional constraints to the optimization problem [2].

In all these activities, the optimization methods emanating from the operations research field play a significant role in the decision-making process associated with UC, ED, and OPF. After outlining these problems, the optimization methods that are currently employed in solving these problems are presented. Mathematically, the optimization methods employed in these formulations are linear programming, mixed-integer programming, decomposition methods and

Classical and Recent Aspects of Power System Optimization. https://doi.org/10.1016/B978-0-12-812441-3.00001-X
© 2018 Elsevier Inc. All rights reserved.

stochastic approaches. These methods constitute a core set of optimization methods that are currently applied [3].

After presenting these optimization problems and solution methods, we will attempt to present some of the emerging challenges we face in current and future power systems. It is believed that the future grid will face some challenging developments such as the smart grid, microgrid, distributed energy resources (DER), and massive amounts of renewable energy sources (RES), to name just a few. These new developments will pose very different kinds of problems to the existing power systems.

In the final section of the chapter, we will present some of these challenges in detail and discuss the optimization methods currently being used to solve these problems. The problems described will be focused on practical applications to the real large-scale systems, including time-coupling constraints, network constraints, size and variation of the control variables, or stochastic nature of renewable resources. Because these are the frontier areas of research with uncertain results, we will also provide some speculations in terms of applications of potential optimization methods. Some innovations that incorporate energy efficiency in to the planning and operation of power systems with high renewables penetration will also be described. New concepts and techniques necessary to manage uncertainty in the models and algorithms that are commonly developed for deterministic conditions will be included as well.

2 KEY SCHEDULING PROBLEMS IN POWER SYSTEM OPERATION

From a general perspective, the power system economic operation involves the operation of generation, transmission, and distribution subsystems with the objective of operating these subsystems efficiently and fulfilling the requirement of maintaining the balance between generation and load at all times. Moreover, the voltage at each system node and the system frequency must remain within prespecified limits. This combination of economic and technical requirements for system operation is the most important task for the system operators. The system operator needs to execute three essential calculations: the UC, the ED, and the OPF. These major decision-making problems are described in Fig. 1 from a temporal perspective.

The next sections will cover these three problems.

2.1 Unit Commitment

The UC problem can be defined as the process of predispatching generation units over a time horizon to satisfy the electricity demand and operational constraints. Typically, a solution to the UC problem is achieved by formulating and solving a specific objective such as production cost minimization [4].

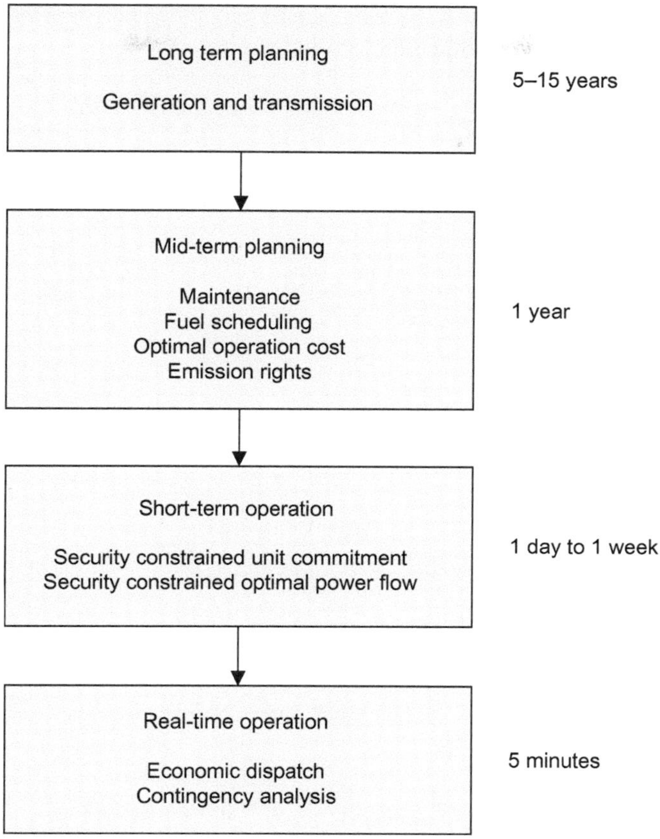

FIG. 1 Temporal perspective—major decision-making problems.

Security-constrained unit Commitment (SCUC) is an extension of the generic UC in which the network constraints are included in the UC problem to ensure that the system is secure after completing this important step. In SCUC formulations, the set of constraints generally include: generation capacity limits; minimum run and down times; ramp up and down rates; operating, regulation, and spinning reserve requirements for multiple zones; fuel constraints by generator, area or system; environmental constraints, such as SO_x and NO_x emissions; interruptible load contracts; hydro reservoir volume limits and targets; discharge limits; natural inflows; maximum spillage; delay times; forbidden zones; energy limits; maintenance schedules; inter-area and inter-zone flow limits, etc. [5].

Nowadays, the optimization techniques applied to these types of problems have been experimental and evolutionary, and as a consequence, become important players in modeling and execution of UC problems.

As a general example, a UC problem can be mathematically represented as a nonlinear optimization formulation as follows:

$$\min_{u,p} F = \sum_{t \in T} \sum_{g \in G} \left(u_{g,t} Cp_{g,t} + s_{g,t} Cup_{g,t} + z_{g,t} Cdown_{g,t} \right)$$

Subject to:
Global constraints
Power balance equation:

$$\sum_{g \in G} u_{g,t} p_{g,t} = D_t \quad \forall t = 1, \ldots, T$$

Reserve capacity:

$$\sum_{g \in G} u_{g,t} \overline{P}_g \geq D_t + R_t \quad \forall t = 1, \ldots, T$$

Local constraints
Unit generation limits:

$$\underline{P}_g \leq p_{g,t} \leq \overline{P}_g \quad \forall t = 1, \ldots, T$$

Ramp limits:

$$p_{g,t} - p_{g,t-1} \leq UR_g \quad \forall t = 1, \ldots, T$$

$$p_{g,t-1} - p_{g,t1} \leq DR_g \quad \forall t = 1, \ldots, T$$

Minimum on/off limits:

$$\left[\boldsymbol{x}_{g,t-1} - \boldsymbol{MUT}_g \right] \left[\boldsymbol{u}_{g,t-1} - \boldsymbol{u}_{g,t} \right] \geq \mathbf{0} \quad \forall \boldsymbol{t} = \mathbf{1}, \ldots, \boldsymbol{T}$$

$$\left[\boldsymbol{x}_{g,t-1} + \boldsymbol{MDT}_g \right] \left[\boldsymbol{u}_{g,t} - \boldsymbol{u}_{g,t-1} \right] \geq \mathbf{0} \quad \forall \boldsymbol{t} = \mathbf{1}, \ldots, \boldsymbol{T}$$

where, D_t is the total system load at period t [MW]; R_t is the total system reserve at period t [MW]; $Cp_{g,\,t}$ is the generation cost [$] of unit g at period t; $Cup_{g,\,t}$ is the startup cost [$] of unit g at period t; $Cdown_{g,\,t}$ is the shutdown cost [$] of unit g at period t; $\boldsymbol{u}_{g,t}$ is the status variable that indicates if unit g is on (1) or off (0) at period t; $s_{g,\,t}$ is the startup status that establishes if unit g goes into service at period t; $z_{g,\,t}$ is the shutdown status that tells if unit g changes from *on* to *off* at period t; $x_{g,\,t}$ is a positive/negative variable that sets the number of periods the unit is on/off at period t; G is the number of generators; T is the total simulation time; $p_{g,\,t}$ is the power generation of unit g at period t [MW]; $\underline{P}_g$ is the minimum power limit of unit g; $\overline{P}_g$ is the maximum power limit of unit g; MUT_g is the minimum time that unit g must remain on when it is set to *on*; $\boldsymbol{MDT}_g$ is the minimum time that unit g must remain *off* when is set to *off*; UR_g is the maximum ramp up limit of unit g and DR_g is the maximum ramp down of unit g [6].

The output from solving this mathematical formulation represents the set of generators that will be set to *on* for each period. This information is given by the value of binary variable $\boldsymbol{u_{g,t}}$.

2.2 Economic Dispatch

Once the UC process is completed, the next step is to solve the ED problem, the solution of which is also used for day-ahead, rescheduling, intraday, and real-time markets. The objective of ED problem is to minimize the total operational cost by taking into consideration the constraints related to the power balance and the unit limits. Mathematically, the ED problem can be described as follows:

$$\min_{p} F = \sum_{g \in UC} Cp_{g,t} \quad \forall t = 1, \ldots, T$$

Subject to:
Power balance equation:

$$\sum_{g \in UC} p_{g,t} = D_t \quad \forall t = 1, \ldots, T$$

Unit generation limits:

$$\underline{P}_g \le p_{g,t} \le \overline{P}_g \quad \forall t = 1, \ldots, T$$

where *UC* is a subset of *G*, representing the number of units that the prior UC problem established as committed [7].

2.3 Optimal Power Flow

The OPF solves the ED problem by adding additional constraints to the optimization problem. The key addition to the ED problem is the inclusion of network constraints. The OPF problem calculates the control and state variables which optimize a selected objective function and guarantee the secure operation of the system. Nowadays, with advances in optimization tools applied to power system problems, several new objectives and constraints are included into a traditional OPF problem. Mathematically, an OPF problem can be formulated as follows:

$$\begin{aligned} &\text{Min} \\ &f(\mathbf{x}, \mathbf{u}) \\ \text{Subject to:} \quad &\mathbf{g}(\mathbf{x}, \mathbf{u}, \mathbf{p}) = \mathbf{0} \quad \Leftrightarrow \quad \boldsymbol{\lambda} \\ &\mathbf{h}(\mathbf{x}, \mathbf{u}, \mathbf{p}) \le \mathbf{0} \quad \Leftrightarrow \quad \boldsymbol{\mu} \end{aligned}$$

where **x** is the state variable vector, **u** is the control variable vector, **p** is the parameter vector, *f* is the objective function, **g** are power balance equations, **h** are the unit and system constraints, **λ** is the Lagrange multipliers vector related to the power balance equation, and **μ** is the Lagrange multipliers vector related to the units and system constraints [8].

2.4 Fuel Scheduling

In general, mainly for thermal units, the cost of fuel represents the majority of operating costs. Restrictions on fuel contracts or the transport of fuels generally mean that the UC must be calculated after considering very restrictive conditions.

Therefore, it is essential to formulate a fuel allocation strategy that allows for the use of long-term fuel consumption constraints within short-term fuel consumption targets in which the objective is to minimize the operational cost. Also, it is important to dispatch generation while optimizing the fuel consumption by considering constraints related to generation, transport, and availability of fuel [9].

In all these activities, the optimization methods from the operations research field play a significant role in the decision-making process associated with ED, UC, OPF, and even fuel scheduling.

3 OPTIMIZATION METHODS

Several optimization techniques have been applied to solve the types of problems described in the previous sections. These methods comprise a broad range of mathematical approaches, including the use of mathematical programming algorithms such as linear and nonlinear programming, dynamic programming, and interior-point methods. Other techniques include artificial intelligence (AI) methods, such as neural networks and fuzzy systems, and evolutionary methods, such as genetic algorithms and the simulated annealing [10].

Fig. 2 illustrates the relationship between the major decision-making tools in power system and the classical optimization problems.

The methods considered in this chapter can be classified as follows:

- Linear programming (LP).
- Mixed-integer programming (MIP).
- Decomposition methods.
- Stochastic programming (SLP).
- Artificial intelligence methods (AI).

These optimization methods can be classified into three main groups: (1) Deterministic methods, (2) stochastic methods, and (3) heuristic and artificial intelligence methods.

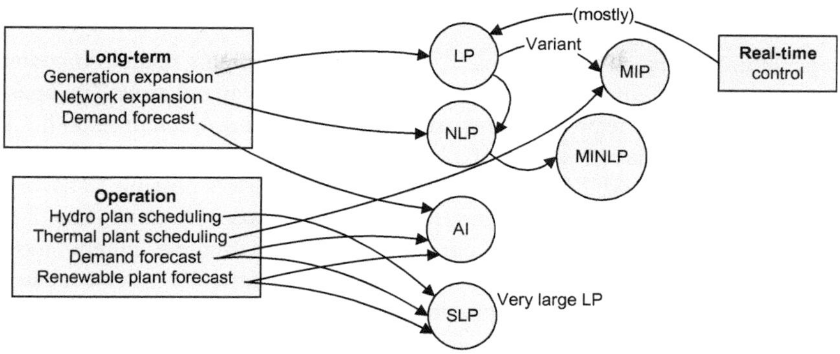

FIG. 2 PES industry and classical optimization problem relationship.

Deterministic methods include linear programming, mixed-integer programming, decomposition methods, and Lagrange relaxation techniques. Stochastic methods include Monte Carlo simulation, chance-constrained programming, Pareto curves, and risk management approaches. Finally, the genetic algorithm, particle swarm optimization method, and evolutionary methods are part of the artificial intelligence methods. Fig. 3 summarizes these methods. Although there are more methods, the most common ones are presented as examples to highlight the complexity of the problems. Several of these methods are further described in the next section.

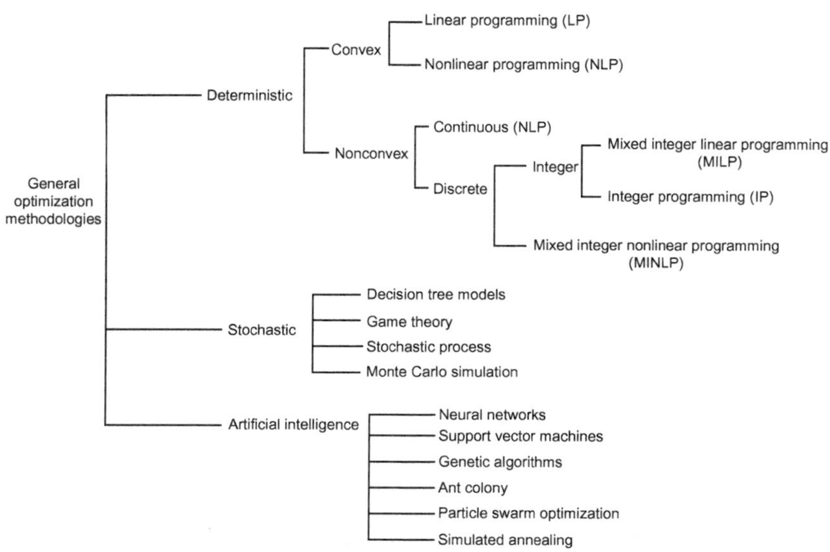

FIG. 3 Optimization classification.

3.1 Linear Programming

In general, the objective functions of optimization problems related to power systems are mainly quadratic. However, they can be linearized around an operating point to make the problem amenable for solving using linear programming techniques. The main advantage of a linear programming approach is the guaranteed solution of a well-formulated problem. Besides, it is possible to solve a large-scale problem using linear programming in a reasonable time. Also, it is the natural way to start solving a complex optimization problem. Finally, most of the commercially available optimization software, particularly those applied to solving power system problems, includes robust and powerful linear programming algorithms. The main algorithms used in the linear programming techniques are the simplex methodology, revised simplex method, sequential linear programming, and interior point techniques [8].

3.2 Mixed-Integer Programming

The implementation of algorithmic solutions based on mixed-integer programming (MIP) began in the early 1960s with the development of two classical methods: (1) Cutting Planes algorithm and (2) branch and bound (B&B) methodology. However, significant innovations related to this method have emerged only during the last 20 years. These innovations and developments have allowed for the application of these methods in real-scale problems. Most of these advances have already been implemented in commercially available programs. Some of the latest advancements are: (1) Significant progress in linear programming algorithms; (2) development of numerical methods for scattered data systems; (3) different types of cuts; (4) preresolution of the problem and nodes; (5) flexible selection of variables, and (6) heuristic techniques [11].

3.3 Decomposition Methods

The decomposition methods were proposed to efficiently solve large-scale optimization problems, such as SCUC. These methods take advantage of the special structure of the problem, by iteratively solving small-scale problems. While these methods are general, their application depends mainly on the particular problem. The decomposition methods can be classified into two forms: (1) when variables are decomposed, or (2) when constraints are decomposed.

The Benders (decomposition) algorithm is usually applied when the restrictions (aka constraints) are decomposed. Hence, the Benders method is mostly used for SCUC-type problems. The Benders method decomposes the SCUC problem into a master problem and a subproblem. The master problem, such as the conventional UC, is represented at the first level of optimization and the signals from the network, commonly referred to as *slices*, are derived from the subproblems at the second level of optimization [12].

There are two primary applications for the Benders method: (1) when variables make the problem much more complex, and (2) when the master problem and the subproblem are of a different nature. For example, a general bi-level optimization problem can be formulated mathematically as follows:

$$\begin{aligned}
&\min_{x_1,x_2} c_1{}^T x_1 + c_2^T x_2 \\
&A_1 x_1 \qquad\qquad\quad = b_1 \\
&B_1 x_1 \;+\; A_2 x_2 \;=\; b_2 \\
&x_1 \quad , \quad x_2 \;\geq\; 0
\end{aligned}$$

where x_1 are the variables considered at the first level and x_2 are the variables considered at the second level. Then, the problem can be expressed as follows:

$$\begin{aligned}
&\min_{x_1,\theta_2(x_1)} \; c^T x_1 + \theta_2(x_1) \\
&A_1 x_1 = b_1 \\
&x_1 \geq 0
\end{aligned}$$

the function $\theta_2(x_1)$ represents the objective function of the following subproblem:

$$\begin{aligned}
&\theta_2(x_1) = \min_{x_2} c_2^T x_2 \\
&A_2 x_2 = b_2 - B_1 x_1 : \pi_2 \\
&x_2 \geq 0
\end{aligned}$$

where π_2 represents the dual variables of the constraints (known as shadow prices).

The dual formulation is:

$$\begin{aligned}
&\theta_2(x_1) = \max_{\pi_2} \, (b_2 - B_1 x_1)^T \pi_2 \\
&A_2^T \pi_2 \leq c_2
\end{aligned}$$

The decomposition method above is presented as an example. For the ED problem, other decomposition techniques exist that have been successfully applied and we encourage the readers to investigate them. For example, dual decomposition, alternating direction method of multipliers, Dantzig-Wolfe reformulation, etc.

3.4 Stochastic Approaches

The stochasticity or uncertainty appears in all the power system problems described above. However, until recently, it was not possible to solve the optimization problems of large-scale systems considering this stochasticity explicitly. In this context, the uncertainty may be due to a lack of reliable data, or information about the future state of variables, such as the future demand, future prices, and the renewable energy contributions. Mathematically, as an example, a linear multistage stochastic problem can be represented as follows:

$$\min_{x_p{}^{w_p}} F = \sum_{p=1}^{P} \sum_{w_p \in \Omega_p} p_p^{w_p} c_p^{w_p^T} x_p^{w_p}$$

Subject to:

$$B_{p-1}^{w_p} x_{p-1}^{w_{p-1}} + A_p^{w_p} x_p^{w_p} = b_p^{w_p} \quad p = 1, \cdots, P$$
$$x_p^{w_p} \geq 0$$
$$B_0^{w_1} \equiv 0$$

where w represents each scenario with probability p [13].

3.5 Artificial Intelligence Methods

Among the artificial intelligence (AI) techniques, the main algorithms applied in power systems are: artificial neural networks, fuzzy logic systems, genetic algorithm, particle swarm optimization, colony optimization, simulated annealing, and evolutionary computing. Some of the distinctive properties of AI methods are: (1) the ability to remember past findings; (2) the methods learn and adapt in their subsequent performances; and (3) the methods can plan their path forward and act intelligently by mimicking human or social intelligence. Improved or hybrid AI methods have been developed by combining the advantages of various search methods [14].

4 APPLICATIONS OF OPTIMIZATION METHODS ON EMERGING CHALLENGES

4.1 Smart Grid Applications

For the optimization problem of a traditional power system, the model is typically based on the estimate of the network's state. However, the optimization problem of a Smart Grid (SG) scenario, in general, will rely on real-time information thanks to the developments of advanced metering infrastructure (AMI) and two-way communications system [15].

One important feature of the Smart Grid is the widespread installation of distributed generations (DG) which include renewable energy sources, microturbines, fuel cells, etc. From the optimization perspective, DGs introduce additional uncertainties into the model. This additional set of uncertainties makes resource scheduling decisions very challenging. For example, it is generally difficult to predict, with any accuracy, near-term or real-time wind speed and wind availability.

The potential deployment of massive DGs suggests that additional optimization areas need to be explored to solve the following problems associated with the strategic placement of DGs in the network: (1) to improve the grid reinforcement, (2) to reduce the losses and the on-peak costs, (3) to improve the voltage profiles, and (4) to improve the security, reliability, and efficiency of the system.

With regards to the model implementation, it is critical to assess the computational performance in solving the power flow problem in SGs. For example, the new communication networks provide faster updates of network information, and the incorporation of phasor measurement units (PMUs) require new analytic tools with more rapid decision-making capabilities.

Volt/Var optimization tools have become popular within the SGs infrastructure. Incorporating variables and constraints related to the reactive part of the model adds another level of complexity. One approach to simplify this problem is to implement the concept of conservative voltage reduction (CVR). The purpose of CVR is to lower the voltage utilization at the end-use consumers such that their energy consumption decreases. Among the potential benefits made possible by this scheme, the most important is peak load reduction, and consequently, the reduction in power supply cost.

New challenges related to the discrete optimization problems arise with Smart Grid networks, particularly related to topological optimization. The reconfiguration of distribution feeders may reduce the system losses, the system cost, and the network congestion. The primary purpose of the feeder reconfiguration is to change the topological structures of distribution feeders by changing the open/closed states of the devices. The key emerging developments in this area are to evaluate how to efficiently solve the underlying discrete optimization problem. Also, load forecast techniques need to be reevaluated in this area, mainly due to their granularity needs. SGs require load forecasts at feeder and housing level to be as accurate as possible.

In summary, it is important to remark that grid optimization applied to SGs strongly relies on the up-to-date technologies in the areas of telecommunication, sensing, and measurements.

4.2 Distributed Energy Resources and Massive Amount of Renewable Energy Resources

Related to the Smart Grid scenario, the increasing penetration of DER will cause significant transformations to power generation systems. The impacts of DER on SGs at the distribution domain are mostly due to the bi-directional power flows because of the installation of small DER units at medium or low voltage levels of the network [16].

Another challenging scenario related to the integration of renewable resources, such as wind, solar, biomass, hydro, etc., is the locations at which these resources will be stored. Most often, the location of these resources is far away from the consumption sites. Thus, it becomes necessary to evaluate and optimize the network facilities, which is extremely expensive.

Development of efficient energy storage devices is another significant challenge. The principal task of the storage devices is to compensate the temporal mismatches between the typically volatile renewable generation, and the energy consumption. With the development of energy storage, it becomes critical to develop optimization tools that address their efficient charge and discharge logic, which can prolong their useful life.

Among the energy storage devices, the electric vehicle (EV) is a unique resource that can be considered as either a distributed generator or a dispersed energy storage device. If the EVs are used appropriately, they can improve electrical power management and increase transportation efficiency. From the electrical power perspective, the EV inclusion in the grid can improve the voltage profiles or reduce peak load. In addition to optimal scheduling, the presence of EVs into the electricity market will require new models to be developed and incorporated into the optimization formulations. For example, such models can include centralized management of charging stations, and handle the inclusion of simultaneous EV chargers.

The optimization tools also need to be ready to incorporate intermittent energy, and be flexible enough for scalability, advanced information management, and improved tools related to predictions. Optimizing resource usage at different locations to maximize revenue is a huge challenge. Besides, Because the amount of data will increase considerably, the incorporation of intelligent big data analytics tools is imperative. Moreover, it is essential to develop sophisticated techniques which will take automated decision making in real-time into consideration when working with large-scale data storage.

4.3 Energy Efficiency

The primary objective of planning in power systems is to define an appropriate strategy for the generation, transmission, and distribution of energy in order to efficiently exploit the system. Nowadays, the level of global investment in energy efficiency is comparable to those in renewable and conventional resources. Global investment in energy efficiency is so important that it has come to be known as the primary fuel or hidden fuel. This statement suggests that energy efficiency is not only a hidden fuel but the first fuel in the world. Therefore, developments regarding energy efficiency play a fundamental role in every energy policy in any country. In a general framework for energy efficiency, optimization applied to the exploitation of electric resources is critical. Specifically, in the power systems area, the energy efficiency can be improved with the addition of renewable generation because the power supply is increased without relying on fuels. Consequently, exploitation of oil resources is reduced, emissions are reduced, and in some countries, the imports of fossil fuels are reduced [17].

In this context, developing innovations that contribute to energy efficiency in the planning and operation of power systems with high wind power penetration is a major challenge. Moreover, new concepts and techniques are needed to manage uncertainty in models and algorithms that are commonly developed for deterministic conditions.

The incorporation of wind power into power systems requires the application of more advanced methodologies to the operation programming problem. These methods include, for example, the dynamic backup reserve allocation and the use of stochastic optimization techniques. Stochastic models present significant computational challenges due to the large scale of the problem. This fact suggests that it is necessary to use decomposition techniques in stochastic problems. Furthermore, it can be useful to take advantage of the units' ability to reschedule generation for reserve allocation.

Therefore, several research proposals are put forward to improve the decomposition methodology that allows the iterative solution of two problems: (1) the first stage problem where generation is scheduled conventionally; and (2) given these results, a second stage problem where generation rescheduling mitigates sudden changes of wind generation.

Decomposition combined with rescheduling is a significant development because it would allow the development of a resolution methodology for an optimization problem which is stochastic mixed-integer and which would otherwise be unsolvable in large-scale power systems.

5 CURRENT CHALLENGES FOR OPTIMIZATION METHODS APPLICABLE TO POWER SYSTEMS

5.1 Time-Coupling Constraints

In general, the time constraints are modeled in a UC problem in which the periods are coupled by unit constraints such as ramp up/down, or minimum on/off restrictions. Other problems, such as the ED problem, are frequently formulated by assuming instantaneous loading for generation units, meaning there are no ramp up/down constraints. Nowadays, the emerging developments in power systems require that even problems previously expressed as static problems need to be formulated as dynamic problems [18].

As an illustration, let us consider the problem of the increasing requirement in operational planning for advanced voltage/VAr management solutions. These new requirements establish more automated scheduling of voltages-VArs, over a near-future time frame which can range from one interval (i.e., 1 h) to several intervals (i.e., 24 or 168 h). That need is exacerbated by the higher uncertainty, such as the penetration of renewable resources. These dispatch solutions are characterized by a multistage problem that includes inter-temporal constraints to secure a soft and realistic operational trajectory over time. As we can see, these complex optimization problems require sophisticated solution techniques. In this particular scenario, the challenge is to implement a sequence of security-constrained optimal power flow (SCOPF) calculations with full AC network models. As a result, the optimal solution needs to provide forward-looking dispatch and control variable schedules over the entire simulation time frame, without violating any network and operational constraints. Another important feature that needs to be evaluated is to allow the execution of a SCOPF for each period in sequence with a definable objective, ramping limits, and tunable movement-inhibiting penalties on specified controls (generator voltages, taps, shunts). It is also necessary to coordinate each SCOPF solution over a time horizon by enforcing time-coupling constraints, while optimizing the equipment settings over a sequence of hours for generator voltage set points by dispatchable reactive power, transformer taps, and shunts.

Concerning the objective functions, new tools must be able to handle different types of objectives. For example, one such objective would be minimizing the movement of control variables from their target values, or minimizing the losses considering the following inter-temporal constraints as a function of the control variables: (1) ramping up and down, (2) minimum on/off time, (3) number of movements at each period, and (4) number of changes over the study horizon.

From the aforementioned challenges, it can be inferred that large-scale problems that include time-coupling constraints are complex, multiperiod, mixed-integer programming (MIP) optimization problems. The solution for this type of complex problem can be found by a decomposition process comprising a sublevel problem, and a master level problem.

In the sublevel problem, the calculation of the optimal network-constrained control variable dispatch for each period can be performed. Then, a mixed-integer problem can be formulated to couple all periods by including constraints to limit control actions between the periods and the network model constraints not only for the base case but also for the contingency case. This problem can be formulated as the master level problem.

In general, this type of problem can potentially have a huge set of critical network constraints (i.e., potentially binding constraints). One of the most-widely used techniques to solve this type of problem is the Benders Decomposition, which can enforce these restrictions efficiently at the master level.

5.2 Network Constraints

Typical network constraints for pre- and all postcontingency scenarios include the bus voltage magnitude limits, the net MVAr interchanges of zones or groups of zones, MVAr reserves of any designated generator groups, weighted sums of the MVAr flows in designated sets of branch groups, voltage magnitude differences between designated pairs of buses, changes in bus voltage magnitudes between the base-case and contingency cases, and branch MVA flows [19].

Considering these constraints in the backdrop of emerging trends in power systems, it becomes necessary to develop more rigorous network optimization methods. Accurate AC network models, both in the pre- and postcontingency states must be included in the problem formulation. Most of the current optimization methods employ only DC network model formulation due to the performance and size issues. In addition, the increasing need to include variables and constraints related to the reactive power into the problem hastens the need to incorporate AC network models, because approximated DC network models for reactive-power scheduling can give incorrect results due to their inherent inaccuracy. Furthermore, they require further work to improve handling contingencies, such as an automatic recursive contingency analysis which allows capturing all system insecurities. This is because correcting existing constraint violations often creates new violations which, if ignored, can produce even worse solutions. Also, additional functions need to be added to handle solution infeasibilities. Solution infeasibility is the most difficult challenge among all other usable engineering results when not all violations are enforceable, because it requires sophisticated analytical and heuristic techniques if practical results are to be obtained. The solutions to this unique problem may include constraint relaxation techniques or incorporation of priority levels for constraints and controls.

From the mathematical point of view, the challenges are related to the nonconvexity nature of the problem caused by nonlinearities of power flow. In particular, these challenges are caused by nonconvexities due to the minimum voltage level and branch flow limits. An additional major challenge comes from the discrete control decisions which make the problem nonconvex.

5.3 Size and Variation of Control Variables

As the size of the power system grows, the number of variables in the optimization problem associated with that system are multiplied. For example, the number of generators in a large-scale real system may be in the thousands and the number of branch constraints modeled in the problem can be in the hundreds or even thousands. Due to the smart-grid development, along with the growing number of DER, future solution methods need to be able to model and find optimal solutions for problems with different types of constraints related to carbon emissions, storage resources, electric vehicles, and renewable energy sources, such as wind farms and PV installations (photovoltaic or solar power system) [20].

Multiplicity and variation of these constraints, as well as those of the state and control variables, can make the original problem too complex and too big to solve. State variables in the power system include bus voltage magnitudes, relative bus voltage angle difference with respect to the reference angle while system control variables are active and reactive power, generator voltage magnitudes, transformer taps, and phase shifters. Simply having too many generators in the system, even if they are small, will easily multiply the number of control and state variables in the optimization problem formulation which will make it very difficult, if not impossible, to solve. These types of models may require modeling constraints as quadratic, which will compound the already complex problem. In addition, it is necessary to incorporate into the model unexpected events associated with changes in system interchanges, generation failures, or device outages; these events are associated with discrete disturbances. Therefore, it becomes necessary to develop computational methods which can handle a huge amount of control variables.

5.4 Stochastic Nature of Renewable Resources

One of the main challenges raised in this chapter is related to the difficulty associated with accurately predicting wind and solar generation. This key challenge suggests the application of a stochastic approach in operating the power system without drastically changing the planning or operation scheme [21].

The main reason for this challenge is that the weather parameters that influence the operation, and hence the output of renewable resources, are difficult to predict. Consequently, the variable and uncertain nature of the wind and solar generation output make it harder to solve the problems related to the optimal operation of the power systems.

One of the solutions to deal with the challenge associated with this uncertain generation is to properly and accurately characterize the uncertainties. Normally, the uncertainty quantities are determined or estimated based on the Probability Density Functions (PDFs) because PDFs portray the most granular form of uncertainty representation. Because PDFs are based on sampling methods, it is necessary to develop techniques that efficiently handle large quantities of sampled data. Typically, the Monte Carlo method is the method that is most-widely used to deal with that problem.

Scenario tree models are another way of solving these problems. Depending on the solution methodology, this type of modeling turns the original problem into different problem formulations among which the most popular are stochastic linear programming, stochastic dynamic programming, stochastic programming with recourse, and stochastic dual dynamic programming. All these techniques are variations of a mathematical method known as multiperiod models with recourse. For this reason, the main challenge is to improve this type of methodology in order to develop better and more accurate modeling of uncertainties.

Pareto curve is another tool that can be employed for these types of problems. Pareto curve does not provide additional information that is helpful for making decisions. Instead, it drives the decision-maker to choose among the best solutions for a given set of data having a single variable without compromising the decision related to another variable. Some companies began to use the Pareto approach to study balancing reserves with the inclusion of a large amount of wind power on the system.

A new development within this type of formulations, which use probabilistic models, makes it possible to explicitly consider the risk levels that were previously assumed. The key challenge that begs further investigation is to ensure an optimal solution can be obtained. With current probabilistic methods, the optimal solutions may not be easily found. As alternatives, robust heuristics and approximation methodologies can be applied to overcome these types of problems.

6 SUMMARY

This chapter summarizes the state-of-the-art related to the application of optimization methods and tools in the planning and operation of power systems to aid in their efficiency. Numerous algorithms used in these optimization methods for solving different power system problems have been described. Several areas that deserve attention and require the utilization of new optimization methods were also mentioned. A significant challenge associated with gradually increasing system sizes tests the limits of the existing algorithms, particularly in real-time applications. Furthermore, the growing penetration of renewable energy resources in the system brings uncertainties to the current problem formulations.

The use of new technologies in dealing with those issues has also been highlighted in the chapter.

REFERENCES

[1] J. Lin, F. Magnago, Electricity Markets: Theories and Applications (978-1119179351), Wiley-IEEE Press, Hoboken, 2017.

[2] A. Wood, B. Wollenberg, G. Sheble, Power Generation, Operation and Control (0471790559), Wiley, Hoboken, 2014.

[3] J. Zhu, Optimization of Power System Operation, second ed., Wiley-IEEE Press, Hoboken, 2015 (9781118854150).

[4] H. Pinto, F. Magnago, S. Brignone, O. Alsac, B. Stott, in: Security constrained unit commitment: Network modeling and solution issues, 2006 IEEE PES Power Systems Conference and Exposition, Atlanta, GA, 2006, , pp. 1759–1766.

[5] Y. Fu, M. Shahidehpour, Z. Li, Security-constrained unit commitment with AC constraints, IEEE Trans. Power Syst. 20 (3) (2005) 1538–1550.

[6] M. Carrion, J.M. Arroyo, A computationally efficient mixed-integer linear formulation for the thermal unit commitment problem, IEEE Trans. Power Syst. 21 (3) (2006) 1371–1378.

[7] G. Kusic, Computer-Aided Power System Analysis (0131645269), Prentice Hall, Englewood Cliffs, 1986.

[8] O. Alsac, J. Bright, M. Prais, B. Stott, Further developments in LP-based optimal power flow, IEEE Trans. Power Syst. 5 (3) (1990) 697–711.

[9] M. Shahidehpour, M. Marwali, Maintenance Scheduling in Restructured Power Systems (0792378725), Kluwer Academic Publishers, Dordrecht, 2000.

[10] N. Rau, Optimization Principles: Practical Applications to the Operation and Markets of the Electric Power Industry (9780470545454), Wiley-IEEE Press, Hoboken, 2003.

[11] G. Morales-España, C.M. Correa-Posada, A. Ramos, Tight and compact MIP formulation of configuration-based combined-cycle units, IEEE Trans. Power Syst. 31 (2) (2016) 1350–1359.

[12] B. Lu, M. Shahidehpour, Unit commitment with flexible generating units, IEEE Trans. Power Syst. 20 (2) (2005) 1022–1034.

[13] Y. Degeilh, G. Gross, Stochastic simulation of utility-scale storage resources in power systems with integrated renewable resources, IEEE Trans. Power Syst. 30 (3) (2015) 1424–1434.

[14] E. Tsoi, K.P. Wong, Artificial intelligence algorithms for short term scheduling of thermal generators and pumped-storage, IEE Proc. Gener. Transm. Distrib. 144 (2) (1997) 193–200.

[15] A. Khodaei, M. Shahidehpour, Microgrid-based co-optimization of generation and transmission planning in power systems, IEEE Trans. Power Syst. 28 (2) (2013) 1582–1590.

[16] L. Che, X. Zhang, M. Shahidehpour, A. Alabdulwahab, A. Abusorrah, Optimal interconnection planning of community microgrids with renewable energy sources, IEEE Trans. Smart Grid 8 (3) (2017) 1054–1063.

[17] P. Pezzini, O. Bellmunt, A. Andreu, Optimization techniques to improve energy efficiency in power systems, Renew. Sust. Energy Rev. 15 (2011) 2028–2041.

[18] H. Pinto, J. Bright, M. Prais, B. Stott, O. Alsac, F. Magnago, in: Advanced voltage/VAr management, FERC Technical Conference, Increasing Real-Time and Day-Ahead Market Efficiency Through Improved Software, June 26–28, 2017, 2017.

[19] H. Pinto, J. Bright, M. Prais, B. Stott, in: Applying voltage/VAr management to support energy markets—practical experience, FERC Conference, Increasing Real-Time and Day-Ahead Market Efficiency through Improved Software, June 25–27, 2012, 2012.

[20] X. Ke, N. Lu, C. Jin, in: Control and size energy storage Systems for Managing Energy Imbalance of variable generation resources, Proceedings of IEEE PES general meeting, Denver, CO, July, 2015, 2015.

[21] M. Milligan, P. Donohoo, M. O'Malley, Stochastic Methods for Planning and Operating Power Systems with Large Amounts of Wind and Solar Power, UTN Libraries Government Documents Department, https://digital.library.unt.edu/ark:/67531/metadc846605/, 2012.

Chapter 2

Application of Robust Optimization Method to Power System Problems

Morteza Nazari-Heris and Behnam Mohammadi-Ivatloo
Faculty of Electrical and Computer Engineering, University of Tabriz, Tabriz, Iran

1 INTRODUCTION

The increasing penetration of renewable energy sources in power systems has introduced uncertainties of power generation output, such as wind turbines and photovoltaic systems. Accordingly, the decision-making process of operation and planning of power systems encounters a level of uncertainty. Optimization problems with uncertainty parameters are handled in different research studies by applying a variety of optimization procedures and uncertainty modeling techniques. A significant sensitivity to parameters may appear in the solutions of optimization problems, which often results to an infeasible solution or sub-optimal solutions [1]. Uncertainty of parameters can be classified in two general types, which include (a) uncertainty resulting from the difference of the real value of a parameter and the value estimated in observations or calculations, and (b) uncertainty of load demand, price of various types of energy, generation availability, and weather interruptions [2]. In traditional approaches, which include information gap decision theory (IGDT) technique [3], probabilistic methods [4], interval based analysis [5], possibilistic procedure [6], hybrid probabilistic and possibilistic methods [7], uncertain parameters of optimization problem have been challenged using probability distribution functions. Such functions have been used to produce scenarios of possible quantities of uncertain parameters, which are utilized to formulate stochastic programming problems for obtaining optimal solutions under uncertainties. However, two main disadvantages are introduced for such methods including (a) difficulties of characterizing some uncertain parameters using distribution functions, and (b) large-scale optimization problems because of the large number of required scenarios.

Classical and Recent Aspects of Power System Optimization. https://doi.org/10.1016/B978-0-12-812441-3.00002-1
© 2018 Elsevier Inc. All rights reserved.

Accordingly, the robust optimization (RO) method is defined as an alternative approach for dealing with uncertainties. This method, which does not require the probability distribution functions of uncertain parameters, is a practical tool for obtaining optimal solutions of optimization problems with uncertain parameters. The RO method aims to solve the optimization problem for finding an optimal solution with two characteristics. First, the obtained solution is feasible for any realization of uncertain parameters and second, it is optimal for worst-case realization of the uncertain parameters.

Recently, the RO method has been applied to handle different series of uncertainties, such as price uncertainty, wind power generation uncertainty, and load uncertainty, which are studied in a variety of publications. In this chapter, a comprehensive review is offered on the employment of the RO method to solve power system optimization problems. Different types of power system problems, which are solved by implementation of the RO method, are classified and studied. In addition, a case study, which is considered as an economic dispatch problem with uncertainty of power market price, is provided to familiarize the readers with an effective approach. The authors aim to analyze the simulation results of the case study to evaluate the performance and ensure the effectiveness of the RO method on power system problems.

The chapter is organized as follows: Section 2 provides a general description of the different approaches implemented to handle uncertainty and considers the advantages and disadvantages of the introduced methods. The RO method definition is provided in Section 3, which is also analyzes the characteristics of the uncertainty modeling method. Section 4 presents the classification of problems handled by robust optimization. The case study is provided in Section 5, which includes the case study definition and analysis of the provided simulation results and impact of the RO method on the optimal solution. Finally, the chapter is concluded and summarized in Section 6.

2 DIFFERENT APPROACHES IMPLEMENTED TO HANDLE UNCERTAINTY OPTIMIZATION

Different uncertainty modeling methods are utilized to handle uncertainties of parameters in power system problems. The general classification of such methods include IGDT technique, probabilistic methods, interval based analysis, possibilistic procedure, hybrid probabilistic and possibilistic methods, and the RO method. In this section, a general definition of the mentioned approaches is offered, which includes an introduction to each method as well as an analysis of the advantages and disadvantages of each method.

2.1 Information Gap Decision Theory (IGDT) Technique

This technique, which was proposed by Yakov ben Haim in 1980, does not require membership function or probability density function (PDF). The difference between what is known and what is vital to be known is the fundamental idea of IGDT technique. Opportunity function and robustness function are two basic functions of such an uncertainty modeling method [8]. This method takes advantage of employment in extreme uncertain conditions; however, complexity can be counted as disadvantage of such uncertainty modeling method. Such an uncertain modeling method is implemented to handle transmission expansion planning [9], distribution of network operations [10], solution of unit commitment problem of CHP plants [11], and planning of wind production [12].

2.2 Probabilistic Methods

This method was first introduced by Dantzig in 1955. It utilizes a determined PDF for each uncertain parameter. For instance, normal distribution function is normally used to model uncertainty of load and Weiball distribution function is utilized to model wind speed in power system problems. Probabilistic methods include Monte Carlo simulation [13], point estimate method [14], and scenario-based decision-making [15].

2.3 Interval Based Analysis

Ramon E. Moore introduced the interval based analysis method in 1966. This method assumes that the uncertain parameters are attained from a known interval. Such an approach is appropriate for parameters when lower and upper bounds are taken from an interval. The range of output variable are defined by using such an interval [16].

2.4 Possibilistic Procedure

The possibilistic procedure was introduced by Lotfi A. Zadeh in 1960. The membership function (MF) is utilized in this procedure for modeling the input parameter, and the MF of output parameters are determined using MF of input parameters [6]. Difficult calculations and a time-consuming process can be numerated as disadvantages of this uncertainty modeling technique.

2.5 Hybrid Probabilistic and Possibilistic Methods

This method, which includes both random and possibilistic parameters, is appropriate to handle objective functions with several variables. Such a method is implemented [17] to evaluate the effect of stochastic renewable and controllable power production on energy losses.

2.6 RO Method

The RO method was first introduced by Soyster in 1973. This uncertainty modeling method is appropriate for conditions when sufficient data are not available. This chapter studies the implementation of the RO method to power system problems. In Section 4, the problems handled by RO method are classified and the research studies published in this area are defined.

3 ROBUST OPTIMIZATION METHOD

The RO method is defined as an alternative modeling method for handling optimization problems with uncertain parameters. Soyster introduced the RO method in 1973. Generally, it is difficult to obtain a realistic PDF of uncertain parameters, however, the state-of-the-art of RO method is adept at solving power system problems with a series of uncertainties. Such uncertainty modeling method, which does not need the PDF or MF of the uncertain parameter, is appropriate for conditions where sufficient data is not available. Accordingly, the RO method facilitates handling uncertainties that are hard to obtain PDF. In the RO method, the optimization problem is optimized for the worst condition of the uncertain parameter.

The optimal solution obtained by implementation of the RO method has two main characteristics. Such a solution is feasible for any uncertain parameters, which only need limited data of the uncertainty parameter including mean, minimum, and maximum amounts of the uncertain parameter. Moreover, the optimal solution provided by the RO method is optimal for all realizations of the uncertain parameter because the worst-case scenario is considered in the decision-making process.

In recent research studies, a significant emphasis is placed on observing the employment of the RO method to solve power system problems with uncertain parameters. The RO method is employed to deal with uncertainties of power generation of renewable energy sources, load demand, and power market price. Uncertain parameters are introduced by so-called uncertainty sets in the RO structure, which defines the available bounds of uncertain parameters quantities. Then, the optimization problem is optimized in the worst-case point of these sets. Later in Section 5, we analyze a case study in which the RO method was implemented to deal with uncertainty of market price.

4 CLASSIFICATION OF PROBLEMS HANDLED BY ROBUST OPTIMIZATION

4.1 Unit Commitment (UC) Problem

A Unit commitment (UC) problem, which aims to obtain a unit commitment schedule with minimum cost of commitment and dispatch, is an important power system problem. Such optimization problem is defined as decision making process accomplished by power system operators in electricity power markets and vertically integrated utilities [18]. A UC problem faces new concerns such as price and demand uncertainties considering the penetration of renewable energy sources such as wind power and price responsive loads. Accordingly, the RO method is introduced as a practical technique to include such uncertainties in the solution of UC problem [19].

In Ref. [20], a two-stage adaptive RO model is introduced to deal with a security-constrained UC problem. The authors implemented the proposed RO method to a large-scale system handled by ISO New England to compare the obtained solution with respect to the current reserve adjustment technique. The comparison ensures the economic advantages and decrement of dispatch costs by implementing the proposed model. The UC problem, which aims to maximize the social welfare, is studied in Ref. [21], by applying a Benders' decomposition technique to obtain the optimal solution. Uncertainties of wind power and demand response scenarios are taken into account in this reference, which also considers the variation of wind power in a determined interval and a differing price-elastic demand curve. The UC robust optimization model for obtaining the optimal commitment schedule in the day-ahead power market is studied in Ref. [22]. In this reference, the impact of wind power uncertainty is considered to obtain the optimal solution in worst case of available wind power, which ensures a reliable power system due to consideration of worst-case scenario. In Ref. [23], contingency-constrained UC problem is handled by employing a robust-based optimization approach, which ensures the optimal solution considering simultaneous outage of generation units. The proposed novel robust-based model is used instead of considering all possible contingency states, which is solved by transforming the bi-level programming to a single programming problem. A distributional RO method is introduced in Ref. [24] for dealing with UC problem considering uncertainties associated with wind power generation. In this reference, a two-stage model is proposed for solving UC problem, in which UC decisions and economic dispatch decisions are considered as here-and-now and wait-and-see methods. In Ref. [25], the authors implemented RO model with uncertainty marginal price (UMP) for the solution of UC problem in a day-ahead market. UMP is introduced in this reference, which considers both price uncertainty and reserve, and is defined as an effective tool for a fair credit of the uncertainties and generation reserve.

4.2 Transmission Network Expansion Planning (TNEP)

The objective of transmission network expansion planning (TNEP) is the minimization of the expansion planning costs of transmission circuits, which satisfies the forecasted load of the system at a determined time horizon [26]. Load demand was first introduced as an uncertain parameter in TNEP problem. The uncertainty associated with renewable energy sources is defined as an interesting filed in the area of such power system problem.

In Ref. [27], the uncertainty model of load and wind power parameters are studied in solving the TNEP problem, where a two-stage RO method is employed. Benders decomposition technique is applied, which aims to minimize expansion plan costs and maximize the curtailment of load and renewable production in the slave program. Requirements of transmission networks for storage systems, including renewable energy sources, is discussed in Ref. [28]. This paper utilized a column-and-constraint generation algorithm and the RO method to prepare the optimal allocation and sizing of storage system in networks. A robust TNEP optimization is performed in Ref. [29], which considers the uncertainties of load demand and renewable energy power generation. Taguchi's orthogonal array testing (TOAT), which chooses representative scenarios from all the available combinations, is implemented in this reference to provide optimal testing scenarios.

4.3 Strategic Bidding of Energy Management System in Electricity Markets

The objective of solving bidding strategies for energy management systems in electricity markets is to maximize the producer profit, which participates in a pool power market [30]. Energy management systems (EMS) are responsible for deriving optimal bidding strategy in day-ahead and real-time power markets. Power market prices and renewable power production are considered as uncertainty parameters in the area of optimal strategic bidding of EMS [31].

An EMS, which handles price-responsive demand, wind power unit, and electrical energy storage unit, is studied in Ref. [32]. RO method is implemented to deal with uncertainties associated with power market price and wind power generation. A robust offering strategy for price-taker producer is proposed in Ref. [33], which considers a confidence price interval in instead of forecasting models. The RO method is implemented in Ref. [34] in an attempt to provide optimal bidding strategies for a price-taker producer, where the minimum and maximum amount of power market price are considered instead of forecasted quantities. Mixed-integer linear programming problem (RMILP) is applied in this reference for partitioning the power price intervals into subintervals, which helps to obtain optimal robust decisions considering price intervals.

4.4 Energy Generation Scheduling in Micro-Grids

Micro-grids (MGs) are appropriate solutions for integrating distributed energy resource (DER) units, and energy storage systems (ESSs), which are capable of meeting energy demands in different types of situations, including when connected to the main grid or isolated from the main grid. The objective of energy generation scheduling in micro-grids is to provide optimal generation scheduling of energy generation units and energy storage systems [35].

A robust generation scheduling of CHP-based MGs is studied in Ref. [36], which considers the uncertainties of net demand, heat demand, and power market price. In this reference, a distribution uncertainty model is applied to show the uncertainty of the net demand and heat demand. Chance constrained approximations and RO method are implemented to transform the problem and obtain the optimal solution. In Ref. [37], the authors solved energy management problem of MGs considering uncertainties of renewable energy sources and load demand. In this reference, the point estimate method is used to model uncertainties associated with wind and solar power production, and the RO method is applied to model load demand uncertainty, which resulted in optimal solutions with less operation risk. This paper aimed to minimize costs related to operation, pollution, and reliability tasks. Optimal generation scheduling of grid-connected MGs considering demand side management is studied in Ref. [38]. In this reference, the RO method is employed to address the uncertainty of power production of renewable energy sources, which used dual decomposition method for decomposing the optimization problem in subproblems. Look-ahead power dispatch of renewable energy sources is studied in Ref. [39], which aims to obtain scheduling of such sources. It is assumed that the control center receives the predicted power output of wind farm, and the allowable power output is calculated by implementing the RO method.

4.5 Economic Dispatch Problem in Hybrid Energy Systems

Hybrid energy systems are defined as the integration of several types of energy generation equipment such as electrical energy generators, electrical energy storage systems, and renewable energy sources. Hybrid energy systems may be utilized in grid-connected mode, isolated from grid, and special aims [40]. Uncertainties associated with renewable power production and load demands are considered in recent studies in the area of scheduling of hybrid energy systems.

The RO method is implemented in Ref. [41] to handle the uncertainties associated with a large-scale intermittent power supply in economic dispatch of a hybrid energy system, which includes wind turbine, photovoltaic system, hydro, and thermal generation plants. In this reference, reliability and economy are considered as the main objectives of the optimization problem, where an adjustable uncertainty budget is proposed for the RO method to obtain a reliable economic solution. The optimal solution for an economic dispatch problem for a multienergy system is studied in Ref. [42] by employment of the RO method. This research considered the realistic interdependency between the electricity and primary energy infrastructures, which provided optimal power dispatch by considering the uncertainty in wind power generation. A multiperiod robust model for economic dispatch problems of a power system with high penetration of wind generation units is proposed in Ref. [43], which contains a two-stage decision-making framework. A rolling-horizon operational structure is proposed in this reference for modeling the real-time ED process, and dynamic uncertainty sets are introduced to study dynamic relationship between uncertainties across decision stages.

5 CASE STUDY

In this section, a case study is considered to evaluate the application of the RO method in solving a power system problem.

Consider a consumer that aims to dispatch a controllable load over 24 h. The forecasted power market price is assumed as uncertain parameters, which are studied employing the RO method. The objective function in deterministic condition without implementation of RO can be stated as:

$$\max. \sum_{t=1}^{24} \left(B - \lambda_{Buy}^{t,u}\right) \times u_t \tag{1}$$

where, B is per unit benefit of power consumption, $\lambda_{Buy}^{t,u}$ is real-time power market price at time t, and u_t is power consumption at time t. The power consumption at each time should be limited to its upper bound, which can be stated as follows:

$$u_t \leq \overline{U}^h \tag{2}$$

where, $\overline{U}^h$ is maximum level of power load. The minimum and maximum total consumption limit during scheduling time horizon should be considered as follows:

$$\begin{aligned} &\sum_{t=1}^{24} u_t \leq \overline{U}^d \\ &\sum_{t=1}^{24} u_t \geq \underline{U}^d \end{aligned} \tag{3}$$

where the minimum and maximum total consumption are shown by $\underline{U}^d$ and $\overline{U}^d$, respectively. The ramp-up and ramp-down limits should also be considered in solving the problem as follows:

$$\begin{aligned} &u_t - u_{t-1} \leq R^U \\ &u_t - u_{t-1} \geq -R^D \\ &u_t \geq 0 \end{aligned} \tag{4}$$

where the ramp-up and ramp-down limits are indicated by R^U and R^D, respectively. The uncertainty of power market price is handled by using RO method. The uncertainty set for power market price is as follows:

$$U_\lambda = \left\{ \begin{array}{l} \lambda_{Buy}^{t,u} \in R^+ : \underline{\Gamma_\lambda} \leq \dfrac{\sum \lambda_{Buy}^{t,u}}{\sum \lambda_{Buy}^{t}} \leq \overline{\Gamma_\lambda}\ \forall t, \\ \lambda_{Buy}^{t,u} \in \left[\underline{\lambda_{Buy}^{t}}, \overline{\lambda_{Buy}^{t}} \right] \quad \forall t \end{array} \right\} \tag{5}$$

The power market price can vary between its lower and upper bounds, which are demonstrated by $\underline{\lambda_{Buy}^{t}}$ and $\overline{\lambda_{Buy}^{t}}$. Uncertainty budgets $\underline{\Gamma_\lambda}$ and $\overline{\Gamma_\lambda}$ are utilized to limit total deviation of the power market price in the scheduling time interval. Objective function with implementation of RO can be stated as:

$$\max.\ \sum_{t=1}^{24} (B \times u_t) - \min \sum_{t=1}^{24} \left(\lambda_{Buy}^{t,u} \times u_t \right) \tag{6}$$

The inner problem can be reformulated equivalently for dealing with min-max structure of Eq. (6). Accordingly, the inner problem can be stated as follows:

$$\begin{aligned} &\min \ \left(\lambda_{Buy}^{t} - z_{Buy}^{t} \hat{\lambda}_{Buy}^{t} \right) \times u_t \\ &\text{s.t. } z_{Buy}^{t} \leq 1 \ : \ \xi_1^t, \ \forall t, \\ &\sum_{t=1}^{T} \left(z_{Buy}^{t} \right) \leq \Gamma : \beta, \\ &z_{Buy}^{t} \geq 0 \end{aligned} \tag{7}$$

where dual variables of the corresponding constraint are shown by ξ_1^t and β. The dual of Eq. (7) can be written as follows:

$$\begin{aligned} &\max \ -\xi_1^t - \Gamma\beta - \\ &\text{s.t. } \xi_1(t) + \beta \geq \hat{\lambda}_h P_{G,Sell}^{t} \\ &\xi_1(t) \geq 0 \\ &\beta \geq 0 \end{aligned} \tag{8}$$

Replacing Eq. (8) in Eq. (6), a single-level problem can be provided as follows:

$$\begin{aligned}
\max\ & \sum_{t=1}^{24}(B \times u_t) - \sum_{t=1}^{24}\left(\lambda_{Buy}^t \times u_t\right) - \xi_1^t - \Gamma\beta - \\
\text{s.t.}\ & u_t \leq \overline{U}^h \\
& \sum_{t=1}^{24} u_t \leq \overline{U}^d \\
& \sum_{t=1}^{24} u_t \geq \underline{U}^d \\
& u_t - u_{t-1} \leq R^U \\
& u_t - u_{t-1} \geq -R^D \\
& u_t \geq 0 \\
& \xi_1(t) + \beta \geq \hat{\lambda}_{Buy}^t u_t \\
& \xi_1(t) \geq 0 \\
& \beta \geq 0
\end{aligned} \tag{9}$$

In this case study, the benefit of power consumption is considered \$100/kWh per hour. The uncertainty budget is considered equal to three in this case study. Considering the forecasted power market prices during a 24-h time period, the prices can vary between the forecasted amount and 30% more than the forecasted amount. In other words, $\overline{\lambda_{Buy}^t}$ is selected as 1.3 the forecasted amount of power market prices. Power market price during scheduling time interval is demonstrated in Fig. 1. As seen in this figure, the maximum power price of the market is related to $t = 14$ h, where the price is equal to \$120/MWh. In addition, the minimum power price, which is related to $t = 24$ h, is equal to \$67/MWh.

The proposed RO method for solving the economic dispatch problem is modeled using General Algebraic Modeling System (GAMS) software [44]. The simulation result is shown in Fig. 2. Consumption of electrical energy during the 24-h period is obtained, which shows it is economical to keep the consumption of the controllable loads between $t = 1$ h and $t = 3$ h, and between $t = 16$ h and $t = 24$ h.

As mentioned previously, the power market price can vary between the forecasted amount and 1.3 times such an amount, which is shown in Fig. 3. As seen in this figure, the consumer purchased power form the market with a price equal to \$75/MWh, which was \$67/MWh before implementation of the RO method. The objective function of the consumer, which includes power consumption

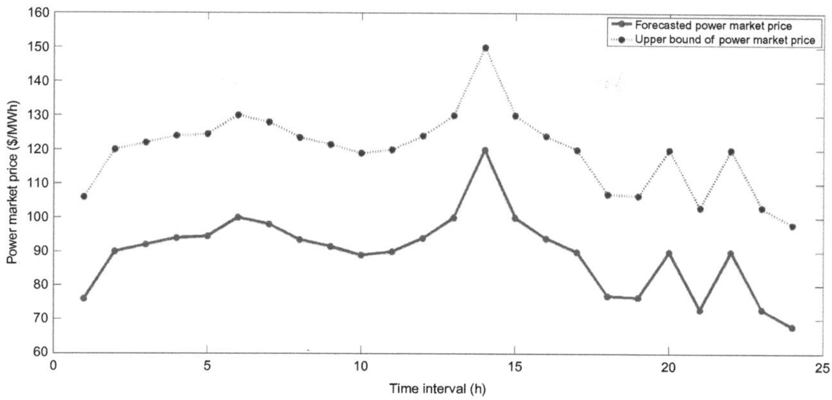

FIG. 1 Power market price.

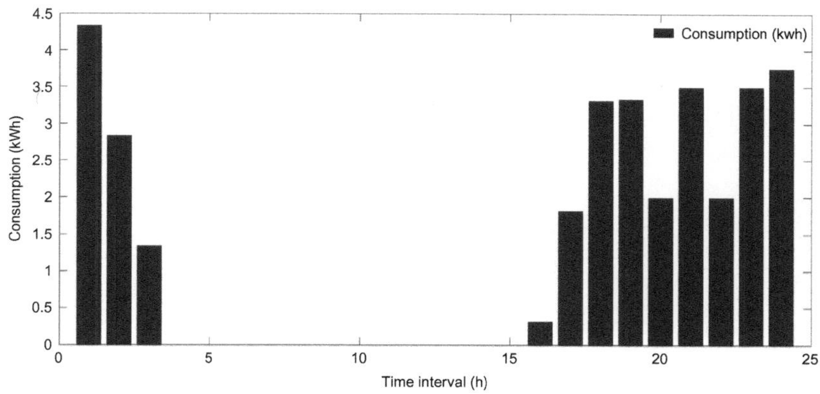

FIG. 2 Consumption of electrical energy (kwh).

cost and benefit obtained by consuming the power, is compared considering the RO method and deterministic condition, where the forecasted power market price is considered as a certain parameter. The deterministic condition is related to the condition where robust budget is zero and there is no uncertainty in power market price. Table 1 shows the comparison between objective function using the RO method and deterministic condition.

6 CONCLUSION

This chapter provided an overview of uncertainty modeling methods in power system problems with a focus on the RO method. The RO method is a practical method to handle the uncertainties associated with a variety of power system parameters including electricity price, power generation of renewable energy

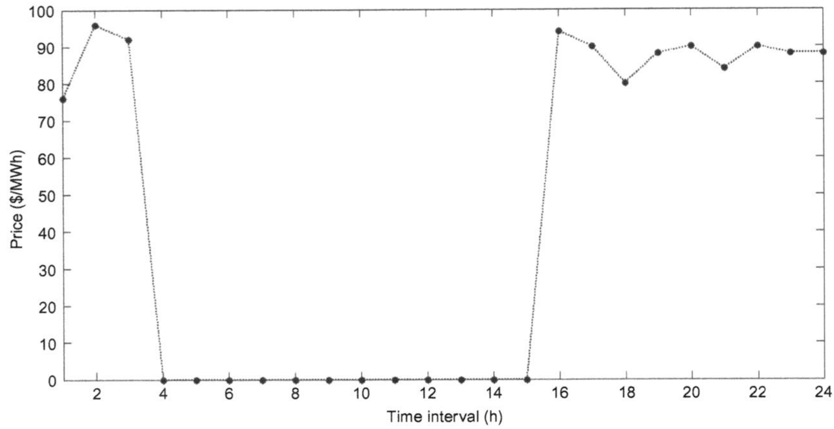

FIG. 3 Price after implementation of RO method.

TABLE 1 Comparison of the Obtained Results in Deterministic Condition and Employment of RO Method

	Implementation of RO Method ($\Gamma = 3$)	Deterministic Condition ($\Gamma = 0$)
Objective function ($)	−436.591	−758.200

sources, and load demands. The power system problems studied by application of the RO method are classified, and a review is prepared for researchers in this area. Finally, the RO method is applied to a case study to familiarize the reader with RO implementation on power system problems. The simulation results are reported, and the obtained solution is analyzed to ensure the performance of the RO method in modeling the uncertainty parameters.

REFERENCES

[1] M. Aien, A. Hajebrahimi, M. Fotuhi-Firuzabad, A comprehensive review on uncertainty modeling techniques in power system studies, Renew. Sust. Energy Rev. 57 (2016) 1077–1089.

[2] Z. Jizhong, Optimization of power system operation, in: IEEE Trans on Power Systems, 2009.

[3] M.A. McCarthy, D.B. Lindenmayer, Info-gap decision theory for assessing the management of catchments for timber production and urban water supply, Environ. Manag. 39 (4) (2007) 553–562.

[4] G.B. Dantzig, Linear programming under uncertainty, Manag. Sci. 1 (3–4) (1955) 197–206.

[5] R.E. Moore, R.B. Kearfott, M.J. Cloud, Introduction to Interval Analysis, SIAM, Philadelphia, 2009.

[6] L.A. Zadeh, Fuzzy sets, Inf. Control. 8 (3) (1965) 338–353.

[7] M. Aien, M. Rashidinejad, M. Fotuhi-Firuzabad, On possibilistic and probabilistic uncertainty assessment of power flow problem: A review and a new approach, Renew. Sust. Energy Rev. 37 (2014) 883–895.

[8] A. Nikoobakht, J. Aghaei, M. Mardaneh, Managing the risk of uncertain wind power generation in flexible power systems using information gap decision theory, Energy 114 (2016) 846–861.

[9] M. Taherkhani, S.H. Hosseini, IGDT-based multi-stage transmission expansion planning model incorporating optimal wind farm integration, Int. Trans. Electr. Energy Syst. 25 (10) (2015) 2340–2358.

[10] A.O. Connell, A. Soroudi, A. Keane, Distribution network operation under uncertainty using information gap decision theory, IEEE Trans. Smart Grid (2016). http://ieeexplore.ieee.org/abstract/document/7546906/.

[11] J. Aghaei, et al., Optimal robust unit commitment of CHP plants in electricity markets using information gap decision theory, IEEE Trans. Smart Grid 8 (5) (2016) 2296–2304.

[12] M. Moradi-Dalvand, B. Mohammadi-Ivatloo, N. Amjady, H. Zareipour, A. Mazhab-Jafari, Self-scheduling of a wind producer based on information gap decision theory, Energy 81 (2015) 588–600.

[13] N. De Nadai, A.Z. de Souza, J. Costa, D. Marujo, F. Portelinha, D. Oliveira, A Monte Carlo simulation approach to determine critical buses in islanded microgrids, in: 5th IET International Conference on Renewable Power Generation (RPG 2016), 2016.

[14] S.M. Mohseni-Bonab, A. Rabiee, B. Mohammadi-Ivatloo, S. Jalilzadeh, S. Nojavan, A two-point estimate method for uncertainty modeling in multi-objective optimal reactive power dispatch problem, Int. J. Electr. Power Energy Syst. 75 (2016) 194–204.

[15] Z. Wang, J. Wang, Self-healing resilient distribution systems based on sectionalization into microgrids, IEEE Trans. Power Syst. 30 (6) (2015) 3139–3149.

[16] L.W. de Oliveira, F.d.S. Seta, E.J. de Oliveira, Optimal reconfiguration of distribution systems with representation of uncertainties through interval analysis, Int. J. Electr. Power Energy Syst. 83 (2016) 382–391.

[17] A. Soroudi, M. Ehsan, A possibilistic–probabilistic tool for evaluating the impact of stochastic renewable and controllable power generation on energy losses in distribution networks—A case study, Renew. Sust. Energy Rev. 15 (1) (2011) 794–800.

[18] V.K. Tumuluru, D.H. Tsang, A two-stage approach for network constrained unit commitment problem with demand response, IEEE Trans. Smart Grid 9 (2) (2016) 1175–1183.

[19] L. Wu, M. Shahidehpour, T. Li, Stochastic security-constrained unit commitment, IEEE Trans. Power Syst. 22 (2) (2007) 800–811.

[20] D. Bertsimas, E. Litvinov, X.A. Sun, J. Zhao, T. Zheng, Adaptive robust optimization for the security constrained unit commitment problem, IEEE Trans. Power Syst. 28 (1) (2013) 52–63.

[21] C. Zhao, J. Wang, J.-P. Watson, Y. Guan, Multi-stage robust unit commitment considering wind and demand response uncertainties, IEEE Trans. Power Syst. 28 (3) (2013) 2708–2717.

[22] R. Jiang, J. Wang, Y. Guan, Robust unit commitment with wind power and pumped storage hydro, IEEE Trans. Power Syst. 27 (2) (2012) 800–810.

[23] A. Street, F. Oliveira, J.M. Arroyo, Contingency-constrained unit commitment with n-k security criterion: a robust optimization approach, IEEE Trans. Power Syst. 26 (3) (2011) 1581–1590.

[24] P. Xiong, P. Jirutitijaroen, C. Singh, A distributionally robust optimization model for unit commitment considering uncertain wind power generation, IEEE Trans. Power Syst. 32 (1) (2017) 39–49.

[25] H. Ye, Y. Ge, M. Shahidehpour, Z. Li, Uncertainty marginal price, transmission reserve, and day-ahead market clearing with robust unit commitment, IEEE Trans. Power Syst. 32 (3) (2016) 1782–1795.

[26] G. Vinasco, M.J. Rider, R. Romero, A strategy to solve the multistage transmission expansion planning problem, IEEE Trans. Power Syst. 26 (4) (2011) 2574–2576.

[27] R. Jabr, Robust transmission network expansion planning with uncertain renewable generation and loads, IEEE Trans. Power Syst. 28 (4) (2013) 4558–4567.

[28] R.A. Jabr, I. Džafić, B.C. Pal, Robust optimization of storage investment on transmission networks, IEEE Trans. Power Syst. 30 (1) (2015) 531–539.

[29] H. Yu, C. Chung, K. Wong, Robust transmission network expansion planning method with Taguchi's orthogonal array testing, IEEE Trans. Power Syst. 26 (3) (2011) 1573–1580.

[30] A.J. Conejo, J. Contreras, R. Espinola, M.A. Plazas, Forecasting electricity prices for a day-ahead pool-based electric energy market, Int. J. Forecast. 21 (3) (2005) 435–462.

[31] H. Pandžić, J.M. Morales, A.J. Conejo, I. Kuzle, Offering model for a virtual power plant based on stochastic programming, Appl. Energy 105 (2013) 282–292.

[32] M. Rahimiyan, L. Baringo, Strategic bidding for a virtual power plant in the day-ahead and real-time markets: a price-taker robust optimization approach, IEEE Trans. Power Syst. 31 (4) (2016) 2676–2687.

[33] L. Baringo, A.J. Conejo, Offering strategy via robust optimization, IEEE Trans. Power Syst. 26 (3) (2011) 1418–1425.

[34] S. Nojavan, B. Mohammadi-Ivatloo, K. Zare, Robust optimization based price-taker retailer bidding strategy under pool market price uncertainty, Int. J. Electr. Power Energy Syst. 73 (2015) 955–963.

[35] M. Nazari-Heris, S. Abapour, B. Mohammadi-Ivatloo, Optimal economic dispatch of FC-CHP based heat and power micro-grids, Appl. Therm. Eng. 114 (2017) 756–769.

[36] R. Wang, P. Wang, G. Xiao, A robust optimization approach for energy generation scheduling in microgrids, Energy Convers. Manag. 106 (2015) 597–607.

[37] S.A. Alavi, A. Ahmadian, M. Aliakbar-Golkar, Optimal probabilistic energy management in a typical micro-grid based-on robust optimization and point estimate method, Energy Convers. Manag. 95 (2015) 314–325.

[38] Y. Zhang, N. Gatsis, G.B. Giannakis, Robust energy management for microgrids with high-penetration renewables, IEEE Trans. Sustain. Energy 4 (4) (2013) 944–953.

[39] W. Wu, J. Chen, B. Zhang, H. Sun, A robust wind power optimization method for look-ahead power dispatch, IEEE Trans. Sustain. Energy 5 (2) (2014) 507–515.

[40] M.S. Adaramola, M. Agelin-Chaab, S.S. Paul, Analysis of hybrid energy systems for application in southern Ghana, Energy Convers. Manag. 88 (2014) 284–295.

[41] C. Peng, P. Xie, L. Pan, R. Yu, Flexible robust optimization dispatch for hybrid wind/photovoltaic/hydro/thermal power system, IEEE Trans. Smart Grid 7 (2) (2016) 751–762.

[42] A. Martinez-Mares, C.R. Fuerte-Esquivel, A robust optimization approach for the interdependency analysis of integrated energy systems considering wind power uncertainty, IEEE Trans. Power Syst. 28 (4) (2013) 3964–3976.

[43] A. Lorca, X.A. Sun, Adaptive robust optimization with dynamic uncertainty sets for multiperiod economic dispatch under significant wind, IEEE Trans. Power Syst. 30 (4) (2015) 1702–1713.

[44] D. K. A. Brooke, and A. Meeraus. (1990). Gams User's Guide, O. A. http://www.gams.com/docs/gams/GAMSUsers, and Guide.pdf.

Chapter 3

Application of New Fast, Efficient-Self adjusting PSO-Search Algorithms in Power Systems' Studies

Adel M. Sharaf*, Hani Mavalizadeh†, Abdollah Ahmadi‡, Foad H. Gandoman§,¶, Omid Homaee† and Heidar A. Shayanfar†
*Engineering and Research Company of SHARAF Energy Systems, Incorporated, Fredericton, NB, Canada, †School of Electrical Engineering, Iran University of Science and Technology, Tehran, Iran, ‡The Australian Energy Research Institute and the School of Electrical Engineering and Telecommunications, The University of New South Wales, Sydney, NSW, Australia, §Research group MOBI—Mobility, Logistics, and Automotive Technology Research Center, Free University of Brussels, Brussels, Belgium, ¶Flanders Make, 3001 Heverlee, Belgium

1 INTRODUCTION

There are many optimization problems in the world and researchers are constantly working to develop efficient algorithms for solving them. One popular approach for solving these problems is the particle swarm optimization (PSO) algorithm.

This chapter presents some efficient self-sealing/self-adjusting, adaptive, error-driven PSO algorithms. The proposed algorithms are based on the modification of inertia factor in the speed and position equations in each iteration of PSO algorithm. The error in each iteration is used to modify the inertia factor to improve the convergence speed and efficiency.

In this chapter, several algorithms are used to modify the weighting factor in PSO algorithm, based on an error in each iteration. These methods are implemented in several test cases, and the results are presented for each method. The obtained results will be compared to a conventional PSO algorithm to show the efficiency of the proposed methods. Following the validation, the best algorithm is selected and utilized to solve a reactive market clearing problem.

Most types of creatures engage in group behaviors. This can be seen in several animal categories, for example, lions, monkeys, and deer. In these situations, the best individuals from the group direct others in the group. There are

Classical and Recent Aspects of Power System Optimization. https://doi.org/10.1016/B978-0-12-812441-3.00003-3
© 2018 Elsevier Inc. All rights reserved.

other types of creatures, for example, flying creatures, fish, and sheep living in gatherings, which don't engage in leader behavior. Every individual can move on the earth with no outer guide. Instead, the individuals can trade the data with their neighbors in the earth. Some way or another, this straightforward connection between particles brings about the more mind-boggling conduct of the group.

To better comprehend the dynamic conduct of PSO calculations [1–4] in [5] a few adjustments were made, and the conduct of the individuals from the group was isolated into the accompanying classes.

- Avoid impacts: Individuals from a group do not collide with each other.
- Set speed: Each part should alter its speed as indicated by their neighbor.
- Retractable focus: Members attempt to move in with their neighbors.

In many papers, optimization algorithms of these simulations have been enlivened [6,7]. Despite other algorithms, PSO does not guarantee a specific answer. One of the best advantages of PSO is that it requires a lower number of parameters to modify.

The PSO calculation indicates some particles are shaped arbitrarily. Every molecule's position and velocity values are models and characterized by a position vector and velocity values, respectively. These particles move as a repeating issue in n-dimensional space to calculate the ideal incentive as a model of evaluation. The algorithm contains a capacity memory for the best position of every particle and a capacity memory for the best circumstance of the considerable number of particles. Therefore, with the experience gained from memory, the particles choose how they move. In every iteration, all particles in n-dimensional space have at last tended to find the general ideal point. Finally, particles, their speed, and their position regarding the absolute best solutions and updates are nearby.

XY axis position illustrates the position of each particle. The velocity is shown by V_x (the velocity of X-axis) and V_y (the velocity of Y-axis). Adjustment of the molecule position is acknowledged by the position and velocity data. This data represents the individual encounters of every particle. Moreover, every particle knows the best value so far in the group (g_{best}) among P_{best}, and every particle knows its best value so far (P_{best}) and its *XY* position. This data represents the knowledge of how the other particles around it have performed. In particular, every particle tries to enhance its position utilizing the accompanying data:

- The current positions (X, Y)
- The current velocities (V_x, V_y)
- The distance between the current location and g_{best}
- The distance between the current location and P_{best}

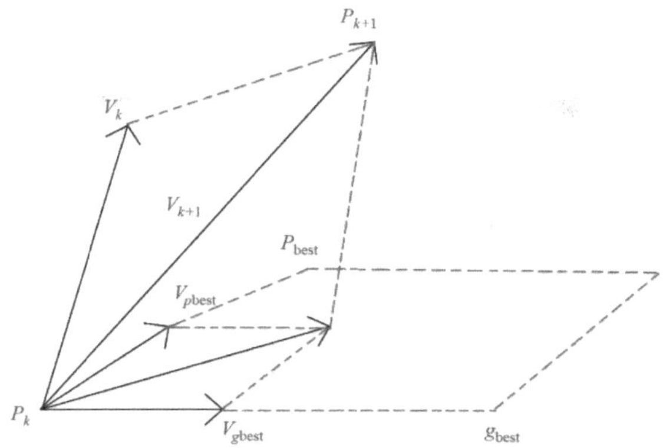

FIG. 1 The adjustment of a looking point by PSO. P_k, is the current position of a particle; P_{k+1}, is its adjusted position; V_K, is its initial velocity; V_{K+1}, is its adjusted velocity; $V_{p_{best}}$, is the velocity considering its p_{best} location; $V_{g_{best}}$, is the velocity considering its g_{best} location.

The principle idea of the PSO strategy lies in quickening every particle toward its g_{best} and P_{best} positions, with an irregular, weighted speeding up at each progression. This appears in Fig. 1.

Utilizing the above idea, a specific velocity, which bit by bit draws near to P_{best} and g_{best}, can be figured. The accompanying condition can alter the present position:

$$V_{new} = WV_{old} + c_1 r_1 (\text{GB} - \text{CP}) + c_2 r_2 (\text{PB} - \text{CP}) \tag{1}$$

where V is the velocity of particle's position change, W is the inertia weight, GB is the global best, PB is the personal best, CP is the current position, r_1 and r_2 are two random variables, C_1 is the global learning coefficient, and C_2 is the personal learning coefficient.

Fig. 2 portrays the general flowchart of the PSO algorithm. The principle ventures in the particle swarm optimization process are depicted by the following:

1. Framework introduced with a populace of arbitrary potential solutions. Every possible arrangement is appointed an irregular "velocity" and is known as a particle. These particles then "flow" through the search space of conceivable solutions
2. Evaluate the wellness of every molecule in the swarm.
3. For each emphasis, contrast every molecule's wellness and its past best wellness (P_{best}) acquired. If the present rate is better than P_{best}, then set P_{best} equal to the present rate and the P_{best} equivalent to the present to the present area in the d-dimensional space.
4. Compare P_{best} of particles with each other and update the swarm global best location with the greatest wellness (g_{best}).

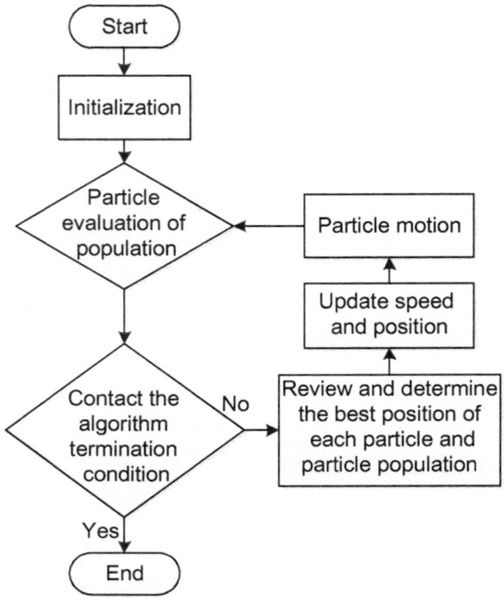

FIG. 2 Search Method of PSO algorithm flowchart.

5. The velocity of each particle is changed (quickened) toward its P_{best} and g_{best}. An arbitrary term weights this increasing speed. Another position in the arrangement space is ascertained for every particle by increasing the value of every segment of the of the particle's position vector.
6. Repeat stages (2)–(5) until convergence is reached based on some desired single or multiple objective criteria.

This chapter is structured as follows: Section 2 presents an overview of PSO algorithm. The proposed Sharaf Accelerated PSO (SAPSO) algorithms are reported in Section 3. The comparison and selection of the best algorithm among other proposed algorithms are presented in Section 4. The next section illustrates the application of proposed algorithms to reactive power market problem, and Section 6 presents the conclusion.

2 OVERVIEW OF PSO ALGORITHM

Over the last three decades, the utilization of optimization algorithms has expanded. Among the algorithms, PSO is known as an exceedingly efficient algorithm. This chapter actualized the PSO algorithm to plan an automatic hydro-power device environmentally friendly power vitality administration plot for hydrogen and island power networks in Ref. [8]. In Ref. [9], a dynamic filter is intended for appropriation system utilizing PSO scheme. Multiobjective PSO [10,11] is likewise utilized as a part of various papers to improve distinctive target works.

Numerous articles on a variety of basic PSO algorithms are provided. However, the primary methods have several noticeable deficiencies [12–14]. The main deficiencies in basic PSO methods can be listed as follows:

- The advantages of velocity clamping (VC) include controlling the development of the molecule by lessening the extent of the progression speed velocity. In any case, in VC if the velocity becomes equal to maximum rate, that will keep on conducting seeks inside a hypercube and it will likely stay in the ideal [12].
- The advantage of inertia weight (IW) is that a larger W at the end of the process will foster the convergence ability. The main disadvantage of IW is that the W has a powerful influence on optimality convergence [12].
- The gain of constriction coefficient (CC) is similar to inertia weight. In this type of basic PSO, the fixed values of the parameters can cause the random fluctuation of particles while the algorithm converges [13].
- The advantages of synchronous and asynchronous (SA) convergence rate is improved [14].

As a result, to enhance the basic PSO problems modification, PSO has introduced [15–19]. In Ref. [20] the application of PSO in smart grids is presented. In Ref. [21], a review of PSO algorithm application to solve the economic dispatch problem in power systems is presented. Also, due to the characteristic of PSO modification, there are several application areas that have developed, such as scheduling, searching, forecasting, feature selection, and classification. Table 1 gives the distribution of search for every function of PSO algorithms (modified PSO).

3 THE DESCRIPTION OF PROPOSED SAPSO ALGORITHMS

3.1 Description of Self-Adjusting Algorithms

In this section, different approaches developed by the First Author are presented to accelerate the convergence of the standard PSO algorithm. All the presented approaches rely on modification of inertia weight in Eq. (1). The value of inertia weight should be adjusted based on the error in each iteration. The general idea is that a smaller error means the particle is probably moving in a correct direction. Therefore, inertia weight must be larger. On the other hand, when the error is increasing, more divergence from the previous direction is required.

TABLE 1 Distribution of Research for Every Function of PSO (Basic and Modified)

	PSO Algorithm									
	Basic				Modified					
Function	*Velocity Clamping*	*Inertia Weight*	*Constriction Coefficient*	*Synchronous and Asynchronous*	*Single Solution of PSO*	*Niching With PSO*	*Constrain Optimized Using PSO*	*Multiobjective Optimization*	*Dynamic Environment of PSO*	*Discrete PSO*
Optimized scheduling								[22]	[23]	[24]
Optimized local search			[14]		[25]	[17]	[18]	[26]	[27]	
Optimized multisearch				[15]	[28]	[29]	[30]	[31]		
Optimized forecasting					[32]			[33]		
Optimized function problem		[13]			[34]	[35]	[36]	[37]	[38]	[39]
Optimized feature selection									[40]	[41]
Optimized classification	[13]				[42]			[43]		[44]
Optimized production rate							[45]	[46]	[47]	

SAPSO Algorithm 1

In SAPSO1 algorithm, the error in kth iteration is defined as the difference between global best in that iteration and its previous iteration. This error is then used to calculate an auxiliary variable for each iteration called ε_k. When the particles are moving in the right direction, the global best is decreasing, this means the value of error becomes negative. As seen in Eq. (3), ξ is normalized using the worst global best, which is the global best in the first iteration. W_0 is the initial inertia weight. It is clear that when the algorithm is moving toward better solution (negative values for ε_k), the inertia weight is reduced, therefore the W_k is always equal or less than W_0. As seen in Eq. (4), the magnitude of change in value of W_k is small in the first iterations. This means larger values for W_k. This is necessary to let the initial population scatter all over the search space. In the later iterations, because the value of k is increased, W_k will have smaller values.

$$\Delta e_k = \mathrm{GB}_k - \mathrm{GB}_{k-1} \tag{2}$$

$$\xi_k = \frac{\Delta e_k}{\max(\mathrm{GB}_k)} \quad \text{for} \quad k = 1, 2, \ldots (k-1) \tag{3}$$

$$W_k = W_0 + W_1 \varphi_k \cdot (K-1)/d \tag{4}$$

where d is a number between 10 and 100. k is the index for iterations. W_0 and W_1 are real numbers between 0 and 1. The appropriate value for these parameters differs for each optimization problem and is tuned by trial and error.

The above formulation is for minimization problems. It can be easily adapted for maximization problems. For a maximization problem, the error in each iteration is divided by the minimum global best.

SAPSO Algorithm 2

In the second SAPSO algorithm, the weight inertia is calculated in each iteration using the following formulation:

$$\Delta e_k = \mathrm{GB}_{k-1} - \mathrm{GB}_{k-2} \tag{5}$$

$$\xi_k = \frac{\Delta e_k}{\max(\mathrm{GB}_k)} \tag{6}$$

$$W_k = W_0 \times (1 + \xi_k) \times (k-1)/d \tag{7}$$

Error is defined in Eq. (5) as the improvement in the global best in term of the fitness function. This error is then used in Eq. (6) to calculate the normalized error called ξ same as SAPSO1 algorithm. The value of W_k is then obtained by Eq. (7). The main difference between SAPSO2 and SAPSO1 algorithms is that in SAPSO1 algorithm, only the velocity is updated by adjusting weight inertia. In the SAPSO2 algorithm, the position is also updated through the following equations:

$$\eta_k = \frac{\mathrm{GB}_{k-1}}{\max(\mathrm{GB}_k)} \tag{8}$$

$$D_k = d_0 \times (1 + \eta_k) \tag{9}$$

$$X_{\mathrm{new}} = X_{\mathrm{old}} \times D_k + V_{\mathrm{old}} \tag{10}$$

where d_0 is a number between 10 and 100. As seen in Eq. (9), a new auxiliary variable is calculated based on global best in the previous iteration. This variable is then used to update the new position of particles in addition to their velocity.

SAPSO Algorithm 3

Another approach for updating the weight inertia is presented in the following formulation:

$$D_k = \mathrm{GB}_k - \mathrm{GB}_{k-1} \tag{11}$$

$$R_k = \sqrt{(\mathrm{GB}_k)^2 + (D_k)^2} \tag{12}$$

$$\Delta e_k = R_k - R_{k-1} \tag{13}$$

$$\xi_k = \frac{\Delta e_k}{\max(\mathrm{GB}_k)} \tag{14}$$

$$W_k = (W_0 + W_1 \times \xi_k) \tag{15}$$

As seen in the formulation of Eqs. (11)–(15), the improvement in global best is used to calculate a new variable called R_k, which is then used to obtain the ξ_k using Eq. (14). The value of ξ_k determines whether inertia weight in each iteration must increase or decrease. It also determines the amount of change for inertia weight in each iteration based on the improvement of the global best in that iteration.

SAPSO Algorithm 4

In the SAPSO4 algorithm, two auxiliary variables called α_k and β_k are introduced. These variables are modified in each iteration as shown in Eqs. (16), (17). The weight factor in PSO algorithm is calculated in each iteration based on α_k and β_k. The procedure is continued until a stopping criterion is reached.

$$\alpha_k = \mathrm{GB}_k - \mathrm{GB}_{k-1} \tag{16}$$

$$\beta_k = \alpha_k - \alpha_{k-1} \tag{17}$$

$$W_k = \frac{W_0 \times (1 + \alpha_k + \beta_k) \times k}{\max(\mathrm{GB}_k)} \tag{18}$$

The improvement in global best in each iteration is α_k. In each iteration, the change in global best improvement (β_k) is also calculated. The weight factor in each iteration is modified based on the value of α_k and β_k as shown in Eq. (18).

3.2 Verification of the Proposed Algorithms

The presented four modified fast convergent error driven PSO algorithms developed by the First Author algorithms are first implemented on several benchmarks to verify their good performance in solving optimization problems. Ackley [48] test case, Griewank [49], and Sphere [50] test cases are selected for comparison purposes. Different versions of SAPSO accelerated PSO algorithms are implemented on the above functions in the remainder of this section.

3.2.1 Sphere Test Case

The Sphere function is shown by the following equation:

$$f_1 = \sum_{i=1}^{n} {x_i}^2 \tag{19}$$

where n is the number of variables. This problem is to minimize the value of f_1. It is clear that the global optimum for this optimization problem is $f_1 = 0$ that occurs when all the variables are zero. The allowable range for x_i is considered between -100 and 100. SAPSO algorithms 1–4 are implemented on the Sphere functions with different dimensions. The results are shown below.

It can be seen in Table 2 that the results obtained by all the presented algorithms are much better in comparison to conventional PSO results for two-dimension Sphere function (i.e., the Sphere function with two variables). The best solution for SAPSO1, SAPSO2, SAPSO3, and SAPSO4 algorithms are 2.04E–80, 7.76E–136, 1.54E–146, 8.40E–320, and 9.31E–308, respectively, for $n = 2$. This shows a considerable improvement in case of the small-scale optimization problem.

The same observation can be made for $n = 10$ and $n = 100$ states. The results presented in Table 2 show that, in this case, the average and the best solution are improved considerably for an equal number of iterations. The best solution for $n = 2$ is obtained by the SAPSO3 algorithm. Moreover, the average of the results obtained in different runs is better in the SAPSO3 algorithm. The results of Table 2 are obtained for 10 separate runs. Mean is the average of obtained values in separate runs, SD is the standard deviation, and best is the best solution. SAPSO3 has produced better results for larger scale problem ($n = 10$ and $n = 100$). The only exception is for $n = 10$, where the SAPSO4 algorithm mean value is slightly better than the mean value of results obtained by the SAPSO3 algorithm.

In Table 2, the maximum iteration was set to 500. To show each algorithm's convergence speed, the stopping criterion of the algorithm is changed. Instead of max iteration, a 1E–8 threshold is set as the stopping criteria, and the number of iterations required for achieving the objective function is less than the threshold shown in Table 3. It should be mentioned that the results shown in Table 3 are the average of 10 separate runs.

The results in Table 3 show a considerable increase in convergence speed. The SAPSO2 algorithm has led to desirable result within a smaller number of generations. The results of Tables 2 and 3 show that the SAPSO2 algorithm has produced better results from both an accuracy and convergence point of view.

TABLE 2 Best Solution, an Average of the Solutions and Standard Deviation of Objective Function for Sphere Function With Different Dimensions

Algorithm		$n=2$	$n=10$	$n=100$
Conventional PSO	Mean	1.43E–74	6.30E–29	2.47E–4
	Best	2.04E–80	1.40E–29	1.81E–4
	SD	2.54E–74	4.48E–29	6.98E–5
SAPSO1	Mean	2.02E–133	6.28E–64	6.38E–22
	Best	7.76E–136	8.85E–65	8.41E–23
	SD	4.33E–133	6.67E–64	2.50E–21
SAPSO2	Mean	1.17E–145	1.54E–86	8.93E–26
	Best	1.54E–146	4.50E–89	8.74E–27
	SD	8.33E–146	2.77E–86	1.23E–25
SAPSO3	Mean	4.59e–312	5.02E–109	7.54E–24
	Best	8.40E–320	1.68E–119	4.99E–26
	SD	6.38e–312	1.00E–108	6.59E–24
SAPSO4	Mean	6.56E–304	1.55E–109	2.59E–16
	Best	9.31E–308	3.12E–118	2.2E–17
	SD	9.53E–304	3.10E–109	3.89E–16

TABLE 3 Average Iterations to Complete for Sphere Function (Tolerance 1E–8)

Number of Variables	Conventional PSO	SAPSO1	SAPSO2	SAPSO3	SAPSO4
$N=2$	29	24	15	24	18
$N=10$	124	115	31	86	106
$N=100$	2149	950	845	1468	1700

The algorithm is superior to different values of problem dimension ranging from the tiny problem ($n=2$) to relatively large-scale problem ($n=100$).

3.2.2 *Ackley Test Case*

The algorithms are further examined using the Ackley function, which is another well-known benchmark for testing optimization algorithms. The objective function can be expressed as follows:

$$F(x,y) = -20 \times e^{-02\times\sqrt{\frac{x^2+y^2}{2}}} - e^{0.5\times(\cos(2\pi x)+\cos(2\pi y))} + e + 20 \qquad (20)$$

The convergence of different algorithms is tested using the Ackley function, and the results are presented in Table 4. The results are the average, often of separate runs. The threshold is considered to be 1E–8, and the number of iterations and the elapsed time are reported.

It is clear from Table 4 that the convergence is accelerated considerably using self-adjusting PSO algorithms. The best performance is obtained by SAPSO2 algorithm, with the same performance for Sphere function. The required iterations to achieve the 1E–8 threshold for the SAPSO2 algorithm is 74.25% less than required iterations for conventional PSO for $n=20$, which is a considerable improvement. Fig. 3 shows the objective function versus iteration of the conventional method and SAPSO2 algorithm. It confirms the results obtained from Table 4 and indicates that the SAPSO accelerated algorithm has reached the optimum solution faster than conventional PSO.

3.2.3 *Griewank Test Case*

Some simulations are performed on the two-dimension Griewank test system to verify the performance of the proposed algorithms. The function is mathematically expressed in Eq. (21):

TABLE 4 Convergence Speed for Ackley Test Case (Tolerance 1E–8)

	$n=2$		$n=10$		$n=20$	
Algorithm	*Iteration*	*Time(s)*	*Iteration*	*Time(s)*	*Iteration*	*Time(s)*
Conventional	59	0.55427	233	1.972538	505	4.114591
SAPSO1	42	0.430867	140	1.239912	335	2.987419
SAPSO2	3	0.077963	55	0.555588	130	1.153908
SAPSO3	3	0.068095	172	1.510101	323	2.757435
SAPSO4	35	0.330774	142	1.218378	291	2.410302

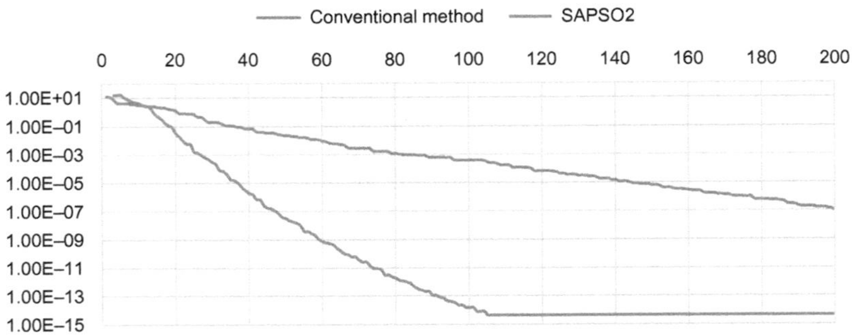

FIG. 3 Comparison between SAPSO2 and Conventional PSO algorithms for Ackley test system.

$$F(x) = 1 + \frac{1}{4000}\sum_{i=1}^{n} x_i^2 - \prod_{i=1}^{n} \cos\left(\frac{x_i}{\sqrt{i}}\right) \tag{21}$$

The results show that the run-time and number of iterations needed to achieve the 1E–8 threshold can be decreased remarkably using a self-adjusting inertia factor. As seen in other test cases used in this chapter, the best results are obtained by the SAPSO2 algorithm (Table 5).

The results show that the use of adaptive inertia factor instead of constant inertia factor can lead to much more convergence speed. The results also revealed that the SAPSO2 algorithm has led to better results among other adaptive algorithms proposed in this chapter. Now that the excellent performance of the SAPSO2 algorithm is verified through the implementation of the algorithm on different benchmarks, the next step is to use it to solve a real engineering problem. In the next subsection, the problem of reactive market clearing [51] is solved using the SAPSO2 algorithm, and the results are compared to the results obtained by conventional PSO algorithm used in Ref. [50].

TABLE 5 Convergence Speed for Griewank Test Case (Tolerance 1E–8)

Algorithm	Iteration	Time
Conventional	495	3.719104
SAPSO1	55	0.499709
SAPSO2	30	0.525535
SAPSO3	92	0.772465
SAPSO4	59	0.524406

4 COMPARISON AND SELECTION OF THE BEST ALGORITHM AMONG OTHER PROPOSED ALGORITHMS

As seen in the previous subsection, the results of each fast converging PSO error-driven algorithms algorithm can be evaluated from two different points of views: convergence speed and accuracy. Some algorithms have good convergence speed in the first iteration but are trapped in local optimum, that is, after initial iterations the improvement of the objective function stops. On the other hand, in some algorithms, the convergence speed is slower, but the possibility of reaching global optimum is higher. For example, in sphere function, the SAPSO3 algorithm provides the most accurate results among other algorithms, but from Table 3 it is clear that SAPSO2 algorithm yields acceptable results in fewer iterations. Therefore, a compromise between convergence speed and accuracy must be made to select the best solution for each objective function. This compromise is made based on the nature of the problem. For example, in long-term planning problems, the accuracy is much more important for the user. In a long-term power system expansion planning problem [51], the planning horizon is between 5 and 20 years. In such problems, the runtime has a small influence on selecting the solution method. On the other hand, in some problems, the problem should be solved in a limited time. A good example is power system economic dispatch problem [52], where the system operator must solve an optimization problem to determine the generation of each generation unit for the day ahead system operation. Therefore, in such problems, a solution with 1-week runtime cannot be acceptable.

In this chapter, a new index is introduced to select the best algorithm for each optimization problem. The index considers both accuracy and convergence speed. For each criterion, two thresholds are defined; the lower threshold and upper threshold. These thresholds are defined by the user based on the nature of the optimization problem, as described earlier in this chapter.

The mathematical formulation can be expressed as follows:

$$\mathrm{SI} = m_1 \times A + m_2 \times B$$

$$\begin{cases} A = 1 & \text{for}\, R \geq R_{\mathrm{th}}^2 \\ A = \dfrac{R_{\max} - R}{R_{\max} - R_{\min}} & \text{for}\ R_{\mathrm{th}}^1 \leq R \leq R_{\mathrm{th}}^2 \\ A = 0 & \text{for}\ R \leq R_{\mathrm{th}}^1 \end{cases} \tag{22}$$

$$\begin{cases} B = 0 & \text{for}\, E \leq E_{\mathrm{th}}^1 \\ B = \dfrac{E_{\max} - E}{E - E_{\min}} & \text{for}\ E_{\mathrm{th}}^1 \leq E \leq E_{\mathrm{th}}^2 \\ B = 1 & \text{for}\ E \geq E_{\mathrm{th}}^2 \end{cases} \tag{23}$$

where m_1 and m_2 are the parameters selected by the user. Higher values of $m_1$show they are more important for runtime and higher values for m_2 indicate that the accuracy is more important for the user. R is the run time in seconds, and E is the error of each solution. The error is calculated here as the difference between each solution compared to the best solution. E_{th}^1 indicates the lower threshold for error. In some problems, a certain degree of accuracy is sufficient and when this threshold is reached, there is no need to put more time to find a better solution. For example, when the global optimum is zero, the system planner may decide to choose a 1E–100 as the lower threshold. E_{th}^2 is the upper limit of the error. This parameter indicates that an error more than the specified value is not acceptable. In other words, the solutions with errors greater than E_{th}^2 should be eliminated regardless of their associated runtime. E_{max} and E_{min} are the maximum and minimum values for error among obtained solutions while R_{max} and R_{min} are the maximum and minimum run time among solutions. R_{th}^1 and R_{th}^2 are the lower and upper thresholds for runtime. As described earlier, in some optimization problems, the runtime should be limited to a certain time and the solutions with more runtime are not acceptable. This can be incorporated in the model by appropriately tuning the R_{th}^2. In most cases, there is no tangible difference between run the times. For example, a solution with 0.02-s runtime is not very different from a solution with a 2 s run time in most cases. R_{th}^1 is used to determine this value.

The presented index is calculated for sphere and Ackley test cases. The algorithm, which obtains the highest SI, excels and is selected as the best algorithm. The results are shown for $m_1 = 1$ and $m_2 = 1$ in Table 6. The dimension of the problem is considered 10. The results show that for both test cases the SAPSO2 algorithm is selected as the best algorithm because it has the highest SI. It should

TABLE 6 SI Index for Sphere and Ackley Test Cases

Test Case	Algorithm	SI
Sphere	Conventional	0
	SAPSO1	0.548387
	SAPSO2	1
	SAPSO3	0.704301
	SAPSO4	0.596774
Ackley	Conventional	0.5
	SAPSO1	0.761236
	SAPSO2	1
	SAPSO3	0.671348
	SAPSO4	0.755618

be noted that best algorithm for each problem is considered based on the nature of the problem and its dimension.

At the end of this subsection, the results are compared with results of another published work to verify the proper performance of the presented algorithms. In Ref. [53], the commodity graphics processing units (GPU) is used to accelerate the PSO with digital pheromones. The results of Ref. [53] for 10-dimension Ackley test case, 50 dimension Griewank function, and 30-dimension sphere function are compared with the presented work in Tables 7–9.

TABLE 7 Comparison With Ref. [53] for Ackley Test Case

	The Objective Function			Average Duration (s)
Algorithm	*Average*	*Small*	*SD*	
Basic PSO [54]	3.54287	0.00299	3.22811	10.26
PSO with Pheromones implemented [54]	0.00143	0.00066	0.00086	8.21
SAPSO1	4.4409E–15	4.4409E–15	0	6.78
SAPSO2	0.0024	4.4409E–15	0.00537	6.39
SAPSO3	4.4409E–14	7.99E–15	3.1777E–15	6.48
SAPSO4	7.99E–15	7.99E–15	0	6.42

TABLE 8 Comparison With Ref. [53] for Griewank Test Case

	The Objective Function			Average Duration (s)
Algorithm	*Average*	*Smallest*	*SD*	
Basic PSO [54]	1.189	1.056	0.133	99.014
PSO with Pheromones implemented [54]	0.008	0.003	0.005	93.801
SAPSO1	9.66E–9	8.91E–9	4.2242E–10	9.8
SAPSO2	3.45E–3	2.11E–15	0.0048	12.91
SAPSO3	2.14E–14	2.78E–15	2.8820E–14	12.83
SAPSO4	8.89E–11	1.13E–12	1.7588E–10	12.6

TABLE 9 Comparison With Ref. [53] for Sphere Test Case

Algorithm	The Objective Function			Average Duration (s)
	Average	*Smallest*	*SD*	
Basic PSO [53]	16.301	0.521	48.692	66.1
PSO with Pheromones implemented [53]	0.084	0.006	0.128	46.46
SAPSO1	4.97E–22	6.74E–24	5.1869E–22	6.22
SAPSO2	3.07E–3	1.90E–4	0.0028	6.25
SAPSO3	1.02E–8	3E–11	1.8065E–7	6.05
SAPSO4	4.26E–11	7.13E–15	7.6468E–11	6.06

It can be seen in Tables 7–9 that the results are much better using the presented algorithms. The results also show that the selection of the best algorithm is completely based on the nature of the problem that is solved. The improvement in both accuracy and runtime is considerable. The runtime for different SAPSO algorithms is not much different. The Tables 7–9 verify the excellent performance of the presented algorithms.

5 APPLICATION OF PROPOSED ALGORITHMS TO REACTIVE POWER MARKET PROBLEM

5.1 A Brief Description of Reactive Market

An appropriate supply of reactive power is one of the most critical requirements of power systems. Reactive power supply impresses total energy loss, stability, and voltage profile of the power system. An appropriate supply of reactive power can efficiently reduce energy loss, increase stability margins, and substantially improve voltage profile. Before the advent of restructuring, the government exclusively ran the generation section. In this circumstance, power systems operators have full authority in the scheduling of power plants. In other words, because power systems operation was one of the governments' responsibilities and all the power plants belonged to governments, power system operators had full authority in the scheduling of active and reactive power of power plants. In these methods, the operation is performed to minimize the total cost of operation.

Restructuring in power systems is conducted with the primary goals of increasing the efficiency, increasing the competition, and determining the real cost of electrical energy production. In restructured power systems, power plants are privatized. Independent system operator (ISO) has the responsibility of appropriately operating the power system. Because power plants are private, the independent system operator does not have full authority in the scheduling of power plants. In this condition, power plants should submit their energy sale offers to the independent system operator. Also, all energy consumers should submit their purchase offers to the independent system operator. Independent system operator clears the energy market according to received offers from energy consumers and energy producers [50].

The inappropriate reactive power supply can lead to voltage collapse in power systems. Recent significant nationwide power systems blackouts were due to the improper supply of reactive power. For example, in 2003 the United States and Canada blackout investigation group reported that inappropriate reactive power supply was one of the reasons for the blackout [54]. In a restructured power system, the primary task of the independent system operator is maintaining system stability and security, by providing ancillary services such as reactive power, spinning reserves, energy balancing, and frequency regulation. Reactive power support needs to have an acceptable voltage profile and allowable transmission lines for power systems. Because reactive power transmission leads to a considerable ohmic loss in transmission lines, therefore, required reactive power in each area should be supplied locally [50,55].

In the restructured power systems, reactive power should be provided by a competitive reactive market. According to Instructions No. 888, issued by Federal Energy Regulatory Commission, voltage control and the reactive power supply is one of the ancillary services that independent system operator should provide in an open-access market [50,56].

Reactive power market mechanisms are not the same in all restructured power systems. In England, generators' offers for reactive power support consist of two components: capacity and utilization. In this market, capacity and utilization components are paid to the selected producers by annual bilateral agreements with National Grid Company (NGC) [57]. In the Nordic energy market, including Norway, Denmark, Finland, and Sweden, because of the immense distance between the generators and loads, reactive power is supplied locally. In fact, reactive power support in the Nordic countries is mandatory. For example, in Norway, only if the generator operating point is outside of the range 0.9 lag to 0.93 lead isthe cost of reactive power paid to them [57]. In Australia, National Electricity Market Management Company (NEMMC) forced generators to supply reactive power in the range of 0.9 lag to 0.93 lead and only if the operating point of a generator is outside of this range will that generator be considered a provider of ancillary services and be eligible to receive money according to the lost opportunity cost [58]. Ref. [58] investigated reactive power management approaches in restructured power systems of different countries.

Cost of reactive power production by synchronous generators is divided into four parts:

1. Availability cost.
2. Losses cost due to injection of reactive power.
3. Losses cost due to absorption of reactive power.
4. Cost of losing the sale opportunity of energy due to injection or absorption of reactive power.

If produced reactive power by a synchronous generator exceeds its reactive power consumption (the amount of reactive power that is needed for appropriate operation of auxiliary equipment including step-up transformers, cooling fans, boiler feed engines, and water circulation system), generators receive money according to the amount of injected reactive power to the network. Otherwise, reactive power output in this region is not considered as ancillary services [50,59].

In most electricity markets, required reactive power is purchased seasonally by long-term contracts. Ref. [60] proposed a day-ahead reactive power market and discussed advantages of these markets compared to long-term contracts.

Ref. [50] has done a comprehensive study on the structure of reactive power market. In the model presented in this reference, same as [57,61,62], the reactive power market is cleared after energy market clearing. All generators should submit their offers for the cost of availability. The offers for the cost are losses due to injection of reactive power, absorption of reactive power, and losing the sale opportunity of energy due to injection or absorption of reactive power. Ref. [50] presents an accurate formulation of the reactive power market.

Before assessing the reactive power market, it is necessary to perform a thorough study of the limitations of the synchronous generator. It should be noted that the energy market is cleared before the reactive power market. Therefore, when reactive power market clearing is performed, the active power output of generators is known. Also, the terminal voltage of synchronous generators is fixed, and equal to network voltages. The primary constraints on reactive power generation by the synchronous generator can be listed as follows [50]:

1. Armature current heating limit.
2. Field current limit.
3. End region heating limit.
4. Steady state stability limit.
5. Limit of volt/Hz.

5.1.1 0 to $Q_{\mathrm{Base},i}$

Q_{Base} is the minimum reactive power required by auxiliary equipment of generator such as step up transformers, boiler engine, cooling fans, etc. In other words, this is the generator internal consumption and will not participate in the ancillary service [50,56,60]. The equation shown in Eq. (24) is used to calculate Q_{Base} [58,63–65]:

$$Q_{\text{Base}} = 0.1 \times Q_{\max} \tag{24}$$

5.1.2 ($Q_{\text{Base},i}$ to $Q_{A,i}$) and ($Q_{C,i}$ to 0)

In this region, the active power output of the generator is independent of reactive power absorbed from or injected to the grid. The cost of losses and cost of availability will be paid to the generator. The range of Q, where the amount of active power is independent of reactive power production can be expressed as follows [50]:

$$Q_{C,i} \le Q_i \le Q_{A,i} \tag{25}$$

where $Q_{A,i}$ is calculated using Eq. (26):

$$\begin{aligned} Q_{A,i} &= -\frac{V_{ti}{}^2}{X_{si}} + \sqrt{\left(\frac{E_i^{\max}}{X_{si}} V_{ti}\right)^2 - P_i^2} \quad \text{if } P_i \le P_i^{\text{rated}} \\ Q_{A,i} &= \sqrt{MVA_i^{\text{rated}2} - P_i^2} \quad \text{if } P_i \ge P_i^{\text{rated}} \end{aligned} \tag{26}$$

where $E_i^{\max}$ is calculated using Eq. (27):

$$E_i^{\max} = \frac{X_{s,i}}{V_{tn,i}} \sqrt{P_i^{\text{rated}2} + \left(Q_i^{\text{rated}} + \frac{V_{tn,i}{}^2}{X_{s,i}}\right)^2} \tag{27}$$

$Q_{C,i}$ is calculated using Eq. (28):

In armature current limitation curve in P-Q page, the active power of the point with $Q = Q_{\min,i}$, is identified as, P_i^a. The value of P_i^a can be mathematically expressed as follows:

$$P_i^a = \sqrt{MVA_i^{\text{rated}2} - Q_{\min,i}{}^2} \tag{28}$$

where $Q_{C,i}$ is obtained by using Eq. (29):

$$\begin{aligned} Q_{C,i} &= \sqrt{MVA_i^{\text{rated}2} - P_i^2} \quad if\ P_i \ge P_i^a \\ Q_{C,i} &= Q_{\min,i} \quad \text{if } P_i \le P_i^a \end{aligned} \tag{29}$$

where $Q_{\min,i}$ is calculated as follows:

$$Q_{\min,i} = -\min\left(\frac{V_{t,i}{}^2}{X_{s,i}}, Q_{ERHL,i}^{\max}\right) \tag{30}$$

5.1.3 ($Q_{A,i}$ to $Q_{B,i}$) and ($Q_{\min,i}$ to $Q_{C,i}$)

To generate the reactive power in this range, the active power generation of the generator must be reduced. This means less active power can be sold by the generator. Therefore, the generator should receive extra pay to compensate for its reduction in active power generation in addition to losses and availability cost. $Q_{B,\,i}$ is the reactive power that the generator is able produce when the active power generation is at minimum level. This can be mathematically expressed as follows:

$$Q_{B,i} = -\frac{V_{ti}{}^2}{X_{si}} + \sqrt{\left(\frac{E_i^{\max}}{X_{si}} V_{ti}\right)^2 - P_{i,\min}^2} \tag{31}$$

The active power in this situation is calculated using the following equation:

$$P_{A,i} = \sqrt{\left(\frac{E_i^{\max}}{X_{si}} V_{ti}\right)^2 - \left(Q_i + \frac{V_{ti}{}^2}{X_{si}}\right)^2} \quad \text{if } Q_i \geq Q_i^{\text{rated}}$$

$$P_{A,i} = \sqrt{MVA_i^{\text{rated}^2} - Q_i^2} \quad \text{if } Q_i \leq Q_i^{\text{rated}} \tag{32}$$

where Q_i^{rated} is calculated as follows:

$$Q_i^{\text{rated}} = \sqrt{MVA_i^{\text{rated}^2} - P_i^{\text{rated}^2}} \tag{33}$$

As mentioned earlier, the results of reactive market clearing can reduce the active power output of some of the generators. This happens when the generator is working within the third performance region. This will result in frequency deviation due to an imbalance between load and generation. This problem should be addressed in the real-time market [55].

Finally, the problem should minimize the total payment function (TPF), to minimize the imposed cost submitted to the ISO. This objective function is formulated as Eq. (34) [50]. where *NU* is the number of generators in the grid; $W_{0,\,i}$, $W_{1,\,i}$, $W_{2,\,i}$, $W_{3,\,i}$, $W_{4,\,i}$ are binary variables which determines the working region of generator and are obtained based on mapping provided in Table 10. ρ_0 is the

TABLE 10 Mapping of Binary Variables to Working Region of the Generators

Parameter					
$W_{0,i}$	$W_{1,i}$	$W_{2,i}$	$W_{3,i}$	$W_{4,i}$	Q
1	1	0	0	0	$Q_C \leq Q \leq 0$
1	0	0	0	1	$Q_{\min} \leq Q \leq Q_C$
1	0	1	0	0	$Q_{\text{base}} \leq Q \leq Q_A$
1	0	0	1	0	$Q_A \leq Q \leq Q_B$

highest accepted offer for availability cost in reactive market. $\rho_1$and ρ_2 are the losses cost resulted from market clearing problem in the regions $Q_{\min} \leq Q \leq 0$ and $Q_{\text{base}} \leq Q \leq Q_B$, respectively. ρ_3 is the price of reduced active power due to reactive generation in region 3. The total reactive power generation for ith synchronous generator is calculated by $Q_{G,\ i} = Q_{1,\ i} + Q_{2,\ i} + Q_{3,\ i} + Q_{4,\ i}$.

$$\begin{aligned}\text{TPF} = \sum_{i=1}^{NU} [&\rho_0 W_{0,i}(Q_{\max,i} - Q_{\min,i}) - \rho_1 W_{1,i} Q_{1,i} \\ &-\frac{1}{2}(\rho_3 W_{4,i}(Q_{C,i}^2 - Q_{4,i}^2)) - \rho_1 W_{4,i} Q_{4,i} \\ &+\rho_2 W_{2,i}(Q_{2,i} - Q_{\text{base},i}) + \rho_2 W_{3,i}(Q_{3,i} - Q_{\text{base},i}) \\ &+\frac{1}{2}(\rho_3 W_{3,i}(Q_{3,i}^2 - Q_{A,i}^2))]\end{aligned} \tag{34}$$

The following constraints should be satisfied in a reactive power market clearing problem. For more details on constraint and objective functions of reactive market clearing problem see Ref. [50].

$$P_{Gi} - P_{Di} = \sum_{j=1}^{n} V_i V_j Y_{ij} \cos(\theta_{ij} + \delta_j - \delta_i) \quad \forall i \tag{35}$$

$$Q_{Gi} - Q_{Di} = -\sum_{j=1}^{n} V_i V_j Y_{ij} \sin(\theta_{ij} + \delta_j - \delta_i) \quad \forall i \tag{36}$$

$$W_{0,i}, W_{1,i}, W_{2,i}, W_{3,i}, W_{4,i} \in \{0, 1\} \quad \forall i \tag{37}$$

$$Q_{G,i} = Q_{1,i} + Q_{2,i} + Q_{3,i} + Q_{4,i} \quad \forall i \tag{38}$$

$$Q_{\min,i} \leq Q_{G,i} \leq Q_{\max,i} \quad \forall i \tag{39}$$

$$W_{1,i} \cdot Q_{C,i} \leq Q_{1,i} \leq 0 \quad \forall i \tag{40}$$

$$W_{2,i} \cdot Q_{\text{base},i} \leq Q_{2,i} \leq W_{2,i} \cdot Q_{A,i} \quad \forall i \tag{41}$$

$$W_{3,i} \cdot Q_{A,i} \leq Q_{3,i} \leq W_{3,i} \cdot Q_{B,i} \quad \forall i \tag{42}$$

$$W_{4,i} \cdot Q_{\min,i} \leq Q_{4,i} \leq W_{4,i} \cdot Q_{C,i} \quad \forall i \tag{43}$$

$$W_{1,i} + W_{2,i} + W_{3,i} + W_{4,i} \leq 1 \quad \forall i \tag{44}$$

$$W_{0,i} . a_{0,i} \leq \rho_0 \quad \forall i \tag{45}$$

$$(W_{1,i} + W_{4,\text{i}}) \cdot m_{1,i} \leq \rho_1 \quad \forall i \tag{46}$$

$$(W_{2,\text{i}} + W_{3,\text{i}}) \cdot m_{2,i} \leq \rho_2 \quad \forall i \tag{47}$$

$$(W_{3,i} + W_{4,i}) \cdot m_{3,i} \leq \rho_3 \quad \forall i \tag{48}$$

$$W_{0,i} = W_{1,i} + W_{2,i} + W_{3,i} + W_{4,i} \quad \forall i \tag{49}$$

$$V_i^{\min} \leq V_i \leq V_i^{\max} \quad \forall i \tag{50}$$

$$|S_{ij}| \leq S_{ij}^{\max} \quad \forall ij \tag{51}$$

$$P_{G,\text{slack}} \leq P_{G,\text{slack}}^{\max} \tag{52}$$

Eqs. (35), (36) show load flow equality constraints. In these equations, n is the number of buses; P_{Di} and Q_{Di} are active load and reactive load at bus i, respectively; P_{Gi} Q_{Gi} are the generated active power and generated reactive power at bus i, respectively; V_i is the voltage magnitude at bus i; δ_i is the angle of V_i; Y_{ij} is element ij of admittance matrix; θ_{ij} is angle associated with Y_{ij}. Eqs. (37)–(44) are related to the generators operation constraints. Determining the market clearing prices are done by satisfying Eqs. (45)–(49). In these equations $a_{0,\,i}$ is the offered cost of availability in dollars; $m_{1,\,i}$ is the offered cost of losses for operation in the region $Q \leq 0$. $m_{2,\,i}$ is the offered cost of losses for the operation in the region $Q_{\text{base}} \leq Q$; and $m_{3,\,i} \cdot Q$ is the offered lost opportunity cost for operation in the regions $Q_A \leq Q \leq Q_B$ and $Q_{\min} \leq Q \leq Q_C$. Network security constraints are considered by Eqs. (50), (51). In these equations $V_i^{\min}$ and $V_i^{\max}$ are minimum and maximum allowable voltage limits at bus i, respectively; $|S_{ij}|$ and $S_{ij}^{\max}$ are the absolute value of the apparent power flowing over the transmission line between the nodes i and j and its allowable limit, respectively. $P_{G,\text{ slack}}$ and $P_{G,\text{ slack}}^{\max}$ are the active power generation of slack bus and its maximum limit, respectively.

Detailed and more comprehensive descriptions of the objective function, constraints, parameters, variables, and formulations of reactive power market clearing are presented in Ref. [50].

5.2 Numerical Results for Reactive Market

The proposed method is examined on the IEEE 24-bus Reliability Test System (RTS) [66]. Table 11 shows the active power output of the generators. The values of Table 11 are the output of energy market clearing problem. The $P_{\min}$ and Q_{Base} of all the generators assumed $0.40 \times (\text{MVA}_{\text{rated}})$ and $0.10 \times Q_{\max}$, respectively.

For load flow, MATPOWER [72] software is used. Each market participant can submit a four element offer. Table 12 shows these offers. Allowable voltage magnitude is limited to $\pm 5\%$range.

All the proposed algorithms are implemented on reactive market problem introduced in Ref. [50]. The best result for each algorithm is shown in Fig. 4:

The CI index is also calculated for reactive power market. Detailed results are shown in Table 13.

It is clear from Table 10 that the best solution obtained by the SAPSO4 algorithm is better because it has the highest CI index. The results are obtained through 10 separate runs, and the average is shown in the table. The solution excels the conventional PSO in both values of the objective function and runs time.

TABLE 11 Generation Scheduling [50]

Bus No.	Unit No.	Active Power (MW)
1	1	10
	2	10
	3	0
	4	60
2	1	10
	2	10
	3	75
	4	75
7	1	0
	2	90
	3	90
13	1	186.51
	2	186.51
	3	186.51
14	1	0
15	1	12
	2	12
	3	12
	4	12
	5	12
	6	140
16	1	140
18	1	400
21	1	400
22	1	0
	2	50
	3	50
	4	50
	5	50
	6	50
23	1	155
	2	0
	3	350

TABLE 12 Generating Units' Reactive Power Offers [50]

Bus No.	Unit No.	*a*0	*m*1	*m*2	*m*3
1	1	0.096	0.86	0.86	0.46
	2	0.094	0.82	0.82	0.45
	3	0.085	0.79	0.79	0.39
	4	0.083	0.82	0.82	0.4
2	1	0.05	0.54	0.54	0.28
	2	0.042	0.42	0.42	0.35
	3	0.069	0.68	0.68	0.39
	4	0.065	0.62	0.62	0.37
7	1	0.075	0.61	0.61	0.43
	2	0.08	0.75	0.75	0.36
	3	0.07	0.65	0.65	0.32
13	1	0.068	0.5	0.5	0.31
	2	0.068	0.5	0.5	0.31
	3	0.068	0.5	0.5	0.31
14	1	0.094	0.81	0.81	0
15	1	0.065	0.6	0.6	0.3
	2	0.05	0.58	0.58	0.25
	3	0.06	0.73	0.73	0.38
	4	0.055	0.61	0.61	0.27
	5	0.052	0.5	0.5	0.26
	6	0.051	0.51	0.51	0.27
16	1	0.05	0.5	0.5	0.3
18	1	0.09	0.85	0.85	0.48
21	1	0.08	0.75	0.75	0.41
22	1	0.042	0.42	0.42	0.17
	2	0.05	0.48	0.48	0.2
	3	0.045	0.42	0.42	0.38
	4	0.048	0.44	0.44	0.35
	5	0.049	0.45	0.45	0.33
	6	0.055	0.46	0.46	0.32
23	1	0.09	0.85	0.85	0.48
	2	0.095	0.89	0.89	0.55
	3	0.086	0.8	0.8	0.45

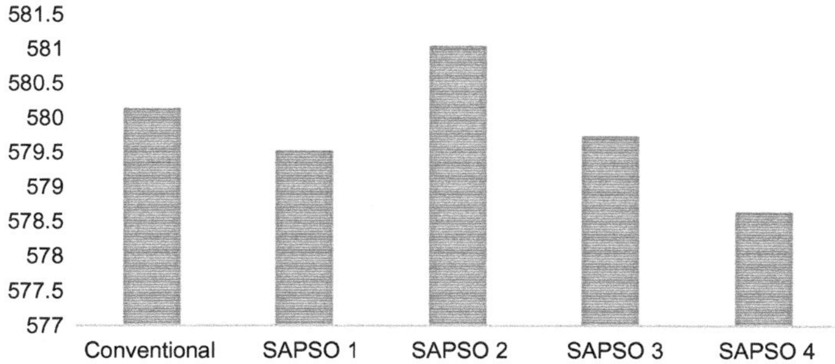

FIG. 4 Comparison between best results obtained for reactive market problem.

TABLE 13 Results for Reactive Power Market Problem

Algorithm	Value	Run Time	CI
Conventional	580.1486	185.2281	0.5705
SAPSO1	579.5306	195.1043	0.6004
SAPSO2	581.0559	223.4959	0
SAPSO3	579.7488	173.3944	0.7717
SAPSO4	578.6509	173.517	0.9988

6 CONCLUSION AND EXTENDED WORK

In this chapter, four new efficient self-sealing/self-adjusting adaptive error-driven modifications for PSO algorithms were presented. In these algorithms, the particles in PSO algorithm move based on an error in each iteration. The algorithms were first implemented on some well-known benchmarks and were compared to conventional PSO algorithm. The results showed considerable improvement of the performance of PSO algorithm. The algorithms then were applied to a real power system problem to test the efficiency of the presented algorithms on a relatively large-scale and highly constrained problem. The results showed good performance of the presented SAPSO algorithms. Future efforts should be made to improve the robustness of the results obtained by the presented SAPSO algorithms. The consideration of uncertainty in power systems is another critical issue that should be addressed when using SAPSO algorithms. The modified PSO fast convergent error driven PSO algorithms are utilized in online control optimal gain searching and online scaling of complex interconnected AC-DC Smart Grid-Renewable Energy Systems as well as optimal selection and sizing of subsystems and super capacitor/battery storage devices and effective coordinated switching and energy sharing of multi source integrated micro grid-storage systems.

REFERENCES

[1] A.M. Sharaf, A.A.A. El-Gammal, Particle swarm optimization PSO: a new tool in power system and electro technology, in: Computational Intelligence in Power Engineering, SCI, vol. 302, 2011, pp. 235–294.

[2] A.M. Sharaf, A.A.A. El-Gammal, An optimal voltage and energy utilization for a stand-alone wind energy conversion scheme WECS based on particle swarm optimization PSO, Int. J. Power Eng. Green Technol. 1 (1) (2010) 51–64. Scientific Publications.

[3] A.M. Sharaf, A.A.A. El-Gammal, in: A novel error driven dynamic tri-loop controller based on multi stage particle swarm optimization-MSPSO for Industrial PMDC Motor Drives, CPE 2009, 6th International Conference-Workshop Compatibility and Power Electronics, May, 2009, pp. 20–22.

[4] A.M. Sharaf, A.A.A. El-Gammal, Particle swarm optimization PSO: A new tool in power system and electro technology, in: B.K. Panigrahi et al., (Ed.), Computational Intelligence in Power Engineering, SCI, vol. 302, Springer-Verlag, Berlin Heidelberg, 2010, pp. 235–294. springerlink.com. Springer-Verlag, Tiergartenstrasse 17, 69121 Heidelberg, Germany.

[5] A. Chatterjee, P. Siarry, Nonlinear inertia weight variation for dynamic adaptation in particle swarm optimization, Comput. Oper. Res. 33 (2006) 859–871.

[6] L.T. Bui, et al., A modified strategy for the constriction factor in particle swarm optimization, in: 2010, ed., Book Series Lecture Notes in Computer Science, vol. 4828, Springer Berlin, Heidelberg, 2010, pp. 333–344.

[7] A.M. Sharaf, A.A.A. El-Gammal, A novel PSO-self regulating micro hydro-fuel cell green energy management scheme for hydrogen and island electricity generation, J. Assoc. Prof. Eng. Trinidad and Tobago 39 (1) (2010) 24–36. ISSN 1000 7924.

[8] A.M. Sharaf, A.A.A. El-Gammal, A novel particle swarm optimization PSO-based dynamic green power filter-compensator for radial distribution systems, Int. J. Power Eng. Green Technol. 1 (1) (2010) 11–37. Scientific Publications.

[9] A.M. Sharaf, A.A.A. El-Gammal, A multi objective multi-stage particle swarm optimization MOPSO search scheme for power quality and loss reduction on radial distribution system, RE&PQJ-7, Renew. Energy Power Qual. J. 7 (2009).

[10] A.M. Sharaf, A.A.A. El-Gammal, Optimal design of hybrid power filter compensator using Multi-Objective Particle Swarm Optimization (MOPSO), Al-Azhar Univ. Eng. Int. Sci. J. (2009). ISSN 1110-6409.

[11] J. Kennedy, R.C. Eberhart, in: Particle swarm optimization, Proceedings of IEEE International Conference on Neural Networks, Piscataway, NJ, 1995, pp. 1942–1948.

[12] R. Eberhart, Y. Shi, in: Comparison between genetic algorithms and particle swarm optimization, Proceedings of the Seventh Annual Conference on Evolutionary Programming, Springer-Verlag, San Diego, 1998, pp. 611–619.

[13] E.O. Wilson, Sociobiology: The New Synthesis, Belknap Press, Harvard, 1975.

[14] M. Imran, R. Hashima, N.E.A. Khalidb, An overview of particle swarm optimization variants, Proc. Eng. 35 (2013) 491–496.

[15] C.W. Reynholds, Flocks, Herds and schools: a distributed behavioral model, Comput. Graph. 21 (4) (1987) 25–34.

[16] L. Bo, et al., An effective PSO-based memetic algorithm for flow shop scheduling, IEEE Trans. Syst. Man Cybern. B Cybern. 37 (2007) 18–27.

[17] S. Changyin, et al., in: Clustering with a weighted sum validity function using a niching PSO algorithm, 2007 IEEE International Conference on Networking, Sensing and Control, 2007, pp. 368–373.

[18] M. Qianzhi, et al., in: Mobile Robot Path Planning with Complex Constraints Based on the Second-Order Oscillating Particle Swarm Optimization Algorithm, Computer Science and Information Engineering WRI World Congress on, 2009, 2009, pp. 244–248.
[19] M.H. Amini, et al., Simultaneous allocation of electric vehicles' parking lots and distributed renewable resources in smart power distribution networks, Sustain. Cities Soc. (2017) 332–342.
[20] G. Abbas, et al., Solution of an Economic Dispatch Problem Through Particle Swarm Optimization: A Detailed Survey—Part II, IEEE Access (2017) 24426–24445.
[21] L. Zhixiong, W. Shaomei, Hybrid Particle Swarm Optimization for Permutation Flow Shop Scheduling, Intelligent Control and Automation, 2006. WCICA 2006. The Sixth World Congress on, 2006, pp. 3245–3249.
[22] S. Chandrasekaran, et al., "Multi-objective particle swarm optimization algorithm for scheduling in flowshops to minimize makespan, total flowtime and completion time variance, in evolutionary computation, 2007. CEC 2007. IEEE Congress on, 2007, pp. 4012–4018.
[23] C.-t. Cheng, et al., Comparison of particle swarm optimization and dynamic programming for large scale hydro unit load dispatch, Energy Convers. Manag. 50 (2009) 3007–3014.
[24] W.C. Yeh, A two-stage discrete particle swarm optimization for the problem of multiple multi-level redundancy allocation in series systems, Expert Syst. Appl. 36 (2009) 9192–9200.
[25] O. Schutze, et al., in: a memetic PSO algorithm for scalar optimization problems, Swarm Intelligence Symposium, 2007. SIS 2007. IEEE, 2007, pp. 128–134.
[26] H. Chen, et al., rfid network planning using a multi-swarm optimizer, J. Netw. Comput. Appl. 34 (3) (2011) 888–901.
[27] S.Z. Zhao, et al., in: Dynamic multi-swarm particle swarm optimizer with local search for large scale global optimization, in evolutionary computation, CEC 2008. (IEEE World Congress on Computational Intelligence). IEEE Congress on, 2008, pp. 3845–3852.
[28] M.R. AlRashidi, M.E. El-Hawary, Hybrid particle swarm optimization approach for solving the discrete OPF problem considering the valve loading effects, IEEE Trans. Power Syst. 22 (2007) 2030–2038.
[29] A. Nickabadi, et al., in: DNPSO: A dynamic niching particle swarm optimizer for multi-modal optimization, in evolutionary computation, CEC 2008. (IEEE world congress on computational intelligence). IEEE Congress on, 2008, pp. 26–32.
[30] M.R. AlRashidi, M.E. El-Hawary, in: Emission-economic dispatch using a novel constraint handling particle swarm optimization strategy, in electrical and computer engineering, CCECE '06. Canadian Conference on, 2006, 2006, pp. 664–669.
[31] M.A. Abido, Multiobjective particle swarm optimization for environmental/economic dispatch problem, Electr. Power Syst. Res. 79 (2009) 1105–1113.
[32] Z. Ruiyou, W. Dingwei, in: Forecasting annual electricity demand using BP neural network based on three sub-swarms PSO, Control and Decision Conference, 2008. CCDC 2008. Chinese, 2008, pp. 1409–1413.
[33] A.C. Briza, P.C. Naval Jr., Stock trading system based on the multi-objective particle swarm optimization of technical indicators on end-of-day market data, Appl. Soft Comput. 11 (1) (2011) 1191–1201.
[34] J. Zhang, et al., in: Multi-sub-swarm particle swarm optimization algorithm for multimodal function optimization, in evolutionary computation, CEC 2007. IEEE Congress on, 2007, 2007, pp. 3215–3220.
[35] R. Brits, et al., Solving systems of unconstrained equations using particle swarm optimization, Systems, Man and Cybernetics IEEE International Conference on, 2002, vol. 3, 2002. 6 pp.

[36] C. Chun-Hong, et al., in: the geometric constraint solving based on memory particle swarm algorithm, in machine learning and cybernetics, Proceedings of 2004 International Conference on, 2004, 4 2004, pp. 2134–2139.
[37] C.K. Goh, et al., A competitive and cooperative co-evolutionary approach to multi-objective particle swarm optimization algorithm design, Eur. J. Oper. Res. 202 (2010) 42–54.
[38] X. Liu, et al., Particle swarm optimization based on dynamic niche technology with applications to conceptual design, Adv. Eng. Softw. 38 (2007) 668–676.
[39] P.Y. Yin, A discrete particle swarm algorithm for optimal polygonal approximation of digital curves, J. Vis. Commun. Image Represent. 15 (2004) 241–260.
[40] X. Yang, et al., A modified particle swarm optimizer with dynamic adaptation, Appl. Math. Comput. 189 (2007) 1205–1213.
[41] A. Unler, A. Murat, A discrete particle swarm optimization method for feature selection in binary classification problems, Eur. J. Oper. Res. 206 (2010) 528–539.
[42] L. Benameur, et al., in: A new hybrid particle swarm optimization algorithm for handling multiobjective problem using fuzzy clustering technique, in computational intelligence, modelling and simulation, CSSim '09. International Conference on, 2009, 2009, pp. 48–53.
[43] B. Alatas, E. Akin, Multi-objective rule mining using a chaotic particle swarm optimization algorithm, Knowl. Based Syst. 22 (2009) 455–460.
[44] W.C. Yeh, et al., A new hybrid approach for mining breast cancer pattern using discrete particle swarm optimization and statistical method, Expert Syst. Appl. 36 (2009) 8204–8211.
[45] S. Sivasubramani, S. Swarup, in: Multi-agent based particle swarm optimization approach to economic dispatch with security constraints in power systems, ICPS '09. International Conference on, 2009, 2009, pp. 1–6.
[46] J. Cai, et al., A multi-objective chaotic particle swarm optimization for environmental/economic dispatch, Energy Convers. Manag. 50 (2009) 1318–1325.
[47] W. Zhenzhen, X. Hancheng, in: Dynamic-probabilistic particle swarm synergetic model: a new framework for a more in-depth understanding of particle swarm algorithms in evolutionary computation, CEC 2008. (IEEE world congress on computational intelligence). IEEE Congress on, 2008, pp. 312–321.
[48] A. Stacey, et al., in: Particle swarm optimization with mutation evolutionary computation, The 2003 Congress on CEC '03, vol. 2, 2003, pp. 1425–1430.
[49] K.M. Bryden, et al., Graph-based evolutionary algorithms, IEEE Trans. Evol. Comput. 10 (2006) 550–567.
[50] O. Homaee, S. Jadid, Investigation of synchronous generator in reactive power market–an accurate view, IET Gener. Transm. Distrib. 8 (2014) 1881–1890.
[51] H. Mavalizadeh, et al., Probabilistic multi-objective generation and transmission expansion planning problem using normal boundary intersection, IET Gener. Transm. Distrib. 9 (2015) 560–570.
[52] V.K. Kalivarapu, Improving solution characteristics of particle swarm optimization through the use of digital pheromones, parallelization, and graphical processing units (GPUs), Retrospective Theses and Dissertations, Paper 15700, 2008.
[53] O. Abedinia, et al., Multiobjective environmental and economic dispatch using imperialist competitive algorithm, Int. J. Tech. Phys. Problems Eng. 4 (2012) 63–70.
[54] C.A. Canizares, et al., Re-defining the reactive power dispatch problem in the context of competitive electricity markets, IET Gener. Transm. Distrib. 4 (2010) 162–177.
[55] K. Bhattacharya, J. Zhong, Reactive power as an ancillary service, IEEE Trans. Power Syst. 16 (2001) 294–300.

[56] FERC Order No. 888: Promotion of wholesale competition through open access non-discriminatory transmission services by public utilities and recovery of stranded costs by public utilities and transmitting utilities, 1996.
[57] A. Rabiee, et al., Reactive power market development considering power system security, Electr. Eng. 92 (2010) 151–164.
[58] J. Zhong, K. Bhattacharya, in: Reactive power management in deregulated electricity markets—a review, IEEE Power Engineering Society Winter Meeting, 2002.
[59] J. Zhong, K. Bhattacharya, Toward a competitive market for reactive power, IEEE Trans. Power Syst. 17 (2002) 1206–1215.
[60] A. Rabiee, et al., A new framework for reactive power market considering power system security, Iran J. Electr. Electron. Eng. 5 (2009) 196–204.
[61] A. Rabiee, et al., Multiobjective clearing of reactive power market in deregulated power systems, Appl. Energy 86 (2009) 1555–1564.
[62] P. Kundur, Power System Stability and Control, McGraw-Hill, New York, 1994.
[63] A.H. Khazali, M. Kalantar, A. Khazali, Fuzzy multi-objective reactive power clearing considering reactive compensation sources, Energy 36 (2011) 3319–3327.
[64] A. Rabiee, H.A. Shayanfar, N. Amjady, Coupled energy and reactive power market clearing considering power system security, Energy Convers. Manag. 50 (2009) 907–915.
[65] Reliability Test System Task Force, The IEEE reliability test system—1996, IEEE Trans. Power Syst. 14 (3) (1999) 1010–1020.
[66] R.D. Zimmerman, C.E. Murillo-Sánchez, R.J. Thomas, MATPOWER: steady-state operations, planning and analysis tools for power systems research and education, IEEE Trans. Power Syst. 26 (1) (2011) 12–19.

FURTHER READING

[67] D. Santos-Martina, S. Arnaltes, J.L. Rodriguez Amenedo, Reactive power capability of doubly fed asynchronous generators, Electr. Power Syst. Res. 78 (2008) 1837–1840.
[68] N. Amjady, A. Rabiee, H.A. Shayanfar, A stochastic framework for clearing of reactive power market, Energy 35 (2010) 239–245.

Chapter 4

Implementation of Radial Movement Optimization (RMO) Algorithm for Solving Economic Dispatch and Fixed Head Hydrothermal Generation Scheduling

M. Vanithasri, R. Balamurugan and L. Lakshminarasimman
Electrical and Electronics Engineering, Annamalai University, Chidambaram, India

NOMENCLATURE

BBO	biogeography based optimization
BSA	backtracking search algorithm
CBA	chaotic bat algorithm
CQGSO	continuous quick group search optimizer
CRO	chemical reaction optimization
DE	differential evolution
DSD	dimensional steepest decline
ED	economic dispatch
EHNN	enhanced hopfield neural network
GA	genetic algorithm
GSO	group search optimizer
HGA	hybrid genetic algorithm
HM	hierarchical method
HMAPSO	hybrid multiagent-particle swarm optimization
IDP	improved dynamic programming
LMO	Levenberg-Marquardt optimization
MDE	modified differential evolution
MTS	modified Tabu search

Classical and Recent Aspects of Power System Optimization. https://doi.org/10.1016/B978-0-12-812441-3.00004-5
© 2018 Elsevier Inc. All rights reserved.

NPSO	new particle swarm optimization
NR	Newton-Raphson
PSM	pattern search method
PSO	particle swarm optimization
QPSO	quantum-inspired particle swarm optimization
RCGA	real coded genetic algorithm
RMO	radial movement optimization
SA	simulated annealing
SHDE	self-tuning hybrid differential evolution
TS	Tabu search
a_i, b_i, and c_i,	fuel cost coefficients of the ith generating unit
B	transmission loss coefficient matrix (MW^{-1})
B_{ij}	loss coefficients
C_1 and C_2	RMO constants
C_k	energy balance constraint in interval k
cp	center point
DR_i	down-ramp limits of the ith generating unit (MW/h)
e_i, and f_i	fuel cost coefficients of valve point effect of generators
$F_i(P_i)$	fuel cost function of the ith thermal generating unit in the system ($/h)
F_sP_{sk}	cost function of each generating unit in interval k
m	number of thermal unit
n	number of hydro unit
N_g	number of generating units in the system
P_D	total power demand of the system (MW)
P_{Dk}	total generation demand in interval k (MW)
P_{hk}^{min} and P_{hk}^{max}	lower and upper generation limits for hydro unit h in interval k
P_i	power generated by the ith thermal generating unit in the system (MW)
P_i^0	previous power output of the ith thermal generating unit (MW)
P_i^{min} and P_i^{max}	minimum and maximum power generation limits of ith generator (MW)
P_L	total system losses (MW)
P_{lossk}	transmission loss in interval k
P_{Lk}	transmission loss in interval k (MW)
P_{sk} and P_{hk}	output of thermal and hydro units in interval k, respectively
P_{sk}^{min} and P_{hk}^{max}	lower and upper generation limits for thermal units in interval k
q_{hk}	rate of water flow from hydro unit h in interval k
T_k	time in interval k
UR_i	upper ramp limits of the ith thermal generating unit (MW/h)

V_h	prespecified volume of water available for generation by hydro unit
V_{ij}	velocity vector
W	inertia weight
W_h	water availability constraint of hydro unit h during the scheduling period
$X_{\max}(j)$ and $X_{\min}(j)$	maximum and minimum limits of jth control variable
λ	Lagrange multiplier
Φ	penalty factor

1 INTRODUCTION

Today, electrical power plays a vital role in our day-to-day. Developments in transportation, industrial, agricultural, entertainment, information, and communication depend heavily on consist access to electrical energy. In other words, the modern economy is entirely reliant on electricity. This led to a surge in the number of power generating stations and a consequent increase in power transmission lines which connect the generating stations to the load centres. Interconnections between generating systems are equally important for a reliable supply of power, which also provide flexibility in system operation [1]. Among the different issues in power systems operation, economic dispatch (ED) and hydro-thermal scheduling (HTS) are found to play a major part.

The ED occupies a prominent place in power systems operation and control. It aims to determine the optimal power outputs with minimum cost in order to meet the load demand subject to satisfying various operational constraints over finite dispatch periods. Similar to most real-world complex engineering optimization problems, the nonlinear and nonconvex characteristics are more prevalent in the ED problem. Therefore, it is possible that computational methods may not yield a global extremum as many local extrema may be encountered, and, in this case, obtaining a truly optimal solution presents a challenge.

In the past decade, the field of global optimization has been very attractive, producing different kinds of deterministic and stochastic algorithms for optimization in the continuous domain. The mathematical methods may not able to provide a global optimal solution for practical economic dispatch problem due to its limitation. Among the stochastic approaches, evolutionary computation offers a number of exclusive advantages: robust and reliable performance, global search capability, etc. These characteristics of evolutionary computation, as well as other supplementary benefits such as ease of implementation, parallelism, no requirement for a differentiable or continuous objective function, etc., make it an attractive choice [2].

Consequently, there have been many studies related to real parameter optimization using evolutionary computation, resulting in many variants such as genetic algorithm (GA), particle swarm optimization (PSO), differential evolution (DE), simulated annealing (SA), evolutionary programming (EP), artificial immune system (AIS), Tabu search (TS), ant colony optimization (ACO), improved particle swarm optimization (IPSO), improved differential evolution (IDE), quantum-inspired evolutionary algorithm (QIEA), artificial bee colony (ABC), integer coded genetic algorithm (ICGA), bacterial foraging (BF), imperialistic competition algorithm (ICA), harmony search algorithm (HSA), Lagrangian relaxation and particle swarm optimization (LR-PSO), straightforward (SF), extended priority list (EPL), advanced three stage approach (ATHS), etc.

Despite the extensive work carried out, developments of consistent techniques to solve the realistic economic dispatch problem are still evolving. A solution to the problem of optimization of load dispatch would be very elaborate and burdensome if calculations are to be made point by point. Hence, a fast method must be invoked for this purpose. In view of nonlinear characteristics of the thermal generating units, there is a need for novel techniques to provide an efficient generation schedule of nonconvex economic dispatch problem.

In this chapter, a recent stochastic optimization method, radial movement optimization is chosen for determining the solutions for economic dispatch problems with nonsmooth fuel cost functions and fixed head hydro thermal scheduling (FHHTS).

2 RADIAL MOVEMENT OPTIMIZATION (RMO)

An efficient tool for global optimization of multivariable complex system, called radial movement optimization, was developed by Ref. [3]. The swarm-based stochastic optimization technique is similar to other evolutionary methods like PSO and DE. The search space in RMO is a vector space wherein a location vector associated with a particle in space proposes a solution to the problem. The movement style of the particle in RMO differentiates it from other optimization methods. The particles get scattered from the center point and are constantly updated at each step in RMO. In a three dimension search space, the particles with different velocities are seen spread along the radii of a sphere. The location of the particle is evaluated by an objective function to return an objective value and, in turn, the location of the best particle is obtained. The unique feature of this method is that the radial movement and updating of particles enhances the exploration and exploitation capabilities to converge on global optimal values.

RMO has an edge over other search techniques as it requires a smaller memory, and can, therefore, carry out better and denser search around the target. This is possible due to the fact that the particle starts moving from an updated point at every iteration and the location and velocity for the particle is not transferred between iteration. Other search techniques, in contrast, are suitable only for small and medium sized problems as the computational cost increases with size.

This technique is also robust in the sense that the global best vector in the updating process presents the starting of algorithm in a local optimum.

The objective function returns a fitness value for the particle at the end of each generation step. The location of the best fitness value for each generation obtained among all evaluated particles is stored as the best solution, R_{best}. Similarly, the best solution obtained among all the generations is stored as G_{best}. Either the number of generations or a definition of a small error for G_{best} from a desired value, or both, serve as the termination criteria depending on the choice of test function. The following section summarizes RMO.

2.1 Step-I: Initialization

RMO algorithm, like any other stochastic optimization method, is triggered by the initialization of the particles in the search space. This is a representation of the solutions to the optimization problem. A nop × nod matrix called X_{ij} demonstrates the location of the particle. Nop represents the number of particles which is fixed by the user and nod represents the number of dimensions, which is usually the number of variables that needs to be obtained and will remain constant during the trials.

The first step in the initialization is the random assignment of all possible locations to particles in the entire dimensional search space of nop × nod matrix.

Eq. (1) shows the location matrix for the algorithm.

$$X_{ij} = \begin{bmatrix} X_{1,1} & X_{1,2} & \cdots & X_{1,n} \\ X_{2,1} & \ddots & \cdots & \vdots \\ \vdots & \cdots & \ddots & \vdots \\ X_{m,1} & X_{m,2} & \cdots & X_{m,n} \end{bmatrix} \tag{1}$$

where $i = 1, 2, 3, \ldots,$ nop and $j = 1, 2, 3, \ldots,$ nod.

A sample of this initialization is given below:

$$X_{ij} = X_{\min(j)} + rand\,(0,1) \times \left(X_{\max(j)} - X_{\min(j)}\right) \tag{2}$$

where $i = 1, 2, 3, \ldots,$ nop and $j = 1, 2, 3, \ldots,$ nod.

$X_{\min(j)}$ and $X_{\max(j)}$ demonstrate the constraints for the jth dimension, which should be determined at the start. The rand (0,1) can be obtained from a normal distribution function, such as the Gaussian Distribution, between 0 and 1. Another initialization method is shown in Eq. (3). The simple random optimization, shown in Eq. (2), and/or in Eq. (3) can be used.

$$\text{step}_j = \frac{X_{\max(j)} - X_{\min(j)}}{\text{nop}} \tag{3}$$

$$\text{for } i = 2, \ldots.\text{nop}$$

Alternatively, the method for initialization of the particle, as described by Eq. (3), shows that effect of the scattering of particles. The particles are scattered by a constant distance and placed uniformly along j^{th} dimension. The placement of the particle is in proportion to the length of the jth dimension and the number of particles, nop. This exploits all locations of the solution space to the fullest and locates the center point through uniform evaluation.

2.2 Step-II: Particle Movement

The next step after obtaining the center point is to sprinkle the particles and make them move along the radii in straight lines from center point (cp). Velocity vector, v_{ij}, a nop*nod random vector determines the movement of the particles and is obtained by the equation as given below:

$$V_{ij} = rand\,(0,1) \times V_{\max(j)} = \begin{cases} V_{\max(j)} = \dfrac{X_{\max(j)} - X_{\min(j)}}{k} \\ i = 1,2,\ldots\ldots,\text{nop} \\ j = 1,2,\ldots\ldots,\text{nod} \end{cases} \tag{4}$$

The coefficient k must be an integer and is carefully chosen. An inertia weight, w, with a descending value decreasing with the runs, is attached to the velocity vector and its relation with the generation run is shown below

$$W_k = W_{\max} - \left(\frac{W_{\max} - W_{\min}}{\text{Generation}_{\max}}\right) \times \text{Generation}_k \tag{5}$$

$$V_{ij}^k = W_k \times rand\,(0,1) \times V_{\max(j)} \tag{6}$$

The minimum and maximum weights, $W_{\min}$ to $W_{\max}$, fixed by the users, drives the effect of the velocity vector, V_{ij}^k on the particle movement. The radial movement of the particles from the centre point, cp is depicted in Ref. [3]. The sphere where the particles are sprinkled has boundaries, which is $V_{\max}$.

2.3 Step-III: Evaluation of the Fitness of the Particle

The particle with best fitness is denoted as the radial best (R_{best}), and its location along with associated fitness value gives the R_{best} particle. Then the update vector updates a new center point location by making use of G_{best} and R_{best} particles. Once the cp is updated the process starts all over again with the sprinkling of the particles from the updated cp and the resulting R_{best} is compared with that of G_{best} of previous generation and thus, the G_{best} gets updated. If R_{best} proves to be a better solution, then it replaces the location of G_{best}. The process continues till the termination criteria are met.

$$cp^{k+1} = cp^k + up \tag{7}$$

$$up = C_1 \times \left(G_{\text{best}} - cp^k\right) + C_2 \times \left(R_{\text{best}} - cp^k\right) \tag{8}$$

C_1 and C_2 are the coefficients that must be set prior to running the algorithm. The two coefficients, C_1 and C_2 of RMO have a direct implication on the values of G_{best} and R_{best}, respectively. These two coefficients allow the user to fix the optimization level as required for specific applications. As far as C_1, the coefficient of G_{best} is concerned, larger is the C_1 if the search space is large and if the problem is nonlinear. Higher values of C_1 increase the pace of convergence but reduce the quantity of optimization and can typically be fixed in the range of 0.4–0.9, but should not exceed unity. C_2, the coefficient of Rbest, can be chosen in line with the range of C_2 but should be greater than C_1. Higher values of C_1 and C_2 are preferred if a fast solution is desired but lower values are usually better for unknown functions.

After updating the *cp*, the sprinkling of the particles will start again from the new *cp*. In the following generation, the value of the G_{best} from the previous generation is updated through comparing with the value of the R_{best} obtained during the sprinkling of the particles in the new generation run. If the R_{best} proposes a better solution than G_{best}, the location of G_{best} must be replaced with R_{best}. The process will continue until the G_{best} reaches a certain defined value or the number of generations reaches its defined maximum value.

3 ECONOMIC DISPATCH

In modern power systems, scarcity of energy resources, increasing power generation costs, and an ever-growing demand for energy necessitate optimal ED. Economic dispatch is a generation allocation problem and defined as the process of calculating the generation of the generating units so that the system load is supplied entirely and most economically, subject to the satisfaction of the constraints.

The main objective of economic dispatch is to reduce the total power generation cost while satisfying various equality and inequality constraints. Traditionally, in ED problems, the cost function for generating units is termed as a quadratic function. Due to physical restriction of power plant components, thermal generating units may have prohibited operating zones lying between their minimum and maximum power outputs. The fuel cost curve of a unit with prohibited operating zones is a discontinuous function. Prohibited zones arise from physical limitations of individual power plant components. The physical limitations may lead to instabilities in operation for certain loads. To avoid these instabilities, the concept of prohibited operating zones has been developed. The power generation of thermal units in prohibited operating areas should be avoided. The presence of prohibited operating zones for individual generators leads to a solution space with disjoint feasible regions.

In the actual operating processes of the generating unit, the operating range of all the committed thermal units are actually restricted by their ramp-rate limits between two operating periods. The generation may increase or decrease with corresponding upward and downward ramp-rate limits of units. In reality, unit input–output curves do not exhibit monotonically increasing shape.

In the traditional economic dispatch problem, valve point effects are ignored. This approximation introduces an inaccuracy into the resulting dispatch. Large steam turbine generators have a number of steam admission valves that are opened in sequence as the unit output is increased. Whenever an additional valve is first opened, the throttling losses increase rapidly and the incremental heat rate rises suddenly. The valve point effects introduce ripples in the heat rate curves of thermal units. The generating units with multivalve steam turbines exhibit great variation in the fuel cost function. The incremental fuel cost function should be modeled in a more practical way by including valve point effects. Therefore, the generator cost functions are represented by quadratic functions with sine components. The superimposed sine components represent the rippling effects produced by the steam admission valve opening. The valve point loading, prohibited operating zones, ramp-rate limits, and other constraints turn the decision space into disjoint subsets, transforming the economic dispatch problem into a hard, nonconvex optimization problem [2].

Traditional dispatch algorithms employ Lagrangian multipliers and require monotonically increasing incremental cost curves. The input-output characteristics of modern units are inherently highly nonlinear because of valve point loadings, ramp-rate limits, etc., and furthermore they may generate multiple local minimum points in the cost function. Classical dispatch algorithms require that these characteristics be approximated; however, such approximations are not desirable as they may lead to suboptimal operation and hence, a huge revenue loss over time [4]. In light of the nonlinear characteristics of the units, there is a demand for techniques that do not have restrictions on the shape of the fuel cost curves. The stochastic algorithms may prove to be very effective in solving nonlinear ED problems without any restrictions on the shape of the cost curves.

3.1 Problem Formulation

3.1.1 Objective Function

The objective of ED is the minimization of thermal unit fuel cost function. It is given by

$$\min F = \sum_{i=1}^{N_g} F_i(P_i) = \sum_{i=1}^{N_g} \left(a_i P_i^2 + b_i P_i + c_i\right) \tag{9}$$

where, N_g, total number of generating units in the system; $F_i(P_i)$, fuel cost of ith generating unit in \$/h; P_i, power generated by the ith generating unit in MW; a_i, b_i, c_i, fuel cost coefficients of ith generator.

3.1.2 ED With Valve Point Effect

Due to the valve point effect, the real input-output cost curves for the generator units are nonconvex. With valve point effect, the objective function becomes,

$$\min\ F=\sum_{i=1}^{N_g}F_i(P_i)=\sum_{j=1}^{N_g}\left(a_iP_i^2+b_iP_i+c_i\right)+\left|e_i\sin\left(f_i\left(P_i^{\min}-P_i\right)\right)\right| \quad (10)$$

where e_i and f_i—fuel cost coefficients of valve point effect of generators.

3.1.3 Optimization Constraints

For ED problems, real power balance criterion and real power generation limits are the equality and inequality constraints. They are represented by.

$$\sum_{i=1}^{N_g}P_i=P_D+P_L \quad (11)$$

$$P_i^{\min}\le P_i\le P_i^{\max} \quad (12)$$

where, P_i, generation of ith generating unit in MW; P_D, total power demand in MW; $P_i^{\min}$ and $P_i^{\max}$, minimum and maximum power generation limits of ith generator; P_L, total transmission losses in MW.

The transmission loss is calculated using Kron's formula (B-coefficients):

$$P_L=\sum_{i=1}^{N_g}\sum_{j=1}^{N_g}P_iB_{ij}P_j+\sum_{j=1}^{N_g}B_{oj}P_j+B_{oo} \quad (13)$$

where P_i and P_j, real power injection at i^{th} and j^{th} generator respectively; B_{ij}, loss coefficients which can be assumed to be constant under normal operating conditions.

3.1.4 ED with Consideration of Prohibited Operating Zones (POZ)

The vibrations in the shaft bearings and/or steam valve operation causes the prohibited zones. The operating zones of unit i is:

$$P_i\in\begin{cases}P_i^{\min}\ \le\ P_i\ \le\ P_{i,1}^{l}\\ P_{i,k-1}^{u}\ \le\ P_i\ \le\ P_{i,k}^{l},\ \ k=2,3,\ldots\ldots\ldots,n\\ P_{i,n_i}^{u}\le\ P_i\le\ P_i^{\max}\end{cases} \quad (14)$$

where n_i is the number of prohibited zones of unit j. k is the index of prohibited zones of a unit i. l and u denote the lower bound and upper bound of the prohibited zone of the generator.

3.1.5 ED with Considerations of Ramp Rate Limits

The physical limitations of starting up and shutting down generators imposes ramp rate limits. The increase and decrease in generation is limited by the following equations

$$P_i - P_i^0 \le UR_i \tag{15}$$

$$P_i^0 - P_i \le DR_i \tag{16}$$

where P_i^0, previous output power of the ith thermal generating unit in MW; UR_i, DR_i, upper ramp and down ramp limit respectively of ith generator.

Combining Eqs. (15), (16) with (12)

$$P_{id} \le P_i \le P_{iu} \tag{17}$$

where

$$P_{id} = \max\left(P_i^{\min}, P_i^0 - DR_i\right) \tag{18}$$

$$P_{iu} = \min\left(P_i^{\max}, P_i^0 + UR_i\right) \tag{19}$$

Combining this with Eq. (10)

$$\min\ F = \sum_{i=1}^{N_g} F_i(P_i) = \sum_{j=1}^{N_g} \left(a_i P_i^2 + b_i P_i + c_i\right) + \left|e_i \sin\left(f_i\left(P_i^{\min} - P_i\right)\right)\right| \tag{20}$$

Subject to $\sum_{i=1}^{N_g} P_i = P_D + P_L.$

$$\max\left(P_i^{\min}, P_i^0 - DR_i\right) \le P_i \le P_{i,1}^l \tag{21}$$

$$P_{i,k-1}^u \le P_i \le P_{i,k}^l, k = 2, 3, \ldots n_j, i = 1, 2, 3, \ldots N_g$$

$$P_{i,n_i}^u \le P_i \le \min\left(P_i^{\max}, P_i^0 + UR_i\right)$$

3.2 Radial Movement Optimization Algorithm for ED Problem

The steps for RMO for parameters estimation of fuel cost functions is given below.

3.2.1 Step-1: Initialization

The initial particles in the search space are randomly generated within their limits. The control variables are real power generation of units and the size of initial population is $m \times n$ matrix. Here m is the number of particles and n is the number of control variables. The initial population of RMO is represented as

$$X_{ij} = \begin{bmatrix} X_{1,1} & X_{1,2} & \cdots & X_{1,n} \\ X_{2,1} & \ddots & \cdots & \vdots \\ \vdots & \cdots & \ddots & \vdots \\ X_{m,1} & X_{m,2} & \cdots & X_{m,n} \end{bmatrix} \tag{22}$$

where $i = 1, 2, 3\ldots.. m$ and $j = 1, 2, \ldots.n$.

3.2.2 Step-2: Generate the Initial Velocity Vector

The velocity vector V_{ij} is randomly generated using the following equation

$$V_{ij} = rand\,(0,1) \times V_{\max(j)}$$

$$V_{\max}(j) = \frac{X_{\max(j)} - X_{\min(j)}}{K} j = 1,2,\ldots.n \tag{23}$$

where, K is an integer number which is chosen judiciously. $X_{\max(j)}$ and $X_{\min(j)}$ are the maximum and minimum limits of jth control variable, respectively. The velocity vector has an inertia weight (W) associated with it, which is chosen as constant in this method.

3.2.3 Step-3: Locate the Center Point

The fitness of each particle in the population vector is estimated using the following equation

$$F_k = \sum_{i=1}^{N_g} F_i(P_i) + \Phi \sum_{i=1}^{N_g} P_i - (P_D + P_L) \tag{24}$$

where, $k = 1, 2, \ldots.m$; Φ, penalty factor and it is positive real number. It is fixed at large value; P_D, total power demand in MW; P_L, total system losses in MW.

The population vector which gives the minimum fitness is chosen as the initial centre point (cp) and it is the G_{best} vector in the initial population.

3.2.4 Step-4: Find the Radial Movement of the Particle

The radial movement of the particles from the centre point, cp is calculated by

$$U_{ij} = W \times V_{ij} + \text{center}\,(j) \quad i = 1,2,\ldots m; j = 1,2,\ldots n \tag{25}$$

W is the inertia weight which is calculated by

$$W = W_{\max} - \left(\frac{W_{\max} - W_{\min}}{Iteration_{\max}} \right) \times Iteration \tag{26}$$

where, $W_{\max} = 1$ and $W_{\min} = 0$.

Here the particles are sprinkled. If any control variable in the population vector U_{ij} violates their maximum or minimum limits then that variable is fixed at their violated limit.

3.2.5 Step-5: Evaluate the Fitness of the Particles

The fitness of each vector in the population U_{ij} is evaluated using Eq. (24). The particle which gives the minimum fitness is selected as radial best (R_{best}) vector.

3.2.6 Step-6: Search Around G_{best}

The convergence of radial movement optimization is improved by generating number of trial vectors around the G_{best} using

$$Y_{ij} = G_{\text{best}} + unifrnd(0,1) * \left(U_{r1j} - U_{r2j} \right) \quad i = 1,2,\ldots.m;\; j = 1,2,\ldots n \tag{27}$$

where r_1 and r_2 are randomly selected vector which are mutually different from each other. The fitness of each vector in Y_{ij} is evaluated and are compared with G_{best}. If the fitness of trial vector is better than G_{best}, trial vector replaced the G_{best} vector. This procedure is repeated for Nt times for each vector in every iteration.

3.2.7 Step-7: Determine the New Center Location

The new center location (Center 1) is updated using the following equation

$$\text{Center}^{K+1} = \text{Center}^{K} + \left(C_1\left[G_{\text{best}} - \text{Center}^{K}\right]\right) + \left(C_2\left[R_{\text{best}} - \text{Center}^{K}\right]\right) \quad (28)$$

where K is the iteration number.

In this study, the values of C_1 and C_2 are chosen as 0.8 and 0.9 respectively and the population size is 50. The fitness of radial best vector (R_{best}) is compared with the fitness of global best (G_{best}) vector. If R_{best} is better than G_{best}, G_{best} is replaced by R_{best}. The iteration is incremented by 1. Steps 2–7 are repeated for predefined iterations.

3.3 Results and Discussion

In this work, the radial movement optimization method is implemented on four different ED problems, that is, 6-unit, 13-unit, 20-unit, and 40-unit systems with complexities are tested. The practical constraints such as ramp rate limits and prohibited zones of units are taken into account. At each test system, under the same evaluation function and individual definition, 20 trials are performed using the RMO method to obtain the best results. The simulation results obtained are compared with other algorithms reported in the literature. The loss coefficients matrix of power system network is employed to determine the transmission line loss. The software is written in MATLAB 2011 and executed in an Intel core i3 CPU, which has 2 GB RAM.

3.3.1 Test System-I

Initially, the RMO algorithm is applied on a small test system, consisting of six generating units with a load demand of 1263 MW. The test system includes transmission loss, prohibited operating zone (POZ), and ramp rate limits. The system data and the B coefficients for transmission loss for 6-unit system are taken from Gaing [5]. The total generations (TG) for power demands of 1263 MW, along with the fuel total cost (TC) and power loss for the six-generating-unit system using RMO are given in Table 1. From the table, it is observed that the optimal total cost for six-generator system using RMO is 15,449.5822 \$/h and the transmission loss is 12.9444 MW. From the simulation results, it is also observed that the generation values satisfy the generation limit constraints and do not fall in the prohibited operating zones.

TABLE 1 Simulation Results of Economic Dispatch for 6-Unit Systems Using RMO

Method (Unit)	P_1 (MW)	P_2 (MW)	P_3 (MW)	P_4 (MW)	P_5 (MW)	P_6 (MW)	TG (MW)	Loss (MW)	TC ($/h)
RMO	447.5010	173.3161	263.4606	139.0630	165.4712	87.1325	1275.9444	12.9444	15,449.5822

A comparison has been made with the results obtained by RMO and other algorithms, that is, PSO [5], GA [5], MDE [6], IDP [7], NPSO [8], DE [9], MTS [10], TS [10], SA [10], RCGA [11], LMO [12], CBA [13] (Data not given). It is seen from the results that the total fuel cost obtained by the RMO is less when compared to most of the recently reported algorithms. The average computation time of RMO is found to be 2.32 s for a 6-unit system. The convergence of solutions with iterations for the RMO algorithm for a typical run is depicted in Fig. 1. From the figure, it is seen that smooth convergence is obtained by the RMO.

3.3.2 Test System-II

The test system-II is a slightly larger system with more nonlinearities. The system consists of 13 generating units meeting a load demand of 1800 MW. It includes valve point loading, which has the effect of introducing several local minima [14], thereby making the cost curve multimodal. Transmission losses are not considered for this case. The system data are obtained from Sinha et al. [4].

The numerical results obtained from RMO are given in Table 2. The optimal cost obtained by RMO is 17,961.0645 $/h. It is also found that the generations satisfy the generation limit constraints. For this system, a comparison has also been made with recent algorithms, that is, DSD [15], PSM [16], DE [17], CBA [13], QPSO [18], HMAPSO [19], SHDE [20], HGA [21]. The results show that the total cost obtained by RMO is less when compared to the above techniques. The average computation time of the RMO for 13-unit system is 4.62 s.

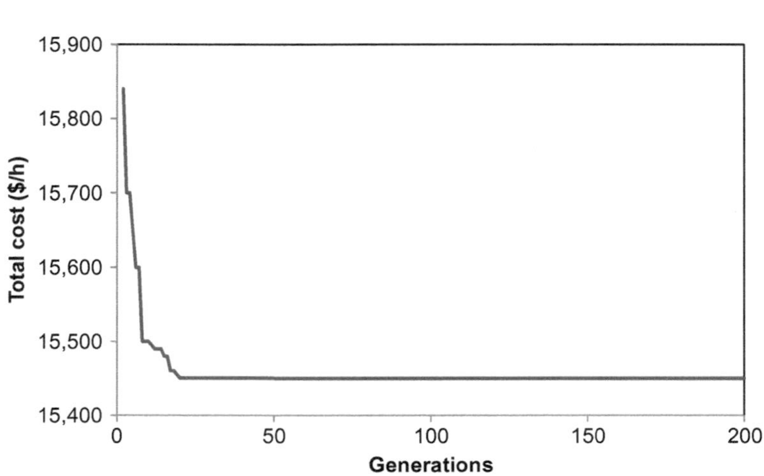

FIG. 1 Convergence characteristics of RMO for 6-unit systems.

TABLE 2 Simulation Results of Economic Dispatch for 13-Unit System Using RMO

Unit	RMO
P_1 (MW)	628.2932
P_2 (MW)	149.6541
P_3 (MW)	222.5471
P_4 (MW)	109.8995
P_5 (MW)	109.8995
P_6 (MW)	109.8995
P_7 (MW)	109.8999
P_8 (MW)	109.8992
P_9 (MW)	60.0000
P_{10} (MW)	40.0000
P_{11} (MW)	40.0000
P_{12} (MW)	55.0000
P_{13} (MW)	55.0000
TC ($/h)	17,961.0645

Fig. 2 shows the graph of the convergence of solutions with iterations for the RMO algorithm for a typical run, for Test System II. It can be seen that smooth convergence is obtained by the RMO.

3.3.3 *Test system-III*

In this test system, the same algorithm is applied on a larger test system consisting of twenty thermal units. The system data are obtained from Su and Lin [22]. This is relatively larger than test system I and II. The load demand of the system is 2500 MW. Transmission losses are included in this system and it is taken from Su and Lin [22] along with the B loss coefficients matrix. To simulate this case, each individual in the population of RMO contains 20 generator power outputs. The experimental results, that is, the optimal generation values, cost and transmission loss are presented in Table 3. From the table it is observed that the results satisfy the system constraints. The optimal cost and the corresponding transmission loss obtained are 62,456.6295 $/h and 91.8894 MW, respectively. It is seen that the all the generation limit constraints are satisfied. The comparison of results of the RMO with other algorithms like HM [22], λ method [22], BBO [23], NR [24], EHNN [25], GSO [25], CQGSO

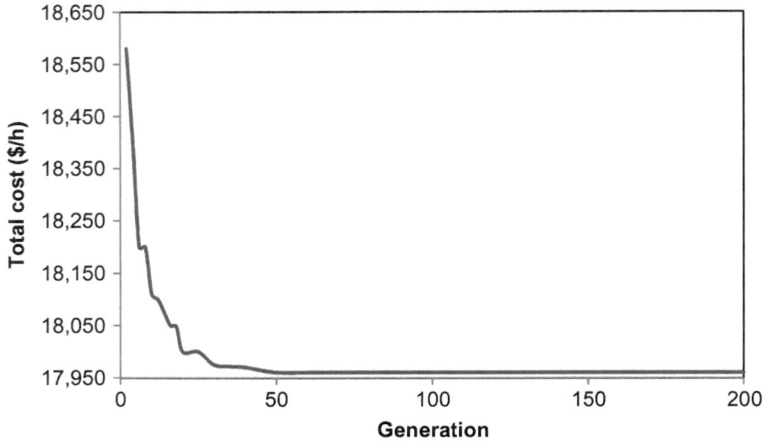

FIG. 2 Convergence characteristics of RMO for 13-unit system.

[25], BSA [26] shows that the total cost obtained by RMO is less. The computation time for the 20-unit system is found to be 5.12 s.

Fig. 3 shows the graph of the convergence of solutions with iterations for the RMO algorithm for a typical run, for 20-unit system. The smooth and early convergence shows the effectiveness of RMO algorithm for solving the 20 unit ED problem.

3.3.4 Test System-IV

This is a large system consisting of 40 generators supplying a demand of 10,500 MW. It includes valve point loading effects, due to which a number of local minima are introduced, thereby making the problem more complex. The system data are obtained from Sinha et al. [4].

The optimal generation values and cost obtained by the simulation results of RMO is presented in Table 4. The optimal cost obtained by RMO for the 40-unit system is 121,412.5462 $/h. It is also noted that the generations satisfy the generation limit and operating constraints. The results obtained are compared with those obtained by several recently reported efficient techniques, that is, QPSO [18], HGA [21], DE [17], CRO [17], CBA [17] BSA [26]. The fuel cost obtained by the RMO is the least when compared to other methods. The comparative results (data not shown) clearly depict that the proposed RMO method is capable of yielding a lower cost than the other algorithms reported in the literature.

The computation time for the 40-unit system by RMO is found to be 9.79 s. Fig. 4 shows the graph of the convergence of solutions with iterations for the RMO algorithm for a typical run, for 40-unit test system. From the figure, it is observed that a smooth convergence is obtained by the RMO.

TABLE 3 Simulation Results of Economic Dispatch for 20-Unit System Using RMO

Unit	RMO
P_1 (MW)	512.5647
P_2 (MW)	169.4251
P_3 (MW)	126.1425
P_4 (MW)	102.1247
P_5 (MW)	113.2541
P_6 (MW)	73.4179
P_7 (MW)	115.1201
P_8 (MW)	116.2221
P_9 (MW)	100.7265
P_{10} (MW)	106.2345
P_{11} (MW)	150.3917
P_{12} (MW)	292.8469
P_{13} (MW)	119.1085
P_{14} (MW)	31.0285
P_{15} (MW)	115.7854
P_{16} (MW)	36.3215
P_{17} (MW)	67.2675
P_{18} (MW)	87.8756
P_{19} (MW)	101.7895
P_{20} (MW)	54.2421
TG (MW)	2591.8894
P_L (MW)	91.8894
TC ($/h)	62,456.6295

4 FIXED HEAD HYDRO THERMAL SCHEDULING

Hydrothermal generation scheduling plays an important role in the operation and planning of a power system. A good generation schedule reduces the production cost, increases the system reliability, and maximizes the energy capability of reservoirs by utilizing the limited water resource [27]. The main

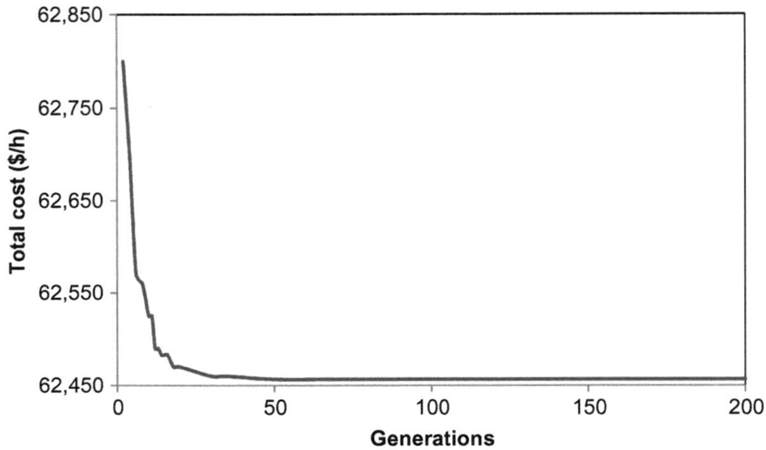

FIG. 3 Convergence characteristics of RMO for 20-unit system.

TABLE 4 Comparison of Simulation Results of ED for 40-Unit System Using RMO

Unit	RMO
P_1 (MW)	110.8021
P_2 (MW)	110.8004
P_3 (MW)	97.3999
P_4 (MW)	179.7331
P_5 (MW)	87.8021
P_6 (MW)	140.0000
P_7 (MW)	259.5882
P_8 (MW)	284.5884
P_9 (MW)	284.5821
P_{10} (MW)	130.0000
P_{11} (MW)	94.0000
P_{12} (MW)	94.0000
P_{13} (MW)	214.7285
P_{14} (MW)	394.2941
P_{15} (MW)	394.2941
P_{16} (MW)	394.2941

TABLE 4 Comparison of Simulation Results of ED for 40-Unit System Using RMO—cont'd

Unit	RMO
P_{17} (MW)	489.1114
P_{18} (MW)	489.1114
P_{19} (MW)	511.2145
P_{20} (MW)	511.2145
P_{21} (MW)	523.2011
P_{22} (MW)	523.2011
P_{23} (MW)	523.2011
P_{24} (MW)	523.2011
P_{25} (MW)	523.2011
P_{26} (MW)	523.2011
P_{27} (MW)	10.0000
P_{28} (MW)	10.0000
P_{29} (MW)	10.0000
P_{30} (MW)	87.8112
P_{31} (MW)	190.0000
P_{32} (MW)	190.0000
P_{33} (MW)	190.0000
P_{34} (MW)	164.7999
P_{35} (MW)	164.3869
P_{36} (MW)	200.0000
P_{37} (MW)	110.0000
P_{38} (MW)	110.0000
P_{39} (MW)	110.0000
P_{40} (MW)	511.2795
TC ($/h)	121,412.5462

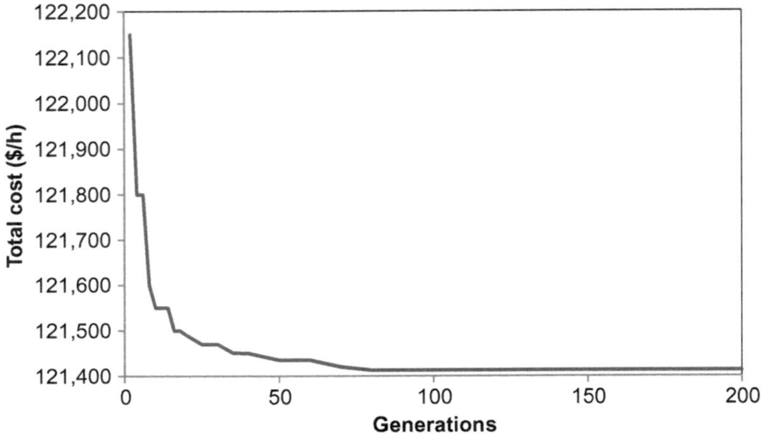

FIG. 4 Convergence characteristics of RMO for 40-unit system.

objective of the short term hydro thermal scheduling problem is to determine the optimal generation schedule of the thermal and hydro units to minimize the total production cost over the scheduling time horizon (typically 1 day or 1 week) subjected to a variety of thermal and hydraulic constraints.

It is a large scale, nonlinear programming problem involving nonlinear objective function and a mixture of complicated constraints [28]. The hydrothermal generation scheduling is mainly concerned with both hydro unit scheduling and thermal unit dispatching. The hydrothermal generation scheduling problem is more difficult than the scheduling of thermal power systems [29]. Because there is no fuel cost associated with hydropower generation, the problem of minimizing the total production cost of hydrothermal scheduling problem is achieved by minimizing the fuel cost of thermal power plants under the constraints of water available for the hydro power generation in a given period of time [30]. The equality constraints are the power balance constraint and the water available constraints for hydro generation. The inequality constraints are due to the generation limits of both thermal and hydro plants [31]. In the past years, conventional methods and stochastic methods have been employed to solve the FHHTS problem. Each of these methods has its own merits and demerits. Hence, there is a need for developing newer techniques to obtain the best solution with minimum computational efforts.

4.1 Problem Formulation

The goal of the economic emission dispatch for the fixed head hydro thermal scheduling problem is to minimize the total fuel cost of all thermal plants while satisfying the system and unit constraints for optimal schedule of hydro and thermal plants of power system. The FHHTS problem scheduled in M subintervals over $j_{\max}$ time intervals is formulated as follows [32]:

$$\text{Minimize } z = \sum_{k=1}^{k_{\max}} \sum_{s=1}^{m} t_k F_S(P_{sk}) \tag{29}$$

Subject to the energy balance equation

$$C_k = \sum_{s=1}^{m} P_{sk} + \sum_{h=1}^{n} P_{hk} - P_{\text{loss}k} - P_{Dk} = 0 \tag{30}$$

for $k = 1, 2 \ldots k_{\max}$.

Water availability constraint

$$W_h = \sum_{k}^{k_{\max}} t_k q_{hk} = V_h \tag{31}$$

for $h = 1, 2 \ldots, n$

with

$$P_{sk}^{\min} \leq P_{sk} \leq P_{sk}^{\max} \tag{32}$$

$$P_{hk}^{\min} \leq P_{hk} \leq P_{hk}^{\max} \tag{33}$$

where, $F_s P_{sk}$, cost function of each generating unit in interval k; P_{sk}, output of thermal unit s in interval k; P_{hk}, output of hydro unit h in interval k; $P_{sk}^{\min}$ $P_{sk}^{\max}$, lower and upper generation limits for thermal unit s in interval k; $P_{hk}^{\min}$ $P_{hk}^{\max}$, lower and upper generation limits for hydro unit h in interval k; $P_{\text{loss}k}$, transmission loss in interval k; P_{Dk}, total generation demand in interval k; T_k, time in interval k; q_{hk}, rate of water flow from hydro unit h in interval k; V_h, prespecified volume of water available for generation by hydro unit h during the scheduling period; C_k, energy balance constraint in interval k; W_h, water availability constraint of hydro unit h during the scheduling period; m, number of thermal unit; n, number of hydro unit.

4.1.1 System Model

The thermal unit cost is represented as a continuously increasing quadratic function of its generation as follows

$$F_s(P_{sk}) = \alpha_{2s} P_{sk}^2 + \alpha_{1s} P_{sk} + \alpha_{0s} \tag{34}$$

The water discharge rate at plant h is a function of the hydro generation of that plant. It is given as

$$q_{hk} = \alpha_{2h} P_{hk}^2 + \alpha_{1h} P_{hk} + \alpha_{0h} \tag{35}$$

The transmission loss is represented by B matrix formulation [33]. The general form is [28]

$$P_{\text{loss}\,k}=\sum_{i=1}^{m+n}\sum_{j=1}^{m+n}B_{ij}P_{ik}P_{jk}+\sum_{i=1}^{m+n}B_{oj}P_{ik}+B_{oo} \tag{36}$$

Lagrange function is [28]

$$L=\sum_{j=1}^{j\max}\left[n_jF\left(P_{sj}\right)+\lambda_j\left(P_{\text{load}j}-P_{Hj}-P_{sj}\right)\right]+\gamma\left[\sum_{j=1}^{j\max}\left[n_jq_j\left(P_{Hj}\right)-q_{\text{Tot}}\right]\right] \tag{37}$$

And for a specific interval $j=k$,

$$\frac{\partial L}{\partial P_{sk}}=0 \tag{38}$$

gives

$$n_k\frac{dF_{sk}}{dP_{sk}}=\lambda_k \tag{39}$$

and

$$\frac{\partial L}{\partial P_{Hk}}=0 \tag{40}$$

gives

$$\gamma n_k\frac{dq_k}{dP_{Hk}}=\lambda_k \tag{41}$$

If the network losses are added to the problem, then at each hour,

$$P_{\text{load}\,j}+P_{\text{loss}\,j}-P_{Hj}-P_{sj}=0 \tag{42}$$

and the Lagrange function becomes

$$\begin{aligned}L=&\sum_{j=1}^{j\max}\left[n_jF\left(P_{sj}\right)+\lambda_j\left(P_{\text{load}j}+P_{\text{loss}j}-P_{Hj}-P_{sj}\right)\right]\\&+\gamma\left[\sum_{j=1}^{j\max}n_jq_j\left(P_{Hj}\right)-q_{\text{Tot}}\right]\end{aligned} \tag{43}$$

With resulting coordination equations (hour k):

$$n_k\frac{dF(P_{sk})}{dP_{sk}}+\lambda_k\frac{\partial P_{\text{loss}k}}{dP_{sk}}=\lambda_k \tag{44}$$

$$\gamma n_k\frac{dq(P_{Hk})}{dP_{Hk}}+\lambda_k\frac{\partial P_{\text{loss}k}}{\partial P_{Hk}}=\lambda_k \tag{45}$$

These equations, along with the constraints, are solved iteratively to obtain the optimum the hydro thermal scheduling.

4.2 RMO Algorithm for Solving Fixed Head Hydro-Thermal Scheduling Problem

In fixed head hydro thermal scheduling problem, fuel cost of thermal units and water discharge are expressed as a quadratic function. The total availability of water for each hydro unit is specified. The total available water should be efficiently utilized for the given scheduling horizon so as to minimize the fuel cost of thermal generating units. The different steps involved in the RMO algorithm for solving the fixed head hydro thermal scheduling problem are detailed below:

4.2.1 Step-I: Initialization

The initial particles of RMO are randomly generated within their limits. In this study, γ is taken as the decision variable. The number of decision variables is equal to the number of hydro plants.

4.2.2 Step-2: Generate the Velocity Vector

The velocity vector V_{ij} is randomly generated using Eq. (4)

4.2.3 Step-3: Locate the Center Point

The real power generations of all thermal and hydro plants over the scheduling horizon are considered as the problem dependent variables and expressed as function of γ. The population vector of RMO is represented as

$$X=\begin{bmatrix} \gamma_{11} & \gamma_{12} & \cdots & \gamma_{1n} \\ \gamma_{21} & \gamma_{22} & \cdots & \gamma_{2n} \\ \vdots & \cdots & \cdots & \vdots \\ \gamma_{m1} & \gamma_{m2} & \cdots & \gamma_{mn} \end{bmatrix} \tag{46}$$

These γ values for the hydro units are randomly generated and for each vector the optimal real power generated is determined using λ method. In λ method, the initial λ value is calculated by neglecting transmission line using

$$\lambda_k=(P_D+P_L)+\sum_{i=1}^{m}\frac{\alpha_{1s}}{2\alpha_{2s}}+\sum_{i=1}^{n}\frac{\lambda\alpha_{1h}}{2\gamma\alpha_{2h}} \tag{47}$$

The optimal generation of thermal and hydro units in each interval are calculated using

$$P_i=\frac{\lambda_k-\alpha_{1s}}{2\alpha_{2s}} \text{ for thermal units; } i=1,2,3\ldots m \tag{48}$$

$$P_i=\frac{\lambda_k-\gamma\alpha_{1n}}{2\gamma\alpha_{2n}} \text{ for hydro units; } i=1,2,3\ldots n \tag{49}$$

The transmission loss is calculated using the optimal generation

For each vector, the above steps 1–3 is repeated till the following constraint is satisfied in k^{th} interval

$$\sum_{g=1}^{m} P_{sk} + \sum_{h=1}^{n} P_{nk} - P_{\text{loss}k} - P_{Dk} = 0 \tag{50}$$

The λ—approach corresponding to each particle provides the optimal thermal and hydro generations at all intervals in the scheduling horizon. Therefore, the fitness of each particle in the population is developed to check only the water availability constraints of the hydro unit. The water discharge of each hydro unit is calculated from Eq. (35) using optimal generation schedule.

The fitness of each particle is evaluated by

$$FIT_i = \frac{1}{1 + \sum_{j=1}^{n} \left| V_j^{av1} - \sum_{k=1}^{k\max} t_k q_{nk} \right|} \tag{51}$$

The vector which gives the maximum fitness is chosen as the initial center point and it is the Gbest vector in the initial population.

4.2.4 Step-4: Find the Radial Movement of the Particle

The radial movement of the particles from the center point is calculated by Eqs. (25), (26). The particles are sprinkled here. If any control variable in the population vector U_{ij} violates their maximum or minimum limits, then that variable is fixed at their violated limit.

4.2.5 Step-5: Evaluate the Fitness of the Particles

The fitness of each vector in the population U_{ij} is evaluated. The particle which gives the minimum error is selected as radial best (R_{best}) vector.

4.2.6 Step-6: Search Around G_{best}

The convergence of radial movement optimization is improved by generating number of trial vectors around the G_{best} using Eq. (27). The fitness of each vector in V_{ij} is evaluated and are compared with G_{best}. If the fitness of trial vector is better than G_{best}, the trial vector replaces the G_{best} vector. This procedure is repeated for Nt times for each vector in every iteration.

4.2.7 Step-7: Determine the New Center Location

The new center location (center 1) is updated using Eq. (28)

For this FHHTS problem, C_1 and C_2 values are fixed as 0.83 and 0.89, respectively, and the population size is 100. The fitness of radial best vector (R_{best}) is compared with the fitness of global best (G_{best}) vector. G_{best} is replaced by R_{best}, when R_{best} is better than G_{best}. The iteration is incremented by 1. The steps 2–7 in the algorithm is repeated for predefined iterations.

4.3 Results and Discussion

In this work, four different fixed head short term hydro thermal scheduling are taken as case studies. The RMO technique is implemented to solve each of the four case studies considered. The algorithm is applied in MATLAB 2011 and the simulation results are presented in Table 5–8. Simulation is performed in an Intel core i3 CPU with 2 GB RAM.

4.3.1 Test System-I

The first test system considered in this study comprises of one thermal and one hydel plant. The details pertaining to thermal plant cost characteristics, hydel plant discharge characteristics, loss coefficients, and availability of water for each hydel plant in the system are given below [32].

$$F_1(P_{1k}) = 0.001991P_{1k}^2 + 9.606P_{1k} + 373.7\ \$/\text{h} \tag{52}$$

$$q_{1k} = 0.0007749P_{2k}^2 - 0.009079P_{2k} + 61.53\ \ \text{M} - \text{ft}^3/\text{h} \tag{53}$$

$$V_1 = 2559.6\,\text{M} - \ \text{ft}^3/\text{h}$$

$$B_{ij} = \begin{bmatrix} 0.00005 & 0.00001 \\ 0.00001 & 0.00015 \end{bmatrix}$$

$$B_{oj} = \begin{bmatrix} 0.0 \\ 0.0 \end{bmatrix}$$

$$B_{oo} = 0.0\,\text{per MW}$$

The total generation demand along with the output of thermal and hydel power plant with the loss are tabulated in Table 5. The gamma value and the best cost obtained by RMO for the test system-I are presented in Table 9.

4.3.2 Test System-II

The second test system considered has one hydro plant and two thermal plants. The details of thermal plant cost characteristics, hydel plant discharge characteristics, loss coefficients, and availability of water for each hydel plant of the systems are [32]:

$$F_1(P_{1k}) = 0.01P_{1k}^2 + 3.0P_{1k} + 15.0\ \ \$/\text{h} \tag{54}$$

$$q_{1k} = 0.0005P_{2k}^2 - 0.03P_{2k} + 0.2\ \ \text{M} - \text{ft}^3/\text{h} \tag{55}$$

$$q_{2k} = 0.0001P_{3k}^2 - 0.06P_{3k} + 0.4\ \ \text{M} - \text{ft}^3/\text{h}$$

$$V_1 = 25.0\text{M} - \text{ft}^3/\text{h}\ \ V_2 = 35.0\text{M} - \text{ft}^3/\text{h}$$

TABLE 5 Optimal Generations Over the Scheduling Period for Test System-I

PD_k (MW)	P_{sk} (MW)	P_{hk} (MW)	P_{loss} (MW)
455	230.5176	236.6293	12.1469
425	202.0664	234.1450	11.2114
415	192.6007	233.3184	10.9191
407	185.0346	232.6578	10.6923
400	178.4189	232.0801	10.4990
420	197.3324	233.7316	11.0640
487	260.9555	239.2870	13.2425
604	373.0476	249.0747	18.1223
665	431.9997	254.2223	21.2220
675	441.6980	255.0692	21.7672
695	461.1237	256.7654	22.8890
705	470.8510	257.6148	23.4658
580	349.9502	247.0579	17.0081
605	374.0111	249.1589	18.1700
616	384.6167	250.0849	18.7016
653	420.3744	253.2072	20.5817
721	486.4352	258.9755	24.4107
740	504.9739	260.5943	25.5682
700	465.9861	257.1900	23.1761
678	444.6094	255.3234	21.9328
630	398.1313	251.2650	19.3963
585	354.7576	247.4777	17.2353
540	311.5745	243.7070	15.2816
503	276.2094	240.6190	13.8284
V total		2559.6 M-ft^3/h	

TABLE 6 Optimal Generations Over the Scheduling Period for Test System-II

PD_k (MW)	P_s (MW)	$P_h(1)$ (MW)	$P_h(2)$ (MW)	P_{loss} (MW)
30	2.6782	18.5841	9.1247	0.3870
33	3.2769	19.8332	10.3367	0.4468
35	3.6765	20.6670	11.1458	0.4892
38	4.2766	21.9193	12.3609	0.5569
40	4.6772	22.7553	13.1720	0.6046
45	5.6806	24.8489	15.2035	0.7330
50	6.6866	26.9480	17.2403	0.8748
59	8.5039	30.7401	20.9198	1.1638
61	8.9089	31.5852	21.7398	1.2339
58	8.3015	30.3179	20.5101	1.1295
56	7.8972	29.4741	19.6913	1.0626
57	8.0993	29.8959	20.1006	1.0958
60	8.7063	31.1626	21.3297	1.1986
61	8.9089	31.5852	21.7398	1.2339
65	9.7202	33.2781	23.3824	1.3808
68	10.3298	34.5502	24.6167	1.4967
71	10.9404	35.8242	25.8529	1.6176
62	9.1116	32.0081	22.1501	1.2698
55	7.6951	29.0525	19.2823	1.0300
50	6.6866	26.9480	17.2403	0.8748
43	5.2789	24.0108	14.3903	0.6801
33	3.2769	19.8332	10.3367	0.4468
31	2.8777	19.0002	9.5285	0.4064
30	2.6782	18.5841	9.1247	0.3870
$V_1 = 25.00$ M-ft^3/h			$V_2 = 35.00$ M-ft^3/h	

TABLE 7 Optimal Generations Over the Scheduling Period for Test System-III

PD_k (MW)	$P_s(1)$ (MW)	$P_s(2)$ (MW)	$P_h(1)$ (MW)	$P_h(2)$ (MW)	P_{loss} (MW)
400	77.61605	136.70832	168.3914	22.91627	5.632
300	61.88352	88.12348	149.1637	4.091418	3.2621
250	53.16140	61.20744	138.0015	0.00000	2.3703
250	53.16140	61.20744	138.0015	0.00000	2.3703
250	53.16140	61.20744	138.0015	0.00000	2.3703
300	61.88358	88.12348	149.1638	4.091417	3.2623
450	85.52748	161.12206	178.0587	32.37615	7.0844
900	158.14320	384.6411	266.8382	119.0057	8.6282
1230	213.1185	553.17071	334.0828	184.3410	54.713
1250	216.5006	563.5179	338.2199	188.3535	56.592
1350	233.4970	615.48970	359.0188	208.5014	66.507
1400	242.0521	641.62539	369.4872	218.6342	71.799
1200	208.0564	537.67948	327.8899	178.3335	51.959
1250	216.5001	563.51794	338.2184	188.3516	56.588
1250	216.5001	563.51794	338.2184	188.3516	56.588
1270	219.8876	573.88068	342.3654	192.3702	58.504
1350	233.4968	615.48977	359.0189	208.5015	66.507
1470	254.0922	678.38615	384.2251	232.8871	79.591
1330	230.0861	605.06358	354.8445	204.4590	64.453
1250	216.5004	563.51785	338.2187	188.3517	56.589
1170	203.0080	522.22285	321.7138	172.3419	49.287
1050	182.9426	460.73999	297.1684	148.5056	39.357
900	158.1431	384.64190	266.8386	119.0061	28.6297
600	109.4450	234.85872	207.2974	60.95082	12.5519
$V_1 = 2500.1$ M-ft^3/h			$V_2 = 2100.1$ M-ft^3/h		

TABLE 8 Optimal Generations Over the Scheduling Period for Test System-IV

Duration (h)	PD_k (MW)	P_s (MW)	P_h (MW)	P_{loss} (MW)
1–12	1200	628.0518	600.8277	28.8795
12–24	1500	628.0518	943.1038	71.1556
V_{total}		100,000.07 M-ft^3/h		

TABLE 9 Gamma Value and Fuel Cost Obtained by RMO

Test System	Gamma Value	Cost ($/d)
1	28.214253	96,024.3345
2	90.668480 48.532780	848.3254
3	9.555120 5.826835	53,051.462
4	2.028377	169,637.5925

$$B_{ij} = \begin{bmatrix} 0.0 & 0.0 & 0.0 \\ 0.0 & 0.001 & 0.0 \\ 0.0 & 0.0 & 0.0005 \end{bmatrix}$$

$$B_{ij} = \begin{bmatrix} 0.0 \\ 0.0 \\ 0.0 \end{bmatrix}$$

$$B_{oo} = 0.0 \text{ per MW}$$

The total generation demand along with the output of thermal and hydel power plant with the loss are presented in Table 6. The gamma value and the best cost obtained by RMO for the test system-II are presented in Table 9.

4.3.3 Test System-III

The third system consists of two thermal and two hydel plants. The details for the test system-III are given below [32].

$$F_1(P_{1k}) = 0.0025P_{1k}^2 + 3.2P_{1k} + 25.0 \ \$/\text{h} \tag{56}$$

$$F_2(P_{2k}) = 0.0008P_{2k}^2 + 3.4P_{2k} + 30.0 \ \$/\text{h} \tag{57}$$

$$q_{1k} = 0.000216P_{3k}^2 - 0.306P_{3k} + 1.98 \quad \text{M} - \text{ft}^3/\text{h} \tag{58}$$

$$q_{2k} = 0.00036P_{4k}^2 - 0.612P_{4k} + 0.936 \quad \text{M} - \text{ft}^3/\text{h} \tag{59}$$

$$V_1 = 2500.0\,\text{M-ft}^3/\text{h} \quad \text{V}_2 = 2100.0\,\text{M-ft}^3/\text{h}$$

$$B_{ij} = \begin{bmatrix} 0.00014 & 0.00001 & 0.000015 & 0.000015 \\ 0.00001 & 0.00006 & 0.00001 & 0.000013 \\ 0.000015 & 0.00001 & 0.000068 & 0.000065 \\ 0.000015 & 0.000013 & 0.000065 & 0.00007 \end{bmatrix}$$

$$B_{oj} = \begin{bmatrix} 0.0 \\ 0.0 \\ 0.0 \\ 0.0 \end{bmatrix}$$

$$B_{oo} = 0.0\,\text{per MW}$$

The total generation demands along with the results obtained from RMO are tabulated in Table 7. The gamma value and the best cost obtained by RMO for the test system-III are presented in Table 9.

4.3.4 Test System-IV

The fourth test system includes one thermal and one hydel plant. The details of thermal plant, hydel plant, and availability of water are given below [32].

$$F_1(P_{1k}) = 0.00184P_{1k}^2 + 9.20P_{1k} + 575\,\$/\text{h} \tag{60}$$

$$q_{1k} = 4.97P_{2k} + 330\,\text{acre} - \text{ft}/\text{h} \tag{61}$$

$$V_1 = 1{,}00{,}000\,\text{acre-ft/h}$$

$$B_{ij} = \begin{bmatrix} 0.0 & 0.0 \\ 0.0 & 0.00008 \end{bmatrix}$$

$$B_{ij} = \begin{bmatrix} 0.0 \\ 0.0 \end{bmatrix}$$

$$B_{oo} = 0.0\,\text{per MW}$$

The total generation demands along with the results obtained from RMO for test system-IV are tabulated in Table 8. The gamma value and the best cost obtained by RMO for the test system-IV are presented in Table 9.

In order to validate the simulation results of FHHTS by RMO, a comparison has been made with other techniques like λ-γ method and optimal gamma based GA [34], a method developed by Rashid and Nor [32]. From the results, it is inferred that the RMO method gives the best solution compared to other methods. These results suggest that the RMO gives better results when compared to the available methods for the test systems I–IV.

Figs. 5–8 show the convergence characteristics of the RMO method used in this work for optimal generation of power over the scheduling period for test system-I–IV, respectively. It can clearly be seen from the graphs that the RMO method convergences after around 25 iterations for test system-I and 40 iterations for test system-IV. This shows the effectiveness of RMO in solving the fixed head hydro thermal scheduling problem.

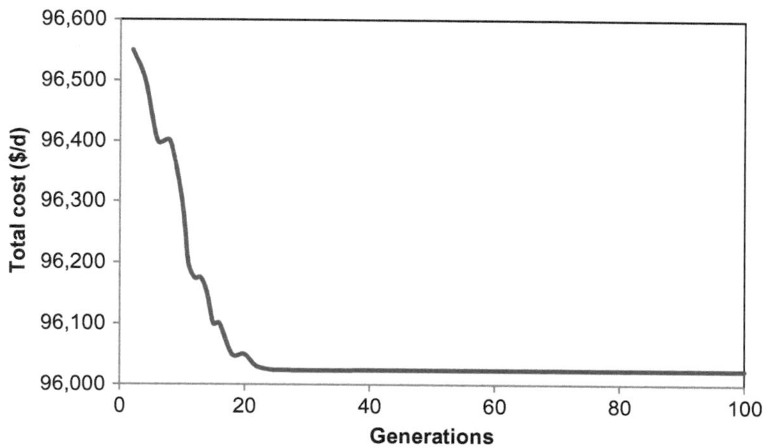

FIG. 5 Convergence characteristics of RMO for test system-I.

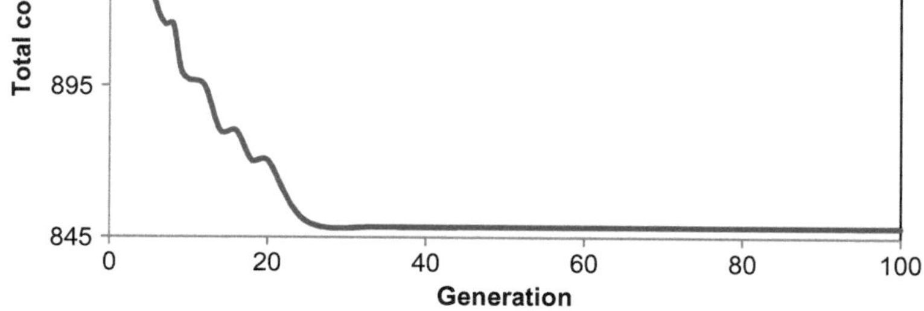

FIG. 6 Convergence characteristics of RMO for test system-II.

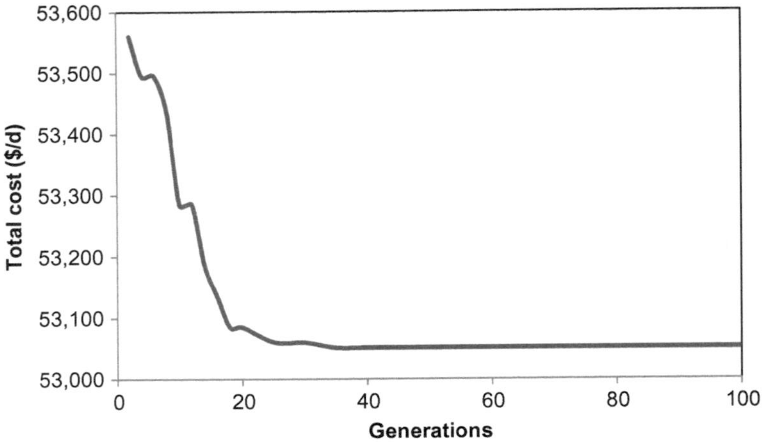

FIG. 7 Convergence characteristics of RMO for test system-III.

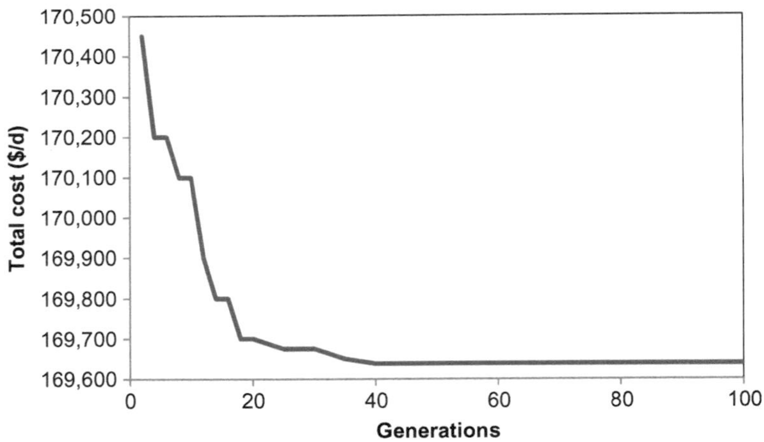

FIG. 8 Convergence characteristics of RMO for test system-IV.

5 CONCLUSIONS

The application of the RMO to different types of nonconvex ED problems and fixed head hydro thermal unit problems are demonstrated in this chapter. Four sample test systems; ED with ramp rate limits and prohibited operating zones (6-unit system), ED with valve point loading effects (13-unit system), ED for 20-unit system, and ED with valve point loading effects (40-unit system) are considered. For alltypes of ED problems, the RMO provides better simulation results compared to most of the other algorithms. Additionally, the RMO converges to the optimum solution in fewer iterations. Thus, the solution obtained through RMO is a better quality solution. It can be concluded that the RMO algorithm is superior to other methods employed previously for solving different types of ED problem in terms of high quality solution and stable convergence characteristic. It can be concluded that the results obtained by radial movement optimization algorithm is the best compared to the solutions obtained by the other methods, that is, $\lambda - \gamma$, optimal gamma-based GA, a method developed by Rashid and Nor. The simulation results reveal the effectiveness of the RMO in searching global or near global optimal solutions to the FHHTS problem. Further, the proposed RMO converges at earlier iteration. Hence, RMO based optimization is a promising technique for solving the FHHTS in power systems. In future, this work may be pursued on the following aspects:

- Application of the RMO algorithm to a security constrained economic dispatch problem that allocates generation in a power system, such that it is in the normal secure state relative to a prespecified contingency list.
- Implementation for the solution of dynamic economic dispatch problems, multiarea economic dispatch problems, and combined economic emission dispatch problems.
- Application of RMO can also be extended to short-term and long-term hydrothermal scheduling problem in a power system.

REFERENCES

[1] V. Kumar, J. Saraswat, A review on optimization techniques for economic load dispatch problem, Int. Res. J. Eng. Technol. 3 (2016) 375–379.

[2] R. Balamurugan, Economic dispatch solution techniques for power system using computational models (Ph.D. Thesis), Annamalai University, India, 2017.

[3] R. Rahmani, R. Yusof, A new simple, fast and efficient algorithm for global optimization over continuous search space problems: Radial Movement Optimization, Appl. Math. Comput. 248 (2014) 287–300.

[4] N. Sinha, R. Chakrabarti, P.K. Chattopadhyay, Evolutionary programming techniques for economic load dispatch, IEEE Trans. Evol. Comput. 7 (2003) 83–94.

[5] Z.L. Gaing, Particle swarm optimization to solving the economic dispatch considering the generator constraints, IEEE Trans. Power Syst. 18 (2003) 1187–1195.

[6] N. Amjady, H. Sharifzadeh, Solution of non-convex economic dispatch problem considering valve loading effect by a new modified differential evolution algorithm, Int. J. Electr. Power Energy Syst. 32 (2010) 893–903.

[7] R. Balamurugan, S. Subramanian, An improved dynamic programming approach to economic power dispatch with generator constraints and transmission losses, J. Electr. Eng. Technol. 3 (2008) 320–330.

[8] I. Selvakumar, K. Thanushkodi, A new particle swarm optimization solution to non-convex economic dispatch problems, IEEE Trans. Power Syst. 22 (1) (2007) 42–51.

[9] W.T. Elsayed, E.F. El-Saadany, A fully decentralized approach for solving the economic dispatch problem, IEEE Trans. Power Syst. 30 (2015) 2179–2189.

[10] S. Pothiya, I. Ngamroo, W. Kongprawechnon, Application of multiple Tabu search algorithm to solve dynamic economic dispatch considering generator constraints, Energy Convers. Manage. 49 (2008) 506–516.
[11] N. Amjady, H. Nasiri Rad, Economic dispatch using an efficient real-coded genetic algorithm, IET Gener. Transm. Distrib. 3 (2009) 266–278.
[12] I. Ciornei, E. Kyriakide, A GAAPI solution for the economic dispatch of generation in power system operation, IEEE Trans. Power Syst. 27 (2012) 233–242.
[13] B.R. Adarsh, T. Raghunathan, T. Jayabarathi, X.-S. Yang, Economic dispatch using chaotic bat algorithm, Energy 96 (2016) 666–675.
[14] M.S.P. Subathra, S.E. Selvan, T.A.A. Victoire, A.H. Christinal, U. Amato, A hybrid with cross entropy method and sequential quadratic programming to solve economic load dispatch problem, IEEE Syst. J. 99 (2015) 1–14.
[15] Q.H.W. Junpeng Zhan, C. Guo, X. Zhou, Economic dispatch with non-smooth objectives-Part II: dimensional, steepest decline method, IEEE Trans. Power Syst. 30 (2015) 722–733.
[16] J.S. Sumait, J.K. Sykulski, A.K. Othman, Solution of different types of economic load dispatch problems using a pattern search method, Electr. Power Compon. Syst. 36 (3) (2008) 250–265.
[17] P.K. Roy, S. Bhui, C. Paul, Solution of economic load dispatch using hybrid chemical reaction optimization approach, Appl. Soft Comput. 24 (2014) 109–125.
[18] K. Meng, H.G. Wang, Z. Dong, K.P. Wong, Quantum inspired particle swarm optimization for valve-point economic load dispatch, IEEE Trans. Power Syst. 25 (2010) 215–222.
[19] R. Kumar, D. Sharma, A. Sadu, A hybrid multi agent based particle swarm optimization algorithm for economic power dispatch, Int. J. Electr. Power Energy Syst. 33 (2011) 115–123.
[20] S.K. Wang, J.P. Chiou, C.W. Liu, Non-smooth/non-convex economic dispatch by a novel hybrid differential evolution algorithm, IET Proc. Gener. Transm. Distrib. 1 (2007) 793–803.
[21] D. He, F. Wang, Z. Mao, A hybrid genetic algorithm approach based on differential evolution for economic dispatch with valve-point effect, Int J. Electr. Power Energy Syst. 30 (2008) 31–38.
[22] C.T. Su, C.T. Lin, New approach with a Hopfield modeling framework to economic dispatch, IEEE Trans. Power Syst. 15 (2000) 541–545.
[23] A. Bhattacharya, P.K. Chattopadhyay, Solving complex economic load dispatch problems using biogeography-based optimization, Expert Syst. Appl. 37 (2010) 3605–3615.
[24] A. Abdelaziz, S. Mekhamer, M. Badr, M. Kamh, Economic dispatch using Hopfield neural network, Electric Power Compon. Syst. 36 (2008) 719–732.
[25] M. Moradi Dalvand, B. Mohammadi Ivatloo, A. Najafi, A. Rabiee, Continuous quick group search optimizer for solving non-convex economic dispatch problems, Electric Power Syst. Res. 93 (2012) 93–105.
[26] M.M. Delshad, N.A. Rahim, Solving non-convex economic dispatch problem via backtracking search algorithm, Energy 77 (2014) 372–381.
[27] D. Yadav, V. Sharma, R. Naresh, Fixed head short term hydro thermal generation scheduling using differential evolution, Asian J. Curr. Eng. Maths 2 (2013) 17–23.
[28] A.J. Wood, B.F. Wollenberg, Power Generation, Operation and Control, second ed., John Wiley and Sons, Hoboken, NJ, 1996.
[29] M.M. Salama, M.M. Elgazar, S.M. Abdelmaksoud, H.A. Henry, Short term optimal generation scheduling of fixed head hydrothermal system using genetic algorithm and constriction factor based particle swarm optimization technique, Int. J. Sci. Res. Publ. 5 (2013) 1–9.
[30] D.P. Kothari, J.S. Dhillon, Power System Optimization, PHI Learning Pvt. Ltd., New Delhi, 2009.

[31] M. Basu, Artificial immune system for fixed head hydrothermal power system, Energy 36 (2011) 606–612.
[32] A.H.A. Rashid, K.M. Nor, An efficient method for optimal scheduling of fixed head hydro and thermal plants, IEEE Trans. Power Syst. 6 (1991) 631–636.
[33] L.K. Kirchmayer, Economic Operation of Power Systems, John Wiley and Sons, Hoboken, NJ, 1958.
[34] J. Sasikala, M. Ramaswamy, Optimal gamma based fixed head hydrothermal scheduling using genetic algorithm, Expert Syst. Appl. 37 (2010) 3352–3357.

Chapter 5

An Intelligent Approach Based on Metaheuristic for Generator Maintenance Scheduling

S. Tamil Selvi*, **S. Baskar**† **and S. Rajasekar**‡
**SSN College of Engineering, Chennai, India, †Thiagarajar College of Engineering, Madurai, India, ‡NEC Laboratories Singapore, NEC Asia Pacific Pte. Ltd, Singapore, Singapore*

1 INTRODUCTION

1.1 Significance of Generator Maintenance Scheduling

The most important function of an electric power system is to provide electric power to its customers at the lowest possible cost with acceptable reliability levels. The prediction of cost of generation is one important aspect of system planning. Reliability analysis is a fundamental tool in the planning of an electric system. These two aspects, economics and reliability often conflict and render power system managers, planners, designers and operators face with a wide range of challenging problems. Power system engineers and managers have been attempting to achieve the highest possible reliability at an affordable cost.

Reliability is assessing the risk of not being able to meet the demand, even at the time of random outages of the units. The ability to deliver the generated energy to the load points is not the only aspect of reliability assessment. In addition, it also assesses sufficient excess capacity of the generation system required to manage random outages of generator units, including for maintenance on the generating facilities, so that random failure of the units is limited. Maintenance of units is essential to reduce the risk of capacity outage and it ensures that the generating units can continue to operate efficiently and reliably in the long run. Proper and regular maintenance improves the availability of units, reduces the risk of energy being unserved, and improves the reliability.

On the other hand, removing generating units from service for maintenance will reduce the available capacity and may increase the system's risk of not being able to meet increasing demand. Well planned maintenance schedule can improve the system reliability, whereas badly planned maintenance schedules would be directly reflected in the risk of energy unserved (Energy Not

Classical and Recent Aspects of Power System Optimization. https://doi.org/10.1016/B978-0-12-812441-3.00005-7
© 2018 Elsevier Inc. All rights reserved.

Supplied—ENS). The operation and planning of power system, including economic dispatch, unit commitment, load dispatch, generation co-ordination, generation expansion planning, and such other planning activities are also highly influenced and directly affected by generator maintenance schedule problem decisions [1].

The maintenance schedule, when not optimized, reduces system reliability and increases the costs of the electric power system [2]. It is important for a power generating utility to decide when generators should go off-line for maintenance [3]. Optimal maintenance scheduling is essential because many short-term, middle-term, and long-term power system costs are directly affected by such maintenance scheduling decisions[1], including unit commitment, fuel scheduling, optimal use of water resources, power system development planning, reliability calculations, generation co-ordination, generation expansion planning, export/import of power and such other planning activities.

Preventive optimal generator maintenance scheduling (GMS) is a challenging task. In this chapter, we discuss creating an optimal outage schedule for any number of generating units of plant, which can result in substantial savings in production cost and reduction in ENS.

1.2 Preventive GMS Approaches

To avoid premature aging and generator failure, which leads to unplanned and costly power outages, it is important to carry out preventive maintenance at regular intervals [4]. The nature of GMS is nonlinear (derived from nonlinear reliability (ENS) and cost link), stochastic (due to demand uncertainty, and forced outage rate (FOR) of the generators), and constrained problem. The job of this GMS is to determine the period for which the units should be taken offline for planned maintenance over the period of stipulated time horizon (one/two years), so the costs involved are minimized, and system constraints are satisfied. It is for this reason that the GMS problem is treated as an optimization problem. The maintenance schedule of each generator must be optimized based on the objective function under a set of constraints. The most typical objectives found in literature are based on either economic criteria, or reliability criteria. GMS constraints are based on the maintenance technology of the generators (describes appropriate duration, sequence, and continuous length of maintenance), availability of man power, predicted load demand, generator type (coal or hydro), material resources, reliability, etc.

1.2.1 Objective Functions

For years, researchers have been searching for an optimal maintenance schedule, which is technically feasible as well as economical. To achieve the solution of this complex task, there are several objective functions which can be implemented. The objective function must be selected in such a way that almost all the maintenance needs of the system are fulfilled.

Reliability Criteria

The GMS problem can be solved by maximizing the system reliability [5]. The quantification of reliability is a vital aspect of reliability assessment for generation systems. Generation system reliability is an important factor in the long-term planning for future system capacity expansions to make sure that the total installed capacity is sufficient to support demand. Generation system reliability is evaluated using different indices. The planning process applies reliability indices as a criterion to decide on investments in additional generation capacities. Reliability evaluations can be deterministic or probabilistic. General deterministic index is reserve margin and this index does not consider the probabilistic nature of system behavior. However, the probabilistic reliability indices are estimated by simulating the actual random behavior of the system. They are loss of load probability (LOLP), loss of load expectation (LOLE), loss of energy probability (LOEP), loss of energy expectation (LOEE), expected energy not served (EENS), loss of load frequency (LOLF) and loss of load duration (LOLD) [6]. Basically, these probabilistic indices are expected values of a random variable. Expectation indices provide valid adequacy indicators which reflect various factors such as system component availability and capacity, load characteristics and uncertainty, system configurations and operational conditions, etc. EENS reflects the true risk and has more physical significance than LOLP or LOLE. [6].

Among all these indices, leveling the reserve generation and leveling the risk LOLP, emphasizing the coordinated deterministic levelized reserve rate method and stochastic levelized risk method, respectively, are considered commonly as reliability criteria for the formulation of objective functions for GMS.

This most common reliability criterion of leveling the reserve over the planning period is done by maximizing the net reserve [7,8]. The effective approach involves the reliability criterion of minimizing the sum of squares of the reserves in each weekly time period, [4,8,9] when there is a large deviation in reserve. However, the main drawback of deterministic reliability objective function approach is that it neglects the randomness of the available generating unit's capacity [10].

The stochastic reliability objective function removes the above defect by considering the random forced outage of the generating units [10]. Wenyuan has discussed a method that includes the aging failures in power system reliability evaluation [11]. Minimization of the annual value of LOLE of power system is an objective function in the period analyzed in [2]. Minimizing the sum of the individual loss of load probabilities for each interval can also be a reliability objective under the conditions of load uncertainty and random forced outages of units [1,8,12]. The objective of the study in literature [13] is to reduce the

LOLP and minimizing the annual supply reserve ratio deviation for a power system, which are considered as a measure of power system reliability. Moreover, in this study the impacts of aging failures of the generators are considered in order to calculate the unavailability of power system which is modeled using the Weibull distribution.

Economic Criteria

Instead of considering reliability as primary objective, literatures pertaining to cost as the main criterion are also seen for the solution of GMS problem. A mathematical model was also proposed to calculate minimum maintenance cost whose failure distribution is considered as Weibull [14]. Each generation producer intends to optimize maintenance scheduling, minimizing fuel costs in the restructured power systems, considering maintenance cost, start-up cost, shut-down cost, production cost, and energy price [15,16]. The most familiar objective found is to minimize the total operating cost, which includes the sum of the energy production cost and maintenance cost [17,18]. Shorter maintenance length results in higher maintenance costs, whereas reduces the generation, possible energy purchase cost, and thereby energy production cost [19]. Production cost minimization requires many approximations or computationally rigorous simulation to create a maintenance schedule. The GMS problem is also solved by minimizing the loss of revenue, wherein the objective is to maximize the expected profit of selling electricity in a competitive environment. Each producer can schedule the optimal maintenance periods of its units by maximizing profit, with the consideration of the production and maintenance costs [20]. The objective function in [21] represents the producer's potential benefit over the entire scheduling horizon, which is calculated as the difference between revenues and costs including three parts: the expected energy-selling profit, the expected renewal cost of the damaged components, and the maintenance cost.

In the GMS problem formulation, the cost criteria can be considered as the primary objective function and the reliability criterion can be met, as constraint. In this way, the GMS problem formulation in [9,22] minimizes the expected energy production cost along the maintenance planning horizon and includes a reliability measure by penalizing nonzero values of EENS. Thus, in the optimization model formulations, dominant objective is considered as a single objective function to be optimized, while the lesser important objectives are incorporated as constraints.

1.3 Solution Methods

The nature of the GMS problem is nonlinear, stochastic, constrained optimization problem. Numerous mathematical programming techniques (MPT) such as successive approximation dynamic programming [23] decomposition approach

[24], integer programming [7], branch and bound technique [8], goal programming [25] benders decomposition approach, [16,26,27] etc., have been used to solve GMS problem. But, due to the nature of nonlinear, nondifferentiable, discontinuous and large-scale generation systems that characterize the GMS problem, conventional methods are not suitable [13].

Most of the earlier works in maintenance scheduling have used techniques that rely heavily on heuristics. Algorithms based on heuristics, [8,28–30] are able to alleviate the limitations of MPT wherein, each generating unit is considered separately for the selection of its optimal outage interval. But the global optimum solution can always be obtained by simultaneously involving all the units in the scheduling process. Due to the coupling constraints, sequential unit scheduling generation often leads to poor local optimum and infeasible solutions.

In recent years, evolutionary algorithms (EAs) have become the effective alternate solution for finding the solution for complex combinatorial problem. Several such methods including metaheuristic-based hybrid approaches [4], genetic algorithm [9,31] tabu search [32], simulated annealing, [22,33] memetic algorithm [17] hybrid evolutionary technique [34], Particle Swarm Optimization (PSO) [13,35–37] etc. have also been used to solve the GMS problem. Survey of literature reveals that, for the nature of GMS problem, genetic algorithm shows rapid convergence and it has more probability of escaping at local optima [9,31]. Hence, in this chapter, Genetic Algorithm (GA) has been applied to solve the GMS optimization problem.

1.4 Objective of the Chapter

Generator Maintenance GMS is one of the large-scale power system optimization problems that needs a software-based technique for the solution. The aim of this chapter is to present review of: (i) formulation of the GMS problem in Section 2; (ii) probabilistic modeling-calculation of objective function, Expected Energy Production Cost (EEPC) and calculation of hard constraint function, EENS for one sample 3-unit system in Section 3; (iii) need for evolutionary algorithms and implementation of GA for solving GMS problem with illustration in Section 4; (iv) results and discussion of optimal GMS for 3-unit, 5-unit and 21-unit test systems using MATLAB in Section 5; and (v) conclusion.

2 PROBLEM FORMULATION

OMS problem is an optimization problem which determines the period for which the units should be removed from service for maintenance over the period of one/two years' time horizon, in order to minimize the cost and increase the system reliability. Among several approaches discussed above, minimization of EEPC and EENS, incorporating the forced outage rate of the generators are often a more crucial factor, among various other cost objectives and reliability index calculations in the planning studies. Thus, in this chapter the problem is formulated as specifying dates at which manpower is to be allocated for an overhaul of every generating unit, such that EEPC is minimized, and simultaneously the overall system security level, that is, EENS is minimum ($\mathcal{E}$), meeting various other system constraints.

Each generating unit has an associated maintenance cost, required maintenance duration, and length of maintenance, a crew, an expected served energy (*EE*), EENS, and EEPC. Power demand forecast is available, with the past experience and future predictions, and a suitable maintenance schedule must ensure that enough units are online over the period to satisfy this demand. The main goal is a valid feasible maintenance schedule that minimizes EEPC and satisfies the EENS constraint.

2.1 Objective Function

At present, power system probabilistic modeling is widely used in utilities. The primary purpose of probabilistic modeling is to simulate the dispatch of generating units and to estimate the production cost. As power systems steadily grow, the accurate representation of production cost in probabilistic modeling is becoming more important. Using single average incremental fuel consumption to present the economic characteristic of a thermal unit usually does not meet the precision requirement [38].

The objective function of the GMS problem under consideration is required to minimize the total EEPC over the planning period, because incorporating the uncertain random outage can only draw out a preventive, planned, and optimal maintenance schedule. The maintenance cost of a generating unit does not change during the planning period and can be regarded as constant in the GMS problem [26]. The values will not influence the search for the optimization operational costs. Therefore, to simplify computation, the maintenance cost of the individual generators is not included in the formulated objective.

The GMS problem can be mathematically formulated as follows [9]:

$$Min,\ EEPC = \sum_{t=1}^{m_t} \sum_{i=1}^{n} a_{it}\, EES_{it}(1 - X_{it}) \tag{1}$$

where,

i = index of generating units
n = total number of units
t = index of periods
m_t = number of planning horizons in weeks
a_{it} = average operating cost of unit "i" at time "t" in \$/MWh
EES_{it} = expected energy served by unit "i" at period "t" in MWh
X_{it} = 1, if unit 'i' is offline for maintenance
0, otherwise.

2.2 Maintenance Constraints

GMS constraints are based on maintenance technology of the generators (describes appropriate duration, sequence, continuous length of maintenance), availability of man power, predicted load demand, material resources, reliability etc.

2.2.1 Window Constraint

It tells the specific time interval during which maintenance on each unit must take place.

$$\sum_{t \in T_i} X_{it} = 1 \tag{2}$$

where,

$$T_i = \{t \in T_i : E_i \leq t \leq L_i - M_i + 1\}$$

T_i is the set of periods, when the maintenance of unit "*i*" start;
L_i is the latest period for unit "*i*" to close the maintenance;
E_i is the earliest period for unit "*i*" to start the maintenance; and.
M_i is the maintenance length of unit "*i*".

2.2.2 Crew Constraint

There are a limited number of maintenance crews, so a limited number of units can be in maintenance simultaneously.

$$\sum_{i \epsilon I_t}^{n} \sum_{k \epsilon S_{it}}^{m_t} X_{ik} M_{ik} \leq AM_t \tag{3}$$

where.

AM_t is the available manpower at period "*t*";
M_{it} is the man power required at period "*t*" by unit "*i*";
S_{it} is set of start time periods "*k*" such that if the maintenance of unit "*i*" starts at period "*k*" that unit will be in maintenance at period "*t*", $S_{it} = k \epsilon T_i$: $t - M_i + 1 \leq k \leq t$;
I_t is the set of units which are allowed to be in maintenance in period "*t*", $I_t = [i : t \in T_i\ ]$.

2.2.3 Duration Constraint

In order for the units to operate in good condition, the units should be maintained for the required duration without interruption consecutively. The schedule to be generated spans a certain number of weeks.

$$\sum_{t=T_i}^{T_i+Y_{i-1}} X_{it} = Y_i \tag{4}$$

where,

Y_i is the maintenance period until the ith unit;
Y_{i-1} is the maintenance period until the $(i-1)$th unit.

2.2.4 Reliability Constraint

The system reliability is described by an acceptable level of EENS. This constraint requires that the system operate within the prescribed level, "$\mathcal{E}$" with the units that remain online.

$$\text{EENS} < \mathcal{E} \tag{5}$$

where, $\mathcal{E}$ is the allowable level of EENS in MWh.

2.2.5 Demand Constraint

The demand of each week should be met by the units that are committed for supplying power.

$$\sum_{i=1}^{n} P_i(1-X_{it}) = D_t \tag{6}$$

where,

D_t is the anticipated demand in MW at period "t";
P_i is the generating capacity of the ith unit.

3 PROBABILISTIC MODELING

The basic task of probabilistic modeling is to calculate (a) the Expected Energy Served (EES) by each generating unit, thereby EEPC, and (b) system reliability index, EENS, which are more meaningful in both system operation and planning [38].

The basic modeling approach for the generating system adequacy assessment [39]:

- Develop model for generation capacity
- Create a load model
- Combine the load model with generation capacity model to obtain a risk model
- Evaluate reliability index

3.1 Methods to Calculate Reliability Indices

To reduce the computational complexity of reliability index calculations for large systems, numerical approximations based on expansions that use the normal probability distribution and its derivatives have been used. These techniques have been variously referred in the power system literatures as follows.

- Analytical methods
- Monte Carlo simulation
- Fourier methods
- Equivalent load duration curve method
- Edge worth approximations
- Cumulant method
- Method of moments
- Gram Charlier expansion
- Segmentation method
- Equivalent Energy Function (EEF) Method

3.2 Equivalent Energy Function Method

Inspired by the idea of Shenk [40], who presented a more accurate, efficient, and new probabilistic modeling approach, Xifan Wang has developed a method based on EEF. This technique enables us to implement convolution and deconvolution by directly using electric energy, so that probabilistic modeling is considerably simplified. EEF approach is not only more efficient, but also more flexible [38]. The main idea of EEF approach is if electric energy consumed in different load level segments are known and can directly be modified when unit failure effects are taken into consideration, then the probabilistic modeling can be carried out easily.

3.2.1 Illustration of EEF Method

In this section, the calculation of EES, EENS, EEPC are discussed in steps, to provide better understanding for the readers [38], for a sample 3-unit test system, whose characteristics are shown in Table 1.

Step (1) Describe the load duration curve by $t=F(x)$. Where,
"x" is the load level in MW;
"t" is the time interval during which the load is equal to "x".

Step (2) Create a generation capacity model and choose a step Δx, greatest common factor of all generating unit capacities.
Capacity, $(C_{i=1,2,3})=[40; 40; 40]$.
So, $\Delta x=40\,\text{MW}$.

TABLE 1 Operating Characteristics of 3-Unit Test System

Hourly Load Duration Curve								
T (H)	1	2	3	4	5	6	7	8
Load (Mw)	20	50	80	90	70	80	60	30

Unit	Capacity (MW)	Forced Outage Rate	Maintenance Window	Duration
1	40	0.1	1–8	2
2	40	0.1	1–8	3
3	40	0.1	1–8	2

Step (3) Describe the outage rate and availability rate of each generating unit. Outage rate, $(q_{i=1,2,3}) = [0.1;\ 0.1;\ 0.1]$.

Availability rate, $(p_{i=1,2,3}) = p_i = 1 - q_i = [0.9;\ 0.9;\ 0.9]$.

Step (4) Segment the load model by step size as in Fig. 1 and note down the area under the individual segments. The area is calculated to be [290; 180; 10] MWH in the segments $[0 - \Delta x;\ \Delta x - 2\Delta x;\ 2\Delta x - 90\,\text{MW}$ (Peak load)] respectively.

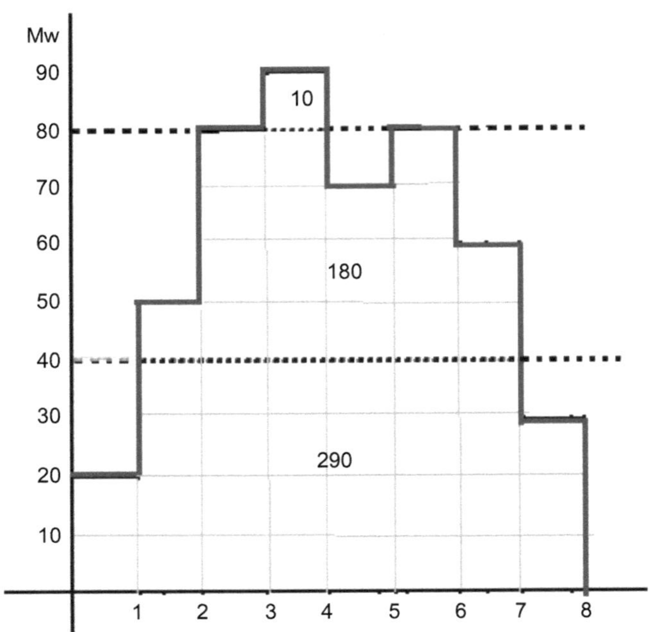

FIG. 1 Load model.

Step (5) Calculate "J" the number of entries for building the energy table (in subsequent steps), which varies from 1 to $J_n + N_E$.

where,

$$Jn = \frac{\text{Total capacities}}{\Delta x}; \quad Jn = 120/40 = 3$$

$$NE = \frac{\text{Peak load}}{\Delta x} + 1; \quad Ne = 90/40 + 1 = 3$$

Therefore "J" varies from 1 to 6 ($Jn + Ne$)

Step (6) Calculate $k_{i=1,\,2,\,3} = C_i/\Delta x$, which is always an integer.

$$k_1 = C_1/\Delta x = 40/40 = 1$$
$$k_2 = C_2/\Delta x = 40/40 = 1$$
$$k_3 = C_3/\Delta x = 40/40 = 1$$

Step (7) Form a table for finding expected values, with 6 "J" entities and enter the calculated area values in $E^0(J)$ column, as found in Table 2.

Step (8) EEF is the energy function that considers the failure effects of committed units. According to the convolution formula [38], Calculate $E^i(J)$ for all values of "i" say, 1, 2, and 3. Let us first commit the unit 1, whose capacity is C_1, and FOR is q_1.

Initially, to calculate $E^1(J)$, for all values of "J" from 1 to 6, formula to be used is.

$$E^1(J) = p_1 * E^0(J) + q_1 * E^0(J - k_1) \tag{7}$$

that is, $E^1(J) = 0.9 * E^0(J) + 0.1 * E^0(J - k_1)$

For, $J = 1$,

- $E^1(1) = 0.9 * E^0(1) + 0.1 * E^0(0)$ [Take, $E^0(0) = E^0(1)$]
  $= 0.9 * 290 + 0.1 * 290 = 290$

TABLE 2 Energy Matrix

J	$E^0(J)$	$E^1(J)$	$E^2(J)$	$E^3(J)$
1	290	290	290	290
2	180	191	200.9	209.81
3	10	27	43.4	59.15
4	–	1	30.6	7.58
5	–	–	0.1	0.45
6	–	–	–	0.01

Similarly, for all other "J" entries

- $E^1(2) = 0.9*180+0.1*290 = 191$
- $E^1(3) = 0.9*10+0.1*180 = 27$
- $E^1(4) = 0.9*0+0.1*10 = 1$
- $E^1(5) = E^1(6) = 0$

Step (9) The load in the interval has been shared by proceeding unit i, when it has been committed. The load expected energy not served (EENS) by the unit 1, becomes,

$$\text{EENS}_1 = \sum_{J>i} E^i(J) = \sum_{J>1} E^1(J) = 191 + 27 + 1 \tag{8}$$

$$\text{EENS}_1 = 219\,\text{MWh}$$

Step (10) The Expected Energy Served (EES) by the unit 1 becomes,

$$\text{EES}_1 = \text{Demand} - \text{EENS}_1 = 480 - 219 = 261\,\text{MWh}$$

Step (11) Next commit 2nd unit, whose capacity is C_2, and FOR is q_2. Calculate $E^2(J)$ for all values of "J" with the formula,

$$E^2(J) = p_2 * E^1(J) + q_2 * E^1(J - k_2) \tag{9}$$

- $E^2(1) = 0.9*290+0.1*290 = 290$
- $E^2(2) = 0.9*191+0.1*290 = 200.9$
- $E^2(3) = 0.9*27+0.1*191 = 43.4$
- $E^2(4) = 0.9*1+0.1*27 = 3.6$
- $E^2(5) = 0.9*0+0.1*1 = 0.1$
- $E^2(6) = 0$

Step (12) The load in the interval has been shared between units "1" and "2", after unit "2" has been committed. The EENS by the unit 2, becomes,

$$EENS_2 = \sum_{J>i} E^i(J) = \sum_{J>2} E^2(J) = 43.4 + 3.6 + (0.1) \tag{10}$$

$$EENS_2 = 47.1\,\text{MWh}$$

Step (13) The expected energy served by the unit 2 becomes,

$$\text{EES}_2 = \text{EENS}_1 - \text{EENS}_2 = 219 - 47.1 = 171.9\text{MWh} \tag{11}$$

Step (14) Next commit 3rd unit, whose capacity is C_3, and FOR is q_3. Calculate $E^3(J)$ for all values of "J" using,

$$E^3(J) = p_3 * E^2(J) + q_3 * E^2(J - k_3) \tag{12}$$

- $E^3(1) = 0.9*290+0.1*290 = 290$
- $E^3(2) = 0.9*200.9+0.1*290 = 209.81$
- $E^3(3) = 0.9*43.4+0.1*200.9 = 59.15$

TABLE 3 Calculation of Cost Objective Function: EEPC

Unit	EES (MWH)	EENS (MWH)	Average Operating Cost ($/MWH)	EEPC ($)
1	261	219	8.2	2140.2
2	171.9	47.1	8.3	1426.77
3	39.06	8.04	8.1	316.386
Total EEPC ($)				3883.356

- $E^3(4) = 0.9*3.6 + 0.1*43.4 = 7.58$
- $E^3(5) = 0.9*0.1 + 0.1*3.6 = 0.45$
- $E^3(6) = 0.9*0 + 0.1*0.1 = 0.01$

Step (15) The load in the interval has been shared between units '1' to '3', after unit '3' has been committed. The EENS by the unit 3, becomes,

$$\text{EENS3} = \sum_{J>i} E^i(J) = \sum_{J>3} E^3(J) = 7.58 + 0.45 + 0.01 = 8.04\,\text{MWh} \quad (13)$$

Step (16) The expected energy served by the unit 3 becomes,

$$\text{EES}_3 = \text{EENS}_2 - \text{EENS}_3 = 47.1 - 8.04 = 39.06\text{MWHh} \quad (14)$$

Step (17) Calculation of EEPC for the three committed units is shown in Table 3.

4 APPLICATION OF GENETIC ALGORITHM FOR GMS

The GMS problem has attracted the interest of many researchers because of its importance and complexity. The no free lunch theorem says that there is no single best optimization algorithm suitable for solving all classes of problems. The purpose of the work is to indicate a suitable optimization technique for solving GMS problems.

4.1 Need for Evolutionary Algorithms

Optimization is the process of making something better. Problems in engineering, economics, and physical and social sciences entail the optimization (minimization or maximization) of an objective function subject to certain constraints (restrictions and trade-offs). Optimization problems can be either unconstrained or constrained; linear or nonlinear; integer or discrete. Unconstrained optimization deals with the optimization of an objective function that is not subject to constraints. Constrained optimization entails the optimization of an objective function subject to certain constraints. Linear programming deals only with linear objective functions and linear constraints. When either set of constraints, objective function, or both are nonlinear, the optimization problem is called as nonlinear programming problem. A mixed-integer programming problem is an optimization problem in which some of the design variables take integer values. A discrete optimization problem is an optimization problem in which some of the design variables must take discrete values that belong to an ordered set of values.

Two different methodologies for the solution of optimization problem are MPTs and MHTs. The best method for solving linear programming problems is the simplex method. Nonlinear programming problems can be solved using direct search and gradient search methods. Methods for solving discrete optimization problems include dynamic programming and branch-and-bound method. The branch-and-bound method is effective in solving mixed-integer programming problems. But, for the complex multimodal GMS problem, an optimal maintenance schedule by MPT is not reliable, because it will obtain optimum with an acceptable tolerance, with the following difficulties.

4.1.1 Drawbacks of MPT

a. The real problem may be so large and complex that an exact mathematical model is not realistic. MPT finds it difficult to model real problems and the problems encountered in calculation. Hence, it uses approximate analysis based on assumptions to avoid this problem. Relationships between input and output variables is usually only approximated and much work is required to generate approximate relationships.

b. When derivative based MPT are applied in solving nonlinear programming problems, the results are often stuck at local optima/premature convergence due to multimodal nature of the search space.

c. It is difficult to handle optimization problems with mixed variables such as integer, real, binary, discrete, and multiple objectives.

d. Problems with complex analytical calculations interacting with look up table, chart, and graphical data cannot be handled accurately by the derivative based methods and an optimal solution is not guaranteed for such problem.

Thus, if there is limited or no knowledge about the behavior of the objective function related to the presence of local minima, the location of feasible region, or the infeasible region in the multidimensional parameter space, it seems advisable to start the MHT, which is a stochastic strategy.

4.1.2 Metaheuristic Technique

Several MHTs have evolved in recent decades which facilitate solving optimization problems that were previously difficult or even impossible to solve. MHTs mimic natural evolutionary principles to perform the search and optimization procedures. These methods, in fact, choose their path through the parameter space using random factor. They can get along with the function value of the objective function only, so that one does not have to bother about the continuity of the objective function or its gradient during the iteration process. This fact enables to find the region of the global optimum with the high probability no matter where the strategy is started from.

Of all the MHTs, GAs are especially simple and effective in the solution of high-complexity, nonconvex, nonlinear, multidimensional, mixed variable, constrained optimization problems, for which a traditional mathematical model is difficult to build, where the nature of the input/output relationship is neither well defined nor easily computable. Hence, this research work has employed GA for finding solution to GMS optimization problem, by minimizing the design objective (EEPC), while simultaneously satisfying all the constraints imposed by specifications.

4.1.3 Genetic Algorithm

EAs are stochastic and their search methods imitate and model some natural phenomena i) survival of the fittest and ii) genetic inheritance. A population of individuals exists in an environment with limited resources. Competition for those resources causes selection of those fitter individuals that are better adapted to the environment. These individuals act as seeds for the generation of new individuals through recombination and mutation. The new individuals have their fitness evaluated and compete (possibly also with parents) for survival. Over time, natural selection causes a rise in the population fitness and the solution gets converged over generations [41]. The simple block diagram of the evolutionary process is shown in Fig. 2.

4.2 Application of GA for GMS

Different type of approaches can be taken in the basic design of a genetic algorithm. Mathematically, the GMS problem can be formulated as an

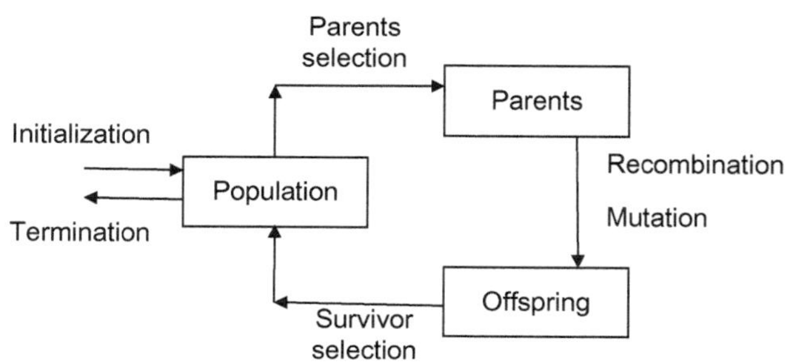

FIG. 2 Block diagram of evolutionary process.

integer programming problem by using integer variables associated with answers to when does maintenance start?", or alternatively by using conventional binary variables associated with the answers to "when does maintenance occur?". However, the first formulation takes care of the constraints on the periods and duration of maintenance and hence the number of unknowns are reduced. Moreover, the answer to the first question automatically provides the answer to the second. Encoding of the problem using an appropriate representation is crucial aspect of the implementation of a GA for solving optimization problem. Hence in this work, the GMS problem is tackled using integer GA rather than traditional binary coding GA, by finding the starting period, that is, when the maintenance should be begun for each generating unit, such that the objective functions are minimized and the constraints are met.

Simple GA has four components that must be designed to solve the problem. They are (i) syntax of the chromosome; (ii) chromosome interpretation; (iii) chromosome evaluation; and (iv) genetic operators. They are explained in detail for GMS as under.

4.2.1 Genetic Representation

Genetic algorithms differing from conventional search techniques start with an initial set of random solutions called population. Each individual in the population, called a chromosome, representing a solution to the GMS problem is represented in integer form. The approach of integer encoding in a genetic structure consists of a string of integers, each one of which indicates the maintenance start period of a unit. For this representation, the string length is equal to the number of units. The genetic structure represented as integer string (chromosome) is:

$$t_1 \quad t_2 \quad t_3 \quad \ldots\ldots \quad t_i \quad \ldots\ldots \quad t_n$$

where, t_i is an integer, indicates the maintenance start period, $(E_i \le t_i \le L_i - M_i + 1)$, for each unit , and "$n$" is the number of units. The number of independent/decision variables in the formulated GMS problem is the number of generators "n".

Suppose if the maintenance window for unit "1", is between 1 and 26 weeks, and the maintenance length is 7 weeks, then t_1 should take an integer value in the interval; $1 \le t_i \le 26 - 7 + 1$ say, 10 indicating the starting week for unit "1" to undergo maintenance. Using MATLAB function randperm, string of integers for this requirement is generated. This type of representation automatically considers the maintenance window constraint and greatly reduce the size of GA search space to.

$$\prod_{i=1}^{n} L_i - M_i - E_i + 1 \tag{15}$$

4.2.2 Initialization

Initial population plays a role in the performance of GA. Because the improvements seen with the heuristic initialization do not persist after first iteration as per the chromosome structure, the initial set of solution called population is randomly generated, as per the formulation, satisfying the maintenance window constraint.

4.2.3 Fitness Function

Evaluation of the chromosome string is accomplished by calculating the objective function. The result of the function calculations is used to calculate the value of the string with respect to all other chromosome strings within the population. This raw value measure is called the string's fitness. The property of fitness influences the GA's performance as GA treats the problem as black box in which the input is the strings and the output is their fitness (F). It is associated directly with the objective function value (EEPC) and the constraints.

(a) *Penalty parameter less constraint handling*
One of the questions of the constrained optimization is how to treat the infeasible individuals throughout the search process commonly called as Constraint Handling Technique (CHT). Penalty parameter less constraint handling is incorporated for treating the complex constraints of GMS problem [42]. This method is preferred in such a way that the different magnitude violation of every constraint is handled properly, and the algorithm can get away from the local optima. In this handling method, infeasible solutions are compared based on their overall constraint violation (CV), whereas the feasible solutions are compared based on their objective function values. The possible three cases that may arise are as follows.
a. In the first case, neither objective function value nor the CV information is necessary, only feasible solution is preferred to any infeasible solution.
b. In the second case, among two feasible solutions, the one with better objective function value alone is considered.
c. In the last case, among two infeasible solutions, the one with smaller CV alone is chosen.

(b) *Fitness function evaluation Using CV*
Overall normalized CV and fitness function are calculated using the following steps.
i. Transform all the constraints equations into inequality form, $h_j(x) \leq d_j$, for $j = 1, 2, \ldots q$.
where, q is the number of constraints, excluding boundary constraints of design variables; (EENS constraint, crew constraint, maintenance boundary constraint, maintenance window constraint, etc.,).

ii. Because the CV, $(h_j(x) - d_j)$ for each constraint 'j' is of different range, calculate the normalized CV using,

$$err_j(x) = \left(h_j(x)/d_j\right) - 1 \tag{16}$$

where, h_j is jth inequality constraint; d_j are constants of constraints; $err_j(x)$ is the normalized constant violation for each constraint

iii. Calculate the overall normalized CV as

$$\text{CV} = \sum_{j=1}^{q} \max\left(0, err_j(x)\right) \tag{17}$$

where, Positive value of $err_j(x)$ denotes the violation of constraint and negative value of $err_j(x)$ shows no violation.

iv. Evaluate the fitness function, $FF(x)$ using penalty parameter less CHT from [42]

$$FF(x) = \left\{ \begin{array}{ll} z(x), & \text{if feasible} \\ z^{max} + \text{CV} & \text{otherwise} \end{array} \right\} \tag{18}$$

where, $z(x)$ is the objective function value (EEPC); z^{max} is the objective function value of the worst feasible solution in the current population.

The fitness of a feasible solution is always fixed and is equal to its objective function value $z(x)$. However, the fitness of the infeasible solutions not only depends on the overall CV, but also on the fitness of the worst feasible solution,z^{max}. If there is no feasible solution in the current population, z^{max} is zero. The infeasible solutions are compared based on their overall normalized CV.

4.2.4 Uniform Integer Crossover and Integer Mutation

Operators used in simple GA cannot be employed for solving GMS problems due to their complexity. Modified genetic operators like uniform integer crossover and integer mutation are applied to GMS problem to avoid settling in false optimum and slower convergence.

Care is required to be taken in the crossover operation that the possibility of arising infeasible offspring from the feasible parents after applying the crossover operation should not be due to the violation of soft constraints like window constraint. In order to reduce the complexity, while crossover the value of gene can be exchanged between the chromosomes in the respective index to furnish new offspring without the window constraint violation like parent chromosome. In this GMS work, uniform integer crossover [41] has been applied and demonstrated in Table 4. A binary mask is generated randomly, wherein binary "1" means bit swapping and binary "0" means bit replicating. The convergence

TABLE 4 Uniform Integer Crossover

Parent 1	3	4	5	6	7
Parent 2	1	2	3	4	5
Mask	1	0	0	1	1
Child 1	1	4	5	4	5
Child 2	3	2	3	6	7

speed of this uniform integer crossover is faster than the single point and two-point crossover.

Mutation is the random alteration of genes in a chromosome. It is believed to help forestall the premature convergence caused by repeated crossover operation. It's a probable random alteration in the string position. If the primitive population by chance lacks one kind of gene which is crucial to converge to the global optimal solution, mutation operation will become absolutely necessary.

In uniform mutation, the genes undergoing mutation are replaced by a randomly generated integer that satisfies the maintenance window of that particular generating unit, in order to obey the soft constraint and avoid its violation after mutation. Also, it is evident [43] that in the integer encoding (representation) of the GMS problem, GA search is limited within the maintenance feasible region and a higher mutation probability increases the exploration for the global minima within this limited region, reducing the chance of premature convergence. Hence in this GMS work, integer mutation has been applied rather than the binary mutation, which is not suitable to produce chromosome without having a violation in window constraint.

4.3 Estimation of Fitness Function from a Chromosome

In GMS problems, all the generating units may not be committed for generation in some Hr/week of the time horizon considered. So, EES, EEPC and EENS must be calculated for each Hr/week, with the units that are not sent for maintenance. This section clearly explains the fitness function evaluation for a chromosome generated randomly by GA, so that one can develop software code in MATLAB to find fitness value for the GMS problem. Once the fitness function is calculated for one solution (chromosome), 'optimtool' helps the beginner to apply GA to solve GMS problem in MATLAB.

Same 3-unit test system considered in Section 3 has been taken for illustration with slight assumption that, hourly Load Duration Curve (LDC) in Fig. 1 is replaced by weekly LDC: Load (MW) Vs time (weeks), considering load remains the same in each week. Finally, EENS and EES (MwWk) will be multiplied by factor 168 (24 h × 7 Days = 168) to get MwH.

Step (1) To Calculate $E^0(J)$.

For the Fig. 1, load model, $E^0(J)$ is calculated for every week, by finding area under curve segment wise, as in Table 5.

TABLE 5 Area Calculation for Each Week of LDC

Week	Segment	$E^0(J)$: Area (MwWk)
1	0–40	20
	41–80	0
	81–90 MW (Peak load)	0
2	0–40	40
	41–80	10
	81–90 MW (Peak load)	0
3	0–40	40
	41–80	40
	81–90 MW (Peak load)	0
4	0–40	40
	41–80	40
	81–90 MW (Peak load)	10
5	0–40	40
	41–80	30
	81–90 MW (Peak load)	0
6	0–40	40
	41–80	40
	81–90 MW (Peak load)	0
7	0–40	40
	41–80	20
	81–90 MW (Peak load)	0
8	0–40	30
	41–80	0
	81–90 MW (Peak load)	0

Step (2) To Find Unit Commitment Schedule.

The genetic structure represented as integer string (chromosome) is denoted by variable 'x'. It indicates the maintenance start up Week of each unit.

Let the chromosome be: $x = [1\ 4\ 7]$.

Maintenance duration in weeks for the three units be: maint_len = [2 3 2].

Then end week of maintenance for each unit becomes: maint_end = x + maint_len - 1.

that is, maint_end = [2 6 8].

Thus, the unit-1 goes for maintenance from week 1, (because, $x(1) = 1$), and continues the maintenance till week 2 (because, maint_end(1) = 2). Similarly, the units '2' and '3' take the maintenance periods of 4th to 6th week and 7th to 8th weeks, respectively. Now, the capacity of those generating units that are committed for generation and those that are not committed (planned for maintenance), in the 8-week planning horizon are displayed in Table 6 as below.

Step (3) To Calculate EENS, EES for Weekly Load.

EES and EENS are calculated using the EEF method and shown in Table 7 for every week.

Step (4) To calculate objective function, EEPC.

For the complete two months horizon, weekly operating costs of energy production for each unit in $/MWH, EES by each unit in MwWk and the EEPC of each unit are calculated below in Table 8 using EEPC = sum(EES. * unit operating cost).

TABLE 6 Unit Commitment Schedule for a Chromosome

	Capacity (MW)		
Week	*Unit-1*	*Unit-2*	*Unit-3*
1	**0**	40	40
2	**0**	40	40
3	40	40	40
4	40	**0**	40
5	40	**0**	40
6	40	**0**	40
7	40	40	**0**
8	40	40	**0**

TABLE 7 EES and EENS Calculation for a Chromosome

Capacity = [0 40 40]; $E^0(J)$ = [20 0 0]; Week = 1; Load = 20 MW			
J	$E^1(J)$	$E^2(J)$	$E^3(J)$
1	20	20	20
2	0	2	3.8
3	0	0	0.2
EES	Unit-1	Unit-2	Unit-3
	0	18	1.8
EENS	0.2		
Capacity = [0 40 40]; $E^0(J)$ = [40 10 0]; Week = 2; Load = 50 MW			
J	$E^1(J)$	$E^2(J)$	$E^3(J)$
1	40	40	40
2	10	13	15.7
3	0	1	2.2
4	0	0	0.1
EES	Unit-1	Unit-2	Unit-3
	0	36	11.7
EENS	2.3		
Capacity = [40 40 40]; $E^0(J)$ = [40 40 0]; Week = 3; Load = 80 MW			
J	$E^1(J)$	$E^2(J)$	$E^3(J)$
1	40	40	40
2	40	40	40
3	4	7.6	10.84
4	0	0.4	1.12
5	0	0	0.04
EES	Unit-1	Unit-2	Unit-3
	36	36	6.84
EENS	1.16		
Capacity = [40 0 40]; $E^0(J)$ = [40 40 10]; Week = 4; Load = 90 MW			
J	$E^1(J)$	$E^2(J)$	$E^3(J)$
1	40	40	40
2	40	40	40
3	13	13	15.7

TABLE 7 EES and EENS Calculation for a Chromosome—cont'd

Capacity = [40 0 40]; $E^0(J)$ = [40 40 10]; Week = 4; Load = 90 MW			
J	$E^1(J)$	$E^2(J)$	$E^3(J)$
4	1	1	2.2
5	0	0	0.1
EES	Unit-1	Unit-2	Unit-3
	36	0	36
EENS	18		
Capacity = [40 0 40]; $E^0(J)$ = [40 30 0]; Week = 5; Load = 70 MW			
J	$E^1(J)$	$E^2(J)$	$E^3(J)$
1	40	40	40
2	31	31	31.9
3	3	3	5.8
4	0	0	0.3
EES	Unit-1	Unit-2	Unit-3
	36	0	27.9
EENS	6.1		
Capacity = [40 0 40]; $E^0(J)$ = [40 40 0];Week = 6; Load = 80 MW			
J	$E^1(J)$	$E^2(J)$	$E^3(J)$
1	40	40	40
2	40	40	40
3	4	4	7.6
4	0	0	0.4
EES	Unit-1	Unit-2	Unit-3
	36	0	36
EENS	8.0		
Capacity = [40 40 0]; $E^0(J)$ = [40 20 0]; Week = 7; Load = 60 MW			
J	$E^1(J)$	$E^2(J)$	$E^3(J)$
1	40	40	40
2	22	23.8	23.8
3	2	4	4
34	0	0.2	0.2

Continued

TABLE 7 EES and EENS Calculation for a Chromosome—cont'd

Capacity = [40 40 0]; $E^0(J)$ = [40 20 0]; Week = 7; Load = 60 MW			
J	$E^1(J)$	$E^2(J)$	$E^3(J)$
EES	Unit-1	Unit-2	Unit-3
	36	19.8	0
EENS	4.2		
Capacity = [40 40 0]; $E^0(J)$ = [40 40 0]; Week = 8; Load = 30 MW			
J	$E^1(J)$	$E^2(J)$	$E^3(J)$
1	30	30	30
2	3	5.7	5.7
3	0	0.3	0.3
EES	Unit-1	Unit-2	Unit-3
	27	2.7	0
EENS	0.3		

Step (5) To calculate fitness function.

For the chromosome, x = [1 4 7], the objective function value EEPC is calculated as,

objfun = Total EEPC = 4087 * 168 = 686,616 $.

The soft constraints, maintenance window constraint, and duration constraint are inherently included in the generation of initial population and offspring itself. But it is desirable to ensure the hard EENS constraint is satisfied. The constraint amount for expected energy not served, EENS in all 8 weeks is found to be.

EENS = [0.2 2.3 1.16 18 6.1 8 4.2 0.3]*168 MWH.
EENS = [33.6386.4194.88 3024 1024.8 1344 705.6 50.4] MwH.

Let the maximum limit for the expected energy not served for each week be: EENS_max = 900MwH. Then, the violation of EENS constraint for the considered chromosome is found using,

EENS_err = (EENS/$EENS_{max}$) − 1.
EENS_err = [−0.962 −0.570 −0.783 2.36 0.1386 0.493 −0.2160 −0.9440]
CV = sum((EENS_err >0).*abs(EENS_err)) = 2.9916.

The solution (chromosome), x = [1 4 7] considered is infeasible, as CV exists and the error is positive. Let the objective function value of the worst feasible chromosome in the current population, z^{max} be 7,50,000.

TABLE 8 Objective Function Calculation from a Chromosome

Week	Average Operating Cost ($/MwH)			Expected Energy Served (EES) in MwWk			Weekly EEPC of each unit × (168)			Weekly EEPC of 3 units × (168)
	Unit1	*Unit2*	*Unit3*	*Unit1*	*Unit2*	*Unit3*	*Unit1*	*Unit2*	*Unit3*	
1	8.2	9.3	10.6	0	18	1.8	0	167.4	19.08	186.48
2	8	9.5	10.7	0	36	11.7	0	342	125.19	467.19
3	8.8	9.7	10.7	36	36	6.84	316.8	349.2	73.188	739.188
4	8	9.6	10.5	36	0	36	288	0	378	666
5	8.7	9.8	10.6	36	0	27.9	313.2	0	295.74	608.94
6	8.6	9.7	10.1	36	0	36	309.6	0	363.6	673.2
7	8.7	9.0	10.3	36	19.8	0	313.2	178.2	0	491.4
8	8.5	9.3	10.2	27	2.7	0	229.5	25.11	0	254.61
Total *EEPC*										4087

Fitness value, $FF = z^{max} + CV = 7{,}50{,}000 + 2.9916 = 7{,}50{,}002.9916$.

If the error is nonpositive/zero, the fitness value would have become the objective function value 686,616 $.

5 RESULTS AND DISCUSSION

Three test systems with different planning periods are used to demonstrate the optimization of GMS problem. In this section, optimized maintenance schedules obtained by genetic algorithm are shown in detail.

5.1 Test System with Three Units

A test system with two-month (eight weeks) horizon is considered in this case study. Table 1 gives the operating characteristics of the generating plants in loading order. Table 8 possess the weekly operating cost in ($/MwH) for all the three units. The maintenance constrains considered for the test system are maintenance window, maintenance duration, and maintenance completion constraints. Apart from these soft constraints, hard constraint EENS is also included, for which the expected energy not supplied in each week must be less than the allowed maximum limit of 8400 Mwh. For the two-month test problem, the load demand (eight weeks) requirement is given by weekly LDC in Table 9, replacing hourly LDC of Table 1.

To implement GA, a population size of 100 and roulette wheel type selection are employed. The number of elite individual is kept as 2. Because GA is a randomized procedure, to prove consistency in obtaining optimal solutions, 20 independent runs are conducted. Program is developed using the MATLAB software package. Crossover and mutation rates are fixed to 0.8 and 0.015, respectively.

The optimization can also be carried out simply using the function 'optimtool' in MATLAB, if readers find difficulty in developing MATLAB code for genetic algorithm. It is sufficient to develop code for evaluating fitness function value for one random chromosome/solution in a mat file and save it as function with extension filename.m in the right path. Type 'optimtool' in the command window for the 'Optimization Tool' window to pop up.

Next, (a.) select the solver as ga-Genetic Algorithm; (b.) give details for problem tab such as, @filename in fitness function and 3 in 'Number of variables'. Specify bounds tab with, Lower as [0.001 0.001 0.001] and Upper as

TABLE 9 Weekly LDC for 3-Unit Test System

T (Week)	1	2	3	4	5	6	7	8
Load (Mw)	20	50	80	90	70	80	60	30

[0.99 0.99 0.99], so that three genes of a chromosome can be generated randomly in between 0 and 1, excluding 0 and 1.

After having entered the GMS details, population size for algorithm may be specified as desired, instead of default number, which is very less (20). Here, in this 3-unit test case, 200 number of chromosomes are generated. Finally, to avoid stalling the program, the maximum number of generations, which can be stopping criterion, is specified as 500 for this case, instead of keeping the default. The rest of the details are not to be altered, because the toolbox takes care of fixing the needed crossover probability, mutation probability, replacement etc. Last, click 'start' for the optimization to begin. The results of optimization have been displayed in Fig. 3 with objective function value and the chromosome solution (in between 0 and 1).

The optimal maintenance schedule (chromosome) which is interpreted solution in maintenance language from [0.793 0.858 0.572] as [6 5 4]. Expected energy served (exp_s) by each unit for 8 weeks, production cost incurred by all the committed units in each week(prod_cost), min objective value (objfun) and weekly EENS (exp_us), zero constraint violation, are captured from the command window of MATLAB and given in Fig. 4.

The optimal generator maintenance schedule for the 3-unit test system of 8-week planning period is found to be [6 5 4], with the minimum production cost of 548,408.448$ as given in detail in Table 10.

5.2 Test System with 5 Units

Test data with three-month (12 weeks) horizon is derived for this case study from [24], where in decomposition approach is used for finding maintenance schedule. Table 11 gives the characteristics of the five generating units. The maintenance constraints considered in this test system are maintenance

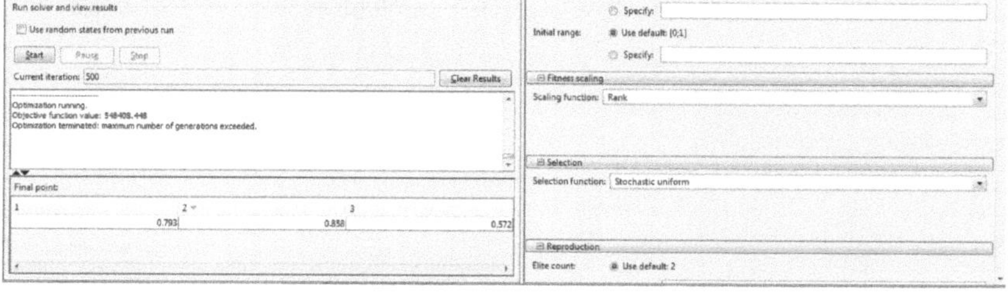

FIG. 3 GMS results for 3unit system in MATLAB optimization tool

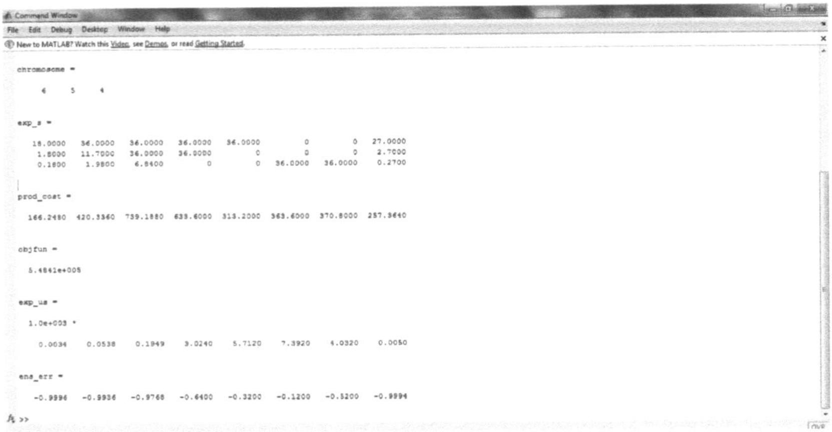

FIG. 4 Summarized optimization results for 3-unit system

TABLE 10 Optimal GMS by GA for 3-Unit System

Solution (X)	[6 5 4]	
Objfun (EEPC)	548,408.448$	
Unit	**Weeks on Maintenance**	**Weeks up**
1	**6**,7	1–5, 8
2	**5**,6,7	1–4, 8
3	**4**,5	1–3, 6–8

TABLE 11 Operating Characteristics of 5 Unit Test System

Unit	Capacity (MW)	Forced Outage Rate	Maintenance Window	Duration
1	150	0.08	1–11	2
2	150	0.09	1–11	2
3	130	0.10	1–11	2
4	90	0.10	1–11	2
5	50	0.15	1–11	2

TABLE 12 Weekly Operating Cost for 5 Unit System in ($/MwH)

Week	Unit-1	Unit-2	Unit-3	Unit-4	Unit-5
1	8.2	9.3	10.6	11.9	12
2	8	9.5	10.7	11.4	12.5
3	8.8	9.7	10.7	11.1	12
4	8	9.6	10.5	11.2	12.3
5	8.7	9.8	10.6	11.3	12.3
6	8.6	9.7	10.1	11.2	12.1
7	8.7	9.0	10.3	11.2	12.4
8	8.5	9.3	10.2	11.1	12.4
9	8.8	9.4	10.5	11.6	12.9
10	8.7	9.6	10.8	11.8	12.8
11	8	9.9	10	11.3	12.9
12	8	9.9	10.4	11.4	12.7

window, maintenance duration, maintenance crew and maintenance completion constraints. Table 12 gives the weekly operating cost in ($/MwH). For the three-month test problem, load demand (12 weeks) requirement is given by weekly LDC as in Table 13. Generating units 1, 2, and 3 are to be maintained by a single crew, meaning the units cannot take up the maintenance simultaneously.

Two cases with different limits for EENS, i.e.., ($\mathcal{E}$) in each week: (a) 400 MwH, (b) 900 MwH, have been analyzed and their effects on the EEPC and optimal GMS results are found. The GMS has been obtained using conventional Decomposition Approach (DA) [24] as well as integer genetic algorithm for the 5-unit system and the results are summarized in Table 14.

For GMS problem, EEPC obtained using DP is found to be more than the one, which is found using GA. Also, it is evident that integer GA can find the optimum cost with less computational time, due to great reduction in search space. The GMS problem is converged around 30th generation by GA to reach the global solution, as shown in Fig. 5. As GA is stochastic, statistical analysis has been done for the 20 independent runs executed, with the case, $\mathcal{E} < 400$ MwH. The mean, best, worst optimal values obtained, standard deviation, and average computational time are reported in Fig. 6.

TABLE 13 Energy Demand (MwH) of Each Weekly LDC for 5-Unit Test System

T (Week)	1	2	3	4	5	6	7	8	9	10	11	12
Energy Demand (MwH)	36,170	38,398	36,170	38,398	36,855	37,707	36,427	37,250	38,379	36,170	36,248	36,855

TABLE 14 GMS Optimization Results for 5-Unit System

	Objective value (EEPC) and Optimal GMS	
$\mathcal{E}$	<400 MwH	<900 MwH
Integer GA	39,968.68.70320 $	3,991,586.8937 $
	[7 11 3 1 11]	[7 11 2 3 7]
Decomposition approach [24]	4,008,450.163509 $	3,998,322.465 $
	[7 10 3 1 5]	[7 11 5 3 11]

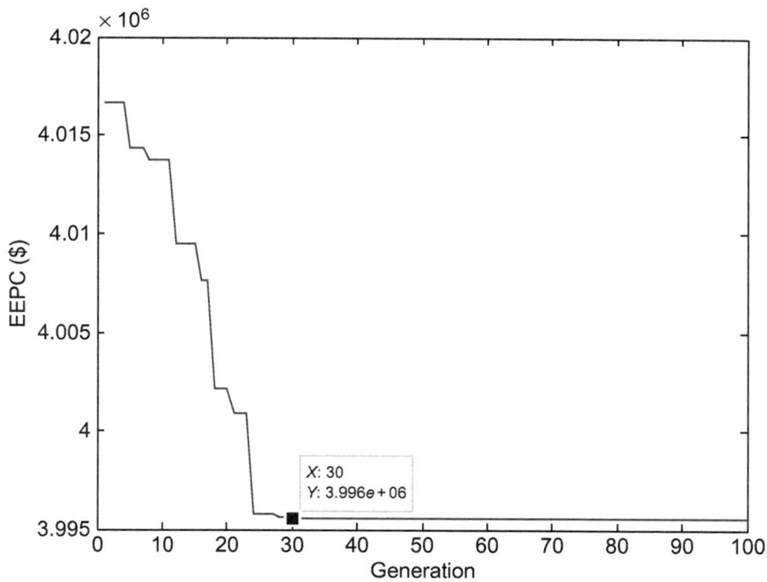

FIG. 5 Convergence plot of GA for $\mathcal{E} < 400$ MwH.

5.3 Test System with 21 Units

This test system is a coal burning utility comprising of 21 units whose operating characteristics has been taken from [19], wherein Dynamic Programming Successive Approximation (DPSA) algorithm is applied for finding maintenance schedule. Generating units range in capacity from 39 to 640 MW. Forced outage

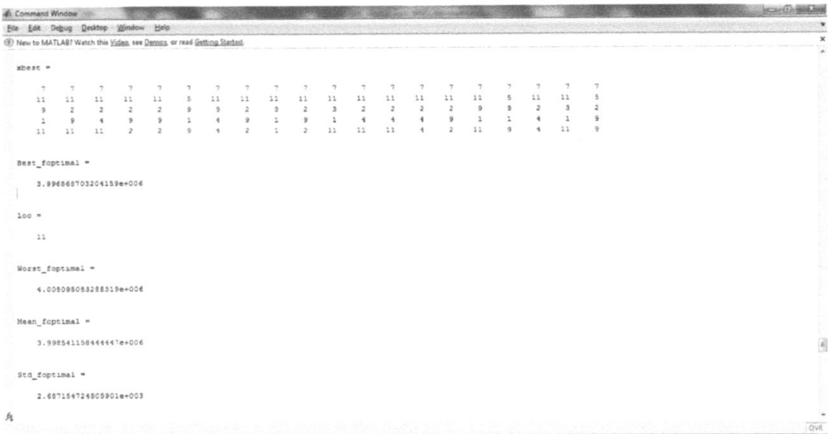

FIG. 6 Statistical analysis of GA for 5-unit system

rate of the units ranges from 0 to 9%. Unit 21 is a hydro unit and its fuel cost is always zero. It is dispatched last due to the limited energy available. Every generating unit is permitted to start the maintenance in any week within the 26-week period. As shown in Table 15, units can begin maintenance either between weeks 1 to 26 or 27 to 52. Thirteen units begin their maintenance in the first half of the year and the remaining eight units' outages are begun in the second half of the year. From the hourly load data of annual load curve [19], the average weekly LDC has been calculated and given in Table 16. The total installed capacity is 5688 MW and the peak load is 4739 MW.

For solving this complex 21-unit GMS problem with DPSA technique, maintenance window, maintenance duration, maintenance completion and load demand constraints are alone considered in [19], by Yamayee without EENS and crew constraints inclusion. In this section, two cases are analyzed to study the complexity of crew constraint on GMS optimization. Case 1 is similar to Case 2 except that case 1 does not consider man power resource constraint. However, the crew constraint [4] is added in Case 2 and the optimal maintenance schedule is needed to satisfy the available man power resource of 20 in each week. The required crew to maintain each unit in the respective outage window is given in Table 15 [4].

The GMS results of 21-unit test system, minimizing EEPC have been obtained with integer genetic algorithm for both the cases and are shown in Tables 17, 18. Case 1 GMS result has been compared with the result of conventional DPSA from [19]. In order to validate the Case 2 study, the optimized schedule obtained with genetic algorithm in Case 2 has been compared with that of Modified Discrete PSO (MDPSO) algorithm from [44], wherein crew constraint is also considered. Result from Table 17 shows that genetic algorithm

TABLE 15 Characteristics of 21-Unit Test System

Unit Number	Capacity (MW)	FOR (%)	Operating Cost ($/MWH)	Outage Window	Duration (Weeks)	Crew Required in Each Week for Maintenance
1	555	8	8.352	1–26	7	10+10+5+5+5+5+3
2	555	8	8.352	27–52	5	10+10+10+5+5
3	180	3.5	8.394	1–26	2	15+15
4	180	0.5	8.394	1–26	1	20
5	640	5	9.884	27–52	5	10+10+10+10+10
6	640	4	9.884	1–26	3	15+15+15
7	640	4	9.884	1–26	3	15+15+15
8	555	7	9.926	27–52	6	10+10+10+5+5+5
9	276	5	10.257	1–26	10	3+2+2+2+2+2+2+2+2+3
10	140	4	10.448	1–26	4	10+10+5+5
11	90	2	10.377	1–26	1	20
12	76	0.5	12.124	27–52	3	10+15+15
13	76	1	12.195	1–26	2	15+15
14	94	4	12.541	1–26	4	10+10+10+10
15	39	4	12.833	1–26	2	15+15
16	188	5	12.921	1–26	2	15+15
17	58	3	14.544	27–52	1	20
18	48	9	14.892	27–52	2	15+15
19	137	6	41.00	27–52	1	15
20	469	4	8.077	27–52	4	10+10+10+10
21	52	0	0	1–26	3	10+10+10

TABLE 16 Weekly Load Duration Curve for 21-Unit Test System

Week	Load (MW)	Week	Load (MW)
1	4739	27	3160
2	4310	28	3109
3	4080	29	3058
4	4029	30	3007
5	3952	31	2981
6	3927	32	2930
7	3875	33	2854
8	3824	34	2802
9	3773	35	2726
10	3748	36	2675
11	3722	37	2598
12	3697	38	2521
13	3651	39	2470
14	3645	40	2419
15	3594	41	2394
16	3569	42	2317
17	3543	43	2291
18	3518	44	2266
19	3492	45	2240
20	3467	46	2215
21	3416	47	2164
22	3390	48	2138
23	3339	49	2087
24	3313	50	2010
25	3262	51	1755
26	3211	52	1500

is able to find the optimal maintenance schedule with less EEPC than DPSA [19] for Case 1 and MDPSO [44] for Case 2. The convergence plot obtained by integer GA for case 1 is shown in Fig. 7. It is also found that in Case 2, when incorporating crew constraint EEPC is found to be higher than the EEPC obtained from Case 1 by GA, for meeting the additional limited man power resource available.

TABLE 17 Comparison of EEPC for Cases With and Without Crew Constraint

Minimization of EEPC Objective			
Algorithm	*Constraint*	*Case (1) Without Crew*	*Case (2) With Crew*
Integer GA	EEPC (Annual in $)	250,375,846.31	254,266,139.422
DPSA [19]	EEPC (Annual in $)	253,400,261.44	–
MDPSO [44]	EEPC (Annual in $)	–	255,183,601.634

TABLE 18 GMS Results for 21-Unit Test System

	Minimization of EEPC Objective			
Unit Number	*DPSA [19] (Without Crew)*	*MDPSO [44] (With Crew)*	*Integer GA (Without Crew)*	*Integer GA (With Crew)*
1	14	17	8	2
2	33	38	48	40
3	12	2	8	15
4	24	1	26	26
5	41	44	42	27
6	7	24	8	9
7	21	14	8	12
8	27	28	47	47
9	10	4	8	16
10	3	8	7	6
11	11	13	26	1
12	33	35	36	42
13	24	5	8	20
14	2	9	20	2
15	12	12	18	24
16	5	1	9	22
17	35	52	27	46
18	35	34	47	32
19	35	27	48	35
20	40	49	45	37
21	10	5	1	17

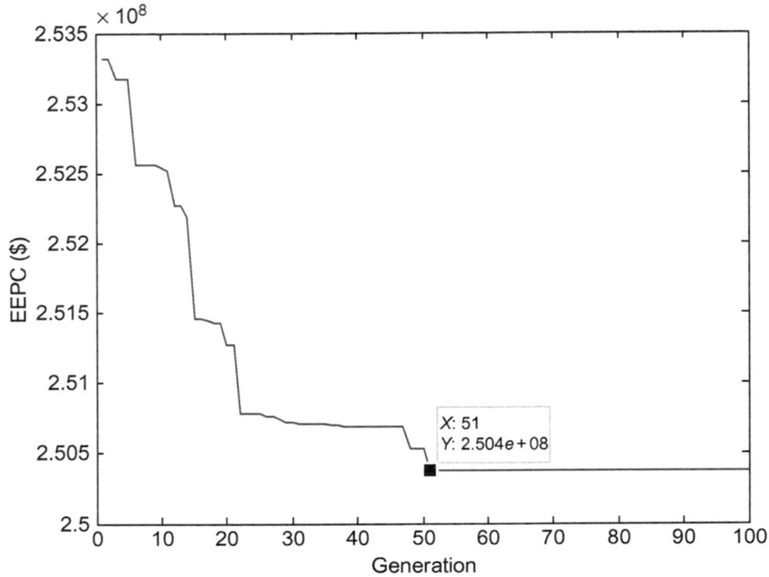

FIG. 7 Convergence plot of GA for case 1 without crew constraint

6 CONCLUSION

The chapter has presented a study of the application of metaheuristic based GA approach to solve power system generator maintenance scheduling problem, taking cost criterion as objective and reliability criterion as the constraint. The application of GA approach has been demonstrated for small, 3-unit test system, medium 5-unit system, and a large 21-unit test problem of generator maintenance scheduling, using an integer GA representation. The appropriate crossover and mutation probabilities of a GA approach are decided after a number of trials, for each of the test systems. The test results of GA are compared with those of MPTs such as decomposition approach and dynamic programming and also with that of stochastic technique MDPSO. The optimal GMS results presented in this chapter show that the GA approach is more robust and suitable than MPTs, irrespective of the size of the GMS problems under consideration. Furthermore, upon statistical analysis, GA is found to have a faster convergence for the complex GMS problem. Generally, MPTs are equipped to find local optimum. But, they can be adapted to seek for a global optimum by considering large number of starting points. The relative merits of evolutionary algorithms and modified MPTs have not yet received required attention in the literature. Thus, it offers a huge potential for future development of hybrid methods intending to combine the advantages of evolutionary algorithms with conventional MPTs. There is also considerable scope for further investigation into solving the complex GMS problem considering reliability criterion as primary objective using hybrid technique.

REFERENCES

[1] J. Endrenyi, et al., The present status of maintenance strategies and the impact of maintenance on reliability, IEEE Trans. Power Syst. 16 (4) (2001) 638–646.

[2] M. Cepin, Assessment of power system reliability: Methods and applications, Springer, London Dordrecht Heidelberg New York, 2011.

[3] K.P. Dahal, K.C. Tan, P.I. Cowling (Eds.), Evolutionary Scheduling, Springer, Berlin Heidelberg New York, 2007.

[4] K.P. Dahal, N. Chakpitak, Generator maintenance scheduling in power systems using metaheuristic-based hybrid approaches, Electric Power Syst. Res. 77 (7) (2007) 771–779.

[5] A.J. Conejo, R. Garcia-Bertrand, M. Diaz-Salazar, Generation maintenance scheduling in restructured power system, IEEE Trans. Power Syst. 20 (2) (2005) 984–992.

[6] R. Billinton, W. Li, Reliability Assessment of Electric Power Systems Using Monte Carlo Methods, Plenum, New York, 1994.

[7] J.F. Dopazo, H.M. Merrill, Optimal generator maintenance scheduling using integer programming, IEEE Transactions on Power Apparatus and Systems 94 (1975) 1537–1545.

[8] G.T. Egan, T.S. Dillon, K. Morsztyn, An experimental method of determination of optimal maintenance schedules in power systems using branch-and-bound technique, IEEE Transactions on Systems, Man and Cybernetics 6 (1976) 538–547.

[9] S. Baskar, P. Subbaraj, M.V.C. Rao, S. Tamilselvi, Genetic algorithms solution to generator maintenance scheduling with modified genetic operators, IEE Proc. Gener. Transm. Distrib. 150 (1) (2002) 56–60.

[10] X. Wang, J.R. Mcdonald, Modern Power System Planning, McGraw-Hill, London, 1994, pp. 247–307.

[11] W. Li, Incorporating aging failures in power system reliability evaluation, IEEE Trans. Power Syst. 17 (3) (2002) 918–923.

[12] H.H. Zurn, V.H. Quintana, Several objective criteria for optimal generator preventive maintenance, IEEE Trans. Power Apparatus Syst. (1977) 984–992. PAS-96 (May/June (3)).

[13] K. Suresh, N. Kumarappan, Generation maintenance scheduling using improved binary particle swarm optimisation considering aging failures, IET Gener. Transm. Distrib. 7 (10) (2013) 1072–1086.

[14] T. Chitra, in: Life based maintenance policy for minimum cost, IEEE Int. Conf. Annual Reliability and Maintainability Symp., Florida, USA, 2003, pp. 470–474.

[15] K.Y. Huang, H.T. Yang, Effective algorithm for handling constraints in generator maintenance scheduling, IEE Proc. Gener. Transm. Distrib. 149 (3) (2002) 274–282.

[16] M.K.C. Marwali, S.M. Shaidehpour, Long-term transmission and generation maintenance scheduling with network, fuel and emission constraints, IEEE Trans. Power Syst. 14 (3) (1999) 1160–1165.

[17] E.K. Burke, A.J. Smith, Hybrid evolutionary techniques for the maintenance scheduling problem, IEEE Trans. Power Syst. 15 (1) (2000) 122–128.

[18] M.Y. El-Sharkh, A.A. El-Keib, Maintenance scheduling of generation and transmission systems using fuzzy evolutionary programming, IEEE Trans. Power Syst. 18 (2) (2003) 862–866.

[19] Z. Yamayee, K. Sidenblad, A computationally efficient optimal maintenance scheduling method, IEEE Trans. Power Apparatus Syst. PAS-102 (1983) 330–338.

[20] J.H. KIM, J.B. PARK, J.K. PARK, B.H. KIM, A new game-theoretic framework for maintenance strategy analysis, IEEE Trans. Power Syst. 18 (2) (2003) 698–706.

[21] C. Feng, X. Wang, F. Li, Optimal maintenance scheduling of power producers considering unexpected unit failure, IET Gener. Transm. Distrib. 3 (5) (2009) 460–471.

[22] J.T. Saraiva, M.L. Pereira, V.T. Mendes, J.C. Sousa, Preventive Generation Maintenance Scheduling—A Simulated Annealing Approach to use in Competitive Markets, (2010).

[23] H.H. Zurn, V.H. Quintana, Generator maintenance scheduling via successive approximation dynamic programming, IEEE Trans. Power Appar. Syst. (1983) 330–338.
[24] J. Yellen, T.M. Al-Khamis, S. Vemuri, L. Lemonidis, A decomposition approach to unit maintenance scheduling, IEEE Trans. Power Appar. Syst. 7 (2) (1992) 726–733.
[25] L.M. Moro, A. Ramos, Goal programming approach to maintenance scheduling of generating units in large scale power systems, IEEE Trans. Power Syst. 14 (3) (1999) 1021–1028.
[26] M.K.C. Marwali, S.M. Shahidehpour, A deterministic approach to generation and transmission maintenance scheduling with network constraints, Electr. Power Energy Syst. 47 (1998) 101–113.
[27] M.K.C. Marwali, S.M. Shahidehpour, Integrated generation and transmission maintenance scheduling with network constraints, IEEE Trans. Power Syst. 13 (3) (1998) 1063–1068.
[28] W.R. Christiaanse, A.H. Palmer, A technique for the automated scheduling of the maintenance of generating facilities, IEEE Trans. Power Appar. Syst. (1) (1972) 137–144.
[29] L.L. Garver, Adjusting maintenance schedules to levelize risk, IEEE Trans. Power Appar. Syst. (5) (1972) 2057–2063.
[30] H. Khatib, Maintenance scheduling of generating facilities, IEEE Trans. Power Appar. Syst. (5) (1979) 1604–1608.
[31] Y. Wang, E. Handschin, A new genetic algorithm for preventive unit maintenance scheduling of power systems, Electr. Power Energy Syst. 22 (2000) 343–348.
[32] I. El-Amin, S. Duffuaa, M. Abbas, A Tabu search algorithm for maintenance scheduling of generating units, Electr. Power Syst. Res. 54 (2000) 91–99.
[33] T. Satoh, K. Nara, in: Maintenance scheduling by using simulated annealing method, Presented at IEEE/PES 1990 summer meeting, Minneapolie, Minnesota, July, 1990, pp. 15–19.
[34] H. Kim, Y. Hayashi, K. Nara, An algorithm for thermal unit maintenance scheduling through combined use of GA, SA and TS, IEEE Trans. Power Syst. 12 (1) (1997) 329–335.
[35] J. Kennedy, R. Eberhart, in: Particle swarm optimization, IEEE Int. Conf. Neural Networks, Perth, Australia, 4 1995, pp. 1942–1948.
[36] T. Logenthiran, D. Srinivasan, in: Particle swarm optimization for unit commitment problem, IEEE Int. Conf. Probabilistic methods Applied to Power Systems, Singapore, 2010, pp. 642–647.
[37] N. Kumarappan, K. Suresh, in: Particle swarm optimization based generation maintenance scheduling using levelized risk method, IEEE Int. Conf. Power System Technology, New Delhi, India, 2008, pp. 1–6.
[38] X. Wang, Equivalent energy function approach to power system probabilistic modeling, IEEE Trans. Power Syst. 3 (3) (1988) 823–829.
[39] R. Billinton, W. Li, Reliability Assessment of Electric Power Systems Using Monte Carlo Methods, Plenum Press, New York, London, 1994.
[40] K.F. Shenk, R.B. MiSl'S, S. Vassos, W. Wen, in: A New Method For Evaluation of Expected Energy Generation And Loss of load Probability, PES. Summer Meeting, 1983.
[41] M. Gen, R. cheng, Genetic Algorithms and Engineering Design, John Wiley and Sons, New York, 1997.
[42] K. Deb, An Efficient Constraint Handling Method for Genetic Algorithms, Comput. Methods Appl. Mech. Eng. 186 (2/4) (2000) 311–338.
[43] K.P. Dahal, J.R. McDonald, Generator maintenance scheduling of electric power systems using genetic algorithms with integer representation, Proceeding of Genetic Algorithms in Engineering Systems: Innovations and Applications, 1997, pp. 456–461.
[44] Y. Yare, G.K. Venayagamoorthy, U.O. Aliyu, Optimal generator maintenance scheduling using a modified discrete PSO, IET Gener. Transm. Distrib. 2 (6) (2008) 834–846.

Chapter 6

Decomposition Methods for Distributed Optimal Power Flow: Panorama and Case Studies of the DC Model

M. Hadi Amini*,[1], Shahab Bahrami[†], Farhad Kamyab[‡], Sakshi Mishra[§], Rupamathi Jaddivada[¶], Kianoosh Boroojeni[||], Paul Weng[#] and Yinliang Xu**

*Department of Electrical and Computer Engineering, Carnegie Mellon University, Pittsburgh, PA, United States, [†]Department of Electrical and Computer Engineering, University of British Columbia, Vancouver, BC, Canada, [‡]Department of Chemical and Process Engineering, University of Surrey, Guildford, United Kingdom, [§]Transmission Planning, American Electric Power, Tulsa, OK, United States, [¶]Electrical Engineering and Computer Science, Massachusetts Institute of Technology, Cambridge, MA, United States, [||]School of Computing and Information Sciences, Florida International University, Miami, FL, United States, [#]University of Michigan-Shanghai Jiaotong University Joint Institute (UM-SJTU JI), Shanghai, China, **Tsinghua-Berkeley Shenzhen Institute (TBSI), Lab of Smart Grid and Renewable Energy, Shenzhen, China

[1]Corresponding author: Email: hadi.amini@ieee.org; Homepage: www.hadiamini.com

1 INTRODUCTION

Electrical power systems provide the necessary infrastructure to generate electricity and deliver it to consumers. Conventionally, electricity is generated in large power plants that use different energy sources such as fossil fuel (e.g., coal, oil, and natural gas), converted fuel (e.g., methane), or geothermal steam [1]. The generated power is transformed through the transmission network at high voltage levels with either direct current (DC) or alternating current (AC). Then, the distribution network enables the transmission of power at medium and low voltage levels to the consumers.

In general, system operators consider the grid-wide generation cost of the operating generators and the system power losses to assess the operation of the power system [2]. Operating conventional power plants is costly due to the high price of their energy sources and maintenance costs. Furthermore,

Classical and Recent Aspects of Power System Optimization. https://doi.org/10.1016/B978-0-12-812441-3.00006-9
© 2018 Elsevier Inc. All rights reserved.

conventional power plants are typically located far from the consumer side, which necessitates using long transmission lines to deliver power from the generators to the consumers. One of the pivotal goals of power grids is to move towards sustainability by deploying renewable resources [3]. Deployment of renewable energy generators such as wind turbines and photovoltaic panels is a promising solution to mitigate the cost of conventional generators and the power losses in transmission networks. Indeed, renewable generators use (almost) free energy sources (e.g., wind and solar) and can be located close to the consumers. Nevertheless, the volatile output power of renewable generators makes it a challenge for system operators to provide an adequate balance between generation and demand [4].

Energy management programs are necessary to enhance the use of the existing and new facilities in transmission systems. They involve exploring and developing tools and techniques for operating the transmission systems in a cost-effective manner. System operators use optimal power flow (OPF) [5,6] analysis to operate the transmission system in an economically efficient manner by reducing the grid-wide generation cost and power losses, while satisfying the operational constraints such as the generation-load balance, bus voltage limits, and the congestion constraints imposed by the transmission lines. A power grid is a time-varying interconnected system, in which the generation is dispatched constantly in response to changes in the load demand [5]. Hence, the OPF program is typically executed on an hourly basis for near real-time system management [7].

The OPF problem in AC grids takes the form of a nonconvex optimization problem and is generally difficult to solve. The nonconvexity of the problem arises from the nonlinear power flow equations and quadratic dependency on the set of bus voltages.

There is a rich body of work on various techniques to solve the OPF problem in AC grids, such as linear approximations [8], mixed-integer linear programming [9], gradient method [10], Newton-Raphson method [11], branch-and-cut method [12], quadratic programming [13], alternating direction method of multipliers (ADMM) [14–16], Lagrange relaxation method [17–19], and metaheuristics [20]. The transition from AC grids to AC-DC grids is accelerating due to the advances in power electronics and the increase of the applications using DC power. Several methods have been proposed to address the AC-DC OPF problem such as heuristic and interior point methods [21,22], the Newton-Raphson method [23], second-order cone programming [24], and sequential approaches [25].

These methods offer no optimality guarantee and may return an infeasible solution to the OPF problem. Jabr [26] was one of the first to introduce a second-order cone relaxation of the power flow equations. Bai et al. [27] presented a semidefinite program (SDP) relaxation for the OPF problem, which was used by Lavaei and Low [28] to show that the SDP relaxation for AC grids has a zero relaxation gap for practical power grids, including the IEEE test power systems. Thereafter, convex relaxations of the OPF problem in AC grids have attracted significant research attention [29–36]. These techniques are guaranteed to determine the global optimal solution when the relaxation gap is zero. Low [29] presented the bus injection and branch flow models of the OPF problem and proved their equivalence. The convex relaxations of the OPF problem based on SDP, chordal extension, and second-order cone programming are studied by Low [30]. Sufficient conditions that guarantee the exactness of these relaxations are provided. Chen et al. [31] proposed a nested optimization approach to decompose the multiperiod joint OPF and EV charging problem into separable subproblems. The decomposed problem is solved using a nonsmooth separable programming technique. Lavaei et al. [32] presented the geometric properties of the set of all vectors of bus power injections satisfying the operating constraints of a radial power network. The convex relaxation of the OPF problem is obtained and a sufficient condition to determine a global optimal solution of the OPF problem is derived. Farivar and Low [33] presented a branch flow model for the analysis of meshed and radial networks. The proposed convex relaxation method is exact for radial networks provided there are no upper bounds on loads or voltage magnitudes. Tan [34] proposed a second-order cone relaxation of the OPF problem for resistive networks. The uniqueness of the optimal solution is characterized for radial and meshed networks. Sojoudi et al. [35] presented a model for the power lines capacity and studied the zero relaxation gap for weak cyclic networks. An upper bound on the rank of the minimum-rank solution of the SDP relaxation is provided. Chen et al. [36] formulated the ACOPF problem as a nonconvex quadratically constrained quadratic program with complex variables and is solved using the SDP relaxation for cases where the relaxation gap is zero. For the case of nonzero relaxation gap, the global optimum solution is determined using a spatial branch-and-bound method. Bahrami et al. [37,38] studied the OPF problem in an AC-DC grid. To this end, the AC-DC OPF problem is transformed into an SDP by using convex relaxation techniques. The zero relaxation gap conditions are determined. It is shown that the conditions can hold in practical AC-DC grids including the IEEE test systems connected to some DC grids.

With renewable energy generators and price responsive load demands in the grid, several challenges need to be addressed in the OPF analysis. First, solving the OPF problem by the independent system operator (ISO) can violate the entities' privacy in a competitive energy market, for example, by revealing the demand information of the load aggregators and cost of the generators. The second challenge is the nonconvexity of the OPF problem, which has been explained before. Even by linearizing the ACOPF problem (e.g., in DCOPF), it can be computationally difficult in large networks, especially when the number of decision variables increases with the participation of the price-responsive load aggregators in the energy market. Third, the uncertainty of the renewable generation puts the generation-load balance at risk, especially when the ISO does not have access to the accurate historical generation data of the privately owned renewable generators.

Instead of solving the nonlinear complex AC power flow equations in the transmission lines, it has been proposed to deploy a linearized power flow model, referred to as DC power flow. DC power flow has several advantages, such as straightforward solution, simple implementation due to matrix representation of the equations, capable of being presented in linear form, obtaining unique solutions, and acceptable accuracy for power flow at high voltage transmission lines [8]. To solve the DC power flow problem, ISO makes the following assumptions: (1) voltage magnitude at all buses is a constant value of 1.0 per unit; (2) the line resistance can be neglected; (3) the voltage phase angle difference of neighboring buses is very small, that is, its cosine and sine are considered as 1 and 0 respectively.

There have been some efforts in the power system operation literature to tackle the above-mentioned challenges. Some of the existing works are concerned with including the power flow constraints in the decentralized system planning procedure using different techniques such as the primal-dual algorithm [39] and convex relaxation [14–18,40–42]. Disfani et al. [39] proposed a distributed DCOPF solution for radial networks using partial prima-dual algorithms. They proved that the obtained solution by the distributed method satisfies the KKT conditions of DCOPF problem for two scenarios: with and without network congestion. Peng and Low [15] presented a convex relaxation approach to deal with power flow problem. Their modified ADMM-based approach customized as a region-based optimization algorithm. Magnusson et al. [16] employed ADMM and sequential convex approximations to solve optimal power flow problem in a distributed fashion. To this end, they decomposed the centralized OPF problem into smaller subproblems that are associated with each bus in the power network. The obtained solutions of these subproblems are coordinated, which increase the scalability of their distributed algorithm. Kargarian et al. [17] proposed a distributed solution to solve the problem of security-constrained unit commitment (SCUC) for large-scale power networks. In their method, the analyzed power system is decomposed into several scalable zones, which may not necessarily include only one bus, and which are connected using tie lines. Each of these zones solved an internal SCUC problem in a parallel fashion and updates its location by communication with other areas. Hug and Andersson [18] introduced a decentralized approach to the optimal power flow problem. To avoid conflicts among neighboring areas in their decentralized method, they proposed a multiarea control scheme. Further, they deployed controllable FACTS devices at each local area to improve the system security. Some other works are concerned with addressing the uncertainty issues by using different techniques, such as chance constrained optimization for optimal power flow under uncertainty [43] and stochastic power system operation by formulating a SCUC problem [44], as well as successive constraint enforcement [45].

Mohammadi [46] provided a comprehensive set of solutions to a variety of energy management problems in power systems considering distributed resources, including optimal energy dispatch problems, optimal power flow control, and security constrained economic dispatch in transmission network. His methods are mainly based on consensus-based distributed optimization that takes advantage of local communication among neighboring agents to find the optimal solution. Mohammadi et al. [47] proposed an asynchronous iterative distributed solution to deal with the DCOPF problem. Their optimization procedure deployed an iterative procedure to solve the first order optimality conditions in a distributed fashion. Further, Kar et al. [48] provided a comprehensive overview of the distributed solutions with applications in the smart power grids. They investigated the distributed solutions that are all based on consensus and innovations to solve three of the foremost energy management problems: state estimation, economic dispatch, and optimal power flow. Mohammadi et al. [49] investigated the role of communication topology among agents in improving the performance of distributed solutions for optimal power flow problem. According to their study, the constrained selection of the additional communication links and the proper utilization of communicated information play a pivotal role in the convergence rate of the distributed algorithm. Kargarian et al. [50] reviewed the decentralized/distributed solutions to the DC optimal power flow (DCOPF) problem. To this end, they studied and compared six decomposition coordination approaches, including ADMM, optimality condition decomposition, consensus + innovations approach, analytical target cascading, proximal message passing, and auxiliary problem principle. Finally, a discussion on their key features was provided based on simulation results.

Several works tackle the DCOPF problem [14,51–55]. Biskas et al. [51] presented a decentralized approach for the DC optimal power flow problem. In their approach, each local transmission system operator is assigned with a computing system to optimize operation of the transmission system for that specific regions. Conejo et al. [52] developed a multiarea optimal power flow using a decentralized method. They used the DC nonlinear OPF model which takes losses into account. Erseghe [14] used ADMM, customized as a region-based optimization procedure, to obtain a distributed implementation of the OPF problem. Mohammadi et al. [53,54] proposed a fully distributed approach to solve the DCOPF problem. Later, they developed an agent-based framework, which can be used for security constrained OPF (SCOPF) [55]. The main improvement of their SCOPF model compared with the basic OPF is considering the contingencies over the network to ensure the acceptable level of security in the network. The distributed solution is similar to the consensus-based distributed solutions developed in Refs. [53,54].

2 DC OPTIMAL POWER FLOW FORMULATION

The optimal power flow (OPF) problem is one of the most important optimization problems of power system operation that calculates the optimum scheduling of the committed generation units [56]. While the unit commitment (UC) problem is solved on a day-ahead basis for 24-h interval; OPF is also solved on an hourly basis and uses the results of UC.

Some studies investigated the application of convex programming to solve this problem [57]. This will lead to an acceptable solution from the ISO's point of view. However, with the advent of novel technologies and demand side management strategies, it is unlikely that the power system operations are smooth without considering the line flow limits. Hence, in this chapter we consider line flow limits, as well as generation limits in the power flow constraints. Power flow study needs to be conducted in order to calculate the power flow in the lines and then check if any of these violate the limits.

In the event of contingency, there might not be enough time to carry out the entire power flow analysis. The analysis of nonlinear equations in AC power flow (ACPF) does not attain the optimal solution quickly. The solution of DCPF can be used as an initial guess instead of performing ACPF for any other system analysis. In such a situation, the DC Power Flow (DCPF) gives a solution, which can be utilized to stabilize the system to ensure security of the system.

In this chapter, we focus on the DC power flow analysis, a simplified version of the power flow analysis. There are numerous advantages of DCPF analysis, which makes it a first choice for an analysis tool by most power engineers. DCPF results in linear constraints, which makes it easy to implement. This linearity in the problem formulation, especially the constraints, makes it more reliable and leads to unique solutions with reasonable accuracy for active power flows. Despite the ease in obtaining the solution, the size of problems in real world power systems may be too large to be solved in a centralized way, which mandates the break-down of one problem to several smaller subproblems that are easier to solve and have a lower computational burden. In this work, we decompose the problem into r subproblems, where each subproblem k represents an area.

More specifically, this chapter focuses on an optimization problem called DC Optimal Power Flow (DCOPF) problem, which is a minimization problem of the following form:

$$\min_{x} f(X) \tag{1a}$$

$$s.t.\ g(X) = 0 \tag{1b}$$

$$h(X) \le 0 \tag{1c}$$

where X is the vector of control variables, $f(X)$ is the objective function, and Eqs. (1b), (1c) are the equality and inequality constraints, respectively. In general, the control variables correspond to the active power outputs.

The objective of DCOPF is to find out the generator's real power outputs at each of the buses, which would minimize the total cost of generation subject to the physical constraints of the system and the feasibility limits. DC power flow is utilized as an approximation to obtain these physical constraints. Hence, in this problem, the control variables are the active power outputs.

To formally express the problem, we introduce the following notations for the parameters and variables:

Ω_{sys}: Set of buses in the system.
Ω_G: Set of buses to which a generator is connected.
Ω_{L_i}: Set of lines in the system incident on the ith bus.
Ω_{tie}: Set of buses correspond to the end points of tie-lines.
Ω_D: Set of buses to which load/demand is connected.
P_{g_i}: Output of generator at bus i.
P_{d_i}: Demand at bus i.
$C_i(P_{g_i})$: Quadratic cost function of generator at bus i.
B_{ij}: The element in the susceptance matrix in the ith row and the jth column.
θ_i: Voltage angle at the ith bus.
F_{ij}: Flow in the line joining the ith bus and the jth bus.
$\Omega_{sys,\ k}$: Set of buses in area (subproblem) k.

Therefore, $X = (P_{g_i}, for\ i \epsilon \Omega_G)$.

- Objective function:

The objective function for the DCOPF problem is the cost of generation of the entire system. The cost of the ith generator is modeled as a quadratic cost function given by Eq. (2).

$$C_i\left(P_{g_i}\right) = c_i + b_i P_{g_i} + \mathrm{a}_i {P_{g_i}}^2 \tag{2}$$

where a_i, b_i and c_i are the cost coefficients of the ith generating unit and a_i is a positive coefficient.

Hence, the objective function in Eq. (1a) is given by Eq. (3).

$$f(X) = \sum_{i \in \Omega_G} C_i\left(P_{g_i}\right)\ \ \$/h \tag{3}$$

- Equality constraints:

By performing the DCPF, the following equations are obtained at each of the buses, which are simple equations balancing the inflow and outflow of real power at each node or bus. The equality constraints in Eq. (1b) can be represented by:

$$P_{g_i} - P_{d_i} - \sum_{j \in \Omega_{L_i}} F_{ij} = 0\ \ \forall i \in \Omega_{sys} \tag{4}$$

In the case of DC power flow, the flow in line F_{ij} is given by Eq. (5).

$$F_{ij} = B_{ij} \times \left(\theta_i - \theta_j\right) \tag{5}$$

Note that the θ_i's are state variables that are completely determined when P_{g_i}'s are found.

- Inequality Constraints:

The DCOPF problem formulation involves limits on the generation and line flows, which can all be considered as inequality constraints. These inequality constraints of Eq. (1c) can be formulated as Eq. (6).

$$P_{g_i}^{min} \leq P_{g_i} \leq P_{g_i}^{max} \quad \forall i \in \Omega_G$$
$$\left|F_{ij}\right| \leq F_{ij}^{max} \quad \forall i, j \in \Omega_{sys} \tag{6}$$

where $P_{g_i}^{min}$, $P_{g_i}^{max}$, F_{ij}^{max} denote the minimum and maximum generation limits of generator i, and the maximum line flow of the line connecting nodes i and j.

- Decomposition into Subproblems

Fig. 1 shows a sample test network that is divided into two areas. These areas are connected via a tie-line. In this chapter, we assume that there is no load or generator at the boundary buses, where the tie-line is located. Tie-line refer to the lines which are connecting two different areas (subproblems).

While we deploy decomposition methods, we need to separately formulate complicating constraints (with variables from different subproblems) and noncomplicating constraints (that only belong to one subproblem). The noncomplicating equality constraints are given by:

$$P_{g_i} - P_{d_i} - \sum_{j \in \Omega_{L_i}} F_{ij} = 0 \quad \forall i \in \Omega_{sys}, i \notin \Omega_{tie} \tag{7}$$

The equality constraint in Eq. (7) is considered noncomplicating because the transmission lines incident to the given bus belongs to a single entity. Thus, it can be assigned as an equality constraint to the subproblem of that entity, upon decomposition. Similarly, noncomplicating inequality constraints can be formulated as Eq. (8). Note that these constraints are being treated as noncomplicating for the same reason the equality constraint is noncomplicating in Eq. (7).

$$P_{g_i}^{min} \leq P_{g_i} \leq P_{g_i}^{max} \quad \forall i \in \Omega_G, i \notin \Omega_{tie}$$
$$\left|F_{ij}\right| \leq F_{ij}^{max} \quad \forall i, j \in \Omega_{sys,k}, i, j \notin \Omega_{tie} \tag{8}$$

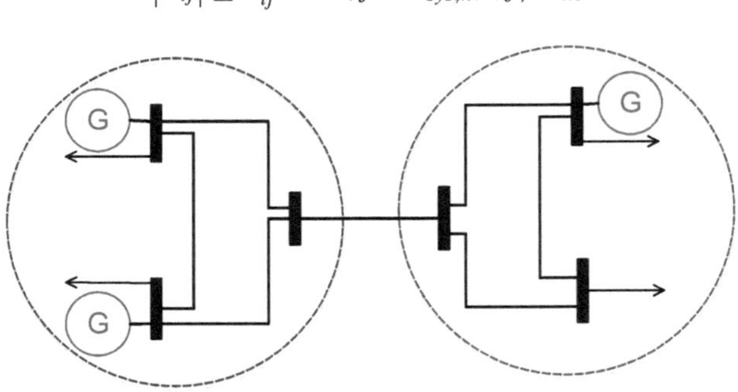

FIG. 1 Illustrative example of a test system decomposed into subsystems (subproblems).

In addition, the complicating equality constraints are given by Eq. (9). These are being considered complicating because the flows from tie-lines are incident on a given bus. In other words, if the whole optimization problem is to be divided into subproblems based on areas/ownerships, then these constraints cannot be assigned to a single subproblem because they involve flows from the lines that are owned by more than one entity.

$$P_{g_i} - P_{d_i} - \sum_{j \in \Omega_{L_i}} F_{ij} = 0, \ \forall i, j \in \Omega_{tie} \tag{9}$$

In addition, the complicating inequality constraints are given by Eq. (10).

$$|F_{ij}| \leq F_{ij}^{max} \ \forall i, j \in \Omega_{tie} \tag{10}$$

The functions defining the complicating equality and inequality constraints will be denoted $g_{int}(X)$ and $h_{int}(X)$, while those for the noncomplicating equality and inequality constraints will be denoted $g_{nc}(X)$ and $h_{nc}(X)$.

Due to the high computational burden, the optimal solution of the entire power system network cannot be obtained efficiently. It also requires that different regions exchange all the data required to solve one centralized optimization problem. Furthermore, more communication links make the system more vulnerable to cyber-attacks. Due to this, it is necessary to divide the entire system into subsystems such that the DCOPF problem can be solved for each of the subsystems separately. The constraints linking different subproblems are then considered and the master problem is solved. There are many techniques discussed in the literature to do the decomposition and solve for the entire system as a whole.

To apply any of the decomposition techniques, there first arises a need to divide the system into subsystems and formulate the DCOPF problem for each of the subsystems. Further, the complicating constraints for the problem is defined.

3 LAGRANGIAN DECOMPOSITION METHOD

The Lagrangian relaxation (LR) decomposition technique is used for solving nonlinear constrained optimization problems. The LR method converts a difficult optimization problem into a problem that is easier to solve and yields a solution that is an approximate solution of the difficult problem. If the complicating constraints can be relaxed, the resulting problem decomposes into several simpler problems, which can then be solved easily [58]. In the LR method, the problem is relaxed by removing the complicating constraints and transferring them to the objective function along with an assigned weight called "Lagrange multiplier". In this way, a Lagrange multiplier penalizes the violation of a complicating constraint in the objective function. The separability of the constraints and objective function is an underlying assumption in the LR decomposition method. The drawback of the LR method is that, if the overall problem is nonconvex, the solution of the dual problem obtained in the LR decomposition method is possibly not the same as for the primal problem.

Once the complicating constraints are identified, the optimization problem is relaxed by moving those complicating constraints from the set of constraints to the objective function. In this way, the Lagrangian function for the DCOPF problem is defined as Eq. (11).

$$\mathcal{L}(X,\lambda,\mu) = f(X) + \lambda^T g_{int}(X) + \mu^T h_{int}(X) \tag{11}$$

where λ and μ are the Lagrange multiplier vectors for the equality and inequality constraints respectively and $g_{int}(X)$ and $h_{int}(X)$ represent the complicating equality and complicating inequality constraints, respectively. Vectors λ and μ are basically the weights assigned to these complicating constraints. When these constraints are violated, the objective function (to be minimized) is penalized by an increased cost.

For fixed λ and μ, the LR primal problem consists of minimizing (11) under the noncomplicating constraints. As function f and the noncomplicating constraints can be decomposed over areas, a subproblem k can be written as follows:

$$\begin{aligned} \text{Subproblem } k: \quad & \min_{x_k} f_k(x_k) + \lambda^T g_{int,k}(x_k) + \mu^T h_{int,k}(x_k) \\ & \text{s.t.} g_{nc,k}(x_k) = 0 \\ & h_{nc,k}(x_k) \le 0 \end{aligned} \tag{12}$$

where f_k, $g_{int,k}$, $h_{int,k}$, $g_{nc,k}$, $h_{nc,k}$ are respectively the cost function, complicating equality function, complicating inequality function, noncomplicating equality function, and noncomplicating inequality function related to subproblem k and $X = (x_k$, for $k = 1 \ldots r)$. Note that because of the decomposition, the functions only depend on the variables of their respective areas.

Under regularity and convexity assumptions, the dual function is defined as shown in Eq. (13).

$$\begin{aligned} \phi(\lambda,\mu) = & \min_X \mathcal{L}(X,\lambda,\mu) \\ & (\text{s.t.})\, g_{nc}(X) = 0 \\ & h_{nc}(X) \le 0 \end{aligned} \tag{13}$$

The subproblems include the constraints of their corresponding area. The formulation of subproblem t is presented in Eq. (14).

$$\begin{aligned} \text{Subproblem } t: \quad & \min_{x_t} f_t(x_t) + \lambda^T g_{int,t}(x_t) + \mu^T h_{int,t}(x_t) \\ & \text{s.t.} g_t(x_t) = 0 \\ & h_t(x_t) \le 0 \end{aligned} \tag{14}$$

3.1 Formulation of the Decomposed System

The primal DCOPF problem is decomposed into r subproblems based on the geography and utility ownership of the power system network. In the following formulation, the kth subsystem parameters and variables are defined in the same way as in the previous section, but with the addition of subscript k. Let $\Omega_{tie,k}$ denote the set of buses in the kth subsystem that has a tie-line incident on it. Applying the LR decomposition, the DCOPF problem can be reformulated for the kth subproblem, $k=\{1,2, \ldots,r\}$, as shown in Eq. (15).

$$\min_{P_{g_i}} \sum_{i \in \Omega_{G,k}} \mathcal{L}(P_{g_i}, \lambda, \mu)$$

$$\text{subject to}: P_{g_i} - P_{d_i} - \sum_{j \in \Omega_{L_i}} F_{ij} = 0; \quad \forall i \in \Omega_{sys,k}, i \notin \Omega_{tie,k}$$

$$P_{g_i}^{min} \le P_{g_i} \le P_{g_i}^{max} \quad \forall i \in \Omega_{G,k}$$

$$|F_{ij}| \le F_{ij}^{max} \forall i,j \in \Omega_{sys,k}, i,j \notin \Omega_{tie,k} \tag{15}$$

In the decomposed formulation, the objective function is composed of the cost functions for all the generators pertaining to the kth region. The decomposition facilitates its solution, and normally allows physical and economical interpretations.

As described in the previous section, the physical constraints comprising of parameters belonging to more than one subproblem are the coupling (complicating) constraints for the power system. If there is a tie-line between the pth region and the qth region, the coupling constraints can be considered as shown in Eq. (16).

$$P_{g_i} - P_{d_i} - \sum_{j \in \Omega_{L_i}} F_{ij} = 0 \quad \forall i \in \Omega_{tie,p}$$

$$P_{g_i} - P_{d_i} - \sum_{j \in \Omega_{L_i}} F_{ij} = 0 \quad \forall i \in \Omega_{tie,q}$$

$$|F_{ij}| \le F_{ij}^{max} \quad \forall i \in \Omega_{tie,p}, j \in \Omega_{tie,q} \tag{16}$$

3.2 Algorithm [58,59]

Here, we present the step-by-step algorithm to apply the Lagrangian relaxation method for the DCOPF problem.

1. The Lagrange Multiplier vector is initialized to zero (flat start)
2. The subproblems are formulated and the relaxed primal problems are solved to obtain the minimizer P_{g_i} and the objective function value at the minimizer is obtained.

3. The values of Lagrange Multipliers are updated using the subgradient method (described in the following section).
4. If the stopping condition is fulfilled, the procedure terminates. Otherwise, the subproblems are solved again through step 2.

3.3 Subgradient Method for Multiplier Update

Let s denote the mismatches of coupling constraints at iteration v.

$$s^{(v)} = \left[g_{int}\left(x^{(v)}\right) \ h_{int}\left(x^{(v)}\right) \right] \tag{17}$$

The multiplier vector is updated as (18) [59,60]:

$$\left[\lambda^{(v+1)} \mu^{(v+1)}\right] = \left[\lambda^{(v)} \mu^{(v)}\right] + \alpha^{v} \frac{s^{(v)}}{\|s^{(v)}\|}, \ if \ \mu^{(v+1)} < 0 => \mu^{(v+1)} = 0 \tag{18}$$

where k^{v}denotes step size that satisfies $\sum_{v=1}^{\infty} \alpha^{v} \rightarrow \infty$, for instance an appropriate choice of α^{v} is $\alpha^{v} = \frac{1}{a+bv}$.

The subgradient method is simple to implement and its computational burden is small. Furthermore, the oscillating behavior makes it very difficult to devise an appropriate stopping criterion. It is typically stopped after a prespecified number of iterations.

4 AUGMENTED LAGRANGIAN DECOMPOSITION METHOD

The Augmented Lagrangian Relaxation (ALR) decomposition technique is utilized for solving nonlinear problems with a decomposable structure, including complicating constraints. This approach has been widely used in power system optimization problems. For instance, Beltran et al. [61] utilized the ALR decomposition to solve the unit commitment problem. It consists of adding a quadratic term to the objective function in order to overcome the issue of the potential suboptimality of the solution for the general decomposition problem. By adding this quadratic term, the objective function becomes locally convex. However, the addition of the quadratic terms makes the corresponding Lagrangian function nonseparable. Consequently, methods such as the alternating direction method of multipliers [14] are used to tackle this difficulty. An interesting study on the application of ALR in multiarea optimal power flow was performed by Conejo et al. [60], who highlighted the advantages of this method. According to Conejo et al., utilizing ALR to solve the OPF problem will improve computation efficiency.

For DCOPF, the problem formulation is described in the previous section by Eqs. (3), (7)–(10). The complicating constraint is denoted by subscript *int*. The augmented Lagrangian function is defined as (19).

$$L_{AL}(X, \lambda, \mu) = f(X) + \lambda^{T} g_{int}(X) + \mu^{T} h_{int}(X)^{T} + \frac{1}{2} \gamma \|g_{int}(X)\|^{2} \tag{19}$$

where λ and μ are the Lagrange multiplier vectors for the equality and inequality constraints respectively and $g_{int}(X)$ and $h_{int}(X)$ are the set of complicating equality and complicating inequality constraints respectively.

Under regularity and convexity assumptions, the dual function is defined as shown in Eq. (20)

$$\begin{aligned} \phi(\lambda, \mu) &= min_X L_{AL}(X, \lambda, \mu) \\ \text{s.t.} g_{nc}(X) &= 0 \\ h_{nc}(X) &\leq 0 \end{aligned} \tag{20}$$

There are r subproblems after decomposing the main optimization problem. Hence, $g(X)=0$ can be expressed as $g_k(x_k)=0$ corresponds to subproblem k. Similarly, the noncoupling inequality constraints are determined by identifying and assigning the inequality constraint related to each subproblem. The coupling equality constraints are included directly in the Lagrange function. Then the ***alternating direction method of multipliers*** is deployed, which fixes some variables during the decomposition procedure. Based on this approach, the subproblems are determined as shown in Eq. (21), assuming that $f(X)=f_1(x_1)+f_2(x)+\ldots+f_r(x_r)$ is combined of r separable objective functions. Here is the sequence of solving subproblems: (1) the first subproblem is minimized, (2) then the second subproblem is solved, (3) the multipliers are updated accordingly based on (22).

Step (1) Formulate the subproblems:

$$\begin{aligned} \text{Subproblem } k: \; \underset{x_k}{min} f_k(x_k) + \lambda^T g_{int,k}(x_k) + \frac{1}{2}\gamma \left\| g_{int}\left(x_1^t, x_2^t, \ldots, x_k, \ldots, x_N^t\right) \right\|^2 \\ \text{s.t.} g_{nc,k}(x_k) = 0 \\ h_{nc,k}(x_k) \leq 0 \end{aligned} \tag{21}$$

Step (2) Update multipliers using $x_1^{t+1}, x_2^{t+1}, \ldots, x_r^{t+1}$:

$$\lambda^{t+1} = \lambda^t + \alpha \cdot g_{int}(X) \tag{22}$$

Because the modeling of the coupling inequality constraints with ALR is not straightforward, the topology of the problem formulation is modified to convert the shared inequality constraint related to the tie-line to two noncoupling inequality constraints in each subproblem separately, that is, the following constraints are added to subproblem k:

$$\left| \sum_{i \in \Omega_{g,k}} P_{gi} - \sum_{i \in \Omega_{sys,k}} P_{di} \right| \leq F_{tie_{max}} \tag{23}$$

where $\Omega_{g,k}$ denoted the set of generation buses at area k. These additional constraints ensure that the tie-line flow does not exceed the limit of thermal rating of the tie-lines.

5 DISTRIBUTED OPTIMIZATION FOR ECONOMIC DISPATCH OF MICROGRIDS

Consensus-based distributed optimization and multiagent-based methods are essential research topics that aim to enable a group of agents to reach the same agreement by exchanging information only with their neighbors [46–50,53–55,62]. The applications include flocking control, attitude alignment of spacecrafts, distributed sensor networks, and optimal power flow problem. Due to the advantages of microgrids, there have been several efforts to increase their utilization in future power systems [63,64]. This high penetration of microgrids raises several challenges for centralized strategies such as computation complexity, communication reliability, and privacy concerns [65]. To address those challenges in the application of the microgrid, a fully distributed control strategy can be considered [66]. In this work, they also tackled DCOPF excluding the constraint related to the line flow limits (Eq. (10). Their method can also be extended to tackle DCOPF by adding another Lagrange multiplier. In this method, agents only exchange information locally to coordinate their decisions with the neighboring nodes. They use the update rule (24) to coordinate the agents at each iteration. To this end, the step size, ε, is defined to control the convergence speed.

$$r_i[t+1] = \sum_{j \in N_i} d_{ij} r_i[t] + \varepsilon P_{D,i}[t] \tag{24}$$

$$P_i[t+1] = \frac{r_i[t+1] - b_i}{a_i} \tag{25}$$

$$P'_{D,i}[t] = P_{D,i}[t] + (P_i[t+1] - P_i[t]) \tag{26}$$

$$P_{D,i}[t+1] = \sum_{j \in N_i} d_{ij} P'_{D,i}[t] \tag{27}$$

where $r_i[t]$ denotes the incremental cost of agent i at iteration t, and $P_{D,i}[t]$ represents the local estimation of the global mismatch between load and generation. Moreover, d_{ij} denotes the communication coefficient.

$$d_{ij} = \begin{cases} 2/(n_i + n_j + 1), & j \in N_i \\ 1 - \sum_{j \in N_i} 2/(n_i + n_j + 1), & i = j \\ 0, & \text{otherwise} \end{cases} \tag{28}$$

This general updating procedure [66] can be rewritten with matrix representations as follows:

$$R[t+1] = D \cdot R[t] + \varepsilon P_D[t] \tag{29}$$

$$P[t+1]=B\cdot R[t+1]+G \tag{30}$$

$$P_D[t+1]=D\cdot P_D[t]+D.(P[t+1]-P[t]) \tag{31}$$

where the column vectors of r_i, $P_{D,i}$, P_i, and $-b_i/a_i$, are respectively denoted by R, P_D, P, and G. Further, we have $B=diag([1/a_1, \ldots, 1/a_n])$.

Xu and Li [66] then introduced the new vector, $[R,P_D]^T$ to represent the above-mentioned update rule in the matrix form as:

$$\begin{bmatrix} R[t+1] \\ P_D[t+1] \end{bmatrix}_{2n\times 1} = \begin{bmatrix} D & \varepsilon I_n \\ DB(D-I_n) & D+\varepsilon DB \end{bmatrix}_{2n\times 2n} \cdot \begin{bmatrix} R[t] \\ P_D[t] \end{bmatrix}_{2n\times 1} \tag{32}$$

Let's define $\mathbf{M}$ as:

$$\mathbf{M}=\begin{bmatrix} D & \varepsilon I_n \\ DB(D-I_n) & D+\varepsilon DB \end{bmatrix}_{2n\times 2n} \tag{33}$$

where $\mathbf{I}_n$ is an n-dimensional identity matrix.

Assume ε is a negligible scalar, the repeated eigenvalue sets of matrix M and matrix D are then identical, because:

$$|\lambda I_{2n}-M|=|\lambda I_n-D|^2+\varepsilon|DB||\lambda I_n-I_n|\approx|\lambda I_n-D|^2 \tag{34}$$

Matrix D is a double-stochastic matrix. Moreover, $D1_n=1_n$ and $(D-I_n)1_n=0_n$ with $1_n=[1, \ldots, 1]^T$ and $0_n=[0, \ldots, 0]^T$. $[1_n, 0_n]^T$ is an eigenvector of M with respect to $\lambda_1=1$.

$$\begin{bmatrix} D & \varepsilon I_n \\ DB(D-I_n) & D+\varepsilon DB \end{bmatrix}_{2n\times 2n} \cdot \begin{bmatrix} 1_n \\ 0_n \end{bmatrix}_{2n\times 1} = \begin{bmatrix} D1_n \\ DB(D-I_n)1_n \end{bmatrix}_{2n\times 1} = \begin{bmatrix} 1_n \\ 0_n \end{bmatrix}_{2n\times 1} \tag{35}$$

According to Yang et al. [67], the system converges to span $[\mathbf{1}_n, \mathbf{0}_n]^T$, as t increases to infinity. Consequently, $r_i[t]$ converges to r^*, which guarantees the optimality, and $P_{D,i}[t]$ disappears, which means the equality constraint is satisfied.

$$\begin{bmatrix} R[\infty] \\ P_D[\infty] \end{bmatrix}_{2n\times 1} = r^* \begin{bmatrix} 1_n \\ 0_n \end{bmatrix}_{2n\times 1} \tag{36}$$

To deal with the inequality constraint, the projection operation is defined as follows [66]:

$$d_{ij}=\begin{cases} P_i^{min}, & \dfrac{r_i[t]-b_i}{a_i}<P_i^{min} \\ \dfrac{r_i[t]-b_i}{a_i}, & P_i^{min}\le\dfrac{r_i[t]-b_i}{a_i}\le P_i^{max} \\ P_i^{max}, & \dfrac{r_i[t]-b_i}{a_i}>P_i^{max} \end{cases} \tag{37}$$

To evaluate the effect of the projection operation on the performance of distributed algorithm in [66], the proposed approach is reformulated as:

$$\begin{bmatrix} R[t+1] \\ P_D[t+1] \end{bmatrix}_{2n\times 1} = \begin{bmatrix} D & \varepsilon I_n \\ D\hat{B}(D-I_n) & D+\varepsilon D\hat{B} \end{bmatrix}_{2n\times 2n} \cdot \begin{bmatrix} R[t] \\ P_D[t] \end{bmatrix}_{2n\times 1} \tag{38}$$

where $\hat{B} = \mathrm{diag}\left(\left[B'_1, \ldots, B'_n\right]\right)$ with

$$B'_i = \begin{cases} 0, \text{if the inequality constraint is violated} \\ \dfrac{1}{a_i}, \text{otherwise} \end{cases} \tag{39}$$

According to the analysis in Ref. [67], the stability of the presented system in Eq. (35) is guaranteed and it converges to span $[\mathbf{1}_n, \mathbf{0}_n]^T$.

6 CONCLUSION AND FUTURE TRENDS

In this chapter, we reviewed distributed/decentralized approaches to solve the optimal power flow (OPF) problem in a power system. To this end, we focused on the widely used DC power flow formulation, which is a simplified power flow model with an acceptable approximation for transmission networks. The main advantage of using the DCOPF formulation is to obtain a linearized model of the power system. We first reviewed the state-of-the-art decomposition techniques to address the scalability of the OPF analysis. Then, we discussed the application of the Lagrangian Relaxation (LR) decomposition and augmented LR in designing decentralized DCOPF algorithms. We also overviewed a consensus-based distributed optimization method to deal with the DCOPF problem in a microgrid. The LR decomposition requires the sharing of the data related to tie-line flows only, that is, voltage magnitudes and angles at the boundary buses. However, a larger amount of data needs to be shared in the case of the ALR decomposition technique, that is, the variables related to the flow of the lines that are incident to the boundary buses in each of the areas.

REFERENCES

[1] J.J. Grainger, W.D. Stevenson, in: Power System Analysis, vol. 621, McGraw-Hill, New York, 1994.

[2] J.D. Glover, M.S. Sarma, T. Overbye, Power System Analysis and Design, fifth ed., Cengage Learning, Boston, MA, 2011.

[3] M.H. Amini, et al., A panorama of future interdependent networks: from intelligent infrastructures to smart cities, in: Sustainable Interdependent Networks, Springer, Cham, 2018, pp. 1–10.

[4] K.G. Boroojeni, M. Hadi Amini, S.S. Iyengar, Smart Grids: Security and Privacy Issues, Springer International Publishing, 2017.

[5] F. Milano, Power System Modelling and Scripting, Springer-Verlag, Berlin, Heidelberg, 2010.

[6] J. Zhu, *Optimization of Power System Operation*, 1st Ed, Wiley, Hoboken, NJ, 2009.

[7] S. Savulescu, Real-Time Stability in Power Systems: Techniques for Early Detection of the Risk of Blackout, second ed., Springer, USA, 2014.

[8] B. Stott, J. Jardim, O. Alsac, DC power flow revisited, IEEE Trans. Power Syst. 24 (3) (2009) 1290–1300.

[9] C.H. Lin, S.Y. Lin, Distributed optimal power flow with discrete control variables of large distributed power systems, IEEE Trans. Power Syst. 23 (3) (Aug. 2008) 1383–1392.

[10] L. Gan, S.H. Low, An online gradient algorithm for optimal power flow on radial networks, IEEE J. Sel. Areas Commun. 34 (3) (Mar. 2016) 625–638.

[11] X. Zhao, Y. Yang, H. Wei, in: An interior point method based on continuous Newton's method for optimal power flow, *Proc. of IEEE Innovative Smart Grid Technologies (ISGT)*, Tianjin, China, May, 2012.

[12] R.A. Jabr, Optimization of AC transmission system planning, IEEE Trans. Power Syst. 28 (3) (2013) 2779–2787.

[13] A.G. Bakirtzis, P.N. Biskas, A decentralized solution to the DC-OPF of interconnected power systems, IEEE Trans. Power Syst. 18 (3) (2003) 1007–1013.

[14] T. Erseghe, Distributed optimal power flow using ADMM, IEEE Trans. Power Syst. 29 (5) (2014) 2370–2380.

[15] Q. Peng, S.H. Low, in: Distributed algorithm for optimal power flow on a radial network, Proc. of IEEE Conf. on Decision and Control, December, 2014, , pp. 167–172.

[16] S. Magnusson, P.C. Weeraddana, C. Fischione, A distributed approach for the optimal power-flow problem based on ADMM and sequential convex approximations, IEEE Trans. Control Netw. Syst. 2 (3) (2015) 238–253.

[17] A. Kargarian, Y. Fu, Z. Li, Distributed security-constrained unit commitment for large-scale power systems, IEEE Trans. Power Syst. 30 (4) (2015) 1925–1936.

[18] G. Hug-Glanzmann, G. Andersson, Decentralized optimal power flow control for overlapping areas in power systems, IEEE Trans. Power Syst. 24 (1) (2009) 327–336.

[19] S. Mhanna, A.C. Chapman, G. Verbic, A fast distributed algorithm for large-scale demand response aggregation, IEEE Trans. Smart Grid 7 (4) (2016) 2094–2107.

[20] M.R. AlRashidi, M.E. El-Hawary, Hybrid particle swarm optimization approach for solving the discrete OPF problem considering the valve loading effects, IEEE Trans. Power Syst. 22 (4) (2007) 2030–2038.

[21] R. Wiget, G. Andersson, in: Optimal power flow for combined AC and multi-terminal HVDC grids based on VSC converters, *Proc. of IEEE Power and Energy Society General Meeting (PES)*, San Diego, CA, July, 2012.

[22] J. Yu, W. Yan, W. Li, C. Chung, K. Wong, An unfixed piecewise-optimal reactive power-flow model and its algorithm for AC-DC systems, IEEE Trans. Power Syst. 23 (1) (2008) 170–176.

[23] A. Pizano-Martinez, C. Fuerte-Esquivel, H. Ambriz-Perez, E. Acha, Modeling of VSC-based HVDC systems for a Newton-Raphson OPF algorithm, IEEE Trans. Power Syst. 22 (4) (2007) 1794–1803.

[24] M. Baradar, M. Hesamzadeh, M. Ghandhari, Second-order cone programming for optimal power flow in VSC-type AC-DC grids, IEEE Trans. Power Syst. 28 (4) (2013) 4282–4291.

[25] C. Liu, B. Zhang, Y. Hou, F.F. Wu, Y. Liu, An improved approach for AC-DC power flow calculation with multi-infeed DC systems, IEEE Trans. Power Syst. 26 (2) (2011) 862–869.

[26] R.A. Jabr, Optimal power flow using an extended conic quadratic formulation, IEEE Trans. Power Syst. 23 (3) (2008) 1000–1008.

[27] X. Bai, H. Wei, K. Fujisawa, Y. Wang, Semidefinite programming for optimal power flow problems, Int. J. Electr. Power Energy Syst. 30 (67) (2008) 383–392.

[28] J. Lavaei, S.H. Low, Zero duality gap in optimal power flow problem, IEEE Trans. Power Syst. 27 (2012) 92–107.

[29] S.H. Low, Convex relaxation of optimal power flow—part I: formulations and equivalence, IEEE Trans. Control Netw. Syst. 1 (1) (2014) 15–27.

[30] S.H. Low, Convex relaxation of optimal power flow—part II: exactness, IEEE Trans. Control Netw. Syst. 1 (2) (Jun. 2014) 177–189.

[31] N. Chen, C.W. Tan, T. Quek, Electric vehicle charging in smart grid: optimality and valley-filling algorithms, IEEE J. Sel. Topics Signal Process. 8 (6) (2014) 1073–1083.

[32] J. Lavaei, D. Tse, B. Zhang, Geometry of power flows and optimization in distribution networks, IEEE Trans. Power Syst. 29 (2) (Mar. 2014) 572–583.

[33] M. Farivar, S.H. Low, Branch flow model: Relaxations and convexification -parts I and II, IEEE Trans. Power Syst. 28 (3) (2013) 2554–2572.

[34] C.W. Tan, D.W.H. Cai, X. Lou, Resistive network optimal power flow: uniqueness and algorithms, IEEE Trans. Power Syst. 30 (1) (Jan. 2015) 263–273.

[35] S. Sojoudi, R. Madani, J. Lavaei, in: Low-rank solution of convex relaxation for optimal power flow problem, *Proc. of IEEE Smart Grid Comm*, Vancouver, Canada, October, 2013.

[36] C. Chen, A. Atamtrk, S.S. Oren, Bound tightening for the alternating current optimal power flow problem, IEEE Trans. Power Syst. 31 (5) (Sept. 2016) 3729–3736.

[37] S. Bahrami, V.W.S. Wong, J. Jatskevich, in: Optimal power flow for AC-DC networks, *Proc. of IEEE Smart Grid Comm*, Venice, Italy, November, 2014.

[38] S. Bahrami, F. Therrien, V.W.S. Wong, J. Jatskevich, Semidefinite relaxation of optimal power flow for AC–DC grids, IEEE Trans. Power Syst. 32 (1) (2017) 289–304.

[39] V.R. Disfani, L. Fan, Z. Miao, in: Distributed DC optimal power flow for radial networks through partial primal dual algorithm, Proc. of IEEE Power & Energy Society General Meeting (PES), 2015, , pp. 1–5.

[40] W. Shi, N. Li, X. Xie, C. Chu, R. Gadh, Optimal residential demand response in distribution networks, IEEE J. Sel. Areas Commun. 32 (7) (2014) 1441–1450.

[41] N. Li, L. Gan, L. Chen, S. Low, in: An optimization-based demand response in radial distribution networks, *Proc. of IEEE Globecom*, Anaheim, CA, 2012.

[42] E. Dall'Anese, H. Zhu, G.B. Giannakis, Distributed optimal power flow for smart microgrids, IEEE Trans. Smart Grid 4 (3) (Sept. 2013) 1464–1475.

[43] H. Zhang, P. Li, Chance constrained programming for optimal power flow under uncertainty, IEEE Trans. Power Syst. 26 (4) (2011) 2417–2424.

[44] M.H. Amini, A. Kargarian, O. Karabasoglu, ARIMA-based decoupled time series forecasting of electric vehicle charging demand for stochastic power system operation, Electr. Power Syst. Res. 140 (2016) 378–390.

[45] R.A. Jabr, Adjustable robust opf with renewable energy sources, IEEE Trans. Power Syst. 28 (4) (2013) 4742–4751.

[46] J. Mohammadi, Distributed Computational Methods for Energy Management in Smart Grids (PhD Dissertation), Carnegie Mellon University, 2016.

[47] J. Mohammadi, G. Hug, S. Kar, in: Asynchronous distributed approach for DC Optimal Power Flow, *Proc. of Powertech*, June 29–July 2, Eindhoven, Netherlands, 2015.

[48] S. Kar, G. Hug, J. Mohammadi, J. Moura, Distributed state estimation and energy Management in Smart Grids: a consensus+innovations approach, IEEE J. Sel. Top. Signal Process. 8 (6) (Dec. 2014) 1022–1038.

[49] J. Mohammadi, G. Hug, S. Kar, in: Role of Communication on the Convergence Rate of Fully Distributed DC Optimal Power Flow, *Proc. of the 5th IEEE International Conference on Smart Grid Communications*, November 3–6, Venice, Italy, 2014.

[50] A. Kargarian, J. Mohammadi, J. Guo, S. Chakrabarti, M. Barati, G. Hug, S. Kar, R. Baldick, Toward distributed/decentralized DC optimal power flow implementation in future electric power systems, IEEE Trans. Smart Grid 99 (2016) 1–21.

[51] P.N. Biskas, A.G. Bakirtzis, N.I. Macheras, N.K. Pasiali, A decentralized implementation of DC optimal power flow on a network of computers, IEEE Trans. Power Syst. 20 (1) (2005) 25–33.

[52] A.J. Conejo, J.A. Aguado, Multi-area coordinated decentralized DC optimal power flow, IEEE Trans. Power Syst. 13 (4) (1998) 1272–1278.

[53] J. Mohammadi, S. Kar, G. Hug, Distributed approach for DC optimal power flow calculations, arXiv preprint arXiv: 1410, 2014, p. 4236.

[54] J. Mohammadi, G. Hug, S. Kar, in: Fully Distributed DC-OPF Approach for Power Flow Control, Power & Energy Society General Meeting, 2015 IEEE, 2015.

[55] J. Mohammadi, G. Hug, S. Kar, Agent-based distributed security constrained optimal power flow, IEEE Trans. Smart Grid 9 (2) (2018) 1118–1130.

[56] A. Mohammadi, M. Mehrtash, A. Kargarian, Diagonal quadratic approximation for decentralized collaborative TSO+DSO optimal power flow, IEEE Trans. Smart Grid (2018), https://doi.org/10.1109/TSG.2018.2796034.

[57] S. Frank, I. Steponavice, S. Rebennack, Optimal power flow: a bibliographic survey I, Energy Syst. 3 (3) (2012) 221–258.

[58] A.J. Conejo, E. Castillo, R. Minguez, R. Garcia-Bertrand, Decomposition Techniques in Mathematical Programming, Springer-Verlag, Berlin, Heidelberg, 2006.

[59] G. Hug, *"18-879 M: Special Topics in ECE: Optimization in Energy Networks"* Course Materials, Spring, Carnegie Mellon University, 2015.

[60] F.J. Nogales, F.J. Prieto, A.J. Conejo, A decomposition methodology applied to the multi-area optimal power flow problem, Ann. Oper. Res. 120 (1–4) (2003) 99–116.

[61] C. Beltran, F. Javier Heredia, Unit commitment by augmented Lagrangian relaxation: testing two decomposition approaches, J. Optim. Theory Appl. 112 (2) (2002) 295–314.

[62] M.H. Amini, B. Nabi, M.-R. Haghifam, in: Load management using multi-agent systems in smart distribution network, Power and Energy Society General Meeting (PES), 2013 IEEE. IEEE, 2013.

[63] S. Parhizi, H. Lotfi, A. Khodaei, State of the art in research on microgrids: a review, IEEE Access 3 (2015) 890–925.

[64] M. Hadi Amini, et al., A comprehensive cloud-based real-time simulation framework for oblivious power routing in clusters of DC microgrids, in: IEEE Second International Conference on DC Microgrids (ICDCM), 2017.

[65] V. Bhattacharjee, I. Khan, A non-linear convex cost model for economic dispatch in microgrids, Appl. Energy 222 (2018) 637–648.

[66] Y. Xu, Z. Li, Distributed optimal resource management based on consensus algorithm in a microgrid, IEEE Trans. Ind. Electron. 62 (4) (2015) 2584–2592.

[67] S. Yang, S. Tan, J. Xu, Consensus based approach for economic dispatch problem in a smart grid, IEEE Trans. Power Syst. 28 (4) (2013) 4416–4426.

FURTHER READING

[68] A.J. Wood, B.F. Wollenberg, Power Generation, Operation, and Control, John Wiley & Sons, New York, 2012.

[69] J. Zhu, Security-constrained economic dispatch, in: Optimization of Power System Operation, John Wiley & Sons, New Jersey, 2010, pp. 141–210.

Chapter 7

Optimal Power Flow Using Recent Optimization Techniques

Mohamed Ebeed*, Salah Kamel† and Francisco Jurado‡
*Department of Electrical Engineering, Faculty of Engineering, Sohag University, Sohag, Egypt,
†Department of Electrical Engineering, Faculty of Engineering, Aswan University, Aswan, Egypt,
‡Department of Electrical Engineering, University of Jaén, Linares, Spain

NOMENCLATURE

a_i, b_i, c_i	cost coefficient of the *i*th generator
d_i, e_i	coefficients reflecting the valve-point effect
F	the objective function
g_i	the equality constraints
G_{ij}, B_{ij}	conductance and susceptance of the admittance matrix
h_j	inequality constraints
K_G, K_Q, K_V, K_S	penalty factors
L_{max}	stability index
L_n	voltage stability local indicator of bus n
NC	number of shunt VAR compensators
NT	number of tap regulating transformers
NG	number of generation buses
NPV	number of PV buses
NPQ	number of load buses
NTL	number of transmission lines
P_{Di}	the active load demand at bus i
P_{G1}^{max}	maximum active power output of slack bus
P_{G1}^{min}	minimum active power output of slack bus
P_{Gi}	generator active power output of generating unit i
P_{Gi}	the generated active power of unit i
P_{G1}	active power generation of slack bus
P_{Gi}^{max}	maximum active power output of *i*th generating unit
P_{Gi}^{min}	minimum active power output of *i*th generating unit

Classical and Recent Aspects of Power System Optimization. https://doi.org/10.1016/B978-0-12-812441-3.00007-0
© 2018 Elsevier Inc. All rights reserved.

QG	the reactive power output of generators
Q_{G1}	active power generation of slack bus
Q_{Ci}	shunt VAR compensation of *i*th shunt compensator (MVAR)
Q_{Ci}^{max}	maximum VAR injection limit of *i*th shunt compensator (MVAR)
Q_{Ci}^{min}	minimum VAR injection limit of *i*th shunt compensator (MVAR)
Q_{Di}	the reactive load demand at bus *i*
Q_{Gi}	the generated reactive power of unit *i*
Q_{Gi}^{max}	maximum reactive power output of *i*th generating unit (MVAR)
Q_{Gi}^{min}	minimum reactive power output of *i*th generating unit (MVAR)
S_{Li}	transmission line loading of *i*th branch (MVA)
S_{Li}^{min}	maximum apparent power flow limit of *i*th branch (MVA)
STL	the apparent power flow in transmission line
T_i	transformer taps settings of *i*th transformer (p.u.)
T_i^{max}	maximum tap settings limit of *i*th transformer (p.u.)
T_i^{min}	minimum tap settings limit of *i*th transformer (p.u.)
u	vector of the control variables
V_{Gi}	voltage magnitude at PV bus *i*
V_{Li}^{max}	maximum load voltage of *i*th bus
V_{Li}^{min}	minimum load voltage of *i*th bus
VD	load bus voltage deviation
VL	the voltage of load bus
x	vector of the state variables
Yij	admittance matrix between bus *i* and bus *j*
$\gamma_i, \beta_i, \alpha_i, \zeta_i, \lambda_i$	coefficients of the *i*th generator emission
δ_{ij}	phase angle difference between buses *i* and *j*
$\omega_1, \omega_2, \omega_3$	weight factors for multiobjective functions

1 INTRODUCTION

Over the past half-century, the optimal power flow (OPF) has gained great attention due to its importance in power system operation. Optimal power flow is considered an important tool for efficient planning and enhancing the operation of electric power systems. The main task of optimal power flow is to determine the best or the most secure operating point (control variables) for certain objective functions while satisfying the system equality and inequality constraints. Various objective functions related to the electric power system can be optimized such as: transmission line losses, total generation cost, FACTS (Flexible Alternating Current Transmission System) cost, voltage deviations, total of power transfer capability, voltage stability, emission of generation units, system security, etc. The system control variables that can be adjusted include the generated active powers, the voltage of generation buses, transformer tap settings and sizing of FACTS devices. Several classical (deterministic) and recent (nondeterministic) heuristic optimization techniques have been proposed to find the solution of optimal power flow problem. Most of the classical optimization techniques use sensitivity analysis and gradient-based methods. It should be pointed out that the classical methods may be trapped in local optimum due to nonlinearity of optimal power flow problems thereby the metaheuristic optimization techniques are widely employed for solving the optimal power flow problems. This chapter presents a comprehensive survey of the recent optimization techniques used to solve optimal power flow problems. The presented optimization techniques are categorized based on their inspirations, such as nature-swarm-inspired methods (particle swarm optimization, cuckoo search algorithm, firefly algorithm, ant colony optimization algorithm, gray wolf optimizer, moth-flame optimizer, etc.), human-inspired algorithms (harmony search, teaching learning based optimization, tabu search, etc.), evolutionary-inspired algorithms (differential evolution, genetic algorithms, etc.), physics-inspired algorithms (simulated annealing , gravitational search algorithm, etc.), and artificial neural networks (ANN).

2 OPF OVERVIEW

OPF was first proposed by Carpentier [1]. Then, several conventional methods have been applied to solve the OPF problem, such as Newton method network flow programming, linear programming, nonlinear programming, quadratic programming, and the interior point [2–18]. The main shortages of classical methods are they are nonsuitable for large and difficult OPF problems which are high nonlinear and multimodal optimization problems, hence these methods may be trapped in local minima.

Recently, many intelligence optimization methods have been developed to solve the optimal power flow (OPF), particularly the nonlinear complex optimization problems. Intelligence optimization methods are based on different concepts such as evolutionary inspired algorithms, human inspired algorithms, natural inspired algorithms and artificial neural networks [19–109]. The conventional and more recent optimization methods that have been applied for solving the OPF problem are depicted in Fig. 1.

In this chapter, a comprehensive survey is presented to illustrate the application of recent optimization algorithms for solving the optimal power flow problem. Organization of the remaining chapter is as follows: Section 3 describes an overview of the optimal power flow problem, the common objective functions and the system constraints. Section 4 presents a brief survey of conventional methods based OPF. Section 5 shows the new applied optimization techniques for optimal power flow. Section 6 is a conclusion of the chapter.

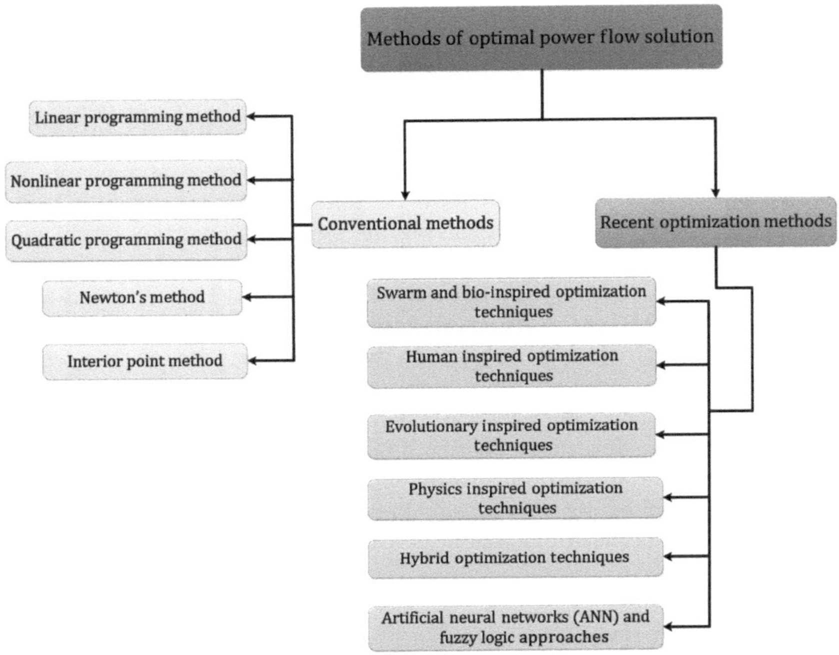

FIG. 1 Solution methods of optimal power flow problem.

3 OPTIMAL POWER FLOW PROBLEM FORMULATION

Solution of the optimal power flow problem aims to optimize an objective function by optimal adjustments of power system control variables with satisfying various equality and inequality constraints. Generally, the optimization problem can be mathematically represented as follows:

$$Min\ F(x,u) \tag{1}$$

Subjected to

$$g_j(x,u)=0 \quad j=1,2,\dots,m \tag{2}$$

$$h_j(x,u)\le 0 \quad j=1,2,\dots,p \tag{3}$$

where F acts the objective function, x is a vector represents the dependent variables (state variables), u is a vector represents the independent (the control) variables, g_j and h_j represent equality and inequality constraints, respectively. m and p are number of equality and inequality constraints, respectively.

The dependent variables (x) in power system can be described as follows:

$$x=\left[P_{G1},V_{L1}\dots V_{L,NPQ},Q_{G,1}\dots Q_{G,NG},S_{TL,1}\dots S_{TL,NTL}\right] \tag{4}$$

where P_{G1} is slack bus power, V_L is voltage of load bus, Q_G is the generator reactive power output, S_{TL} is the apparent power flow in transmission line, *NPQ* is number of load buses, *NG* is number of generation buses, and *NTL* is number of transmission lines.

The independent variables u of power system can be described as follows:

$$u=[P_{G,2}\ldots P_{G,NG}, V_{G,1}\ldots V_{G,NG}, Q_{C,1}\ldots Q_{C,NC}, T_1\ldots T_{NT}] \quad (5)$$

where P_G is output active power of generator, V_G is voltage of generation bus, Q_C is the injected reactive power of shunt compensator, T is tap setting of transformer, *NC* is number of shunt compensator units, and *NT* is the number of transformers.

3.1 Objective Functions

3.1.1 Single Objective Functions

The optimization techniques can be applied for optimal power flow solution where the most common objective functions can be presented as follows:

(1) Quadratic fuel cost

The objective function is the quadratic equation of total generation fuel cost which formulated as follows:

$$F_1=\sum_{i=1}^{NG} F_i(P_{Gi})=\sum_{i=1}^{NPV}\left(a_i+b_iP_{Gi}+c_iP_{Gi}^2\right) \quad (6)$$

where, F_i is the fuel cost of the ith generator. a_i, b_i, and c_i are the cost coefficients of ith generator.

(2) Quadratic cost with valve-point effect

Practically, the quadratic fuel costs of some generation units are nonsmooth functions due to influences of the steam admission from their control valves which call valve-point loading effect. Steam admission will lead to occurrence of ripples in fuel cost. The valve-point loading effect is considered by adding a sine term to the fuel cost as follows:

$$F(x,u)=\sum_{i=1}^{NG} F_i(P_{Gi})=\sum_{i=1}^{NG}\left(a_i+b_iP_{Gi}+c_iP^2{}_{Gi}\right)+\left|d_i sin\left(e_i\left(P_{Gi}^{min}-P_{Gi}\right)\right)\right| \quad (7)$$

where, d_i and e_i are the fuel cost coefficients of the ith unit with valve-point effects.

(3) Piecewise quadratic cost functions

Practically, the thermal generation unit may have several fuel sources such as coal, natural gas, and oil. Hence, the fuel cost functions of these units will be nonconvex problem and it is split as piecewise quadratic cost functions for different fuel types as follows:

$$F(P_{Gi}) = \begin{cases} a_{i1} + b_{i1}P_{Gi} + c_{i1}P_{Gi}^2 & P_{Gi}^{min} \le P_{Gi} \le P_{G1} \\ a_{i2} + b_i P_{Gi2} + c_i P_{Gi}^2 & P_{G1} \le P_{Gi} \le P_{G2} \\ \cdots & \\ a_{ik} + b_i P_{Gik} + c_i P_{Gi}^2 & P_{Gi(k-1)} \le P_{Gi} \le P_{Gi}^{max} \end{cases} \quad (8)$$

(4) Voltage profile improvement

The third objective function is minimizing the voltage deviations of load buses from a specified voltage. The required objective function can formulate as follows:

$$F_3 = VD = \sum_{i=1}^{NPQ} |(V_i - 1)| \quad (9)$$

(5) Voltage stability enhancement

Voltage stability index (L) is an important value proposed by Kessel and Glavitsch to indicate system stability [110]. This value is calculated for all load buses, which vary between 0 (no load case) to 1 maximum loading point (voltage collapse case) hence, the minimizing of this value keeps the system far away from voltage collapse of power system. However, this objective function can be formulated as follows:

$$F_4 = min(L_{max}) = min(max(L_n)) \quad n = 1, 2, \ldots, NPQ \quad (10)$$

where, L_n is the voltage stability index of bus nth that can be calculated as follows:

$$L_n = \left| 1 - \sum_{i=1}^{NPV} F_{ij} \frac{V_i}{V_j} \right| \quad (11)$$

where, V_i is the voltage of ith generator bus and V_j is the voltage of load bus. F_{ij} can be obtained from system Y_{bus} matrix as follows:

$$\begin{bmatrix} I_G \\ I_L \end{bmatrix} = \begin{bmatrix} Y_{GG} & Y_{GL} \\ Y_{LG} & Y_{LL} \end{bmatrix} \begin{bmatrix} V_G \\ V_L \end{bmatrix} \quad (12)$$

where I_G, I_L are the complex currents and V_G, V_L are the complex voltages vectors at the generator and load buses. Y_{GG}, Y_{LL}, Y_{GL}, Y_{LG} are submatrices of system Y_{bus} by making some manipulation in Eq. (12)

$$\begin{bmatrix} V_L \\ I_G \end{bmatrix} = \begin{bmatrix} Z_{LL} & F_{LG} \\ K_{GL} & Y_{GG} \end{bmatrix} \begin{bmatrix} I_L \\ V_G \end{bmatrix} \quad (13)$$

where, $F_{ij} = [F_{LG}] = -[Y_{LL}]^{-1}[Y_{LG}]$.

(6) Real power losses minimization

The required objective function is minimizing the active power loss which can be formulated as follows:

$$P_{loss} = \sum_{i=1}^{NTL} G_{ij}\left(V_i^2 + V_j^2 - 2V_iV_j\cos\delta_{ij}\right) \tag{14}$$

where, G_{ij} the conductance of a transmission, NTL is number of transmission lines, and δ_{ij} is phase difference of voltages.

(7) Emission minimization

Reducing the pollution of the environment can be achieved by minimizing the emission gases from thermal power plants. The objective function for these emission gases can be formulated as follows:

$$E_m = \sum_{i=1}^{NG} \gamma_i P_{Gi}^2 + \beta_i P_{Gi} + \alpha_i + \zeta_i exp(\lambda_i P_{Gi}) \tag{15}$$

where, $\gamma_{i,}$ $\beta_{i,}$ α_i, ζ_i, and λ_i are the emission coefficient of the ith generator.

3.1.2 Multiobjective Functions

The main purpose of solving multiobjective problems is to optimize several independent objective functions simultaneously.

The multiobjective problem can be defined as follows:

$$Min\ F(x,u) = [F_1(x,u), F_2(x,u)\ldots, F_i(x,u)] \tag{16}$$

where "i" is the number of objective functions. Solving the multiobjective functions can be accomplished by Pareto optimization method or weight factors as:

$$F(x,u) = \omega_1\left(\sum_{i=1}^{NG}\left(a_i + b_iP_{Gi} + c_iP^2{}_{Gi}\right)\right) + \omega_2\left(\sum_{i=1}^{NPQ}|(V_i - 1)|\right) \tag{17}$$

$$F(x,u) = \omega_1\left(\sum_{i=1}^{NG}\left(a_i + b_iP_{Gi} + c_iP^2{}_{Gi}\right)\right) + \omega_2\left(\sum_{i=1}^{NTL} G_{ij}\left(V_i^2 + V_j^2 - 2V_iV_j\cos\delta_{ij}\right)\right) \tag{18}$$

$$F(x,u) = \omega_1\left(\sum_{i=1}^{NG}\left(a_i + b_iP_{Gi} + c_iP^2{}_{Gi}\right)\right) + \omega_2(max(L_n)) \tag{19}$$

$$F(x,u)=\omega_1\left(\sum_{i=1}^{NPQ}|(V_i-1)|\right)+\omega_2\left(\sum_{i=1}^{NTL}G_{ij}\left(V_i^2+V_j^2-2V_iV_j\cos\delta_{ij}\right)\right)$$

$$+\omega_3(\max(L_n)) \tag{20}$$

where, ω_1, ω_2, and ω_3 are weight factors which are selected based on relative important of one objective to others. Usually, the values of weight factors are selected as follows:

$$\sum_{i=1}^{n}\omega_n=0 \tag{21}$$

3.1.3 Constraints

The transmission system has several constraints which can be categorized as follows:

(1) *Equality constraints*

The equality constrains represent the balanced load flow equations as follows:

$$P_{Gi}-P_{Di}=|V_i|\sum_{j=1}^{NB}|V_j|\left(G_{ij}\cos\delta_{ij}+B_{ij}\sin\delta_{ij}\right) \tag{22}$$

$$Q_{Gi}-Q_{Di}=|V_i|\sum_{j=1}^{NB}|V_j|\left(G_{ij}\cos\delta_{ij}+B_{ij}\sin\delta_{ij}\right) \tag{23}$$

where, P_{Gi} and Q_{Gi} are the generated active and reactive power at bus i, respectively. P_{Di} and Q_{Di} are the active and reactive load demand at bus i, respectively. G_{ij} and B_{ij} are the conductance and susceptance between bus i and bus j, respectively.

(2) *Inequality constrains*

The inequality constrains can be classified as follows:

I. Generators active power output

$$P_{Gi}^{min}\le P_{Gi}\le P_{Gi}^{max}\quad i=1,2,\ldots,NG \tag{24}$$

II. Generators bus voltages

$$V_{Gi}^{min}\le V_{Gi}\le V_{Gi}^{max}\quad i=1,2,\ldots,NG \tag{25}$$

III. Generators reactive power output

$$Q_{Gi}^{min}\le Q_{Gi}\le Q_{Gi}^{max}\quad i=1,2,\ldots,NG \tag{26}$$

IV. Transformer tap settings

$$T_i^{min} \le T_i \le T_i^{max} \quad i = 1, 2, \ldots, NT \tag{27}$$

V. Shunt VAR compensator

$$Q_{Ci}^{min} \le Q_{Ci} \le Q_{Ci}^{max} \quad i = 1, 2, \ldots, NC \tag{28}$$

VI. Apparent power flow in transmission lines

$$S_{Li} \le S_{Li}^{min} \quad i = 1, 2, \ldots, NTL \tag{29}$$

VII. Voltage magnitude of load buses

$$V_{Li}^{min} \le V_{Li} \le V_{Li}^{max} \quad i = 1, 2, \ldots, NPQ \tag{30}$$

The dependent control variables can be easily considered in an optimization solution by using the quadratic penalty formulation of objective function which can be expressed as follows:

$$F_g(x,u) = F_i(x,u) + K_G(\Delta P_{G1})^2 + K_Q \sum_{i=1}^{NPV} (\Delta Q_{Gi})^2$$

$$+ K_V \sum_{i=1}^{NPQ} (\Delta V_{Li})^2 + K_S \sum_{i=1}^{NTL} (\Delta S_{Li})^2 \tag{31}$$

where, K_G, K_Q, K_V, and K_S are the penalty factors, these values are high positive ΔP_{G1}, ΔQ_{Gi}, ΔV_{Li} and ΔS_{Li} are penalty terms which can be defined as follow:

$$\Delta P_{G1} = \begin{cases} (P_{G1} - P_{G1}^{max}) & P_{G1} > P_{G1}^{max} \\ (P_{G1} - P_{G1}^{min}) & P_{G1} < P_{G1}^{min} \\ 0 & P_{G1}^{min} < P_{G1} < P_{G1}^{max} \end{cases} \tag{32}$$

$$\Delta Q_{Gi} = \begin{cases} (Q_{Gi} - Q_{Gi}^{max}) & Q_{Gi} > Q_{Gi}^{max} \\ (Q_{Gi} - Q_{Gi}^{min}) & Q_{Gi} < Q_{Gi}^{min} \\ 0 & Q_{Gi}^{min} < Q_{Gi} < Q_{Gi}^{max} \end{cases} \tag{33}$$

$$\Delta V_{Li} = \begin{cases} (V_{Li} - V_{Li}^{max}) & V_{Li} > V_{Li}^{max} \\ (V_{Li} - V_{Li}^{min}) & V_{Li} < V_{Li}^{min} \\ 0 & V_{Li}^{min} < V_{Li} < V_{Li}^{max} \end{cases} \tag{34}$$

$$\Delta S_{Li} = \begin{cases} (S_{Li} - S_{Li}^{max}) & S_{Li} > S_{Li}^{max} \\ (S_{Li} - S_{Li}^{min}) & S_{Li} < S_{Li}^{min} \\ 0 & S_{Li}^{min} < S_{Li} < S_{Li}^{max} \end{cases} \tag{35}$$

4 CONVENTIONAL OPTIMIZATION METHODS FOR OPTIMAL POWER FLOW

The conventional methods were based on linearized objective functions and these methods apply sensitivity analysis and gradient based optimization algorithms. The conventional methods that have been applied for the OPF problem are illustrated as follows.

4.1 Linear Programming Method

Linear programming (LP)-based method is used to linearize nonlinear power system optimization problems. This method is reliable and has a good convergence characteristic, however the main shortage is it could be trapped in local minima. Moreover, errors could appear, especially in constraints, due to rounding by digital computers [5–7].

4.2 Nonlinear Programming Method

The nonlinear programming (NLP) method is more accurate compared to linear programs where it can be applied for the nonlinear objective functions and constraints. The NLP techniques are based on reduced gradient method utilizing the Lagrange multiplier or use the penalty function optimization approach. In this method, the first partial derivatives of the equations (the reduced gradient) is used to choose a search direction in the iterative procedure. The merits of this method are it can be applied in a large-scale system while the drawbacks of this method are that some system components are not considered [8–12].

4.3 Quadratic Programming Method

The quadratic programming (QP) can be considered a special form of nonlinear programming method where the objective function is quadratic, and the constraints are in linear form. This method doesn't require the determination of the gradient steps. The obtained solution using the QP method is more accurate compared with LP and NLP and it can be applied in ill conditions, moreover it has fast convergence characteristic [13,14].

4.4 Newton's Method

Newton's method is commonly used in power flow problems based on the creation of the Lagrangian or decomposition approach by applying second-order partial derivatives (the Hessian). The advantages of Newton's method are that it can be applied for different optimal power flow problems, it has fast convergence characteristics, and it can handle inequality constraints efficiently. The disadvantage of Newton's method is that its convergence characteristics are very sensitive for the initial condition [15–18].

4.5 Interior Point Method

Interior point (IP) method is usually applied for large problems and has been employed to solve the optimal power problem. However, the results obtained using IP result are better and need less iteration compared with LP, and it can be trapped in local minima [19,20].

5 RECENT OPTIMIZATION METHODS FOR OPTIMAL POWER FLOW

The OPF problem is a multimodal, nonlinear, or nonconvex problem; hence, application of conventional methods is not always suitable and can't guarantee a global solution. To overcome the shortages of conventional methods, several heuristics optimization techniques have been developed to solve the OPF problem. The merits of recent optimization techniques can be summarized as follows:

(1) These methods can be applied in small and large-scale systems
(2) High reliability to obtain the optimal solutions
(3) These methods rarely suffer from stagnations or trapped in local minima solutions
(4) These methods converge rapidly to the optimal solution compared to conventional methods.

In this section, we offer a comprehensive survey of the application of the recent (nondeterministic) heuristic optimization techniques for optimal power flow. The heuristic optimization algorithms that applied for OPF problem can be classified based on inspiration methods as follows:

5.1 Swarm and Bio-inspired Optimization Techniques

Nature-inspired and bio-inspired optimization techniques are inspired from motion and searching behavior of swarms of animals or birds for food sources. Table 1 summarizes the nature-inspired techniques that have been applied to the OPF problem.

5.2 Human-Inspired Optimization Techniques

Several optimization techniques mimic human behavior, especially in thinking or decision-making. The human-inspired algorithms that have been employed to solve the OPF problem are listed in Table 2.

5.3 Physics-Inspired Optimization Techniques

Physics-inspired algorithms are conceptualized from physics laws or natural phenomena in space. The applied physics-inspired optimization methods for OPF are listed in Table 3.

TABLE 1 Summary of the Literature Review Regarding Nature Inspired Algorithms for OPF Problem

Algorithm	Objective Function	System	Year	Ref.
Improved artificial bee colony	Fuel cost & fuel cost with valve effect & P_{loss}	IEEE 30-bus, IEEE 118-bus	2017	[22]
Moth swarm algorithm	Fuel cost & fuel cost with valve effect & emission & L-index & P_{loss} & piecewise cost & VD	IEEE 30-bus, IEEE-57, IEEE-118 bus	2017	[24]
Adaptive partitioning flower pollination	Fuel cost & P_{loss} & VD	IEEE 30-bus, IEEE 57-bus	2016	[27]
Glow-worm swarm optimization	Fuel cost & emission	IEEE 30-bus, Indian 75-bus	2016	[28]
Best-guided artificial bee colony algorithm	Fuel cost	IEEE 30-bus, IEEE 57-bus	2016	[30]
Oppositional krill herd algorithm	Fuel cost & fuel cost with valve effect emission & P_{loss}	IEEE 30-bus, IEEE-57	2016	[32]
Chaotic krill herd algorithm	Fuel cost & fuel cost with valve effect & P_{loss} & VD	IEEE-26-, IEEE 57-bus	2015	[36]
Improved group search optimization	Fuel cost with valve effect	26-bus, IEEE 30-bus, IEEE 118-bus	2015	[39]
Chaotic artificial bee colony algorithm	Fuel cost & transient stability	IEEE 30-bus, New England 39-bus systems	2015	[41]
Adaptive clonal selection	Fuel cost & P_{loss} & L-index & VD	IEEE 30-bus	2014	[45]

TABLE 1 Summary of the Literature Review Regarding Nature Inspired Algorithms for OPF Problem—cont'd

Algorithm	Objective Function	System	Year	Ref.
Multihive bee foraging algorithm	Fuel cost with valve effect & emission & P_{loss}	IEEE 30-bus	2014	[47]
Artificial bee colony algorithm	Fuel cost & fuel cost with valve effect & emission & L-index & P_{loss} & piecewise cost & VD	IEEE 9-bus, IEEE 30-bus, IEEE 57-bus	2013	[50]
Moth-flame algorithm	L-index & P_{loss} & VD	IEEE 30-bus	2016	[53]
Modified artificial bee colony algorithm	Fuel cost with valve effect & emission & P_{loss} & piecewise cost & VD	IEEE 30-bus, IEEE 118-bus	2013	[55]
Modified shuffle frog leaping algorithm	Fuel cost & emission	IEEE 30-bus	2011	[60]
Chaotic invasive weed optimization algorithms	Fuel cost & fuel cost with valve effect & fuel cost considering the prohibited zones & piecewise cost	IEEE 30-bus	2014	[65]
Particle swarm optimization	Fuel cost & L-index & piecewise cost & VD	IEEE 30-bus	2002	[68]
Group search optimization	Fuel cost with valve effect & emission & L-index & piecewise cost & VD	IEEE 30-bus, 57-bus, 118-bus	2016	[72]
Gray wolf optimizer	Fuel cost with valve effect & P_{loss} & Q_{loss}	IEEE 30-bus, 118-bus	2016	[74]
Moth-flame optimizer	Fuel cost & L-index & P_{loss} & VD	IEEE 30-bus	2016	[77]
Stud krill herd algorithm	Fuel cost & fuel cost with valve effect & emission & L-index & P_{loss}	IEEE 14-bus, IEEE 30-bus, IEEE 57-bus	2016	[78]
Artificial bee colony algorithm	Fuel cost & fuel cost with valve effect & piecewise cost	IEEE 30-bus	2014	[82]

Continued

TABLE 1 Summary of the Literature Review Regarding Nature Inspired Algorithms for OPF Problem—cont'd

Algorithm	Objective Function	System	Year	Ref.
Enhanced ant colony optimization	Fuel cost & emission	IEEE 30-bus, IEEE 118-bus	2012	[86]
Biogeography-based optimization	Fuel cost & fuel cost with valve effect & VD & L-index	IEEE 30-bus	2010	[89]
The modified flower pollination	Fuel cost & P_{loss} VD	IEEE 30-bus	2017	[92]
Modified honey bee mating optimization	Fuel cost & fuel cost with valve effect	IEEE 14, IEEE 30-bus, IEEE 118-bus	2011	[96]
Particle swarm	Fuel cost & emission & L-index & P_{loss}	IEEE 30-bus	2015	[99]

TABLE 2 Summary of the Literature Review Regarding Human Inspired Algorithms for OPF Problem

Algorithm	Objective Function	System	Year	Ref.
Biogeography-based optimization	Fuel cost & emission & L-index & P_{loss} & VD	IEEE 30-bus, IEEE 57-bus	2015	[35]
Imperialist competitive algorithm	Fuel cost & Emission & P_{loss} & VD	IEEE 30-bus, IEEE 57-bus	2014	[40]
The league championship algorithm	Fuel cost & fuel cost with valve effect & emission & L-index & P_{loss} & VD	Algerian power system network	2014	[42]

TABLE 2 Summary of the Literature Review Regarding Human Inspired Algorithms for OPF Problem—cont'd

Algorithm	Objective Function	System	Year	Ref.
Modified imperialist competitive algorithm	Fuel cost & P_{loss} & VD	IEEE 57-bus	2014	[44]
Teaching learning based optimization	Fuel cost & L-index & P_{loss} & piecewise cost & VD	IEEE 30-bus, IEEE 118-bus	2014	[46]
Quasioppositional teaching learning-based optimization	Fuel cost with valve effect & emission & L-index & P_{loss}	IEEE 30-bus, Indian utility 62-bus, IEEE 118-bus	2014	[48]
Modified teaching-learning based optimization	Fuel cost with valve effect & emission	IEEE 30-bus, IEEE 57-Bus	2013	[50]
Improved harmony search algorithm	Fuel cost with valve effect	6-bus, 14-bus, 30-bus, 57-bus,118-bus	2013	[54]
Tabu search algorithm	Fuel cost & fuel cost with valve effect & VD	30-bus	2002	[71]
Symbiotic organisms search algorithm	Fuel cost & fuel cost with valve effect & fuel cost considering the prohibited zones	IEEE 30-bus	2016	[87]
Improved teaching-learning	Fuel cost & fuel cost with valve effect & emission & piecewise cost & VD	IEEE 30-bus, IEEE 57-bus	2015	[100]

TABLE 3 Summary of the Literature Review Regarding Physics Inspired Algorithms for OPF Problem

Algorithm	Objective Function	System	Year	Ref.
Improved colliding bodies optimization algorithm	Fuel cost & fuel cost with valve effect & emission & L-index & P_{loss} & piecewise cost & VD	IEEE 30-bus, IEEE 57- IEEE 118-bus	2016	[33]
Opposition based gravitational search algorithm	Fuel cost & fuel cost with valve effect & emission & L-index & P_{loss} & piecewise cost & VD	IEEE 30-bus	2015	[38]
Black-hole-based optimization approach	Fuel cost & L-index & P_{loss} & VD	30-bus, Algerian 59-bus system	2014	[43]
Gravitational search algorithm	Fuel cost & fuel cost with valve effect & L-index & P_{loss} & piecewise cost & VD	IEEE 30-bus, 57-bus	2012	[58]
Simulated annealing	Fuel cost	IEEE 6-bus, IEEE 30-bus	2003	[67]
Improved electromagnetism-like mechanism method	Fuel cost & L-index & P_{loss} & Q_{loss} & piecewise cost & VD	IEEE 30-bus, IEEE 57-bus	2016	[37]

5.4 Evolutionary-Inspired Optimization Techniques

The evolutionary optimization algorithms are derived from the mechanics of natural selection and genetics or living organism or creatures. The evolutionary-based optimization techniques that applied for OPF problem are listed in Table 4.

5.5 Hybrid Optimization Techniques

Several hybrid optimization algorithms have been proposed to capture merits of several techniques and to gain superior results compared to singular technique. Hybrid optimization techniques that have been applied for OPF are listed in Table 5.

TABLE 4 Summary of the Literature Review Regarding Evolutionary Inspired Algorithms for OPF Problem

Algorithm	Objective Function	System	Year	Ref.
Enhanced self-adaptive differential evolution	Fuel cost & emission& L-index & P_{loss}	IEEE 30-bus, IEEE 57-bus	2017	[21]
Improved evolutionary algorithm	Fuel cost & emission	IEEE 30-bus, IEEE 57-bus	2017	[23]
Backtracking search optimization	Fuel cost & fuel cost with valve effect emission & L-index & piecewise cost & VD	IEEE 30-bus, IEEE 57-bus, and IEEE 118-bus	2016	[25]
Modified evolutionary algorithm based decomposition	Fuel cost emission & L-index & VD & P_{loss}	IEEE 30-bus	2016	[26]
Differential search algorithm	Fuel cost & emission & L-index & P_{loss} & VD	IEEE 9-bus, IEEE 30-bus IEEE 57-bus	2016	[29]
Enhanced charged system search algorithm	Fuel cost with valve effect & piecewise cost with valve effect	IEEE 30-bus, IEEE 118-bus	2012	[57]
Multiagent-based differential evolution approach	Fuel cost & fuel cost with valve effect & piecewise cost with valve effect	6-bus, IEEE 30-bus	2012	[59]
Differential evolution algorithm	Fuel cost & VD & L-index & piecewise cost	IEEE 30-bus	2010	[61]
Enhanced genetic algorithm	Fuel cost & L-index & P_{loss}	IEEE 30-bus	2010	[62]
Modified differential evolution algorithm	Fuel cost & piecewise cost & fuel cost with valve effect	6-bus, IEEE30-bus	2008	[63]
Evolutionary programming	Fuel cost	IEEE 30-bus	2004	[64]
Differential search algorithm	Fuel cost & emission & L-index & P_{loss} & VD	IEEE 30-bus	2014	[69]

Continued

TABLE 4 Summary of the Literature Review Regarding Evolutionary Inspired Algorithms for OPF Problem—cont'd

Algorithm	Objective Function	System	Year	Ref.
Improved evolutionary programming	Fuel cost & piecewise cost & fuel cost with valve effect	IEEE 30-bus	2014	[70]
Adapted genetic algorithm	Fuel cost & VD	IEEE 30-bus	2012	[75]
Evolutionary programming	Fuel cost & piecewise cost	IEEE 30-bus, Indian utility 62-bus	2005	[76]
Differential evolution	Fuel cost with valve effect emission & L-index & piecewise cost & P_{loss}	IEEE 30-bus, IEEE 118-bus	2012	[79]
Backtracking search algorithm	Fuel cost & fuel cost with valve effect & fuel cost considering the prohibited zones	IEEE 30-bus	2015	[81]
Differential evolution	Fuel cost & fuel cost with valve effect & L-index & P_{loss}	IEEE 30-bus	2017	[83]
Forced initialized differential evolution algorithm	Fuel cost & VD & L-index & P_{loss}	IEEE 30-bus, IEEE 57-bus	2016	[88]
Enhanced genetic algorithm	Fuel cost	IEEE 30-bus, IEEE RTS96	2002	[90]
Evolutionary programming	Fuel cost & piecewise cost & fuel cost with valve effect & VD	IEEE 30-bus	1999	[91]
Genetic-algorithm	Fuel cost & severity index	IEEE 30-bus, IEEE 118-bus	2005	[93]
Improved genetic algorithms	Fuel cost	IEEE 30-bus	1997	[97]
Genetic algorithm	Fuel cost	IEEE 6-bus	2004	[98]
Faster evolutionary algorithm	Fuel cost & fuel cost with valve effect	IEEE 30-bus, IEEE 118-bus, IEEE 300-bus	2014	[101]

TABLE 5 Summary of the Literature Review Regarding Hybrid Inspired Algorithms for OPF Problem

Algorithm	Objective Function	System	Year	Ref.
Fuzzy logic system with harmony search algorithm	Cost & severity index	IEEE 30 bus, IEEE 57 bus, IEEE 118 bus	2016	[31]
Particle swarm optimization with an aging leader and challengers algorithm	Fuel cost & fuel cost with valve effect & P_{loss} & VD	IEEE 30 bus, IEEE 118 bus	2016	[34]
Artificial bee colony algorithm with quantum theory	Fuel cost & fuel cost with valve effect	IEEE 30-bus, IEEE 118-bus	2015	[37]
Fuzzy particle swarm optimization and Nelder-Mead algorithm	Fuel cost & fuel cost with valve effect & emission & L-index & piecewise cost & VD	IEEE 30-bus	2014	[49]
Fuzzy evolutionary and swarm optimization	Fuel cost	IEEE 30-bus	2013	[52]
Particle swarm optimization and the shuffle frog leaping algorithms	Fuel cost with valve effect & piecewise cost with valve effect & emission	IEEE 30-bus, IEEE 57-bus, IEEE 118-bus	2013	[56]
Imperialist competitive algorithm and teaching learning algorithm	Fuel cost & fuel cost with valve effect & fuel cost considering the prohibited zones & piecewise cost	IEEE 30-bus, IEEE 57-bus	2014	[66]
GA-fuzzy and PSO-fuzzy	Fuel cost	IEEE 30-bus	2014	[80]
Genetic evolving ant direction	Fuel cost & fuel cost with valve effect & piecewise cost	IEEE 30-bus, IEEE 118-bus	2012	[84]

Continued

TABLE 5 Summary of the Literature Review Regarding Hybrid Inspired Algorithms for OPF Problem—cont'd

Algorithm	Objective Function	System	Year	Ref.
Hybrid genetic algorithm and particle swarm	Fuel cost & fuel cost with valve effect	IEEE 30 bus	2017	[85]
Harmony search algorithm and an ant system	P_{loss} & VD & performance index	IEEE 30-bus and Taiwan Power Company 345 kV simplified systems	2014	[94]
GA with SA	Fuel cost & P_{loss} & security margin index & emission	IEEE 30-bus, IEEE 118-bus	2003	[95]
Shuffle frog leaping algorithm and simulated annealing	Fuel cost & fuel cost with valve effect & fuel cost considering the prohibited zones	IEEE 30-bus	2012	[102]

5.6 Artificial Neural Networks (ANN) and Fuzzy Logic Approach

Artificial neural networks (ANNs) are computational methods mimic the operation of biological neural networks while the fuzzy set theory is a natural and suitable tool to represent inexact relations. The optimizations methods based on ANNs and fuzzy logic approach are listed in Table 6.

TABLE 6 Summary of the Literature Review Regarding ANNs and Fuzzy Approach Inspired Algorithms for OPF Problem

Algorithm	Objective Function	System	Year	Ref.
Neural network	P_{Loss}	22-bus system	1997	[103]
Artificial neural network	Fuel cost & maximize the voltage stability margin	IEEE 30-bus	2003	[104]
Fuzzy logic model	Fuel cost & emission	IEEE 14-bus	2010	[105]

TABLE 6 Summary of the Literature Review Regarding ANNs and Fuzzy Approach Inspired Algorithms for OPF Problem—cont'd

Algorithm	Objective Function	System	Year	Ref.
Fuzzy logic approach	Fuel cost	IEEE 14-bus	1997	[106]
Fuzzy mathematical programming	P_{Loss}	IEEE 14-bus	1988	[107]
Hopfield neural network	Minimum deviations in real power generations and loads at buses	IEEE 6-bus	1996	[108]
Fuzzy linear programming	Fuel cost & maximizing the generation reserve	IEEE 5-bus, IEEE 14-bus	2005	[109]

6 CONCLUSIONS

This chapter has presented a survey related to the optimal power flow (OPF) problem. However, the chapter has described the following:

- OPF problem formulation, including the common objective functions of power system, control variables and operating constraints.
- The conventional methods that have been employed to solve OPF problem, including the pros and cons of these methods.
- The recent optimization techniques that have been applied for OPF. In addition, the recent optimization algorithms have been categorized based on inspiration methods such as evolutionary, human, nature, bio-inspired, and physics inspired techniques.

However, the recent optimization techniques have a superiority than the conventional techniques due to the following merits:

- These methods can be applied in both small and large-scale systems
- High reliability to obtain the optimal solutions
- These methods rarely suffer from stagnations or trapped in local minima solutions
- These methods converged rapidly to the optimal solution compared with conventional methods.

REFERENCES

[1] J. Carpentier, Contribution a l'etude du dispatching economique, Bull. Soc. Franc. Elec. 3 (1962) 431–447.

[2] J.A. Momoh, R. Adapa, M. El-Hawary, A review of selected optimal power flow literature to 1993. I. Nonlinear and quadratic programming approaches, IEEE Trans. Power Syst. 14 (1999) 96–104.
[3] J.A. Momoh, M. El-Hawary, R. Adapa, A review of selected optimal power flow literature to 1993. II. Newton, linear programming and interior point methods, IEEE Trans. Power Syst. 14 (1999) 105–111.
[4] J. Zhu, Optimization of Power System Operation, vol. 47, John Wiley & Sons, Hoboken, 2015.
[5] D. Wells, Method for economic secure loading of a power system, Proc. Inst. Electr. Eng. 115 (1968) 1190–1194.
[6] B. Stott, E. Hobson, Power system security control calculations using linear programming, part I, IEEE Trans. Power Appar. Syst. 5 (1978) 1713–1720.
[7] C. Shen, M. Laughton, Power-system load scheduling with security constraints using dual linear programming, Proc. Inst. Electr. Eng. 117 (1970) 2117–2127.
[8] C. Shen, M. Laughton, Determination of optimum power-system operating conditions under constraints, Proc. Inst. Electr. Eng. (1969) 225–239.
[9] A.M. Sasson, Combined use of the Powell and Fletcher-Powell nonlinear programming methods for optimal load flows, IEEE Trans. Power Appar. Syst. 10 (1969) 1530–1537.
[10] A.H. El-abiad, F.J. Jaimes, A method for optimum scheduling of power and voltage magnitude, IEEE Trans. Power Appar. Syst. 4 (1969) 413–422.
[11] H.W. Dommel, W.F. Tinney, Optimal power flow solutions, IEEE Trans. Power Appar. Syst. 10 (1968) 1866–1876.
[12] A.M. Sasson, Decomposition techniques applied to the nonlinear programming load-flow method, IEEE Trans. Power Appar. Syst. 1 (1970) 78–82.
[13] G. Contaxis, C. Delkis, G. Korres, Decoupled optimal load flow using linear or quadratic programming, IEEE Trans. Power Syst. 1 (1986) 1–7.
[14] N. Nabona, L. Freris, Optimisation of economic dispatch through quadratic and linear programming, Proc. Inst. Electr. Eng. 120 (1973) 574–580.
[15] G.A. Maria, J. Findlay, A Newton optimal power flow program for Ontario Hydro EMS, IEEE Trans. Power Syst. 2 (1987) 576–582.
[16] A. Monticelli, W.-H. Liu, Adaptive movement penalty method for the Newton optimal power flow, IEEE Trans. Power Syst. 7 (1992) 334–342.
[17] D.I. Sun, B. Ashley, B. Brewer, A. Hughes, W.F. Tinney, Optimal power flow by Newton approach, IEEE Trans. Power Appar. Syst. 10 (1984) 2864–2880.
[18] S.-D. Chen, J.-F. Chen, A new algorithm based on the Newton-Raphson approach for real-time emission dispatch, Electr. Power Syst. Res. 40 (1997) 137–141.
[19] K. Ponnambalam, V. Quintana, A. Vannelli, in: A fast algorithm for power system optimization problems using an interior point method, Power Industry Computer Application Conference, 1991. Conference Proceedings, 1991, pp. 393–400.
[20] J. Momoh, R. Austin, R. Adapa, E. Ogbuobiri, in: Application of interior point method to economic dispatch, IEEE International Conference on Systems, Man and Cybernetics, 1992, pp. 1096–1101.
[21] H. Pulluri, R. Naresh, V. Sharma, An enhanced self-adaptive differential evolution based solution methodology for multiobjective optimal power flow, Appl. Soft Comput. 54 (2017) 229–245.
[22] W. Bai, I. Eke, K.Y. Lee, An improved artificial bee colony optimization algorithm based on orthogonal learning for optimal power flow problem, Control. Eng. Pract. 61 (2017) 163–172.

[23] X. Yuan, B. Zhang, P. Wang, J. Liang, Y. Yuan, Y. Huang, et al., Multi-objective optimal power flow based on improved strength Pareto evolutionary algorithm, Energy 122 (2017) 70–82.

[24] A.-A.A. Mohamed, Y.S. Mohamed, A.A. El-Gaafary, A.M. Hemeida, Optimal power flow using moth swarm algorithm, Electr. Power Syst. Res. 142 (2017) 190–206.

[25] A. Chaib, H. Bouchekara, R. Mehasni, M. Abido, Optimal power flow with emission and non-smooth cost functions using backtracking search optimization algorithm, Int. J. Electr. Power Energy Syst. 81 (2016) 64–77.

[26] J. Zhang, Q. Tang, P. Li, D. Deng, Y. Chen, A modified MOEA/D approach to the solution of multi-objective optimal power flow problem, Appl. Soft Comput. 47 (2016) 494–514.

[27] B. Mahdad, K. Srairi, Security constrained optimal power flow solution using new adaptive partitioning flower pollination algorithm, Appl. Soft Comput. 46 (2016) 501–522.

[28] S.S. Reddy, C.S. Rathnam, Optimal power flow using glowworm swarm optimization, Int. J. Electr. Power Energy Syst. 80 (2016) 128–139.

[29] K. Abaci, V. Yamacli, Differential search algorithm for solving multi-objective optimal power flow problem, Int. J. Electr. Power Energy Syst. 79 (2016) 1–10.

[30] H. Jadhav, P. Bamane, Temperature dependent optimal power flow using g-best guided artificial bee colony algorithm, Int. J. Electr. Power Energy Syst. 77 (2016) 77–90.

[31] K. Pandiarajan, C. Babulal, Fuzzy harmony search algorithm based optimal power flow for power system security enhancement, Int. J. Electr. Power Energy Syst. 78 (2016) 72–79.

[32] A. Mukherjee, V. Mukherjee, Solution of optimal power flow with FACTS devices using a novel oppositional krill herd algorithm, Int. J. Electr. Power Energy Syst. 78 (2016) 700–714.

[33] H. Bouchekara, A. Chaib, M.A. Abido, R.A. El-Sehiemy, Optimal power flow using an improved colliding bodies optimization algorithm, Appl. Soft Comput. 42 (2016) 119–131.

[34] R.P. Singh, V. Mukherjee, S. Ghoshal, Particle swarm optimization with an aging leader and challengers algorithm for the solution of optimal power flow problem, Appl. Soft Comput. 40 (2016) 161–177.

[35] A.R. Kumar, L. Premalatha, Optimal power flow for a deregulated power system using adaptive real coded biogeography-based optimization, Int. J. Electr. Power Energy Syst. 73 (2015) 393–399.

[36] A. Mukherjee, V. Mukherjee, Solution of optimal power flow using chaotic krill herd algorithm, Chaos, Solitons Fractals 78 (2015) 10–21.

[37] X. Yuan, P. Wang, Y. Yuan, Y. Huang, X. Zhang, A new quantum inspired chaotic artificial bee colony algorithm for optimal power flow problem, Energy Convers. Manag. 100 (2015) 1–9.

[38] A.R. Bhowmik, A.K. Chakraborty, Solution of optimal power flow using non dominated sorting multi objective opposition based gravitational search algorithm, Int. J. Electr. Power Energy Syst. 64 (2015) 1237–1250.

[39] Y. Tan, C. Li, Y. Cao, K.Y. Lee, L. Li, S. Tang, et al., Improved group search optimization method for optimal power flow problem considering valve-point loading effects, Neurocomputing 148 (2015) 229–239.

[40] M. Ghasemi, S. Ghavidel, M.M. Ghanbarian, M. Gharibzadeh, A.A. Vahed, Multi-objective optimal power flow considering the cost, emission, voltage deviation and power losses using multi-objective modified imperialist competitive algorithm, Energy 78 (2014) 276–289.

[41] K. Ayan, U. Kılıç, B. Baraklı, Chaotic artificial bee colony algorithm based solution of security and transient stability constrained optimal power flow, Int. J. Electr. Power Energy Syst. 64 (2015) 136–147.

[42] H. Bouchekara, M. Abido, A. Chaib, R. Mehasni, Optimal power flow using the league championship algorithm: a case study of the Algerian power system, Energy Convers. Manag. 87 (2014) 58–70.
[43] H. Bouchekara, Optimal power flow using black-hole-based optimization approach, Appl. Soft Comput. 24 (2014) 879–888.
[44] M. Ghasemi, S. Ghavidel, M.M. Ghanbarian, H.R. Massrur, M. Gharibzadeh, Application of imperialist competitive algorithm with its modified techniques for multi-objective optimal power flow problem: a comparative study, Inf. Sci. 281 (2014) 225–247.
[45] B.S. Rao, K. Vaisakh, Multi-objective adaptive clonal selection algorithm for solving optimal power flow considering multi-type FACTS devices and load uncertainty, Appl. Soft Comput. 23 (2014) 286–297.
[46] H. Bouchekara, M. Abido, M. Boucherma, Optimal power flow using teaching-learning-based optimization technique, Electr. Power Syst. Res. 114 (2014) 49–59.
[47] H. Chen, M.L. Bo, Y. Zhu, Multi-hive bee foraging algorithm for multi-objective optimal power flow considering the cost, loss, and emission, Int. J. Electr. Power Energy Syst. 60 (2014) 203–220.
[48] B. Mandal, P.K. Roy, Multi-objective optimal power flow using quasi-oppositional teaching learning based optimization, Appl. Soft Comput. 21 (2014) 590–606.
[49] M. Joorabian, E. Afzalan, Optimal power flow under both normal and contingent operation conditions using the hybrid fuzzy particle swarm optimisation and Nelder–Mead algorithm (HFPSO–NM), Appl. Soft Comput. 14 (2014) 623–633.
[50] M.R. Adaryani, A. Karami, Artificial bee colony algorithm for solving multi-objective optimal power flow problem, Int. J. Electr. Power Energy Syst. 53 (2013) 219–230.
[51] A. Shabanpour-Haghighi, A.R. Seifi, T. Niknam, A modified teaching–learning based optimization for multi-objective optimal power flow problem, Energy Convers. Manag. 77 (2014) 597–607.
[52] S. Kumar, D. Chaturvedi, Optimal power flow solution using fuzzy evolutionary and swarm optimization, Int. J. Electr. Power Energy Syst. 47 (2013) 416–423.
[53] M. Ebeed, S. Kamel, H. Youssef, Optimal setting of STATCOM based on voltage stability improvement and power loss minimization using Moth-Flame algorithm, 2016 Eighteenth International Middle East Power Systems Conference (MEPCON), 2016, pp. 815–820.
[54] N. Sinsuphan, U. Leeton, T. Kulworawanichpong, Optimal power flow solution using improved harmony search method, Appl. Soft Comput. 13 (2013) 2364–2374.
[55] A. Khorsandi, S. Hosseinian, A. Ghazanfari, Modified artificial bee colony algorithm based on fuzzy multi-objective technique for optimal power flow problem, Electr. Power Syst. Res. 95 (2013) 206–213.
[56] M.R. Narimani, R. Azizipanah-Abarghooee, B. Zoghdar-Moghadam-Shahrekohne, K. Gholami, A novel approach to multi-objective optimal power flow by a new hybrid optimization algorithm considering generator constraints and multi-fuel type, Energy 49 (2013) 119–136.
[57] T. Niknam, R. Azizipanah-Abarghooee, M.R. Narimani, Reserve constrained dynamic optimal power flow subject to valve-point effects, prohibited zones and multi-fuel constraints, Energy 47 (2012) 451–464.
[58] S. Duman, U. Güvenç, Y. Sönmez, N. Yörükeren, Optimal power flow using gravitational search algorithm, Energy Convers. Manag. 59 (2012) 86–95.
[59] S. Sivasubramani, K. Swarup, Multiagent based differential evolution approach to optimal power flow, Appl. Soft Comput. 12 (2012) 735–740.

[60] T. Niknam, M. Rasoul Narimani, M. Jabbari, A.R. Malekpour, A modified shuffle frog leaping algorithm for multi-objective optimal power flow, Energy 36 (2011) 6420–6432.
[61] A.A. El Ela, M. Abido, S. Spea, Optimal power flow using differential evolution algorithm, Electr. Power Syst. Res. 80 (2010) 878–885.
[62] M.S. Kumari, S. Maheswarapu, Enhanced genetic algorithm based computation technique for multi-objective optimal power flow solution, Int. J. Electr. Power Energy Syst. 32 (2010) 736–742.
[63] S. Sayah, K. Zehar, Modified differential evolution algorithm for optimal power flow with non-smooth cost functions, Energy Convers. Manag. 49 (2008) 3036–3042.
[64] P. Somasundaram, K. Kuppusamy, R.K. Devi, Evolutionary programming based security constrained optimal power flow, Electr. Power Syst. Res. 72 (2004) 137–145.
[65] M. Ghasemi, S. Ghavidel, E. Akbari, A.A. Vahed, Solving non-linear, non-smooth and non-convex optimal power flow problems using chaotic invasive weed optimization algorithms based on chaos, Energy 73 (2014) 340–353.
[66] M. Ghasemi, S. Ghavidel, S. Rahmani, A. Roosta, H. Falah, A novel hybrid algorithm of imperialist competitive algorithm and teaching learning algorithm for optimal power flow problem with non-smooth cost functions, Eng. Appl. Artif. Intell. 29 (2014) 54–69.
[67] C. Roa-Sepulveda, B. Pavez-Lazo, A solution to the optimal power flow using simulated annealing, Int. J. Electr. Power Energy Syst. 25 (2003) 47–57.
[68] M. Abido, Optimal power flow using particle swarm optimization, Int. J. Electr. Power Energy Syst. 24 (2002) 563–571.
[69] M. Varadarajan, K.S. Swarup, Solving multi-objective optimal power flow using differential evolution, IET Gener. Transm. Distrib. 2 (2008) 720–730.
[70] W. Ongsakul, T. Tantimaporn, Optimal power flow by improved evolutionary programming, Electr. Power Compon. Syst. 34 (2006) 79–95.
[71] M. Abido, Optimal power flow using tabu search algorithm, Electr. Power Compon. Syst. 30 (2002) 469–483.
[72] M. Basu, Group search optimization for solution of different optimal power flow problems, Electr. Power Compon. Syst. 44 (2016) 606–615.
[73] H.R. El-Hana Bouchekara, M.A. Abido, A.E. Chaib, Optimal power flow using an improved electromagnetism-like mechanism method, Electr. Power Compon. Syst. 44 (2016) 434–449.
[74] A.A. El-Fergany, H.M. Hasanien, Single and multi-objective optimal power flow using grey wolf optimizer and differential evolution algorithms, Electr. Power Compon. Syst. 43 (2015) 1548–1559.
[75] A.-F. Attia, Y.A. Al-Turki, A.M. Abusorrah, Optimal power flow using adapted genetic algorithm with adjusting population size, Electr. Power Compon. Syst. 40 (2012) 1285–1299.
[76] R. Gnanadass, P. Venkatesh, N.P. Padhy, Evolutionary programming based optimal power flow for units with non-smooth fuel cost functions, Electr. Power Compon. Syst. 33 (2004) 349–361.
[77] I.N. Trivedi, P. Jangir, S.A. Parmar, N. Jangir, Optimal power flow with voltage stability improvement and loss reduction in power system using moth-flame optimizer, Neural Comput. Appl. 1 (2016) 1–16.
[78] H. Pulluri, R. Naresh, V. Sharma, A solution network based on stud krill herd algorithm for optimal power flow problems, Soft. Comput. (2016) 1–18.
[79] M. Abido, N. Al-Ali, Multi-objective optimal power flow using differential evolution, Arab. J. Sci. Eng. (2012) 37.
[80] S. Kumar, D. Chaturvedi, Optimal power flow solution using GA-fuzzy and PSO-fuzzy, J. Inst. Eng. (India): Ser. B 95 (2014) 363–368.

[81] U. Kılıç, Backtracking search algorithm-based optimal power flow with valve point effect and prohibited zones, Electr. Eng. 97 (2015) 101–110.
[82] S.S. Jadon, J.C. Bansal, R. Tiwari, H. Sharma, Artificial bee colony algorithm with global and local neighborhoods, Int. J. Syst. Assur. Eng. Manag. 8 (2014) 1–13.
[83] S.S. Reddy, P. Bijwe, Differential evolution-based efficient multi-objective optimal power flow, Neural Comput. Appl. (2017) 1–14.
[84] K. Vaisakh, L. Srinivas, K. Meah, Genetic evolving ant direction PSODV hybrid algorithm for OPF with non-smooth cost functions, Electr. Eng. Arch. Elektrotech. 95 (2012) 1–15.
[85] A. Gacem, D. Benattous, Hybrid genetic algorithm and particle swarm for optimal power flow with non-smooth fuel cost functions, Int. J. Syst. Assur. Eng. Manag. 8 (2017) 146–153.
[86] V. Raviprabakaran, R.C. Subramanian, Enhanced ant colony optimization to solve the optimal power flow with ecological emission, Int. J. Syst. Assur. Eng. Manag. 5 (2012) 1–8.
[87] S. Duman, Symbiotic organisms search algorithm for optimal power flow problem based on valve-point effect and prohibited zones, Neural Comput. Appl. (2016) 1–15.
[88] A.M. Shaheen, R.A. El-Sehiemy, S.M. Farrag, Solving multi-objective optimal power flow problem via forced initialised differential evolution algorithm, IET Gener. Transm. Distrib. 10 (2016) 1634–1647.
[89] A. Bhattacharya, P. Chattopadhyay, Application of biogeography-based optimisation to solve different optimal power flow problems, IET Gener. Transm. Distrib. 5 (2011) 70–80.
[90] A.G. Bakirtzis, P.N. Biskas, C.E. Zoumas, V. Petridis, Optimal power flow by enhanced genetic algorithm, IEEE Trans. Power Syst. 17 (2002) 229–236.
[91] J. Yuryevich, K.P. Wong, Evolutionary programming based optimal power flow algorithm, IEEE Trans. Power Syst. 14 (1999) 1245–1250.
[92] E. Barocio, J. Regalado, E. Cuevas, F. Uribe, P. Zúñiga, P.J.R. Torres, Modified bio-inspired optimisation algorithm with a centroid decision making approach for solving a multi-objective optimal power flow problem, IET Gener. Transm. Distrib. 11 (2016).
[93] D. Devaraj, B. Yegnanarayana, Genetic-algorithm-based optimal power flow for security enhancement, IEE Proc. Gener. Transm. Distrib. 152 (2005) 899–905.
[94] C.-M. Huang, Y.-C. Huang, Hybrid optimisation method for optimal power flow using flexible AC transmission system devices, IET Gener. Transm. Distrib. 8 (2014) 2036–2045.
[95] D.B. Das, C. Patvardhan, Useful multi-objective hybrid evolutionary approach to optimal power flow, IEE Proc. Gener. Transm. Distrib. 150 (2003) 275–282.
[96] T. Niknam, M. Narimani, J. Aghaei, S. Tabatabaei, M. Nayeripour, Modified honey bee mating optimisation to solve dynamic optimal power flow considering generator constraints, IET Gener. Transm. Distrib. 5 (2011) 989–1002.
[97] L.L. Lai, J. Ma, R. Yokoyama, M. Zhao, Improved genetic algorithms for optimal power flow under both normal and contingent operation states, Int. J. Electr. Power Energy Syst. 19 (1997) 287–292.
[98] M. Osman, M.A. Abo-Sinna, A. Mousa, A solution to the optimal power flow using genetic algorithm, Appl. Math. Comput. 155 (2004) 391–405.
[99] S. Kahourzade, A. Mahmoudi, H.B. Mokhlis, A comparative study of multi-objective optimal power flow based on particle swarm, evolutionary programming, and genetic algorithm, Electr. Eng. 97 (2015) 1–12.
[100] M. Ghasemi, S. Ghavidel, M. Gitizadeh, E. Akbari, An improved teaching–learning-based optimization algorithm using Lévy mutation strategy for non-smooth optimal power flow, Int. J. Electr. Power Energy Syst. 65 (2015) 375–384.
[101] S.S. Reddy, P. Bijwe, A. Abhyankar, Faster evolutionary algorithm based optimal power flow using incremental variables, Int. J. Electr. Power Energy Syst. 54 (2014) 198–210.

[102] T. Niknam, M.R. Narimani, R. Azizipanah-Abarghooee, A new hybrid algorithm for optimal power flow considering prohibited zones and valve point effect, Energy Convers. Manag. 58 (2012) 197–206.
[103] T.T. Nguyen, in: "Neural network optimal-power-flow" Australia, Proceedings of the 4th International Conference on Advances in Power System Control, Operation and Management, APSCOM-97, Hong Kong, November, 1997.
[104] B. Venkatesh, Online ANN memory model–based method for unified OPF and voltage stability margin maximization, Electr. Power Compon. Syst. 31 (2003) 453–465.
[105] V.C. Ramesh, X. Li, Optimal power flow with fuzzy emissions constraints, Electr. Mach. Power Syst. 25 (1997) 897–906.
[106] V.C. Ramesh, X. Li, A fuzzy multiobjective approach to contingency constrained OPF, IEEE Trans. Power Syst. 12 (1997) 1348–1354.
[107] Y. Terasawa, S. Iwamoto, Optimal power flow solution using fuzzy mathematical programming, Electr. Eng. Jpn. 108 (1988) 46–54.
[108] S. Ghosh, B.H. Chowdhury, Security-constrained optimal rescheduling of real power using Hopfield neural network, IEEE Trans. Power Syst. 11 (1996) 1743–1748.
[109] A.A. Abou El-Ela, et al., Optimal preventive control actions using multi-objective fuzzy linear programming technique, Electr. Power Syst. Res. 74 (2005) 147–155.
[110] P. Kessel, H. Glavitsch, Estimating the voltage stability of a power system, IEEE Trans. Power Deliv. 1 (1986) 346–354.

Chapter 8

Optimal Conductor Selection of Radial Distribution Feeders: An Overview and New Application Using Grasshopper Optimization Algorithm

Sherif M. Ismael*, Shady H.E. Abdel Aleem†, Almoataz Y. Abdelaziz‡ and Ahmed F. Zobaa§

**Engineering for the Petroleum and Process Industries (Enppi), Cairo, Egypt, †Higher Institute of Engineering, Cairo, Egypt, ‡Ain Shams University, Cairo, Egypt, §Brunel University London, Uxbridge, Middlesex, United Kingdom*

1 INTRODUCTION

In recent years, researchers have focused on conserving electric energy in line with increasing the generated power to cater for the continuing growth in electrical demands. Power loss reduction is considered one of the most important ways of conserving generated energy. The significant power loss in the overall power system networks are concentrated in the distribution systems because they are usually characterized by their reduced voltage levels, low lines' resistances, and high lines' currents [1–3]. Many methodologies have been explored in the literature for power loss reduction, such as capacitor sitting, distributed generation (DG) allocation, network reconfiguration, and optimal conductor selection. Optimal selection of the conductor sizes of a distribution network is a technoeconomic problem as it considers many constraints, such as the specified voltage limits for the different buses, current carrying capacity of the feeders, loading profile alteration, and load growth, as well as other economic considerations that concern with cost of losses and conductors, and interest and depreciation rates. One can notice that there are almost no published materials that analyzes what the optimal selection of conductors in distribution systems over the past 60 years reveals. To redress this gap, this chapter provides a comprehensive survey of the optimal conductor's selection problem presented in a

Classical and Recent Aspects of Power System Optimization. https://doi.org/10.1016/B978-0-12-812441-3.00008-2
© 2018 Elsevier Inc. All rights reserved.

tabular form as a guide for researchers interested in this problem as presented in Table 1. The optimal conductor selection problem can be categorized into two main approaches, the first approach is based on conventional analytical algorithms and the second relies on intelligent meta-heuristic algorithms. The literature survey is performed based on the database found in the IEEE Xplore digital library, IET digital library, ScienceDirect, and the research available in Google Scholar. About 34 publications were chosen from these databases. Only the articles written in the English language were considered. The research performed in this chapter has considered a time span of 60 years, from 1955 through 2017. This literature is organized in chronological order according to the year of publication, and a systematic overview is performed.

Further, this chapter also presents a novel approach based on a recent meta-heuristic algorithm, the grasshopper optimization (GO) algorithm, [38] to solve the optimal conductor selection problem of radial distribution networks. From the literature survey, it is clear that application of GO algorithm has not been discussed so far to solve the optimal conductor selection problem in radial distribution systems. This encourages utilizing this new algorithm for the problem. A practical conductors' library containing 29 conductor types are introduced based on data of conductors that comply with the BS-50182 [39,40]. This rich library explores a wider search space and ensures finding the most optimal set of conductors that satisfy the economic objective functions while complying with the constraints. A constrained objective function is employed to minimize overall cost and comply with the system voltage limits and the conductors' current carrying capacities. Practical aspects are considered, such as available market conductors' utilization, as well as load growth considerations. The proposed GO is applied to two different systems; the first one is a small-scale 26-bus system and the second is a large-scale 85-bus system. The obtained results are compared with other results available in the literature and showed the effectiveness of the proposed algorithm in reducing the network losses, maximizing the overall saving, while maintaining the specified constraints with high annual load growth scenario.

Organization of the remaining chapter is as follows: Section 2 presents the problem formulation and constraints for the optimal conductor selection problem. Section 3 presents an overview of the GO algorithm. Section 4 presents the simulation results and discussion of results. Finally, the conclusion is given in Section 5.

2 PROBLEM FORMULATION

2.1 Objective Function

In an optimal conductor selection problem, the goal is to choose the conductor size and type from a set of available inventories such that the total cost is minimized while satisfying nonlinear constraints on voltage limits of the buses and

TABLE 1 Comprehensive Survey of the Optimal Conductor Selection Problem

Refs.	Year	Category / Inspiration	Algorithm	No. of Conductors	Test System	Objective Function	Load Growth Considered?	Remarks
[4]	1955	Analytical	–	7	Nil	Cost ($)	✓	Uniformly distributed loads were assumed for all the feeders, which is not the actual case in practical distribution networks
[5]	1982	Analytical	Dynamic programming	3	14-bus	Conductor area (Rs)	✓	Nonuniform distribution of loads was considered
[6]	1985	Analytical	Computational programming	3	4-bus 14-bus	Cost (Rs)	✓	A simple and direct solution procedure was proposed for solving the conductors' selection problem

Continued

TABLE 1 Comprehensive Survey of the Optimal Conductor Selection Problem—cont'd

Refs.	Year	Category / Inspiration	Algorithm	No. of Conductors	Test System	Objective Function	Load Growth Considered?	Remarks
[7]	1988	Analytical	Computational programming	40	250-bus	Cost ($)	×	Regulators were used as a solution for bus voltage enhancement No details were provided for the considered conductor set
[8]	1995	Analytical	–	4	Nil	Cost ($)	✓	Economic research was presente
[9]	1999	Analytical	Computational programming	6	52-bus	Cost ($)	×	A computer-based procedure is proposed to solve the conductors' selection problem from economic and technical aspect

[10]	2000	Analytical	Integer programming	8	33-bus	Cost (¥)	✓	A practical procedure is introduced for optimal conductor selection considering the nonuniform loading
[11]	2002	Analytical	Computational programming	3	26-bus 32-bus	Cost ($)	×	An analytical algorithm is proposed based on radial network load flow to select the optimal conductor size
[12]	2002	Analytical	Computational programming	5	18-bus	Cost ($)	✓	It was concluded that enlarging the conductor selection set provides better economic results
[13]	2003	Evolutionary Algorithms (Nature)	Evolutionary Programming	4	16-bus	Cost (Rs)	×	Voltage deviation index was proposed to quantify the voltage limit violation

Continued

TABLE 1 Comprehensive Survey of the Optimal Conductor Selection Problem—cont'd

Refs.	Year	Category / Inspiration	Algorithm	No. of Conductors	Test System	Objective Function	Load Growth Considered?	Remarks
[14]	2005	Analytical	Computational programming	5	27-bus	Cost (Rs)	✓	Economic research was presented considering load growth profile
[15]	2006	Analytical	Computational programming	4	26-bus	Cost ($)	✓	Different load modeling was considered such as constant power, constant current, constant impedance, exponential or composite load
[16]	2006	Evolutionary Algorithms (Nature)	Evolutionary strategy (ES)	6	8-bus	Cost	×	ES were applied to solve the optimal conductor selection problem

[17]	2008	Analytical	Partial enumeration	9	20-bus 123-bus	Cost (Rs)	✓	Practical aspects were considered, such as non-uniform loading, load growth, load factor and diversity in load peaks
[18]	2008	Evolutionary Algorithms (Nature)	Fuzzy logic	4	26-bus 32-bus	Cost (Rs)	✓	Fuzzy Evolutionary Programming (FEP) was applied to solve the optimal conductor selection problem
[19]	2009	Swarm Algorithms (Nature)	Discrete particle swarm optimization (DPSO)	5	26-bus 32-bus	Cost (Rs)	×	DPSO was applied to solve the optimal conductor selection problem and the proposed algorithm was a practical indicant system was tested

Continued

TABLE 1 Comprehensive Survey of the Optimal Conductor Selection Problem—cont'd

Refs.	Year	Category / Inspiration	Algorithm	No. of Conductors	Test System	Objective Function	Load Growth Considered?	Remarks
[20]	2009	Evolutionary Algorithms (Nature)	Genetic Algorithms (GA)	4	28-bus	Cost (Rs)	×	It was concluded that the GA can obtain higher total cost savings and higher loss reduction compared to the conventional method
[21]	2009	Evolutionary Algorithms (nature)	Genetic Algorithms (GA)	5	27-bus	Cost ($)	×	Combination of optimal capacitor placement and optimal conductor sizes selection were examined
[22]	2010	Evolutionary Algorithms (Nature)	Computational programming	5	22-bus	Cost (Rs)	×	An analytical method for conductor selection based on the cost of energy losses were proposed

[23]	2010	Evolutionary Algorithms (Nature)	Genetic Algorithms (GA)	4	30-bus	Cost ($)	×	The problem of optimal conductor selection was examined in presence of wind power penetration
[24]	2011	Evolutionary Algorithms (Nature)	Genetic Algorithms (GA)	4	13-bus	Cost (Rs)	×	It is concluded that GA yields better results and less computational time when compared to conventional method
[25]	2011	Physical Algorithms (Human)	Harmony search with differential operator (HSDE)	8	16-bus 85-bus	Cost (Rs)	×	HSDE was applied to solve the optimal conductor selection problem
[26]	2012	Swarm Algorithms (Nature)	Selective particle swarm optimization (SPSO)	5	27-bus	Cost ($)	×	Combination of optimal capacitor placement and optimal conductor sizes selection were examined.

Continued

TABLE 1 Comprehensive Survey of the Optimal Conductor Selection Problem—cont'd

Refs.	Year	Category / Inspiration	Algorithm	No. of Conductors	Test System	Objective Function	Load Growth Considered?	Remarks
[27]	2013	Analytical	Mixed integer linear programming (MILP)	4	50-bus 200-bus 600-bus	Cost ($)	×	MILP was applied to solve the optimal conductor selection problem
[28]	2014	Swarm Algorithms (Nature)	Particle swarm optimization (PSO)	3	60-bus	Cost ($)	×	Combination of optimal capacitor placement and optimal conductor sizes selection were examined It is concluded that PSO yields better results when compared to conventional method
[29]	2014	Evolutionary Algorithms (Nature)	Modified Differential Evolution (MDE)	4	32-bus	Cost (Rs)	✓	MDE was applied to solve the optimal conductor selection problem

[30]	2014	Physical Algorithms (Human)	Imperialism competitive algorithm (ICA)	4	69-bus	Cost ($)	×	It is concluded that ICA yields better results when compared to the conventional method and PSO
[31]	2015	Swarm Algorithms (Nature)	Bacterial foraging optimization algorithm (BFA)	4	69-bus	Cost ($)	×	It is concluded that BFA yields better results when compared to conventional method and ICA algorithm
[32]	2015	Swarm Algorithms (Nature)	Ant colony optimization (ACO)	4	10-bus 118-bus	Cost ($)	×	Multiobjective problem was presented to search for optimum substation location, substation sizing, feeder type and route

Continued

TABLE 1 Comprehensive Survey of the Optimal Conductor Selection Problem—cont'd

Refs.	Year	Category / Inspiration	Algorithm	No. of Conductors	Test System	Objective Function	Load Growth Considered?	Remarks
[33]	2016	Analytical	Branch-wise minimization (BWM)	13	69-bus 85-bus	Cost ($)	×	BWM was applied to solve the optimal conductor selection problem
[34]	2016	Swarm Algorithms (Nature)	Discrete particle swarm optimization (DPSO)	5	26-bus 32-bus	Cost (Rs)	×	DPSO was applied to solve the optimal conductor selection problem
[35]	2016	Evolutionary Algorithms (Nature)	Differential Evolution (DE)	9	19-bus 34-bus	Real power loss	×	Combination of optimal capacitor placement and optimal conductor sizes selection were examined for an unbalanced radial system

[36]	2017	Swarm Algorithms (Nature)	Crow search algorithm (CSA)	8	16-bus 85-bus	Cost (Rs)	×	CSA was applied to solve the optimal conductor selection problem
[37]	2017	Population-based algorithm	Sine-Cosine algorithm (SCA)	20	45-bus	Cost ($)	✓	SCA was applied to solve the optimal conductor selection problem of a real Egyptian distribution system

maximum ampacities of the conductors. The cost consists of the annual cost of energy losses (*CL*) and the capital investment cost of the conductor (*CC*). It can be formulated for a branch j with type c conductor as follows:

$$Cost\,(j,c) = CL(j,c) + CC\,(j,c) \tag{1}$$

The annual cost of the loss in branch j with c type conductor (*CL*) is given as:

$$CL(j,c) = P_{loss}\,(j,c) \times \left[k_p + (k_e \times LSF \times T)\right] \tag{2}$$

where $P_{loss}\,(j,c)$ is the active power loss of the jth branch with cth conductor under peak load conditions, k_p is the cost of peak demand power loss ($/kW), k_e is the cost of energy losses (($//kWh), T is the number of hours per year (8760 h), and *LSF* is the loss factor which is defined as the ratio of energy loss in the system during a given period to the energy loss that could result if the system peak loss had persisted throughout that period. *LSF* is expressed in terms of the load factor (*LF*), the average load to the peak demand in a given period, as follows [15,25]:

$$LSF = 0.2\,(LF) + 0.8(LF)^2 \tag{3}$$

Thus, the total annual cost of the energy loss (CL^{total}) can be expressed as follows:

$$CL^{total} = \sum_{j}^{b}\sum_{c}^{n} P_{loss}\,(j,c) \times \left[k_p + (k_e \times LSF \times T)\right] \tag{4}$$

where b is the total number of branches in the system and n is the number of the available conductors' types.

The capital investment cost of the conductor (*CC*) is defined in the terms of the annual depreciation of the capital cost of the jth branch with the cth conductor, thus:

$$CC\,(j,c) = IDF \times l\,(j) \times IC\,(c) \tag{5}$$

so that,

$$IDF = \frac{i(i+1)^Y}{(i+1)^Y - 1} \tag{6}$$

where *IDF* is the interest and depreciation factor, i is the interest rate value, Y is the planned lifetime (years), $l(j)$ is the length of branch j (km), $A(c)$ is the cross-section area of the cth conductor (mm^2), and $IC(c)$ is the investment cost of the c type conductor per unit area per unit length ($/km).

Thus, the objective function (OF) for the optimal selection of types and sizes of the conductors can be expressed as:

$$OF = \min\left(CL^{total} + CC^{total}\right) \tag{7}$$

where CC^{total} is the total capital investment cost of all branches.

2.2 Constraints

The considered constraints can be expressed, as follows:

- *Bus voltage constraint*

$$U_{min}(m) \leq |U(m,c)| \leq U_{max}(m) \quad \forall(m \in k, c \in n) \tag{8}$$

where m is the bus number, k is the total number of buses in the system. $U_{min}(m)$ and $U_{max}(m)$ are the minimum and maximum voltage at bus m, respectively. They are considered as 0.9 p.u. and 1.1 p.u., respectively.

- *Conductor current carrying capacity constraint*

$$|I(j,c)| \leq I_{max}(c) \quad \forall(j \in b, c \in n) \tag{9}$$

where $I(j,c)$ is the current flowing in conductor j of type c, and $I_{max}(c)$ is the maximum current carrying capacity of the cth type conductor.

2.3 Load Growth

Practically, optimal conductor sizes selection should consider the existing loads in addition to possible load growth for a certain planning period [15]. This period is usually determined by the feeder ability to accept load growth without violating the node voltage or the branch's current constraints. During the "predetermined" planning period, the load growth can be considered as an annual growth rate in proportion to the connected loads in the base year (first year). Mathematically, the active and reactive power loads at the Nth year is given by:

$$P_L(N) = \begin{cases} P_L(1) \times (1+g)^N & \text{for } N = 1,2,3,\ldots M \\ P_L(1) \times (1+g)^M & \text{for } N = M+1,\ldots F \end{cases} \tag{10}$$

$$Q_L(N) = \begin{cases} Q_L(1) \times (1+g)^N & \text{for } N = 1,2,3,\ldots M \\ Q_L(1) \times (1+g)^M & \text{for } N = M+1,\ldots F \end{cases} \tag{11}$$

where $P_L(1)$ and $Q_L(1)$ are the active and reactive load power in the first year, respectively. Similarly, $P_L(N)$ and $Q_L(N)$ are the active and reactive load power in the Nth year, respectively. g is the annual load growth rate, M is the planning period up to the condition that the feeder is not capable of accepting load growth (breaking year), and F is the planned lifetime of the feeder.

Eqs. (8) and (9) of the proposed constraints shall be checked initially in the base year, and then the load growth shall be calculated according to Eqs. (10) and (11).

TABLE 2 Numerical Values of the Used Parameters

Parameter	Value
k_p (\$/kW)	1.04
k_e (\$/ kWh)	0.029
LF	0.4
LSF	0.2
i (%)	8
Y (years)	25
IDF	0.1
g (%)	10

Further, the loading pattern shall increase gradually until violation of any of the constraints occurs. In this condition, the breaking year M is reached and the maximum duration that the feeder can handle the load growth without violating the preset constraints is determined. Beyond the breaking year, the optimally selected conductors can accept additional load growth until the end of their lifetime, but with the support of external compensators such as the shunt capacitors, DGs, and energy storage schemes. Table 2 gives the numerical values of the parameters used in the formulation of the objective function and the constraints. The available conductor types and their specifications [40] are presented in Table 3. The conductors' library used in [11,25] is presented in Table 4 to be compared with the results obtained using the proposed GO algorithm.

3 OVERVIEW OF GRASSHOPPER OPTIMIZATION ALGORITHM

Grasshopper optimization (GO) algorithm is a nature-inspired meta-heuristic optimization algorithm that was developed based on mimicking the hunting behavior of grasshoppers [38]. Grasshoppers are considered one of the most harmful pests due to their damage to crop production and agriculture. The life cycle of the grasshoppers starts with an egg phase, followed by a nymph phase, and further ends with an adult phase. The social interaction between grasshoppers is categorized into three zones, repulsion zone, comfort zone, and attraction zone. Fig. 1 shows the life cycle and the social interactions between grasshoppers.

$$X_i = (r_1 S_i) + (r_2 G_i) + (r_3 A_i) \tag{12}$$

TABLE 3 Electrical Specifications of the Used ACSR Conductors

Conductor Type	$A(mm^2)$	$R(\Omega/km)$	$X(\Omega/km)$	$I_{max}(A)$	$IC(\$/km)$
1	6.5	2.718	0.374	70	90
2	13	1.374	0.355	120	170
3	16	1.098	0.349	130	210
4	20	0.9116	0.345	150	260
5	25	0.6795	0.339	175	340
6	30	0.5449	0.335	200	420
7	40	0.4565	0.353	250	500
8	42	0.3977	0.327	270	540
9	45	0.3841	0.327	257	590
10	48	0.3656	0.329	260	630
11	50	0.3434	0.328	270	770
12	55	0.302	0.327	290	760
13	65	0.2745	0.315	305	820
14	80	0.2193	0.282	395	1010
15	80	0.2214	0.268	380	1040
16	80	0.2221	0.271	385	1130
17	95	0.1844	0.266	425	1370
18	110	0.1589	0.261	470	1590
19	130	0.1375	0.256	510	1840
20	140	0.1223	0.252	560	2060
21	160	0.1102	0.25	590	2290
22	185	0.08989	0.224	665	2810
23	225	0.07771	0.24	745	3240
24	250	0.07213	0.242	775	3600
25	260	0.06786	0.237	800	3713
26	260	0.06915	0.237	812	3270
27	300	0.0611	0.235	850	4130
28	300	0.06125	0.237	825	3620
29	325	0.05517	0.235	880	4030

TABLE 4 Electrical Specifications of the used ACSR Conductors in Refs. [11,25].

Conductor Type	A(mm^2)	R(Ω/km)	X(Ω/km)	I_{max}(A)
A (Squirrel)	12.90	1.3760	0.3896	115
B (Gopher)	15.91	1.098	0.3100	138
C (Weasel)	19.55	0.9108	0.3797	150
D (Ferret)	25.87	0.6795	0.2980	180
E (Rabbit)	32.26	0.5441	0.3673	208
F (Mink)	37.32	0.4565	0.2850	226
G (Beaver)	42.07	0.3841	0.2795	250
H (Raccoon)	48.39	0.3657	0.3579	270

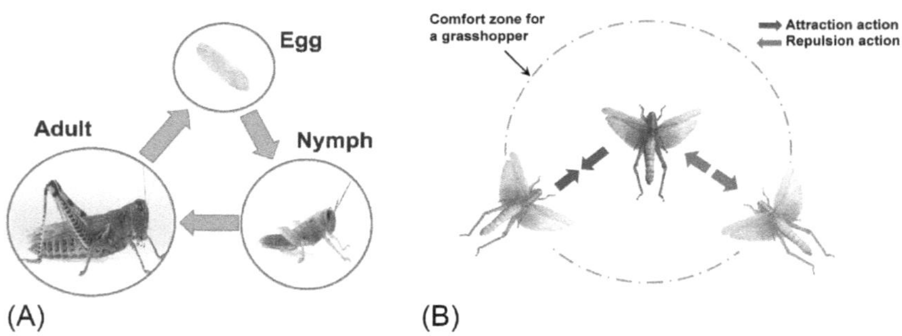

FIG. 1 (A) The life cycle of a grasshopper, (B) the social interaction between various grasshoppers.

GO depends on three factors, namely social interaction between grasshoppers, gravity force, and downwind advection. The mathematical behavior of GO algorithm as a swarming-based approach [38] is given later.

where X_i defines the position of the ith grasshopper, S_i is the social interaction, G_i is the gravity force on the ith grasshopper, A_i represents wind advection, and r_1, r_2, r_3 are random numbers in [0,1]. S_i depends on the social forces between two grasshoppers where a repulsion force prevents collisions over a short length scale and an attraction force exists for aggregation purpose (to form a swarm). It can be defined as:

$$S_i = \sum_{\substack{j=1 \\ j \neq i}} s\left(d_{ij}\right) \vec{d}_{ij} \tag{13}$$

where d_{ij} is the distance between the ith and the jth grasshopper, calculated as $d_{ij} = |X_j - X_i|$, s is a function to define the strength of social forces, and $\overrightarrow{d}_{ij}$ is a unit vector from the ith to the jth grasshopper.

The strength of social forces s, which defines the social forces (attraction and repulsion) of grasshoppers, is calculated as follows:

$$s(d_{ij}) = f\, e^{\frac{-d_{ij}}{l_{att}}} - e^{-d_{ij}} \tag{14}$$

where f indicates the intensity of attraction and l_{att} is the attractive length scale.

The gravitational force on the ith grasshopper, G_i is defined as:

$$G_i = -v \times \overrightarrow{e_g} \tag{15}$$

$\overrightarrow{e_g}$ where v is the gravitational constant and is a unity-vector toward the center of earth.

In addition, the wind advection effect on the motion swarm (A_i) in (12) can be is defined as:

$$A_i = u \times \overrightarrow{e_w} \tag{16}$$

where u is a constant that represents the wind drift and $\overrightarrow{e_w}$ is a unity-vector in the direction of wind.

Substituting S_i, G_i, and A_i, in (12), the grasshopper behavior can be represented as follows:

$$X_i = \sum_{\substack{j=1 \\ j \neq i}}^{N_g} s\left(|x_j - x_i|\right) \frac{x_j - x_i}{d_{ij}} - \left(v \times \overrightarrow{e_g}\right) + \left(u \times \overrightarrow{e_w}\right) \tag{17}$$

where N_g is the number of grasshoppers, which is equivalent to the number of search agents.

The previous equation is modified to be implemented for optimization problems and for enhancing the capability global searching of the algorithm. The modified equation is presented as follows:

$$X_i^d = c\left[\sum_{\substack{j=1 \\ j \neq i}}^{N_g} c\,\frac{ub_d - lb_d}{2}\, s\left(\left|x_j^d - x_i^d\right|\right) \frac{x_j - x_i}{d_{ij}}\right] + \overrightarrow{T_d} \tag{18}$$

where ub_d is the upper bound in the Dth dimension, lb_d is the lower bound in the Dth dimension, is the best solution found so far, and c is an adaptive coefficient that decreases linearly for shrinking the comfort zone, repulsion zone and attraction zone. In addition, the adaptive coefficient factor c is utilized to enhance the search capability of GO as follows:

$$c = c_{max} - z\frac{c_{max} - c_{min}}{Z} \tag{19}$$

GO includes four parameters (c_{max}, c_{min}, f, and l_{att}) that must be adjusted. In this work, the number of search agents is set to 30, c_{max} and c_{min} are set to 1 and 0.00004, respectively, f is set to 0.5, l_{att} is set to 1.5, and the maximum number of iterations is set to 500.

The GO is proposed in this work due to the intelligent behavior of grasshoppers in the food hunting process [38]. The GO has multiple advantages, such as having few parameters to be set, faster convergence capability, and higher sensitivity compared to the widely known meta-heuristic algorithms. Due to these advantages, it has been recently employed to solve many engineering problems [41–43]. For more details about GO, readers can refer to Ref. [38].

4 SIMULATION RESULTS AND DISCUSSIONS

In this chapter, 29 conductor types are used. The numerical values of the used parameters are summarized in Table 3. Load flow analysis based on a forward-sweep algorithm described in [36,44] is performed, and the corresponding fitness function is calculated using the GO. In order to verify the effectiveness of the proposed GO algorithm two radial distribution networks are examined in MATLAB platform, the first one is a small-scale 26-bus system and the second is a large-scale 85-bus system.

4.1 26-Bus System

The 26-bus system configuration shown in Fig. 2 has a base voltage and apparent power of 11 kV and 10 MVA, respectively. The substation bus voltage (bus 1) is 1 p.u. and the line and load data for this system were obtained from Ref. [34]. Practical conductors' library is introduced based on actual manufacturer data [40]. The available conductor types and their electrical specifications

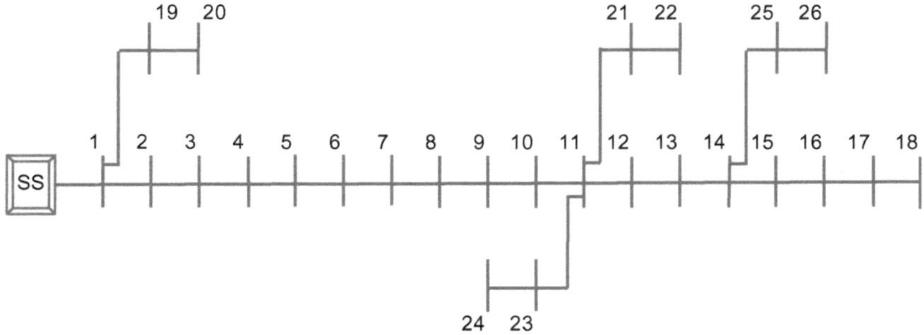

FIG. 2 Configuration of the 26-bus radial distribution network.

TABLE 5 Comparison of Conductor Selection Results for the 26-Bus System

Branch No.	From Bus	To Bus	Original System	GA [11]	DPSO [34]	Proposed GO
1	1	2	C	F	F	29
2	2	3	C	F	F	28
3	3	4	C	F	F	26
4	4	5	C	F	F	26
5	5	6	C	F	F	28
6	6	7	C	F	F	26
7	7	8	C	F	F	23
8	8	9	C	F	F	26
9	9	10	C	F	F	22
10	10	11	C	F	F	26
11	11	12	C	F	F	26
12	12	13	C	F	F	14
13	13	14	C	F	F	21
14	14	15	C	F	F	18
15	15	16	C	F	F	15
16	16	17	C	F	F	8
17	17	18	C	F	D	3
18	2	19	C	F	D	2
19	19	20	C	F	A	14
20	11	21	C	F	F	15
21	21	22	C	F	A	12
22	11	23	C	F	F	1
23	21	24	C	F	A	3
24	14	25	C	F	F	12
25	25	26	C	F	A	7

are presented in Table 2. The obtained GO results are compared to those obtained using GA [11] and discrete particle swarm optimization (DPSO) [34] and the comparison is presented in Tables 5 and 6. It is noticed that the network active power losses optimized by the proposed GO are reduced to 11.2197 kW, which is less than other published results; this means the power

TABLE 6 Comparison of Final Results for the 26-Bus System

Variables	Original System	After Optimal Conductor Selection		
		GA [11]	*DPSO [34]*	*Proposed GO*
U_{min} at node 18	0.931	0.9586	0.9589	0.98542
Power loss (kW)	154.6	76.44	73.8526	11.2197
Loss reduction (%)	–	50.55	52.223	92.7
Optimal cost (Rs)*	504,082	253,419	246,510	157,092
Net saving (%)	–	49.73	51.1	68.84

**USD ($) was exchanged to Indian Rupee (Rs.) to allow comparison with results of Refs. [11,34], based on an exchange rate of 1$ equals 63.60 Rs.*

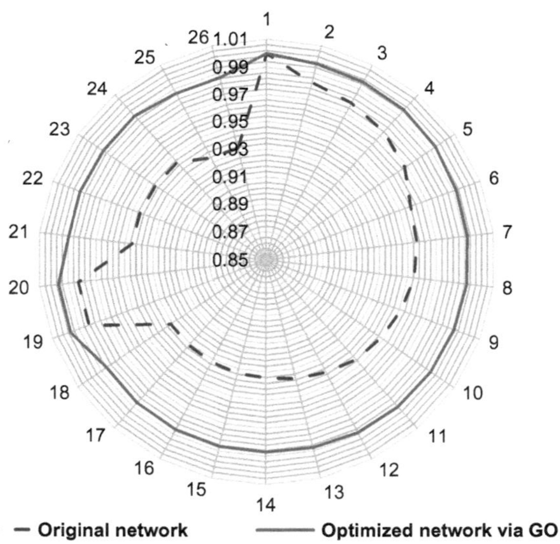

FIG. 3 Bus voltages of the original 26-bus network and the optimized network by GO.

loss is reduced by 92.7% compared to the original network loss. The optimal annual cost obtained by the proposed algorithm is 157,092 $/year, which will save about 68.84% of the original network cost. The bus voltage values of the original network and that optimized using the proposed algorithm are presented in Fig. 3. Fig. 4 shows the current flow in branches compared to their maximum limits.

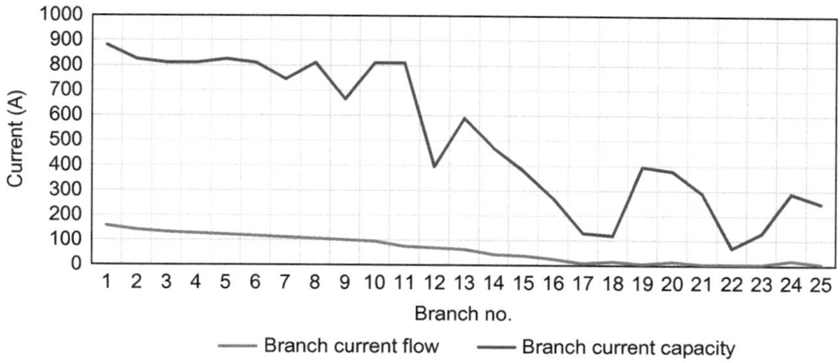

FIG. 4 Branch currents compared to their maximum limits for the 26-bus system.

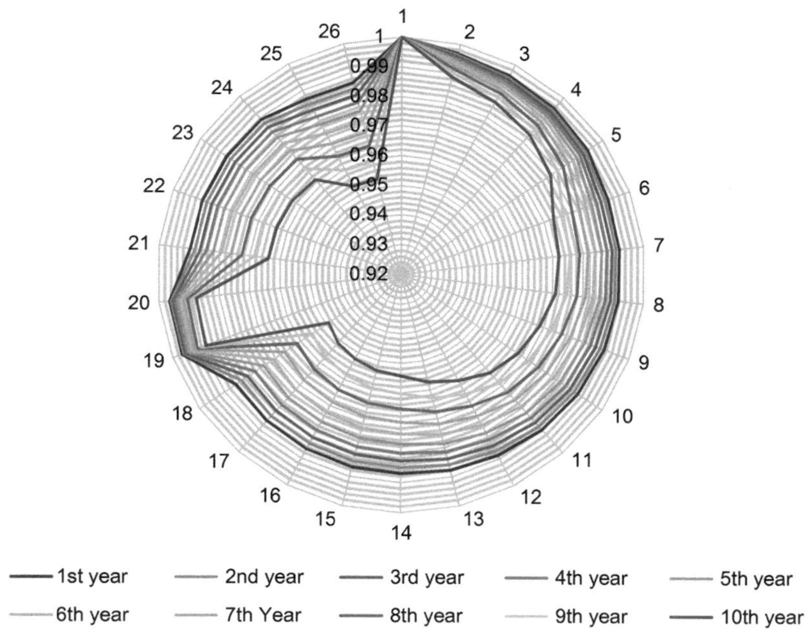

FIG. 5 Voltage profile over a 10-year span considering for the 26-bus system.

The load growth results are presented in Figs. 5 and 6 showing the voltage profile and branch current flows of the 26-bus system over a 10-year span. From the presented figures, one can notice that the proposed optimal conductor selection algorithm succeeded in minimizing the overall system costs and keeping the bus voltages and branch currents within the prescribed limits not only for the current load profile but also for a 10-year span with high annual load growth factor of 15%.

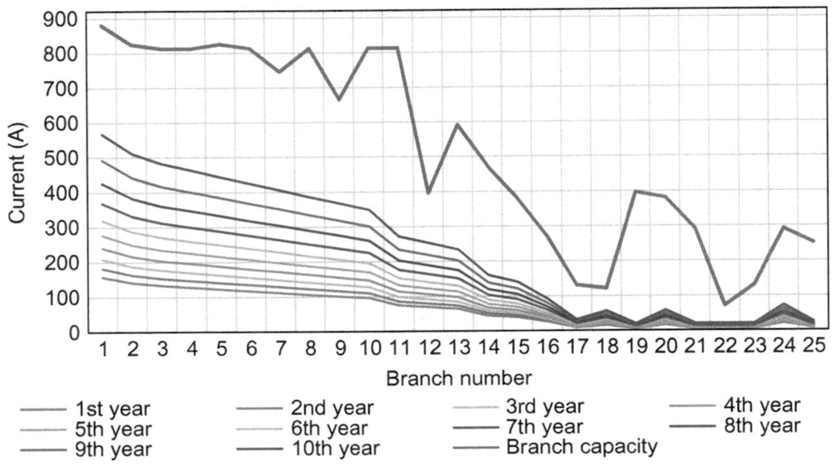

FIG. 6 Branch currents over a 10-year span for the 26-bus system

As presented in Fig. 5, after the seventh year, the voltage of bus 18 begins to reach its minimum allowable limit (0.95 p.u.). Therefore, other compensation techniques should be adopted to allow for more load growth without violating the system limits.

4.2 85-Bus System

The 85-bus system configuration is shown in Fig. 7, the base voltage and apparent power of this system are 11 kV and 100 MVA, respectively. The substation bus voltage (Bus 1) is 1 p.u. and the line and load data for this system are initiated in Ref. [44] and developed in Ref. [25]. The obtained GO results are compared to those obtained using BWM [33], HSDE in Ref. [25], and CSA in Ref. [36], and the comparison is presented in Tables 7 and 8. It is noticeable that the network active power losses optimized by the proposed algorithm are reduced to 52.88 kW, which is less than other published results; this means the power loss is reduced by 83.25% compared to the original network loss. The optimal annual cost obtained by the proposed algorithm is 6054.04 $/year, which will save about 64.64% of the original network cost. The bus voltage values of the original network and that optimized using the proposed GO algorithm are presented in Fig. 8. Fig. 9 shows the current flow in branches compared to their maximum limits.

The response of the GO algorithm is presented in Fig. 10. The load growth results are presented in Figs. 11 and 12 showing the voltage profile and branch current flows of the 85-bus system over a 7-year. From the presented figures, one can notice that the proposed optimal conductor selection algorithm succeeded in minimizing the overall system costs and keeping the bus voltages

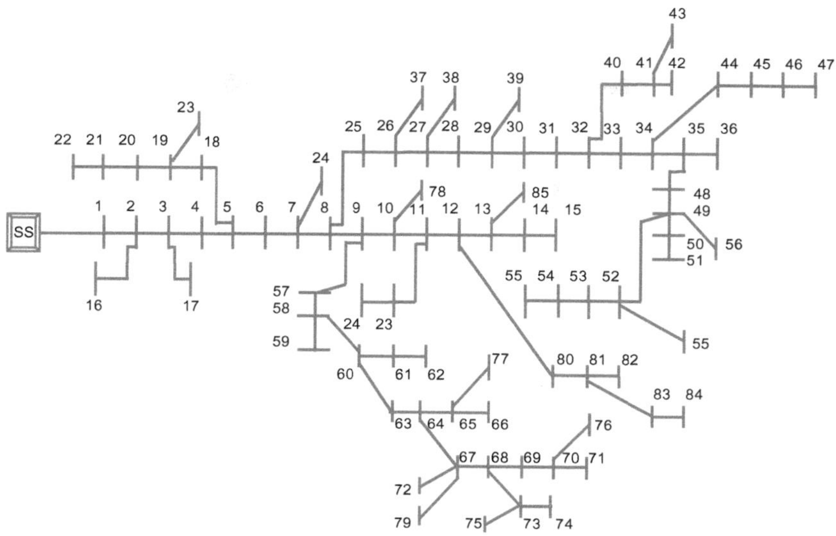

FIG. 7 Configuration of 85-bus radial distribution network.

and branch currents within the prescribed limits not only for the current load profile but also for a 7-year span with high annual load growth factor of 10%. As presented in Fig. 11, after the seventh year, the voltage of Bus 54 begins to reach its minimum allowable limit (0.9 p.u.) Therefore, other compensation techniques should be adopted to allow higher load growth without violating the system limits.

5 CONCLUSIONS

Optimal distribution system planning and proper selection of conductors can reduce electrical power losses while complying with the technical operational limits and economic aspects. In this chapter, a comprehensive survey of the optimal conductor selection problem is presented. The GO algorithm is proposed to solve the problem of optimal selection of conductors of radial distribution networks. The proposed algorithm is then applied to two test systems. The achieved results are compared with other methods available in the literature and show the effectiveness of the proposed algorithm over the other published methods. A recent conductors' library is used. One can notice that the proposed set of conductors explored a wider search space and succeeded in finding the most optimal set of conductors that satisfy the objective functions while complying with the preset constraints with high annual load growth scenarios.

TABLE 7 Comparison of Conductor Selection Results for the 85-Bus System

Branch No.	From Bus	To Bus	BWM [33]	HSDE [25]	CSA [36]	Proposed GO
1	1	2	F	H	F	28
2	2	3	F	H	D	28
3	3	4	F	H	A	28
4	4	5	F	H	F	29
5	5	6	F	H	B	26
6	6	7	E	H	A	29
7	7	8	E	H	A	26
8	8	9	E	H	D	20
9	9	10	E	G	D	13
10	10	11	B	H	G	9
11	11	12	B	E	B	6
12	12	13	B	C	H	6
13	13	14	B	A	H	1
14	14	15	B	A	F	1
15	2	16	B	H	C	1
16	3	17	B	G	B	2
17	5	18	B	E	E	6
18	18	19	B	F	H	5
19	19	20	B	A	H	3
20	20	21	B	B	B	1
21	21	22	B	A	G	1
22	19	23	B	C	H	2
23	7	24	B	C	A	1
24	8	25	B	G	G	21
25	25	26	B	H	C	14
26	26	27	B	G	D	14
27	27	28	B	H	C	15
28	28	29	B	H	A	15
29	29	30	B	G	G	12

TABLE 7 Comparison of Conductor Selection Results for the 85-Bus System—cont'd

Branch No.	From Bus	To Bus	BWM [33]	HSDE [25]	CSA [36]	Proposed GO
30	30	31	B	H	A	17
31	31	32	B	H	C	15
32	32	33	B	E	B	10
33	33	34	B	H	A	13
34	34	35	B	F	E	2
35	35	36	B	H	E	1
36	26	37	B	A	A	1
37	27	38	B	C	A	2
38	29	39	B	G	E	2
39	32	40	B	B	A	2
40	40	41	B	B	B	1
41	41	42	B	F	B	2
42	41	43	B	B	D	2
43	34	44	B	H	C	3
44	44	45	B	D	B	1
45	45	46	B	G	A	1
46	46	47	B	D	C	1
47	35	48	B	E	G	6
48	48	49	B	A	E	3
49	49	50	B	D	C	2
50	50	51	B	D	C	1
51	48	52	B	D	H	7
52	52	53	B	B	C	1
53	53	54	B	A	A	1
54	52	55	B	G	A	1
55	49	56	B	D	C	7
56	9	57	B	H	B	15
57	57	58	B	H	F	14

Continued

TABLE 7 Comparison of Conductor Selection Results for the 85-Bus System—cont'd

Branch No.	From Bus	To Bus	BWM [33]	HSDE [25]	CSA [36]	Proposed GO
58	58	59	B	A	B	1
59	58	60	B	E	B	12
60	60	61	B	D	A	1
61	61	62	B	E	A	1
62	60	63	B	H	D	12
63	63	64	B	G	H	13
64	64	65	B	H	G	1
65	65	66	B	A	H	3
66	64	67	B	G	A	6
67	67	68	B	D	D	5
68	68	69	B	E	C	2
69	69	70	B	E	D	2
70	70	71	B	A	F	1
71	67	72	B	H	B	1
72	68	73	B	B	E	3
73	73	74	B	H	H	1
74	73	75	B	A	E	2
75	70	76	B	C	F	4
76	65	77	B	A	C	1
77	10	78	B	C	H	1
78	67	79	B	A	G	1
79	12	80	B	D	E	4
80	80	81	B	B	E	2
81	81	82	B	G	G	1
82	81	83	B	C	H	1
83	83	84	B	F	A	1
84	13	85	B	A	C	1

TABLE 8 Comparison of Final Results for the 85-Bus System

Variables	Original System	After Optimal Conductor Selection			
		BWM [33]	*HSDE [25]*	*CSA [36]*	*Proposed GO*
U_{min} at node 54	0.8713	0.9458	0.9296	0.8900	0.9514
Power loss (kW)	315.84	134.26	256.96	254.152	52.88
Loss reduction (%)	–	57.51	18.64	19.53	83.25
Optimal cost (Rs)*	1,106,732	485,040	929,035	927,786.37	391,030.44
Net saving (%)	–	56.17	16.06	16.17	64.64

**USD ($) was exchanged to Indian Rupee (Rs.) to allow comparison with results of Refs. [25,44], based on exchange rate of 1$ equals 63.60 Rs.*

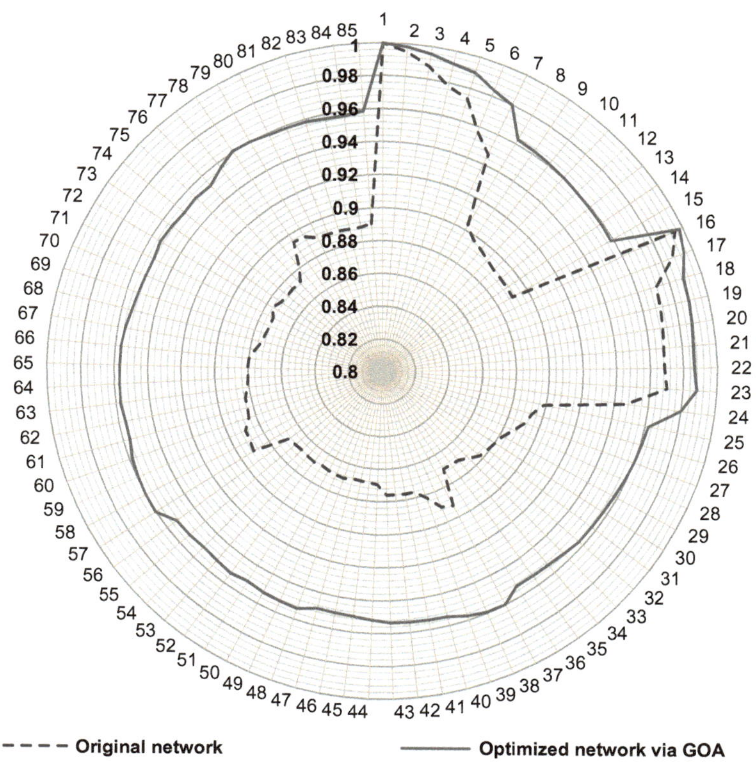

FIG. 8 Bus voltages of the original 85-bus network and the optimized network by GO.

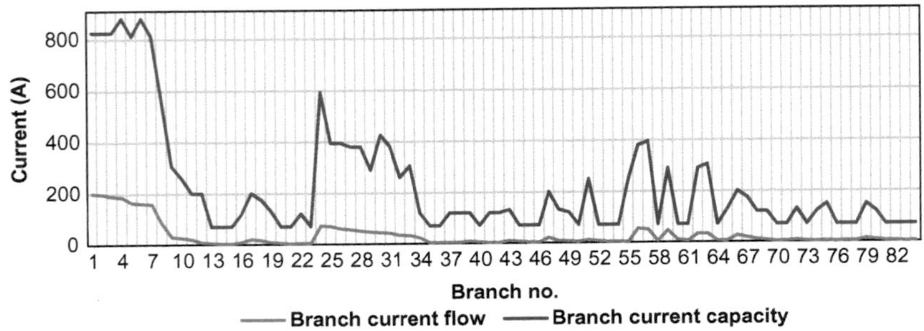

FIG. 9 Branch currents compared to their maximum limits for the 85-bus system.

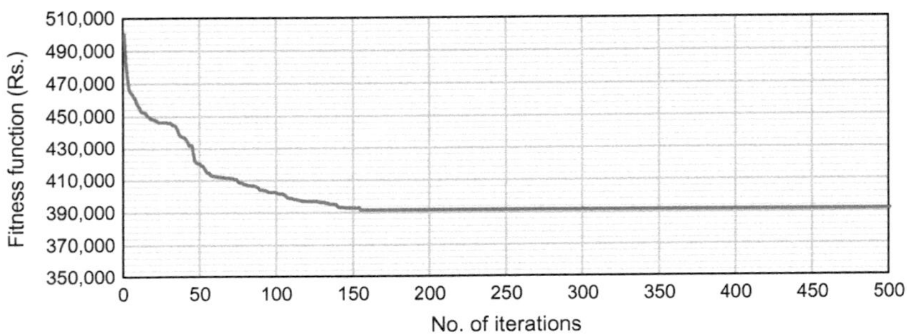

FIG. 10 The variation of fitness function with number of iterations for 85-bus system.

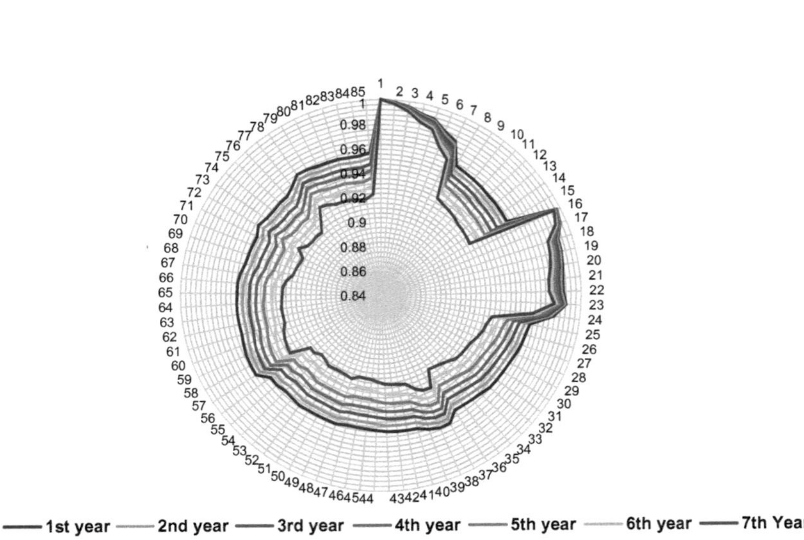

FIG. 11 Voltage profile over a 7-year span for the 85-bus system.

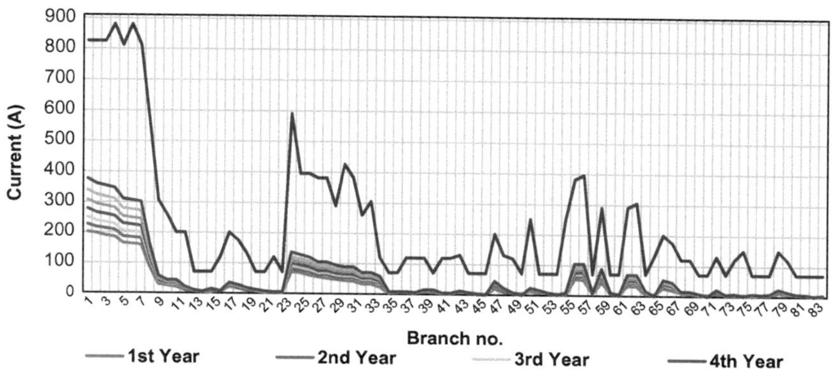

FIG. 12 Branch currents over a 7-year span for the 85-bus system.

REFERENCES

[1] P. Paliwal, N.P. Patidar, R.K. Nema, Planning of grid integrated distributed generators: a review of technology, objectives and techniques, Renew. Sust. Energ. Rev. 40 (2014) 557–570.

[2] E.S. Ali, S.M. Abd Elazim, A.Y. Abdelaziz, Improved harmony algorithm and power loss index for optimal locations and sizing of capacitors in radial distribution systems, Int. J. Electr. Power Energy Syst. 80 (2016) 252–263.

[3] S. Sakar, M.E. Balci, S.H.E. Abdel Aleem, A.F. Zobaa, Integration of large-scale PV plants in non-sinusoidal environments: considerations on hosting capacity and harmonic distortion limits, Renew. Sust. Energ. Rev. 82 (2018) 176–186, https://doi.org/10.1016/j.rser.2017.09.028.

[4] A.W. Funkhouser, R.P. Huber, A method for determining ACSR conductor sizes for the distribution systems, AIEE Trans. PAS 74 (3) (1955) 479–484.

[5] M. Ponnavaikko, K.P. Rao, An approach to optimal distribution system planning through conductor gradation, IEEE Trans. PAS 101 (6) (1982) 1735–1742.

[6] P.N. Rao, An extremely simple method of determining optimal conductor sections for radial distribution feeders, IEEE Trans. Power App. Syst. 104 (6) (1985) 1439–1442.

[7] H.N. Tram, D.L. Wall, Optimal conductor selection in planning radial distribution systems, IEEE Trans. Power Syst. 3 (1) (1988) 200–206.

[8] S.M. Leppert, A. Allen, in: Conductor Life Cycle Cost Analysis, 39th Rural Electric Power Conference, Nashville, USA, Apr. (1995), 1995.

[9] S.J. Islam, M.R.A. Ghani, in: Economical optimization of conductor selection in planning radial distribution networks, IEEE Transmission and Distribution Conference, New Orleans, USA, Apr. (1999), 1999.

[10] Z. Wang, H. Liu, D.C. Yu, et al., A practical approach to the conductor size selection in planning radial distribution systems, IEEE Trans. Power Del. 15 (1) (2000) 350–354.

[11] S. Sivanagaraju, N. Sreenivasulu, M. Vijayakumar, et al., Optimal conductor selection for radial distribution systems, Electr. Power Syst. Res. 63 (2) (2002) 95–103.

[12] S. Mandal, A. Pahwa, Optimal selection of conductors for distribution feeders, IEEE Trans. Power Syst. 17 (1) (2002) 192–197.

[13] R. Ranjan, A. Chaturvedi, P.S. Solanki, et al., in: Optimal conductor selection of radial distribution feeders using evolutionary programming, Convergent technology for the Asia-Pacific region conference (TENCON), Bangalore, India, 456–459, Oct. (2003), 2003.

[14] Falaghi, H., Ramezani, M., Haghifam, MR. et al.: Optimal selection of conductors in radial distribution systems with time varying load. IET 18th International Conference and Exhibition on Electricity Distribution, CIRED, Turin, Italy, Jun. (2005).

[15] S. Satyanarayana, T. Ramana, G.K. Rao, et al., Improving the maximum loading by optimal conductor selection of radial distribution systems, Electr. Pow. Compo. Sys. 34 (7) (2006) 747–757.

[16] F. Mendoza, D. Requena, J.L. Bemal-agustin, et al., in: Optimal conductor size selection in radial power distribution systems using evolutionary strategies, IEEE Transmission & Distribution conference and exposition, 2006.

[17] D. Kaur, J. Sharma, Optimal conductor sizing in radial distribution systems planning, Int. J. Elect. Pow. Energy Syst. 30 (4) (2008) 261–271.

[18] M. Sreedhar, N. Visali, S. Sivanagaraju, et al., A novel method for optimal conductor selection of radial distribution feeders using fuzzy evolutionary programming, Int. J. Electr. Power Eng. 2 (1) (2008) 6–10.

[19] S. Sivanagaraju, J.V. Rao, Optimal conductor selection in radial distribution system using discrete particle swarm optimization, World J. Model. Simul. 5 (2) (2009) 96–104.

[20] A.L. Devi, A. Shereen, Optimal conductor selection for radial distribution networks using genetic algorithm in SPDCL, AP—a case study, J. Theor. Appl. Inf. Technol. 5 (6) (2009) 674–685.

[21] M. Vahid, N. Manouchehr, A. Jamaleddin, et al., in: Combination of optimal conductor selection and capacitor placement in radial distribution systems for maximum loss reduction, IEEE International Conference on Industrial Technology, Gippsland, Australia, 2009.

[22] M.R. Raju, K.R. Murthy, K. Ravindra, et al., in: Optimal conductor selection for agricultural distribution system-a case study, IEEE International Conference on Intelligent and Advanced Systems, ICIAS, Kuala Lumpur, Malaysia, 2010.

[23] H. Falaghi, S. Chanan, Optimal conductor size selection in distribution systems with wind power generation, in: W. Lingfeng, S. Chanan, K. Andrew (Eds.), Wind Power Systems, first ed., Springer Berlin Heidelberg, 2010, , pp. 25–51.

[24] M. Thenepalle, A comparative study on optimal conductor selection for radial distribution network using conventional and genetic algorithm approach, Int. J. Comput. Appl. 17 (2) (2011) 6–13.

[25] R.S. Rao, K. Satish, S.V.L. Narasimham, Optimal conductor size selection in distribution systems using the harmony search algorithm with a differential operator, Electr. Pow. Compo. Sys. 40 (1) (2011) 41–56.

[26] M.K. Tamer, V.G. Alexander, in: Optimal conductor selection and capacitor placement for loss reduction of radial distribution systems by selective particle swarm optimization, Seventh International Conference on Computer Engineering & Systems (ICCES), Cairo, Egypt, 2012.

[27] J.F. Franco, M.J. Rider, M. Lavorato, et al., Optimal conductor size selection and reconductoring in radial distribution systems using a mixed-integer LP approach, IEEE Trans. Power Syst. 28 (1) (2013) 10–20.

[28] M.M. Legha, F. Ostovar, L.M. Mohammad, Combination of optimal conductor selection and capacitor placement in radial distribution systems using PSO method, Iraqi J. Electr. Electr. Eng. 10 (1) (2014) 33–41.

[29] M.K. Belal, Conductor selection optimization in radial distribution system considering load growth using MDE algorithm, World J. Model. Simul. 10 (3) (2014) 175–184.

[30] S. Haidar, M.M. Legha, in: Optimal Conductor Selection in Radial Distribution Using Imperialism Competitive Algorithm and Comparison with PSO Method, 6th International Conference for Scientific Computing to Computational Engineering, Athenes, Greece, 2014.

[31] M.M. Legha, H. Noormohamadi, A. Barkhori, Optimal conductor selection in radial distribution using bacterial foraging algorithm and comparison with ICA method, WALIA J. 31 (s3) (2015) 37–43.

[32] J. Shu, L. Wu, B. Han, et al., Enhanced multi-dimensional power network planning based on ant colony optimization, Int. Trans. Electr. Energ. Syst. 25 (7) (2015) 1204–1222.

[33] A.R. Abul'Wafa, Multi-conductor feeder design for radial distribution networks, Electr. Power Syst. Res. 140 (2016) 184–192.

[34] S. Manikandan, S. Sasitharan, J.V. Rao, et al., in: Analysis of Optimal Conductor Selection for Radial Distribution Systems Using DPSO, 3rd International Conference on Electrical Energy Systems (ICEES), Chennai, India, 2016.

[35] P. Samal, S. Mohanty, S. Ganguly, in: Simultaneous Capacitor Allocation and Conductor Sizing in Unbalanced Radial Distribution Systems Using Differential Evolution Algorithm, National Power Systems Conference (NPSC), Bhubaneswar, India, 2016.

[36] A.Y. Abdelaziz, A. Fathy, A novel approach based on crow search algorithm for optimal selection of conductor size in radial distribution networks, Eng. Sci. Technol. Int. J. 20 (2) (2017) 391–402.

[37] S.M. Ismael, S.H.E. Abdel Aleem, A.Y. Abdelaziz, in: Optimal selection of conductors in Egyptian radial distribution systems using sine-cosine optimization algorithm, 19th International Middle-East Power Systems Conference, MEPCON' 2017, 2017.

[38] S. Mirjalili, A. Lewis, The grasshopper optimization algorithm, Adv. Eng. Softw. 95 (2016) 51–67.

[39] B.S. EN, Standard: 50182-2001.: Conductors for Overhead Lines, in: Round Wire Concentric Lay Stranded Conductors, 2001.

[40] ACSR Conductor catalogue, http://www.linttop.com/uploads/ACSR_Wires1.Pdf, Online, [accessed 14 Jan. 2018].

[41] A. Tharwat, E.H. Houssein, M.M. Ahmed, A.E. Hassanien, T. Gabel, MOGOA algorithm for constrained and unconstrained multi-objective optimization problems, Appl. Intell. (2017), https://doi.org/10.1007/s10489-017-1074-1.

[42] S.Z. Mirjalili, S. Mirjalili, S. Saremi, H. Faris, I. Aljarah, Grasshopper optimization algorithm for multi-objective optimization problems, Appl. Intell. (2017), https://doi.org/10.1007/s10489-017-1019-8.

[43] J. Wu, et al., Distributed trajectory optimization for multiple solar-powered UAVs target tracking in urban environment by adaptive grasshopper optimization algorithm, Aerosp. Sci. Technol. 70 (Supplement C) (2017) 497–510, https://doi.org/10.1016/j.ast.2017.08.037.

[44] D. Das, D. Kothari, A. Kalam, Simple and efficient method for load flow solution of radial distribution network, Int. J. Electr. Power Energy Syst. 15 (5) (1995) 335–346.

Chapter 9

Classical and Recent Aspects of Active Power Filters for Power Quality Improvement

Mihail Hristov Antchev
Section Power Electronics, Faculty of Electronic Engineering and Technology, Technical University-Sofia, Sofia, Bulgaria

1 INTRODUCTION

There are several methods to increase the energy efficiency of an AC source and to improve its quality. (1) Development of power electronic converters with increased power factor K_P regarding the source. (2) Development of special electronic devices to improve the power factor of the installed electrical equipment. (3) Development of special electronic devices to improve the quality of the supplying voltage of "sensitive" consumers. Active power filters (APF) belong to the last two stated groups. Their development starts in the 1970s [1,2]. The significant progress in this domain became possible after the creation of the IGBT-s (insulated gate bipolar transistors) and DSP-s (digital signal processors) [3–5]. The following problems of the worsen quality of the electrical energy could be solved by means of these filters [6–12]:

- Removal of the harmonics of the consumed by the source current and/or removal of the harmonics of its voltage;
- Compensation of the reactive power for the main frequency (existence of phase difference between the first harmonics of the consumed current and the source voltage);
- Presence of negative sequence and main frequency components (unbalanced components) of the consumed current;
- Presence of zero sequence and main frequency components (current in the neutral wire);
- Presence of Flicker-effect (low frequency modulation of the electrical energy flow).

Classical and Recent Aspects of Power System Optimization. https://doi.org/10.1016/B978-0-12-812441-3.00009-4
© 2018 Elsevier Inc. All rights reserved.

Depending on their type, the APF eliminates one or several of the presented problems. As each power electronic devices, active power filters are composed of power circuit, control system, and sensors to follow the main parameters, related to the operation of the filter [13].

2 TYPES OF ACTIVE POWER FILTERS

The most popular classification of APF's is: shunt, series and hybrid active power filters. There are also combinations of active and passive filter and combination of APF's. Fig. 1 presents the block diagram of shunt APF. Generally, the load is nonlinear and its apparent power S is composed of active power P, reactive power Q, and distortion power D. The shunt filter should compensate these two last powers, that is, the system filter-load should behave as active load.

By permanent following of the load current I_L, the APF should generate current I_F, which should complete I_L to a sinusoidal current I_S in phase with the source voltage. For this purpose, sensors follow the load current and one of the two other currents—the filter's current I_F or the source current I_S. The compensation of the higher harmonics of the current I_L could be done also by series filter (Fig. 2) [14,15]. In this case the current of the load I_L is followed, and the power circuit generates voltage U_F, which eliminates the higher harmonics of the current.

Series filter is more often used as a solution to eliminate the distortion of the quality of the source voltage (Fig. 3). This is the so called dynamic voltage restorer. Usually this solution follows the waveform of the load voltage U_L or the source voltage U_S. The filter generates such a voltage U_F, which completes the source voltage to a sinusoidal voltage which supplies the load.

Fig. 4 shows a combination of the classic method—APF and shunt passive filters. It is applied for loads with relatively constant distortion power D. In that

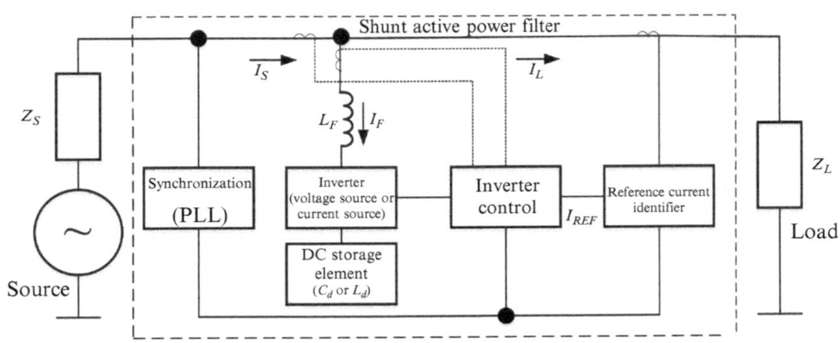

FIG. 1 Shunt active power filter—block diagram.

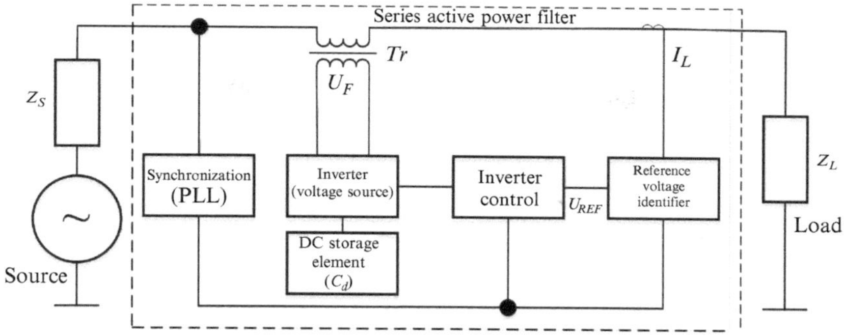

FIG. 2 Series active power filter—block diagram.

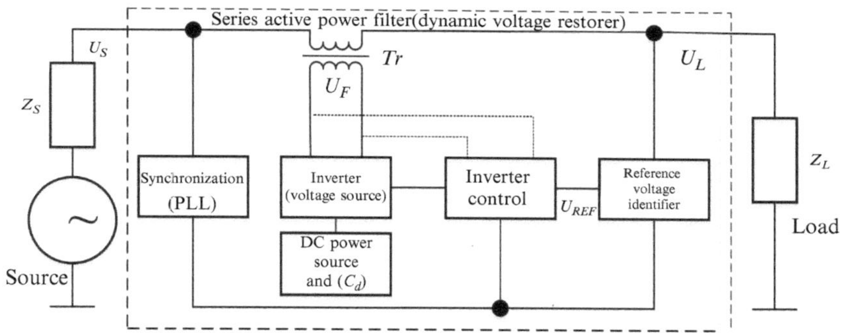

FIG. 3 Series active power filter (dynamic voltage restorer)—block diagram.

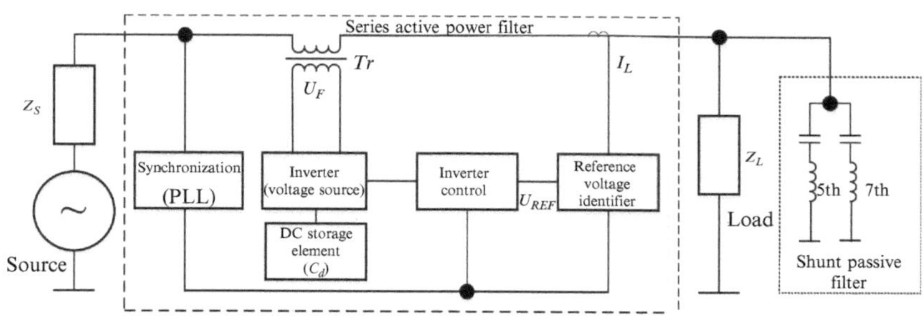

FIG. 4 Combination of a series active power filter and a shunt passive filter—block diagram.

case, the harmonics fluctuations are lower. The passive filter compensates the main part and the compensation of the dynamic variations is done by the active filter.

The series connection of shunt APF and passive filter is called hybrid APF (Fig. 5) [16,17].

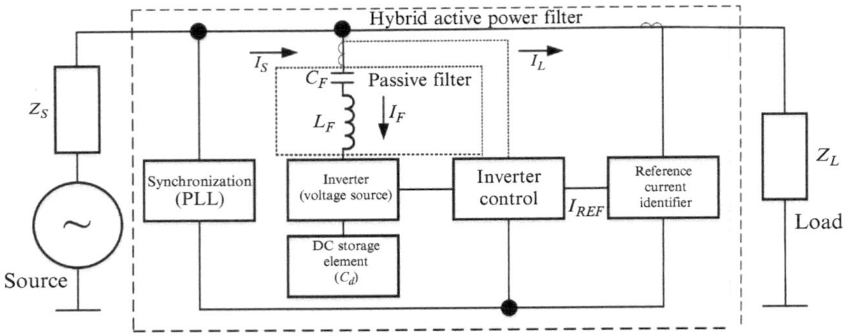

FIG. 5 Hybrid active power filter (series connection of a shunt APF and a passive filter).

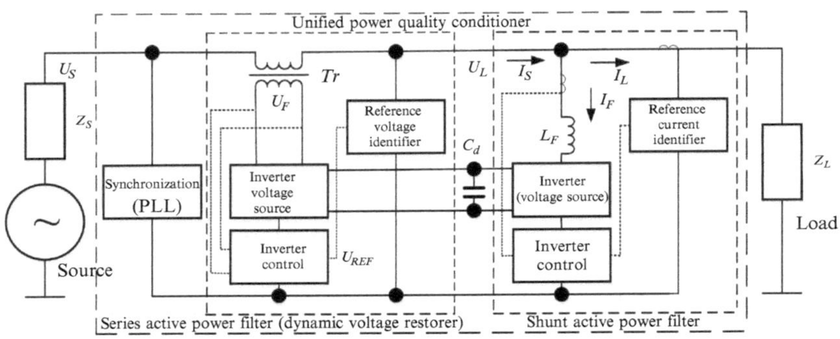

FIG. 6 Unified power quality conditioner—block diagram.

Fig. 6 presents the use of a combination of two APF's [18–20]. The series APF operates as dynamic voltage restorer, and then the load is supplied with the voltage U_L which is of better quality than the source voltage U_S. The shunt filter compensates the distortion power D and the reactive power Q of the load, and then improves the power factor regarding the source [12,21,22]. All the presented cases require synchronization of the control systems of the active power filters and the source voltage. It is done by means of a special block, which often contains phase locked loop (PLL).

Sometimes the active power filters can operate as uninterruptible power supply [23,24]. They are used in the power distribution systems [25] and systems with renewable energy sources [26,27].

3 POWER CIRCUITS FOR REALIZATION OF ACTIVE POWER FILTERS

Power circuits can be based on current source inverter (with inductance L_d as energy storage element) or on voltage source inverter (with capacitor C_d as energy storage element). The second type is more widely used. The difference with the classic voltage source inverter is that they do not contain direct voltage source U_d, but capacitor C_d as energy storage element on the direct current side, and that is why they are sometimes called "DC Capacitor Based Energy Storage System." Their operation is described by the single phase bidirectional converter presented in Fig. 7 [28]. By means of this figure is explained the possibility to change the direction of the current i_F, as well the energy flow. Transistors switch is a high frequency switch and the study is done for the first harmonics of the main parameters. Their vector diagram is shown on Fig. 8. From Fig. 7 follows:

$$I_{F1} = \frac{U - U_{C1}}{j \cdot \omega \cdot L_F} \tag{1}$$

The active power is defined by the following equation

$$P = U \cdot I_{F1} \cdot \cos \varphi \tag{2}$$

And as per Fig. 8

$$U_{L_F1} \cdot \cos \varphi = \omega \cdot L_F \cdot I_{F1} \cdot \cos \varphi = U_{C1} \cdot \sin \delta \tag{3}$$

so

$$P = \frac{U^2}{\omega \cdot L_F} \left(\frac{U_{C1}}{U} \cdot \sin \delta \right) \tag{4}$$

The reactive power is defined by the following equation

$$Q = U \cdot I_{F1} \cdot \sin \varphi \tag{5}$$

And as per Fig. 8

$$U_{C1} \cdot \cos \delta = U = U_{L_F1} \cdot \sin \varphi = U - \omega \cdot L_F \cdot I_{F1} \cdot \sin \varphi \tag{6}$$

so

$$Q = \frac{U^2}{\omega \cdot L_F} \left(1 - \frac{U_{C1}}{U} \cdot \cos \delta \right) \tag{7}$$

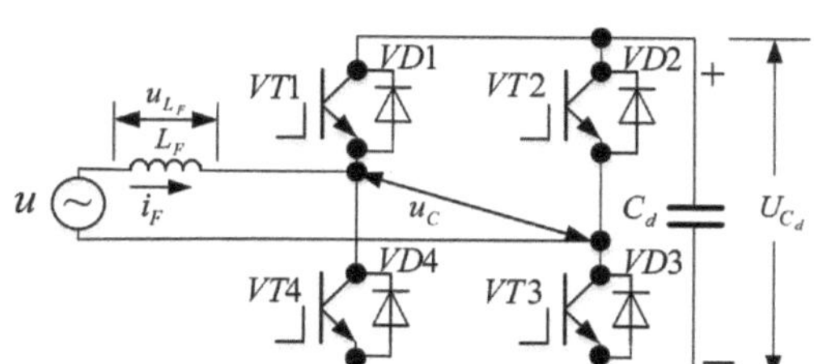

FIG. 7 Single phase bidirectional converter—power circuit.

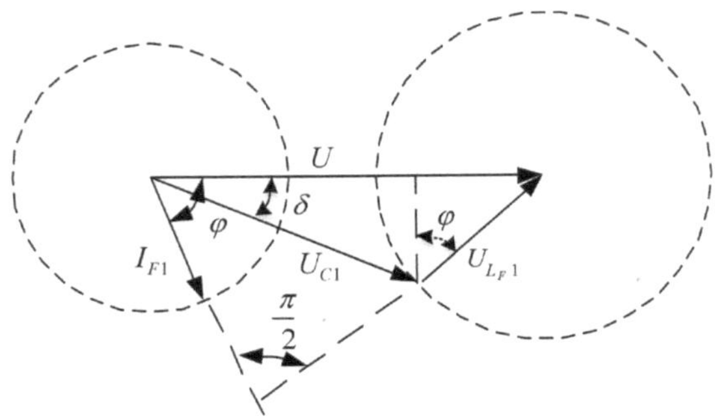

FIG. 8 Single phase bidirectional converter—vector diagram.

From Eqs. (4), (7), it is obvious that if the value of the inductance L_F is fixed and for any value of the voltage of the source U, the desired values of P and Q can be obtained by varying the input voltage of the converter U_{C1} and the phase difference angle regarding the source voltage δ.

Fig. 8 shows how will change U_{C1} if I_{F1} should be kept invariable for different values of the angle φ. Fig. 9 shows vector diagrams for two equal values for the current I_{F1} in rectifier mode (Fig. 9A) phase difference 0 and $cos\ \varphi = 1$, and inverter mode (Fig. 9B) phase difference π and $cos\ \varphi = -1$.

Fig. 10 shows current and voltage diagrams of the circuit of Fig. 7 corresponding to rectifier mode operation of Fig. 9A. It can be observed that the current i_F follows the reference curve (in this case with sinusoidal waveform) by switching of the power semiconductor devices. When the voltage U_C is 0, the current i_F increases, and when U_C is equal to the capacitor voltage U_{C_d}—the current decreases.

In the study of Fig. 10 uses a unipolar pulse-width modulation. Fig. 11 presents the equivalent circuits of the bipolar pulse width modulation and operation in the positive half period (Fig. 11A), or in the negative one (Fig. 11B) of the source voltage. They help describe how one should switch the transistors, so that the current i_F could follow a reference curve for the current of the filter. The description of the operation mode uses the bridge power circuit of Fig. 7. By means of closed loop system for automatic regulation of the voltage U_{C_d} it is maintained at a higher value than the maximum value of the source voltage. The current of the inductance L_F is followed by means of a sensor, the output signal of which is sent to the control system and is compared with the reference curve for the current of the active power filter.

When the control impulse to turn on the couple of diagonally connected $VT1$ and $VT3$ is sent ($VT1$ and $VT3$ or $VD1$ and $VD3$ conduct the current), the applied voltage polarity U_{C_d} is the one without brackets. When the other couple of $VT2$ and $VT4$ or $VD2$ and $VD4$ is on, the applied polarity of the voltage U_{C_d} is the one in the brackets.

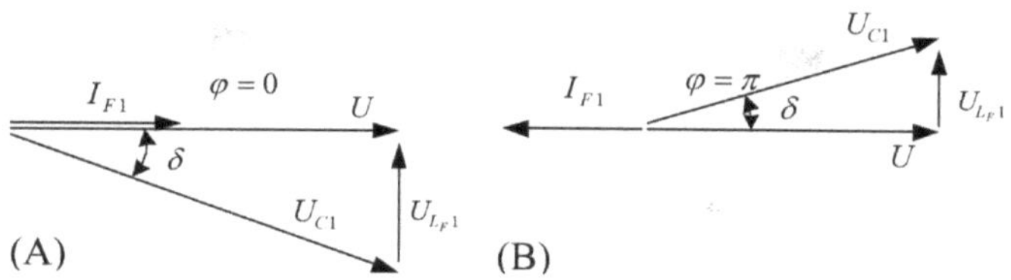

FIG. 9 Single phase bidirectional converter: (A) rectifier mode and (B) inverter mode.

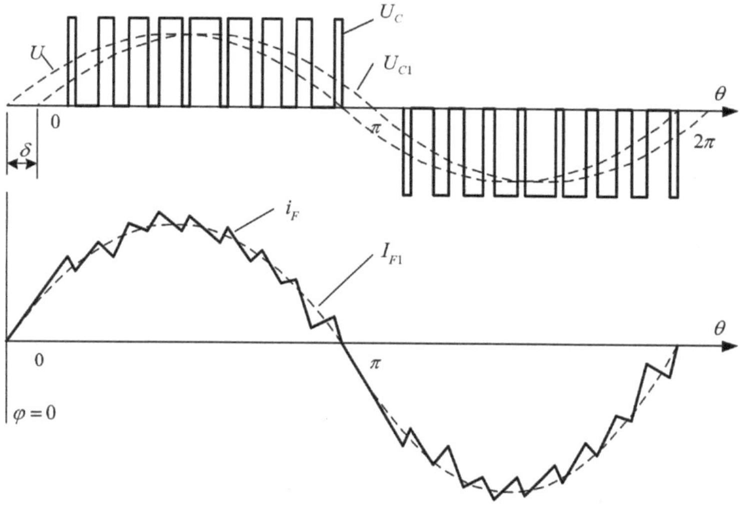

FIG. 10 Single phase bidirectional converter—waveforms.

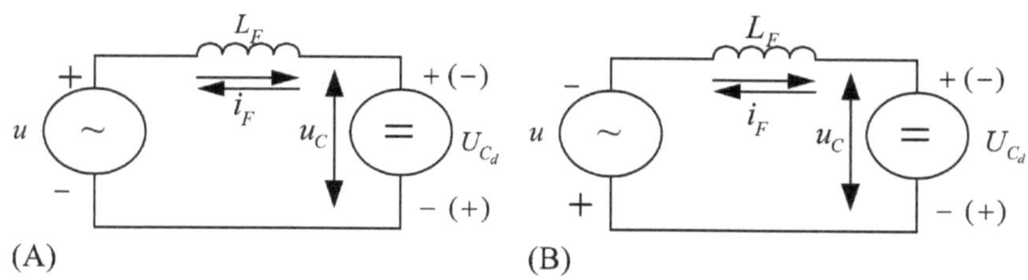

FIG. 11 Single phase bidirectional converter—equivalent circuits: (A) for positive half period, (B) for negative half period.

From Fig. 11A it is obvious that if the current i_F flow is from left to right (rectifier mode), at the switching on of the transistors *VT*2 and *VT*4 this current should increase (*VT*2 and *VT*4 conduct the current) and at the switching on of the transistors *VT*1 and *VT*3 it should decrease (*VD*1 and *VD*3 conduct the current). During the positive half period the current i_F could also flow from right to left (inverter mode) as the value of U_{C_d} is higher than the maximum value of the source voltage. Then, when switching on $VT1$ and $VT3$ the absolute value of the current will increase ($VT1$ and $VT3$ conduct the current) and with the switching on of $VT2$ and $VT4$, its absolute value will decrease ($VD2$ and $VD4$ conduct the current). Some identical analysis could be done for the negative half period of the supplying voltage (Fig. 11B). If the flow of the current i_F was from right to left (rectifying mode) and the transistors $VT1$ and $VT3$ are switched on, then the absolute value of this current will increase ($VT1$ and $VT3$ conduct the current), and when $VT2$ and $VT4$ are switched on its absolute value will decrease ($VD2$ and $VD4$ conduct the current). If the current flow was from left to right (inverter mode), then at the switching on $VT2$ and $VT4$ it will increase ($VT2$ and $VT4$ conduct the current) and at the switching on of $VT1$ and $VT3$ it will decrease ($VD1$ and $VD3$ conduct the current).

Let us look at two consecutive moments of time, corresponding to the switching on of two different transistor couples ($VT1$,$VT2$ or $VT2$,$VT4$) and two instantaneous values of the current i_F before and after these moments. From this study, we can make the conclusion: If these instantaneous values are compared with their signs (positive if the current flow is from left to right and negative if the current flow is form right to left), so independently of the operation mode (inverter or rectifier), the switching on of the couple of transistors $VT2$ and $VT4$ causes an increase of the current, and the switching on of the couple of transistors $VT1$ and $VT3$ causes a decrease. This conclusion is the basis of the follow of the reference curve for the current i_F with certain hysteresis. This happens with continuous switching of the couples of transistors, depending whether it is necessary that the instantaneous value of the current should increase or decrease in comparison with the reference value.

Besides the studied full-bridge circuit of Fig. 7, the realization of single phase active power filters also used half-bridge (Fig. 12A) and three-level of the voltage u_C circuits (Fig. 12B).

Fig. 13 shows the circuit of the three-phase bridge bidirectional converter, used for the realization of three phase active power filters with AC source without neutral.

When there is a neutral point N , it is used the circuit of Fig. 14 [29]. The circuit of Fig. 15 has reduced number of power semiconductor devices.

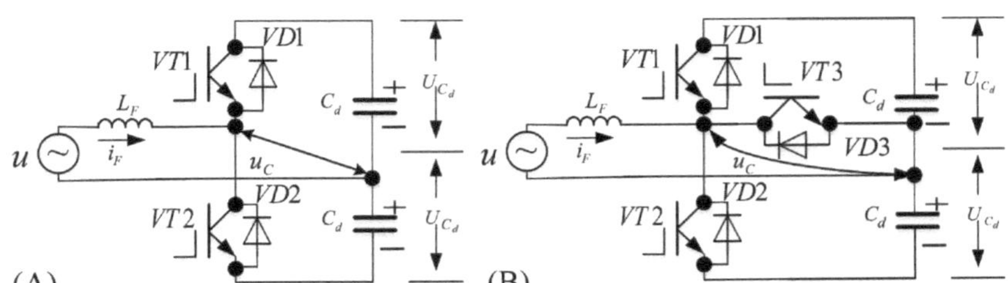

FIG. 12 Single phase bidirectional converters: (A) half bridge and (B) half bridge and zero level.

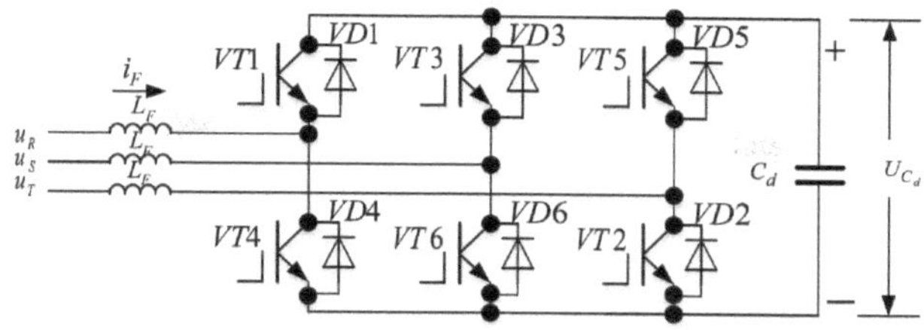

FIG. 13 Three phase bidirectional converter.

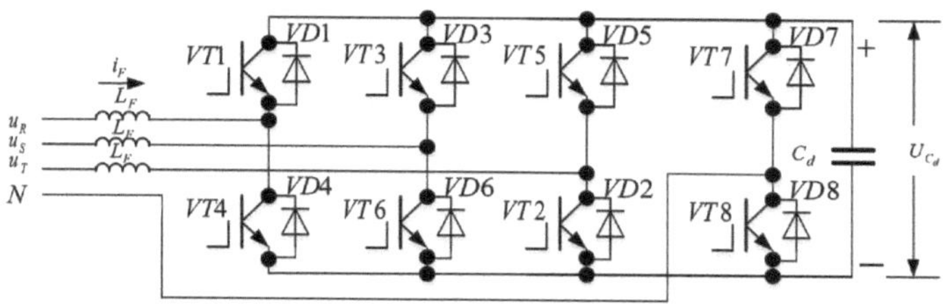

FIG. 14 Three phase bidirectional converter with neutral point.

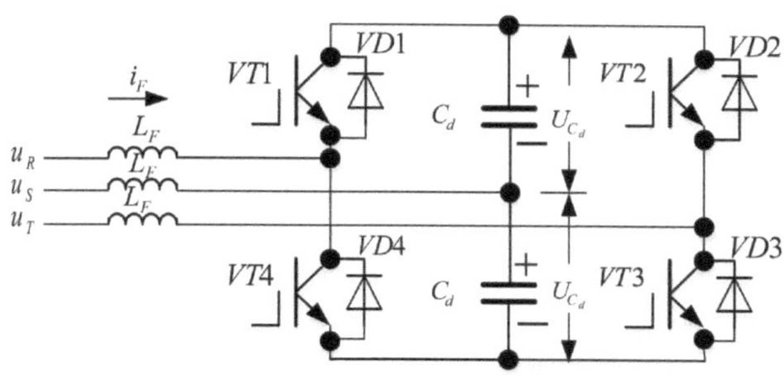

FIG. 15 Three phase bidirectional converter with reduced number of power semiconductor devices.

The realization of active power filters also uses the so-called multilevel inverters as in Fig. 16 [30]. The most frequently used are inverters with clamped diodes (Fig. 16A) and cascaded single-phase full-bridge inverters (Fig. 16B). For the circuit of Fig. 16A the voltage of the lower end of each of the inductances with regards to the neutral point of the source can have three levels: $\frac{U_d}{2}$,0 and $-\frac{U_d}{2}$. The number of the states for one cell of Fig. 16B is three: $-U_d$, 0 and U_d. In a series connection of p cells, the voltage of the phase (the lower end of each of the inductances regarding point N) is defined as per:

$$U_{RN} = \sum_{i=1}^{p} (S_{Ri} - 1) \cdot U_{di} \tag{8}$$

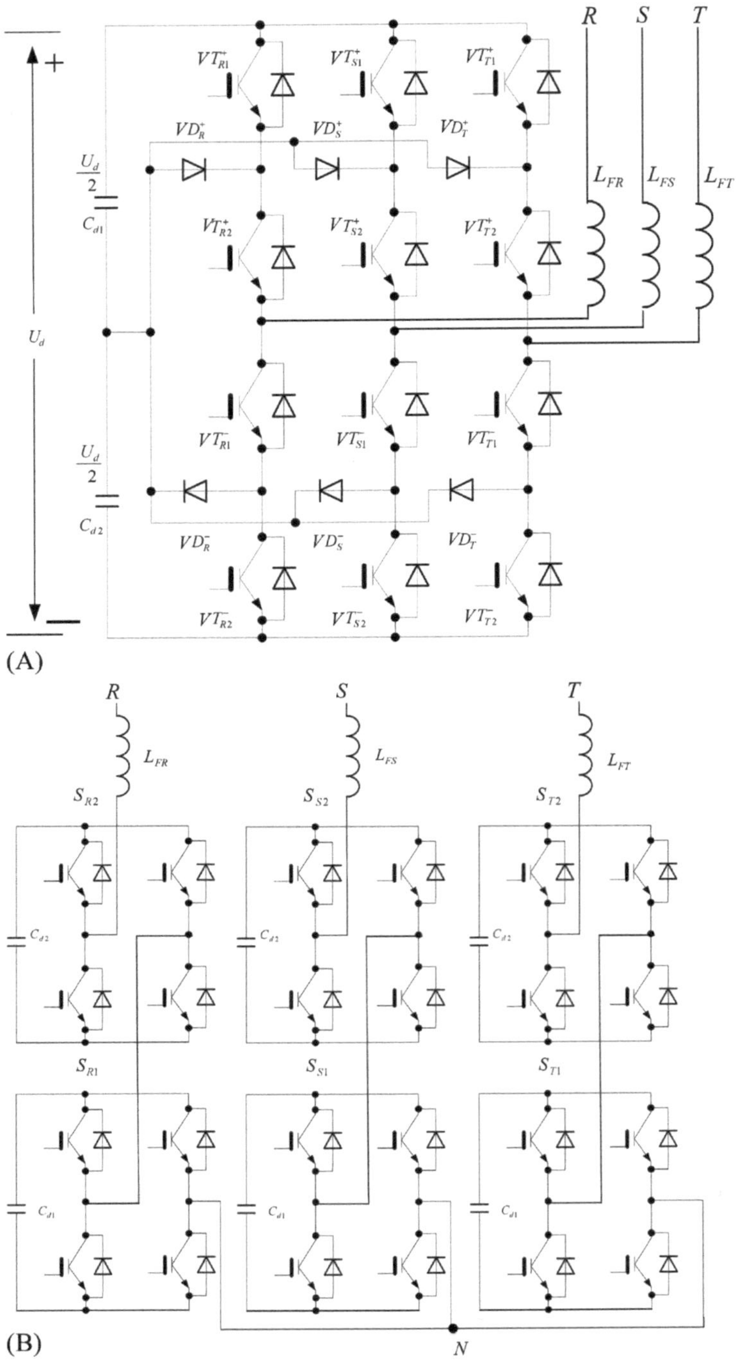

FIG. 16 Three phase multilevel converters: (A) neutral point clamped diodes and (B) cascaded.

S_{Ri} is the number of states for one cell, p is the number of cells, and U_{di} is the supplying voltage of the cell. If the supplying voltages of all the cells have the same value equal to U_d, then the common number of modulation levels is:

$$n = 3 + 2 \cdot (p - 1) \tag{9}$$

Then, the number of the modulation levels of the circuit of Fig. 16B is five.

The following three examples show the connection of some of the studied power circuits for realization of active power filters according to some of the block diagrams presented in part one. Fig. 17 presents three phase shunt active power filter realized from the circuits of Figs. 1 and 13.

Fig. 18 presents the realization of dynamic voltage restorer using the circuits of Figs. 3 and 13. It shows the necessity of additional DC power source. Usually, it is obtained from the voltage of the three-phase source by means of a rectifier, but it is also possible to obtain it from a renewable energy source.

By combination of the circuits of Figs. 17 and 18 is realized unified power quality conditioner (Figs. 6 and 19).

A main issue with the design of shunt active power filters is the definition of the effective value of the filter current I_F. The studied case is of a single-phase filter. The known values are the apparent power of the load S, its harmonics coefficient K_H and its $cos\ \varphi$. In the following studies, the elements of the active power filter are supposed to be ideal, meaning there are no active power losses in these elements. First, only the compensation of the distortion power D is studied, meaning only the compensation of the higher harmonics which the filter should eliminate as in Fig. 20. The nonsinusoidal waveform of the current of the nonlinear load is described by the coefficient of the

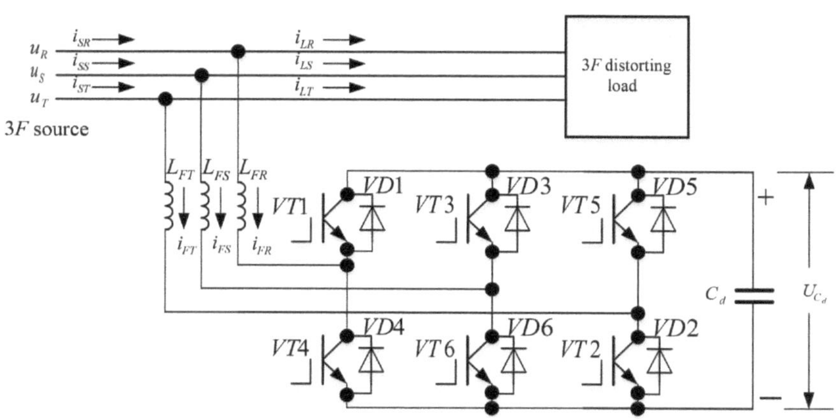

FIG. 17 Three phase shunt active power filter—connection diagram.

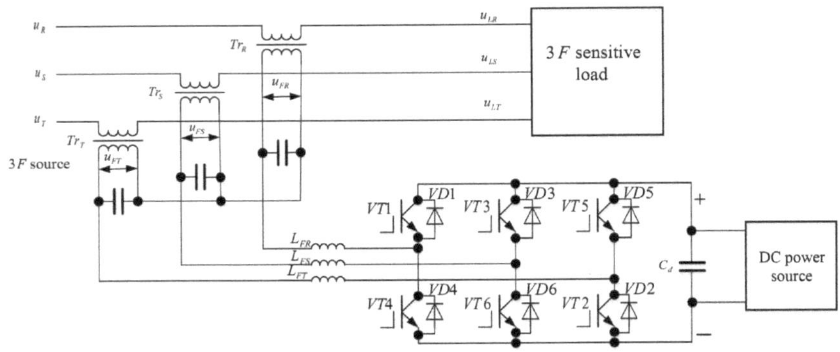

FIG. 18 Three phase series active power filter (dynamic voltage restorer)—connection diagram.

distortion of the current waveform υ or the harmonics coefficient K_H, which are related as follows:

$$\upsilon = \frac{I_1}{I}; \quad \mathrm{K}_H = \frac{\sqrt{\sum_{n=2}^{\infty} I_n^2}}{I_1}; \quad \upsilon = \frac{1}{\sqrt{1+K_H^2}} \tag{10}$$

The effective value of the current of the nonlinear load I_L is:

$$I_L = \sqrt{I_1^2 + \sum_{n=2}^{\infty} I_n^2} = \sqrt{\frac{\sum_{n=2}^{\infty} I_n^2}{K_H^2} + \sum_{n=2}^{\infty} I_n^2} = \sqrt{\sum_{n=2}^{\infty} I_n^2} \cdot \sqrt{\frac{1+K_H^2}{K_H^2}} \tag{11}$$

In the ideal case, when the APF does not consume any active power, its current contains only higher harmonics

$$I_F = \sqrt{\sum_{n=2}^{\infty} I_n^2} \tag{12}$$

If we consider Eqs. (8), (9):

$$I_L = I_F \cdot \sqrt{\frac{1+K_H^2}{K_H^2}} = I_F \cdot \sqrt{\frac{1}{1-\nu^2}} \tag{13}$$

Then the apparent power, consumed by the load, is presented as:

$$S = U_S \cdot I_L = U_S \cdot I_F \cdot \sqrt{\frac{1+K_H^2}{K_H^2}} = U_S \cdot I_F \cdot \sqrt{\frac{1}{1-\upsilon^2}} \tag{14}$$

For a single-phase active power filter

$$\frac{S}{U_S} = I_F \cdot \sqrt{\frac{1+K_H^2}{K_H^2}} \tag{15}$$

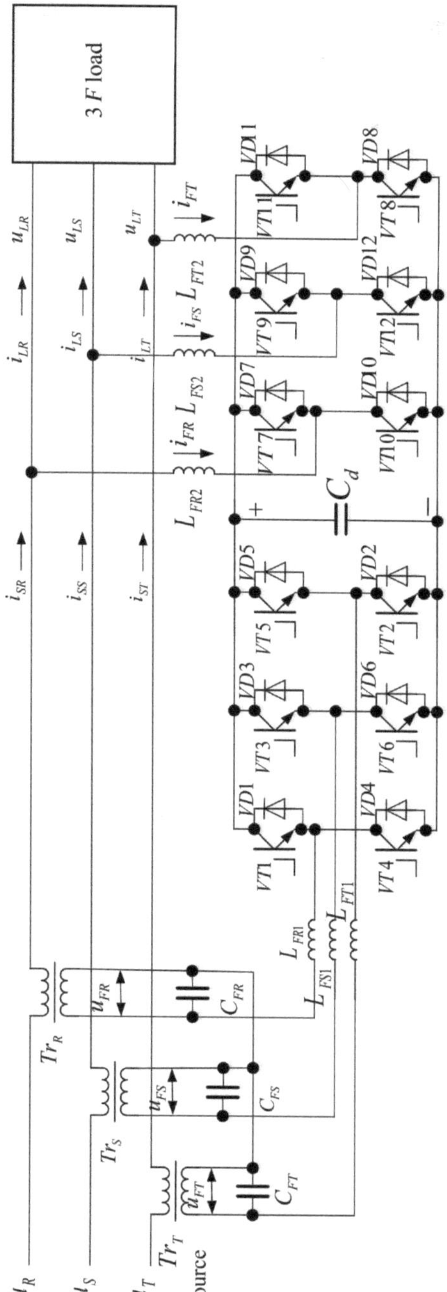

FIG. 19 Three phase unified power quality conditioner—connection diagram.

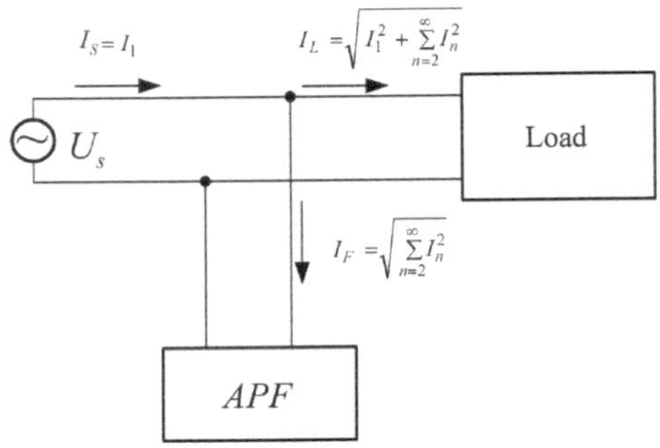

FIG. 20 Shunt active power filter—compensation only of the distortion power D.

For three-phase active power filter

$$\frac{S}{\sqrt{3} \cdot U_{LL}} = I_F \cdot \sqrt{\frac{1 + K_H^2}{K_H^2}} \tag{16}$$

In Eq. (15) S is single phase apparent power and U_S is the effective value of the phase voltage. In Eq. (16) S is three phase apparent power and U_{LL} is the effective value of the line to line voltage. In both cases I_F is the effective value of the filter current per phase. As per Eqs. (15), (16) in Fig. 21 are shown 3D graphic relations. The relations (15) and (16) can be used for the following purposes:

- When the source voltage is U_S or U_{LL}, the apparent power S and the harmonics coefficient of the load current K_H are known, it is possible to find

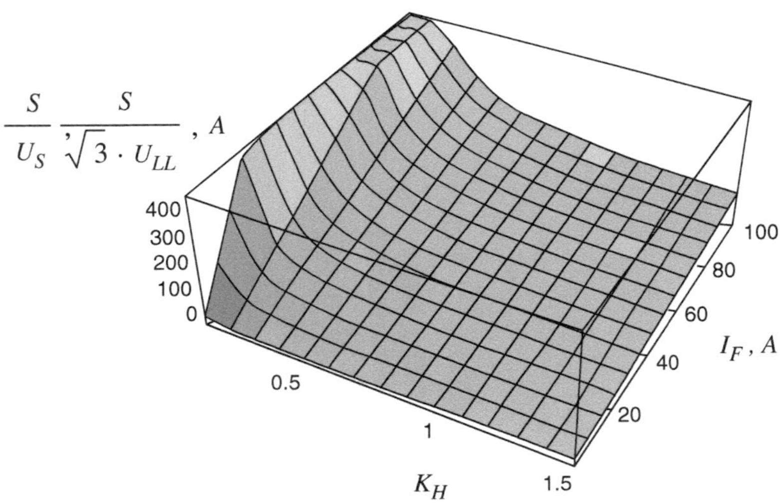

FIG. 21 Shunt active power filter—3D-graphics in accordance with expressions (15) and (16).

the value of the filter I_F necessary for the compensation of this harmonics coefficient.

- For some filters with a known value of the current I_F and voltage U_S or U_{LL}, it is possible to find what value of the harmonics coefficient K_H it can compensate for some apparent power of the load S.

On Fig. 22, these relations are shown in a plane for several typical values of the current of active power filter.

Next case is the compensation only of reactive power Q. It is presented in Fig. 23 for the first harmonics of the currents. The following expression is valid:

$$I_{SM1} \cdot sin\ \omega \cdot t = I_{LM1} \cdot sin\,(\omega \cdot t - \varphi) + F(\omega \cdot t, \varphi) \tag{17}$$

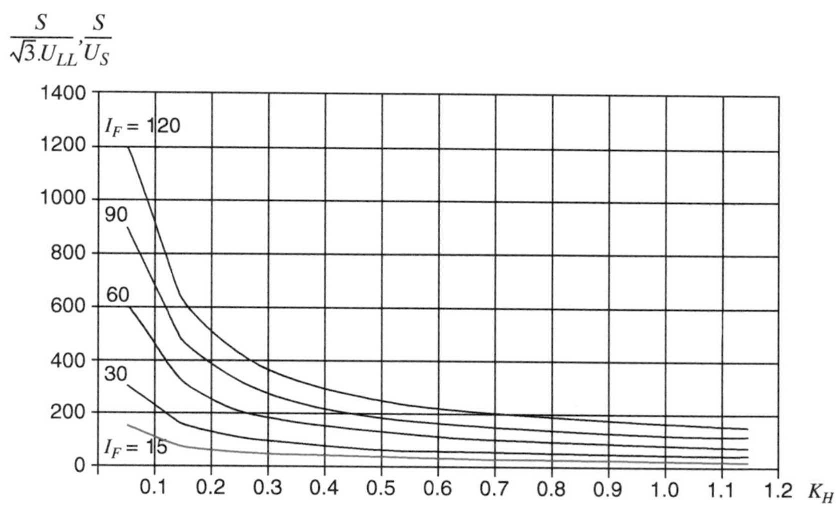

FIG. 22 Shunt active power filter—graphics in accordance with expressions (15) and (16).

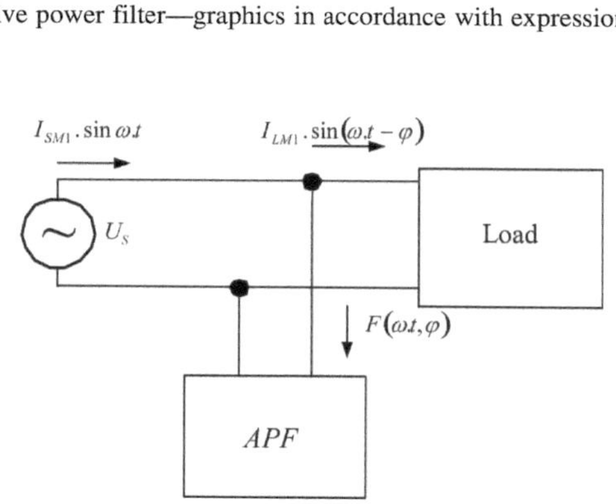

FIG. 23 Shunt active power filter—compensation only of the reactive power Q.

If we assume there are no active power losses in the active power filter, then from the equality of the active powers before and after the switch on of the filter it follows that:

$$I_{SM1} = I_{LM1} \cdot cos\ \varphi \tag{18}$$

If we replace Eq. (18) in Eq. (17) and solve, we can find the function of variation of the current of the active power filter:

$$F(\omega \cdot t, \varphi) = I_{LM1} \cdot sin\ \varphi \cdot cos\ \omega \cdot t = I_{FM1} \cdot cos\ \omega \cdot t \tag{19}$$

Then the effective value of the filter current is defined as per:

$$I_{F1} = I_{L1} \cdot sin\ \varphi \tag{20}$$

In this case for a known value of the first harmonic of the load current I_{L1} (from S and U_S is defined I and from known υ from Eq. 10 is defined I_{L1}) and the angle φ (when $cos\ \varphi$ is known), from Eq. (20) we can find the value of the active power filter current, which should fluctuate in the time as per Eq. (19) . If we have a look on the general case of compensation of the reactive power Q and the distortion power D, then the square of the effective current value of the active power filter should be equal to the sum of the squares of the effective current values for both cases. When we replace from Eqs. (13), (20) we obtain:

$$\frac{\dfrac{I_L^2}{1 + K_H^2}}{K_H^2} + I_{L1}^2 \cdot sin^2\ \varphi = I_F^2 \tag{21}$$

After a transformation and taking in consideration (8) we have:

$$\frac{I_F}{I_L} = \sqrt{\frac{1}{1 + K_H^2}(K_H^2 + 1 - cos^2\ \varphi)} \tag{22}$$

Fig. 24 shows the graphic relations as per Eq. (22).

The main issue for the series active power filter, connected as dynamic voltage restorer, is the definition of the apparent power of the filter, S_F when the apparent power of the load S and the harmonics coefficient of the source voltage k_H are known. Fig. 25 shows the connection of a consumer and a DVR to one phase of the supplying grid. In this case, the series active power filter is not used to correct the power factor, so that the effective values of the current I before and I' after the switch on of the filter will be approximately equal. Then the active power losses in the filter are neglected and:

$$S \approx S' \tag{23}$$

The effective value of the source voltage is defined as:

$$U = \sqrt{U_1^2 + \sum\nolimits_{n=2}^{\infty} U_n^2} \tag{24}$$

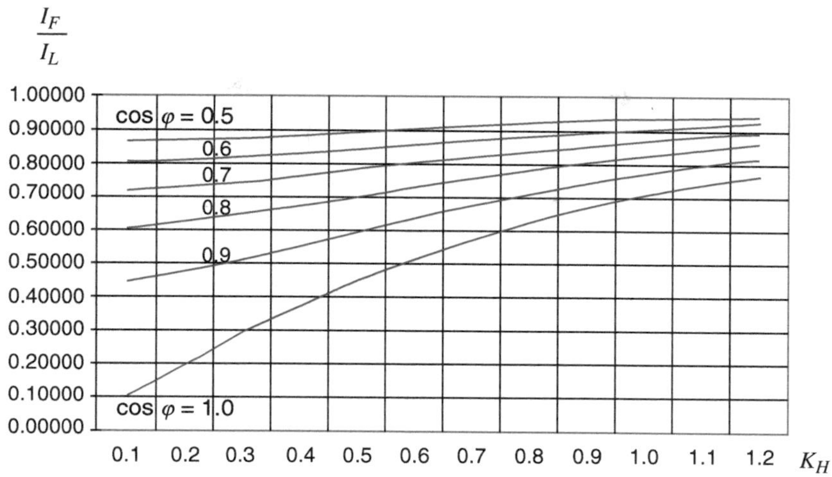

FIG. 24 Shunt active power filter—graphics in accordance with expression (22).

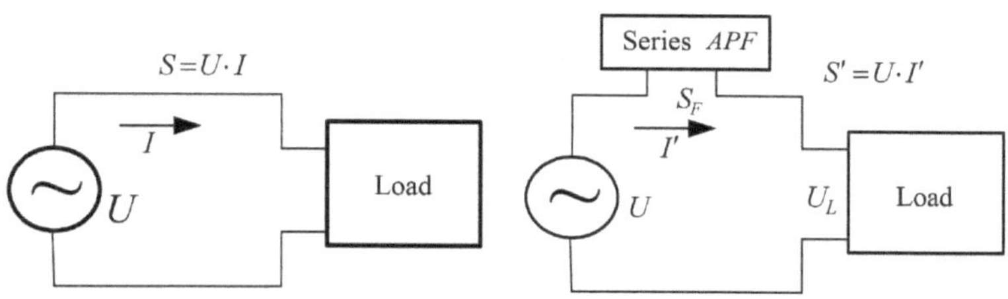

FIG. 25 Series active power filter (dynamic voltage restorer)—connection with load.

The series active power filter eliminates the higher harmonics of the source voltage and the load voltage is sinusoidal. Let us accept, that its effective value is equal to a preliminary defined one.

$$U_L = U_{1R} = const. \tag{25}$$

If the effective values of the output voltage of DVR, the preliminary defined values of the first harmonic and nth harmonic are U_F, U_{1R}, and U_n, then from the operation mode in this case it follows that:

$$U_F^2 = (U_{1R} - U_1)^2 + \sum_{n=2}^{\infty} U_n^2 \tag{26}$$

Let us determine with k the ratio between the effective values of the reference voltage (the preliminary defined one) and the first harmonic of the source voltage:

$$k = \frac{U_{1R}}{U_1} \tag{27}$$

Then from Eqs. (26), (27) it follows that:

$$U_F = U_1 \cdot \sqrt{(k-1)^2 + k_H^2} \tag{28}$$

Then the ratio of the apparent power of DVR S_F and the apparent power of the load S will be:

$$\frac{S_F^2}{S^2} = \frac{U_1^2\left[(k-1)^2 + k_H^2\right] \cdot I'^2}{\left[U_1^2 + \sum_{n=2}^{\infty} U_n^2\right] \cdot I^2} = \frac{(k-1)^2 + k_H^2}{1 + k_H^2} \tag{29}$$

On Figs. 26 and 27 are presented the graphics in accordance with the following expressions for this case:

$$\frac{S_F}{S} = \sqrt{\frac{(k-1)^2 + k_H^2}{1 + k_H^2}} \tag{30}$$

$$\frac{U_L}{U} = \frac{U_{1R}}{U} = k \cdot \sqrt{\frac{1}{1 + K_H^2}} \tag{31}$$

4 ACTIVE POWER FILTERS CONTROL SYSTEMS

From the operation mode of the active power filters follows the main specific functions which their control system should have: (1) definition of the reference

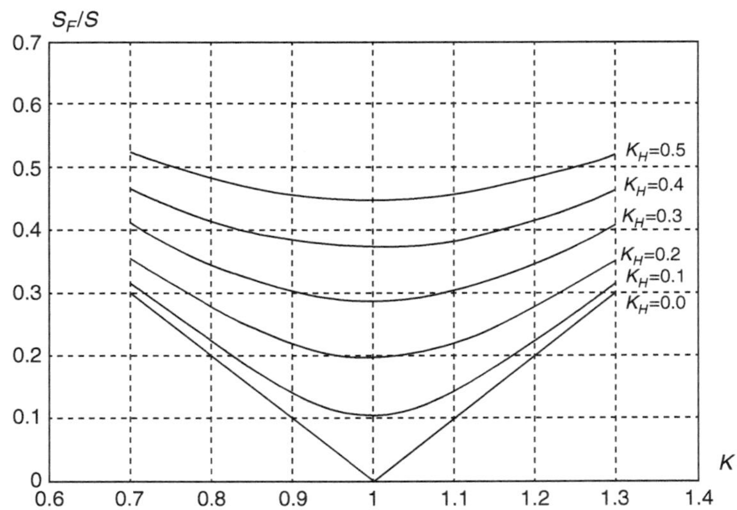

FIG. 26 Series active power filter (dynamic voltage restorer—graphics in accordance with expression (30).

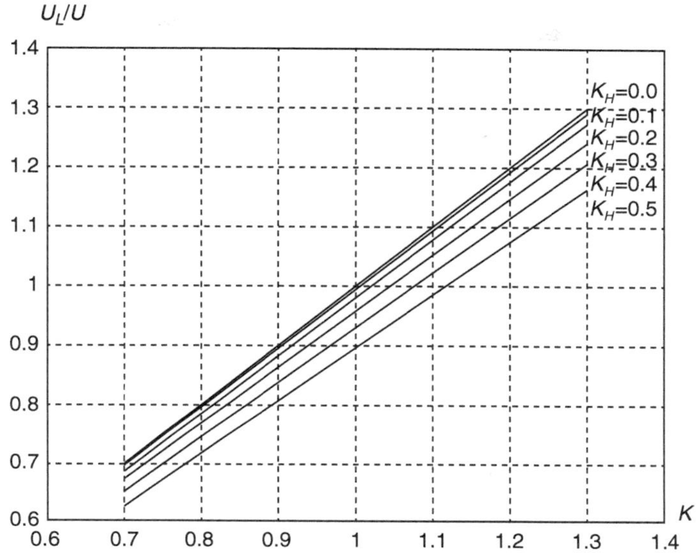

FIG. 27 Series active power filter (dynamic voltage restorer—graphics in accordance with expression (31).

curve for the active power filter current $i_{F_{REF}}$ or for source current $i_{S_{REF}}$, (2) formation of active power filters current curve $i_{F_{REF}}$ or of the source current $i_{S_{REF}}$, in accordance with the reference curve, and (3) stabilization of the voltage U_d of the energy storage capacitor C_d.

4.1 Definition of the Reference Curve of the Active Power Filter Current $i_{F_{REF}}$ or of the Source Current $i_{S_{REF}}$

In case of variations of the type and power of the load, a reference curve, which should be followed, should be defined permanently. The possibilities are the curve of the active power filter current $i_{F_{REF}}$ or the curve of the system active power filter-load (of the source $i_{S_{REF}}$).

High frequency filtering method. A high pass filter eliminates the main frequency of the load current and retains only its harmonics, which are used for the forming of a reference signal for the active power filter current $i_{F_{REF}}$. This method generates a big error in the phases and amplitudes of the harmonics. It is very sensitive to a high frequency noise.

Low frequency filtering method. A low pass filter is used the higher harmonics of the load current and it is obtained the reference signal with the main frequency $i_{S_{REF}}$. The reference signal is obtained by subtraction of this current from the load current $i_{F_{REF}} = i_L - i_{S_{REF}}$. This method has the same disadvantages as the previous one—big errors in the phase and the amplitude of the harmonics. There are other filtering methods [31].

Fourier transforms method. It is based on an analysis in the frequency domain using a simple and quick Fourier transform. Calculations are extracted for the higher harmonics of the load current i_L and on this basis, a reference curve forms for the filter current $i_{F_{REF}}$ [32,33].

PQ-theory. For control of three phase shunt active power filter, the "Akagi-Nabae theory" is often used. It is also known as "theory of the instantaneous powers" or "*p-q* theory" [6,34–36]. The studied system is three-phase, three-wire system source-load. It uses a power invariant transformation, which is different from the Clarke-transform. The three phase system voltages and currents from the *R-S-T* coordinate systems are transformed in one plane $\alpha-\beta$ coordinate systems by means of the following expressions:

$$\begin{bmatrix} u_\alpha \\ u_\beta \end{bmatrix} = \sqrt{\frac{2}{3}} \begin{bmatrix} 1 & -\frac{1}{2} & -\frac{1}{2} \\ 0 & \frac{\sqrt{3}}{2} & -\frac{\sqrt{3}}{2} \end{bmatrix} \cdot \begin{bmatrix} u_R \\ u_S \\ u_T \end{bmatrix} \tag{32}$$

$$\begin{bmatrix} i_\alpha \\ i_\beta \end{bmatrix} = \sqrt{\frac{2}{3}} \begin{bmatrix} 1 & -\frac{1}{2} & -\frac{1}{2} \\ 0 & \frac{\sqrt{3}}{2} & -\frac{\sqrt{3}}{2} \end{bmatrix} \cdot \begin{bmatrix} i_R \\ i_S \\ i_T \end{bmatrix} \tag{33}$$

The definitions "real instantaneous power *p*" and "imaginary instantaneous power *q*" are used. The power *p* is defined in the two coordinate systems:

$$p = u_R \cdot i_R + u_S \cdot i_S + u_T \cdot i_T = u_\alpha \cdot i_\alpha + u_\beta \cdot i_\beta \tag{34}$$

The power *q* is defined only in the plain system:

$$q = u_\alpha \cdot i_\beta - u_\beta \cdot i_\alpha \tag{35}$$

The expressions (14) and (15) can be presented also as:

$$\begin{bmatrix} p \\ q \end{bmatrix} = \begin{bmatrix} u_\alpha & u_\beta \\ -u_\beta & u_\alpha \end{bmatrix} \cdot \begin{bmatrix} i_\alpha \\ i_\beta \end{bmatrix} \tag{36}$$

As in the system $\alpha-\beta$ the power *p* is presented as a multiplication of voltages and currents with the same coordinates, it is measured in "watt" [*W*]. Besides *q* is presented as a multiplication of voltages and currents, it is different from *p*, as they do not have the same coordinates. It is measured in "imaginary watt" [*IW*]. By multiplying both sides of Eq. (36) with the reverse matrix of the voltages it follows:

$$\begin{bmatrix} i_\alpha \\ i_\beta \end{bmatrix} = \begin{bmatrix} u_\alpha & u_\beta \\ -u_\beta & u_\alpha \end{bmatrix}^{-1} \cdot \begin{bmatrix} p \\ q \end{bmatrix} \tag{37}$$

This expression can be presented as:

$$\begin{bmatrix} i_\alpha \\ i_\beta \end{bmatrix} = \begin{bmatrix} u_\alpha & u_\beta \\ -u_\beta & u_\alpha \end{bmatrix}^{-1} \cdot \begin{bmatrix} p \\ 0 \end{bmatrix} + \begin{bmatrix} u_\alpha & u_\beta \\ -u_\beta & u_\alpha \end{bmatrix}^{-1} \cdot \begin{bmatrix} 0 \\ q \end{bmatrix} = \begin{bmatrix} i_{\alpha p} \\ i_{\beta p} \end{bmatrix} + \begin{bmatrix} i_{\alpha q} \\ i_{\beta q} \end{bmatrix} \tag{38}$$

Considering Eqs. (14), (18) then:

$$\begin{bmatrix} p_\alpha \\ p_\beta \end{bmatrix} = \begin{bmatrix} u_\alpha \cdot i_\alpha \\ u_\beta \cdot i_\beta \end{bmatrix} = \begin{bmatrix} u_\alpha \cdot i_{\alpha p} \\ u_\beta \cdot i_{\beta p} \end{bmatrix} + \begin{bmatrix} u_\alpha \cdot i_{\alpha q} \\ u_\beta \cdot i_{\beta q} \end{bmatrix} \tag{39}$$

Then:

$$p = p_\alpha + p_\beta = u_\alpha \cdot i_{\alpha p} + u_\beta \cdot i_{\beta p} + u_\alpha \cdot i_{\alpha q} + u_\beta \cdot i_{\beta q} \tag{40}$$

By replacing Eqs. (38), (39):

$$p = \frac{u_\alpha^2}{u_\alpha^2 + u_\beta^2} \cdot p + \frac{u_\beta^2}{u_\alpha^2 + u_\beta^2} \cdot p + \frac{-u_\alpha \cdot u_\beta}{u_\alpha^2 + u_\beta^2} \cdot q + \frac{u_\alpha \cdot u_\beta}{u_\alpha^2 + u_\beta^2} \cdot q \tag{41}$$

The sum of the last two members is 0. Then:

$$p = u_\alpha \cdot i_{\alpha p} + u_\beta \cdot i_{\beta p}; \qquad 0 = u_\alpha \cdot i_{\alpha q} + u_\beta \cdot i_{\beta q} \tag{42}$$

From the presented dependencies, the following conclusions can be obtained:

- The first equation gives the active power in three-phase system expressed by its components in the coordinate system $\alpha - \beta$.
- The second equation determines that the sum of these components is 0, which leads to the conclusion that they compensate each other and are equal to the reactive power in the three-phase system, which can also be expressed by its components in the coordinate system $\alpha - \beta$.

The idea of the Akagi-Nabae theory is that it shows the components $i_{\alpha q}$, $i_{\beta q}$ of the currents i_α, i_β, that should be eliminated by the active power filter. The main block diagram that shows the realization of the method is presented in Fig. 28. The synchronization with the system of supplying voltages is on one of these voltages—in this case u_R—by means of phase locked loop (PLL). The signals $sin\ \theta$ and $cos\ \theta$ are necessary for the direct and reverse Park transforms. The three phase currents of the load are followed—i_{LR}, i_{LS}, i_{LT}, and from them are obtained the components i_p, i_q. First from Eq. (33) are obtained i_α, i_β, which then are transformed by Park direct transform:

$$\begin{bmatrix} i_p \\ i_q \end{bmatrix} = \sqrt{\frac{2}{3}} \cdot \begin{bmatrix} \cos\theta & \sin\theta \\ -\sin\theta & \cos\theta \end{bmatrix} \cdot \begin{bmatrix} 1 & -\frac{1}{2} & -\frac{1}{2} \\ 0 & \frac{\sqrt{3}}{2} & -\frac{\sqrt{3}}{2} \end{bmatrix} \cdot \begin{bmatrix} i_{LR} \\ i_{LS} \\ i_{LT} \end{bmatrix} \tag{43}$$

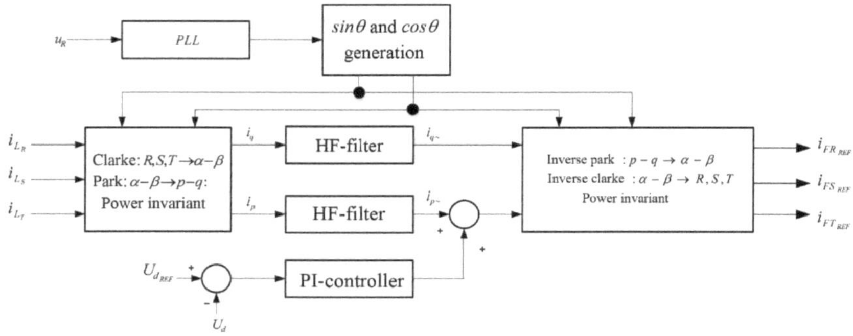

FIG. 28 Realization of three phase active power filter control in accordance with Akagi-Nabae theory.

The different harmonics of the phase currents which interact with the phase voltages, but in the $\alpha - \beta$ coordinates system, by means of Eq. (36) will define different components of the powers p, q, and then taking in consideration (37) and (43), also the components of the currents i_p, i_q. The first harmonics of the phase currents, which are in phase with the phase voltages, describe unipolar curves (similar to those after a full bridge rectification) containing a constant component. So, the constant components of the currents of Eq. (43) correspond to the system active power. The higher harmonics of the phase currents, which interact with the phase voltages, will define the variable components of the currents from Eq. (43). These are the components which should be compensated by the active power filter. That is why both currents are filtered by high pass filters (usually the cutoff frequency is 10Hz), on the outputs of which are obtained only the variable components $i_{p\sim}$, $i_{q\sim}$. In addition, there is the part for stabilization of the voltage U_d of the capacitor C_d, as the output signal of the PI-regulator forms the final component of $i_{p\sim}$ for the next block. In the next block is realized a reverse $p-q$ transformation (reverse Park transform), to find out the reference curves for the filter currents in accordance with the expression below:

$$\begin{bmatrix} i_{FR_{REF}} \\ i_{FS_{REF}} \\ i_{FT_{REF}} \end{bmatrix} = \sqrt{\frac{2}{3}} \cdot \begin{bmatrix} 1 & 0 \\ -\frac{1}{2} & \frac{\sqrt{3}}{2} \\ -\frac{1}{2} & -\frac{\sqrt{3}}{2} \end{bmatrix} \cdot \begin{bmatrix} \cos\theta & -\sin\theta \\ \sin\theta & \cos\theta \end{bmatrix} \cdot \begin{bmatrix} i_{p\sim} \\ i_{q\sim} \end{bmatrix} \tag{44}$$

Modified PQ-theory. In the modified PQ theory, the active and reactive powers are defined respectively as:

$$p = u_R \cdot i_R + u_S \cdot i_S + u_T \cdot i_T \tag{45}$$

$$q = u'_R \cdot i_R + u'_S \cdot i_S + u'_T \cdot i_T \tag{46}$$

where u'_R, u'_S, u'_T have 90 degrees phase shift respectively with u_R, u_S, u_T. For the three-phase system, it is valid that:

$$i_T = -(i_R + i_S) \tag{47}$$

Then from Eqs. (20), (21), (22) follows:

$$\begin{bmatrix} p \\ q \end{bmatrix} = \begin{bmatrix} u_R - u_T & u_S - u_T \\ u'_R - u'_T & u'_S - u'_T \end{bmatrix} \cdot \begin{bmatrix} i_R \\ i_S \end{bmatrix} \tag{48}$$

If we multiply both sides of Eq. (48) by the reverse matrix of the voltages, we obtain:

$$\begin{bmatrix} i_R \\ i_S \end{bmatrix} = \frac{1}{\Delta} \begin{bmatrix} u'_S - u'_T & u_T - u_S \\ u'_T - u'_R & u_R - u_T \end{bmatrix} \cdot \begin{bmatrix} p \\ q \end{bmatrix} \tag{49}$$

where the determinant is $\Delta = (u_R - u_T) \cdot (u'_S - u'_T) - (u_S - u_T) \cdot (u'_R - u'_T)$.

Using the same criteria as in the PQ-theory and defining p by means of Eq. (34), the reference curves for the source currents can be calculated:

$$\begin{bmatrix} i_{SR_{REF}} \\ i_{SS_{REF}} \end{bmatrix} = \frac{1}{\Delta} \begin{bmatrix} u'_S - u'_T & u_T - u_S \\ u'_T - u'_R & u_R - u_T \end{bmatrix} \cdot \begin{bmatrix} p \\ 0 \end{bmatrix} \tag{50}$$

$$i_{ST_{REF}} = -(i_{SR_{REF}} - i_{SS_{REF}}) \tag{51}$$

The reference curves of the filter currents can be obtained as a difference of these currents and the load currents.

Method of the synchronous reference frame. In this method the system of three phase load currents i_{LR}, i_{LS}, i_{LT} is transformed in two phase active component i_{Ld} and reactive component i_{Lq} with $\frac{\pi}{2}$ phase shift. These components form a rotating synchronous frame with straight sequence of the three-phase voltage system [37]. The following transformation is used:

$$\begin{bmatrix} i_{Ld} \\ i_{Lq} \\ 0 \end{bmatrix} = \sqrt{\frac{2}{3}} \begin{bmatrix} \cos\theta & \cos\left(\theta - \frac{2\cdot\pi}{3}\right) & \cos\left(\theta + \frac{2\cdot\pi}{3}\right) \\ \sin\theta & \sin\left(\theta - \frac{2\cdot\pi}{3}\right) & \sin\left(\theta + \frac{2\cdot\pi}{3}\right) \\ \frac{1}{\sqrt{2}} & \frac{1}{\sqrt{2}} & \frac{1}{\sqrt{2}} \end{bmatrix} \cdot \begin{bmatrix} i_{LR} \\ i_{LS} \\ i_{LT} \end{bmatrix} \tag{52}$$

Similarly, the three-phase voltage system u_R, u_S, u_T is transformed into u_d, u_q.

The instantaneous active and reactive powers are defined as:

$$p_L = \frac{3}{2}(u_d \cdot i_{Ld} + u_q \cdot i_{Lq}); \qquad q_L = \frac{3}{2} \cdot (u_q \cdot i_{Ld} - u_d \cdot i_{Lq}) \tag{53}$$

When the voltages synchronous frame coincides with the axis R of the three-phase system, the component u_q with phase difference $\frac{\pi}{2}$ is equal to 0. Then:

$$p_L = \frac{3}{2} u_d \cdot i_{Ld}; \qquad q_L = -\frac{3}{2} \cdot u_d \cdot i_{Lq} \tag{54}$$

From the stated above it follows that, the regulation of the active and reactive powers can be done with independently regulation of the components i_{Ld}, i_{Lq} respectively.

Each of these components contains constant and alternative parts, as only the constant part of i_{Ld} ($i_{Ld=}$) defines the active power, that is, $i_{Lq}=0$:

$$i_{Ldq} = i_{Ldq=} + i_{Ldq\sim} \tag{55}$$

The circuit to be implemented is similar to the one in Fig. 28, but the transformation in the first block is done according to Eq. (52). The angle θ and the trigonometric functions are obtained from the PLL block. The constant part of i_{Ld} ($i_{Ld=}$) is removed with a filter, and the reference curves for the source currents are obtained in the last block according to the following transformation:

$$\begin{bmatrix} i_{SR_{REF}} \\ i_{SS_{REF}} \\ i_{ST_{REF}} \end{bmatrix} = \sqrt{\frac{2}{3}} \cdot T^{-1} \cdot \begin{bmatrix} i_{Ld=} \\ 0 \\ 0 \end{bmatrix} \tag{56}$$

where:

$$T^{-1} = \begin{bmatrix} \cos\theta & \cos\left(\theta - \frac{2\cdot\pi}{3}\right) & \cos\left(\theta + \frac{2\cdot\pi}{3}\right) \\ \sin\theta & \sin\left(\theta - \frac{2\cdot\pi}{3}\right) & \sin\left(\theta + \frac{2\cdot\pi}{3}\right) \\ \frac{1}{\sqrt{2}} & \frac{1}{\sqrt{2}} & \frac{1}{\sqrt{2}} \end{bmatrix}^{-1} \tag{57}$$

Modified method of the synchronous reference frame. The values of u_α, u_β and $i_{L\alpha}$, $i_{L\beta}$ are defined with Clarke's transformation. The necessary trigonometric functions for Park's transformation, from which are defined i_{Ld}, i_{Lq} by $i_{L\alpha}$, $i_{L\beta}$, are obtained from:

$$\cos\theta = \frac{u_\alpha}{\sqrt{u_\alpha^2 + u_\beta^2}}; \quad \sin\theta = \frac{u_\beta}{\sqrt{u_\alpha^2 + u_\beta^2}}; \quad -\sin\theta = -\frac{u_\beta}{\sqrt{u_\alpha^2 + u_\beta^2}} \tag{58}$$

$$\begin{bmatrix} i_{Ld} \\ i_{Lq} \end{bmatrix} = \begin{bmatrix} \cos\theta & \sin\theta \\ -\sin\theta & \cos\theta \end{bmatrix} \cdot \begin{bmatrix} i_{L\alpha} \\ i_{L\beta} \end{bmatrix} \tag{59}$$

From Eqs. (58), (59) it follows that:

$$i_{Ld} = \frac{u_\alpha \cdot i_{L\alpha} + u_\beta \cdot i_{L\beta}}{\sqrt{u_\alpha^2 + u_\beta^2}} = \frac{p_{L\alpha\beta}}{\sqrt{u_\alpha^2 + u_\beta^2}} = i_{Ld=} + i_{Ld\sim} \tag{60}$$

And again, only the constant part of i_{Ld} is used:

$$i_{Ld=} = \left(\frac{p_{L\alpha\beta}}{\sqrt{u_\alpha^2 + u_\beta^2}} \right)_= \tag{61}$$

The reference curves for the source currents in a $\alpha - \beta$ coordinate system are obtained from the following expression:

$$\begin{bmatrix} i_{S\alpha_{REF}} \\ i_{S\beta_{REF}} \end{bmatrix} = i_{Ld=} \cdot \frac{1}{\sqrt{u_\alpha^2 + u_\beta^2}} \begin{bmatrix} u_\alpha \\ u_\beta \end{bmatrix} \tag{62}$$

Power factor equal to 1. After the switching on of the active power filter, the source currents should be in phase and with the same waveform as the source voltages:

$$[i_{S_{REF}}] = K \cdot [u] \tag{63}$$

Then the consumed active power will be:

$$p = [i_{S_{REF}}] \cdot [u] = K \cdot [u] \cdot [u] = K \cdot \left[u_0^2 + u_\alpha^2 + u_\beta^2 \right] \tag{64}$$

If we ignore the losses in the filter, then $p = p_L$ and the constant K is defined from the expression here below:

$$K = \left(\frac{p_{L\alpha\beta} + p_{L0}}{\left(u_0^2 + u_\alpha^2 + u_\beta^2 \right)} \right)_= \tag{65}$$

Then the reference curves for the source currents in a $\alpha - \beta$ coordinate system could be defined from:

$$\begin{bmatrix} i_{S0_{REF}} \\ i_{S\alpha_{REF}} \\ i_{S\beta_{REF}} \end{bmatrix} = \frac{p_{L\alpha\beta} + p_{L0}}{\left(u_0^2 + u_\alpha^2 + u_\beta^2 \right)_=} \cdot \begin{bmatrix} u_0 \\ u_\alpha \\ u_\beta \end{bmatrix} \tag{66}$$

Equivalent sine wave method. This method is a variation of the previously presented method. It is suitable for single phase and symmetrical three phase systems and loads [38]. The instantaneous values of one of the source voltages and its respective load current are transformed in discrete values.

In each period is defined the active power P per phase and the effective value of the voltage U:

$$P \approx \frac{1}{N} \cdot \sum_{i=1}^{N} u_i \cdot i_i; \qquad U \approx \sqrt{\frac{1}{N} \sum\nolimits_{i=1}^{N} u_i^2} \tag{67}$$

where N is the number of the discrete values.

If the power factor of the system active power filter-load is approximately equal to 1 and the losses in the filter are neglected, the maximum value of the source current can be defined from:

$$I_{SM_{REF}} \approx \frac{P \cdot \sqrt{2}}{U} \tag{68}$$

By multiplication of this value with *sin θ*, obtained for example by means of phase locked loop for the phase voltage, is obtained the reference curve for the source current for the respective phase:

$$i_{S_{REF}} = I_{SM_{REF}} \cdot sin\ \theta \tag{69}$$

The reference curves for the other phase currents are defined with their respective phase shift.

Elimination of the harmonics, the reactive power and the imbalance. This method is similar to the method with power factor equal to 1, but only the first harmonics of the voltages with direct sequence are considered:

$$[i_{S_{REF}}] = K \cdot [u_1^+] \tag{70}$$

Then the total active power of the system filter-load is:

$$p = [i_{S_{REF}}] \cdot [u] = K \cdot [u] \cdot [u_1^+] = K \cdot \left[u_\alpha \cdot u_{1\alpha}^+ + u_\beta \cdot u_{1\beta}^+\right] \tag{71}$$

The constant K is defined from the expression:

$$K = \left(\frac{p_{L\alpha\beta} + p_{L0}}{\left(u_{1\alpha}^{+2} + u_{1\beta}^{+2}\right)} \right)_{=} \tag{72}$$

The reference curves for the source current in $\alpha - \beta$ coordinate system are defined from:

$$\begin{bmatrix} i_{S\alpha_{REF}} \\ i_{S\beta_{REF}} \end{bmatrix} = \left(\frac{p_{L\alpha\beta} + p_{L0}}{\left(u_{1\alpha}^{+2} + u_{1\beta}^{+2}\right)} \right)_{=} \cdot \begin{bmatrix} u_{1\alpha}^+ \\ u_{1\beta}^+ \end{bmatrix} \tag{73}$$

Methods with tracking of the voltage of the capacitor C_d. There are methods which use only the fluctuations of the voltage of the storage element as information for the balance of the powers, without extraction of a reference signal for the current. The tracking values are the source voltage, the voltage U_d of the filter capacitor C_d and the source current. The source voltage is used for generation of reference sinusoid with amplitude 1 and in phase with it. The fluctuations of the voltage of the filter capacitor are used as information for the balance of the powers. In other cases, the generated reference sinusoid in phase with the source voltage is proportional to its maximum value.

Other methods. There are other methods which are less developed and used, including flow-based method [39], regulation in open loop system [40], direct current control [41], and odd harmonic digital repetitive control [42]. Special solutions are applied for unbalanced systems and nonlinear loads [43–46].

4.2 Forming of Curve $i_{F_{REF}}$ of the Active Power Filter Current or $i_{S_{REF}}$ of the Source Current, Corresponding to the Reference Curve

This part presents the methods of forming of the control signals for the power semiconductor devices of the active power filter. Should be obtained a curve $i_{F_{REF}}$ of the active power filter current or $i_{S_{REF}}$ of the source current, corresponding to the reference curve.

Pulse-width modulation control. The operation is described in Fig. 29.

The difference between the reference and the real value of the current defines the input signal of the PI-regulator. The output signal of the regulator U_C is compared to a voltage with triangle waveform U_T and frequency f_T. The switching of the power semiconductor devices is done by the pulse width modulation method with constant frequency $f_{SW}=f_T$. The advantages of this method are the constant frequency and the low distortions of the obtained curve for the current [47].

Hysteresis current control. The switch of the power devices in this method provides the follow-up of the reference curve with a certain hysteresis. It could be constant or adaptive hysteresis. Fig. 30 shows the operation with hysteresis current control of the power circuit of Fig. 7.

The adaptive hysteresis (variation of H of the comparator) provides more accurate tracking of the reference curve. In both cases, a disadvantage is the

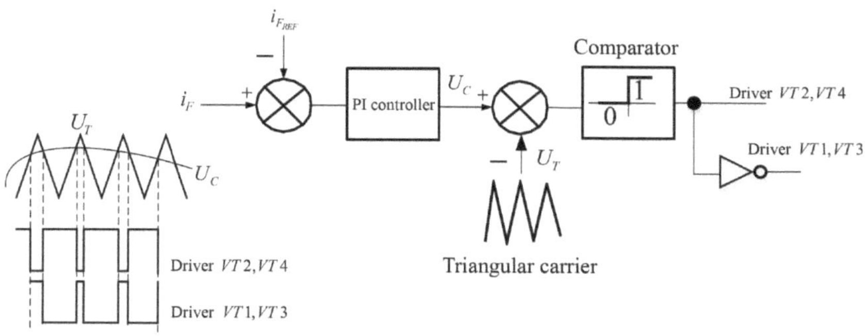

FIG. 29 Realization of pulse width modulation control.

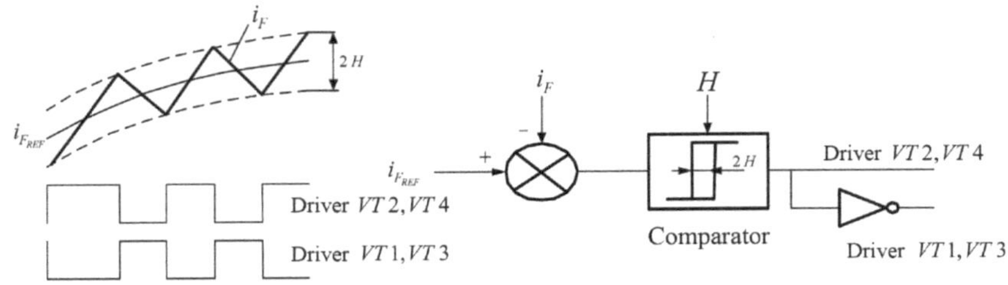

FIG. 30 Realization of hysteresis current control.

variable switching frequency of the power semiconductor devices. Fig. 31 presents hysteresis current control with maximum frequency limitation and adaptive hysteresis, but with higher pulsations around reference curve [48]. For the operation with constant hysteresis of the circuit of Fig. 29 the average frequency of switching of the semiconductor devices for half-period is defined by [38]:

$$f_{SW_{AV}} \approx \frac{U_d}{4 \cdot H \cdot L_F}\left[1-\left(\frac{U_S}{U_d}\right)^2\right] \tag{74}$$

Current control with delta modulation. This control method is similar to the hysteresis current control with switching frequency limitation as shown in Fig. 31. The delta modulation control is presented in Fig. 32. The comparator used is without hysteresis. The advantage of this method, compared to hysteresis current control, is constant switching frequency. The disadvantage, compared to the pulse width modulation control, is the more important distortions of the current curve.

One-cycle control. The power semiconductor devices are controlled with constant switching frequency f_{SW}. In each switching period, there are two switching states. The first state for each phase is for the period $.T_{SW}$, when switched on the upper transistors for the respective phase of the circuit of Fig. 17—*VT*1, *VT*3, *VT*5. The second state is for the period $(1-D)$. T_{SW},

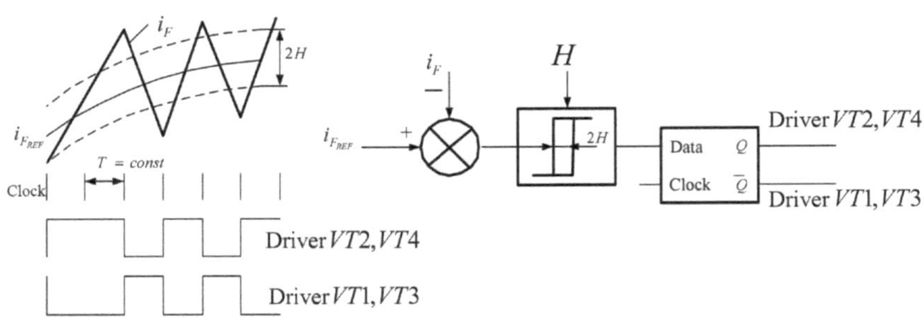

FIG. 31 Realization of hysteresis current control with switching frequency limitation.

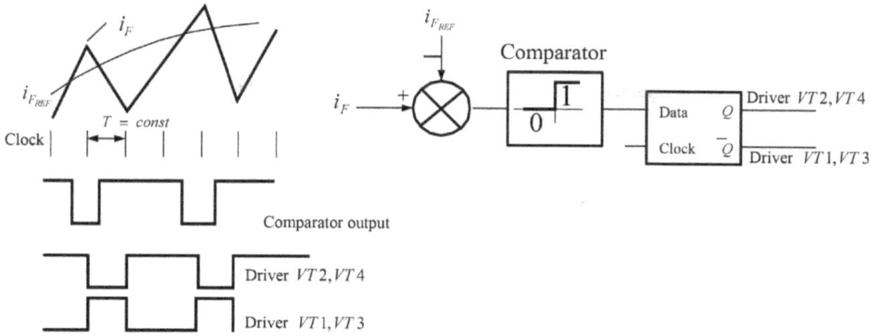

FIG. 32 Realization of delta modulation control.

when switched on the lower transistors for the respective phase of the circuit of Fig. 17—*VT*4, *VT*6, *VT*2. $D = \frac{t_{ON}}{T_{SW}}$ is the duty ratio, and t_{ON} is the time of switching on of the upper transistors. The dependence between the value of the voltage of the filter capacitor $- U_d$ and the instantaneous value u_S of the phase voltage of the source at the moment of the respective switching is:

$$U_d = \frac{1}{2 \cdot D - 1} \cdot u_S \tag{75}$$

If the power factor is 1, the system active power filter-load should be equivalent to a resistor R_E, that is,

$$u_S = R_E \cdot i_S \tag{76}$$

If we replace Eq. (76) in Eq. (75), and multiply both sides with R_S (the resistance of the source) after some transformations we obtain:

$$(2 \cdot D - 1) \cdot u = R_S \cdot i_S \tag{77}$$

where $u = \frac{R_S}{R_E} \cdot U_d$. Expression (77) allows the definition of the duty ratio D in each switching period according to the instantaneous value of the sinusoidal current of the source i_S.

Sliding mode control. The idea of this method is the tracking of a specific trajectory of one of the currents $i_{F_{REF}}$ or $i_{S_{REF}}$ or the voltage u_{LREF} [49]. Then it can be used the sliding mode control, used also for other cases—DC-DC converters, electric motors control, etc. The basis of this control is the forcing of one or more variables (often but not mandatory, state variables) to follow a specific trajectory, called sliding surface. If the power factor of the system filter-load is 1, the current, consumed from the source, should follow the curve of its voltage:

$$i_{S_{REF}} = K \cdot u_S \tag{78}$$

where K is a coefficient, based on the active power, necessary for the load. If we suppose that the waveform is sinusoidal, the form of i_S could be found by the method of the equivalent sinusoid from Eq. (69). The most frequently used form of description of the sliding surface is:

$$s = i_S - i_{S_{REF}} = i_S - K \cdot u_S = 0 \tag{79}$$

If the source current is on the sliding surface, then there is equality. In order to guarantee, that the system can stay on this surface, it is necessary to find the suitable control, that should satisfy the following inequality:

$$s \cdot \dot{s} \leq 0 \tag{80}$$

for all values, that the current from the source i_S could have.

From the last inequality, it follows that if $s > 0$, then $\dot{s} < 0$ and if $s < 0$, then $\dot{s} > 0$. From Eq. (79) it follows, that the sign of $\dot{s}$ could depend on the sign of the differential of the current $\dot{i}_S$, as the differential of the current $i_{S_{REF}}$ is defined. As $i_S = i_F + i_L$, then the sign of the differential of the current $\dot{i}_S$ depends on the sign of the differential of the current of the filter i_F. This is done by means of switching the respective couple of transistors as it was described for the equivalent circuits of Fig. 11.

There are some other types of sliding surfaces: in $d-q$ plain; by use of a system of differential equations and state variables (i_{FR}, i_{FS}, i_{FT}, u_d).

Predictive current control. Predictive current control has some advantages such as accurate tracking of the reference curve for the current in a wide frequency range, based on mathematical model [50]. To provide the stability of the system, the parameters of this model should be recognizable (by means of possible measurement) which is possible in the case of active power filters. From Fig. 7 could be defined the following differential equation for single phase active power filter, which is valid for one of the phases of the three-phase filter of Fig. 17:

$$\frac{di_F}{dt} + \frac{R_F}{L_F} \cdot i_F = \frac{u - u_C}{L_F} \tag{81}$$

where R_F is the resistance in the phase (in series with L_F).

Supposing that

$$\frac{di_F}{dt} \approx \frac{i_F[k+1] - i_F[k]}{T_S} \tag{82}$$

$T_S = const.$ is the duration of the interval, $i_F[k+1]$, $i_F[k]$ is the current of the filter in $[k+1]$ and $[k]$ interval. Then from Eqs. (81), (82) it follows that:

$$i_F[k+1] = i_F[k] \cdot a + (u[k] - u_C[k]) \cdot b \tag{83}$$

where a and b are constants:

$$a = \left(1 - \frac{R_F}{L_F} \cdot T_S\right); \qquad b = \frac{T_S}{L_F} \tag{84}$$

If during one interval T_S the values of the source voltage $u[k]$ and the value of the converter voltage $u_C[k]$ are supposed constant, respectively equal to $U[k]$ and $U_C[k]$ (the mean values of the voltages in the $[k]$ interval), then Eq. (83) is transformed to:

$$i_F[k+1] = i_F[k] \cdot a + (U[k] - U_C[k]) \cdot b \tag{85}$$

The error in the values of the reference $i_{F_{REF}}[k+1]$ and the real $i_F[k+1]$ value of the current of the filter in the $[k+1]$ interval will be:

$$\Delta i_F[k+1] = i_{F_{REF}}[k+1] - i_F[k+1] \tag{86}$$

As the objective is that this error should be 0:

$$i_{F_{REF}}[k+1] = i_F[k+1] \tag{87}$$

Then from Eqs. (85), (87) follow:

$$U_C[k] = U[k] + \frac{1}{b} \cdot \left(i_{F_{REF}}[k+1] - i_F[k] \cdot a\right) \tag{88}$$

From Fig. 10 and from the description of the operation of the single-phase filter from Fig. 7 it is obvious that the mean value of the voltage in the $[k]$ interval $U_C[k]$ could be defined by the times of switch on of the couples of transistors, supposing that $U_d = const.$

The previous equation also defines the output voltage for each phase of the three-phase active power filter in the $[k]$ interval, as the target function (as close as possible to 0) could be:

$$\begin{aligned} G[k+1] = {} & \left(i_{FR_{REF}}[k+1] - i_{FR}[k+1]\right)^2 + \left(i_{FS_{REF}}[k+1] - i_{FS}[k+1]\right)^2 \\ & + \left(i_{FT_{REF}}[k+1] - i_{FT}[k+1]\right)^2 \end{aligned} \tag{89}$$

In each interval is defined the vector $\begin{bmatrix} U_{CR} \\ U_{CS} \\ U_{CT} \end{bmatrix}$, which minimizes the target function, and to the transistors of the filter are applied the respective signals to obtain it according to the used modulation (sinusoidal PWM, Space Vector PWM, etc.).

Deadbeat control. Similar to the previous method, this one is based on a differential Eq. (81) and the expression (83). Supposing that $R_F \approx 0$, then $a = 1$ and:

$$i_F[k+1] = (U[k] - U_C[k]) \cdot \frac{T_S}{L_F} + i_F[k] \tag{90}$$

In the case of deadbeat control the algorithm for definition of the voltage of the converter in the $[k+1]$ interval is [51]:

$$U_C[k+1] = \frac{L_{FM}}{T_S}\left(i_{F_{REF}}[k] - i_F[k]\right) + 2 \cdot U[k] - U_C[k] \tag{91}$$

where L_{FM} is modeled inductance, $U[k]$ is measured or predicted. The predicted value of the source voltage could be defined by means of Eq. (90):

$$e[k-1] = U_C[k-1] + \frac{L_{FM}}{T_S} \cdot (i_F[k] - i_F[k-1]) \tag{92}$$

The defined predicted value is used in Eq. (85), where $U[k]$ is replaced with $e[k-1]$ [52].

Fuzzy logic control and genetic algorithm control. The fuzzy logic is also used to follow the reference curve for the current [52,53]. It is also possible on the base of fuzzy logic to define the value of the adaptive hysteresis, according to the difference between the reference and actual values of the current—this is the so called fuzzy logic hysteresis control of shunt and series filters [54]. An example of definition of the input variables of the fuzzy controller in series active power filters (Fig. 3) is presented in Fig. 33.

The inputs of the fuzzy controller are the error of the voltage $e[k]$ and the variation of the error $ce[k]$. They are defined as follows:

$$e[k] = u_{LREF}[k] - u_L[k]; \qquad ce[k] = e[k] - e[k-1] \tag{93}$$

where $u_L[k]$ is the voltage of the load in the moment $k \cdot T_S$, and $u_{LREF}[k]$ is the reference voltage at this moment. T_S defines the duration of one interval, and k is the number of the interval. The outputs of the controller are the values of the turn on time of the couple of transistors for increase or decrease of the voltage, according to the sign and the value of the input variables. The genetic algorithm is based on Darwin's theory of natural selection and the work of Mendel in the domain of the genetic heritage—the stronger organism is the most probable to survive in the conditions of a competitive environment. By calculation and combination of multiple possible solutions of a problem, the best one could be found. There are some applications of the genetic algorithms to control active power filters [55].

4.3 Stabilization of the Voltage U_d of the Energy Storage Capacitor C_d

As the active power filter does not give active power to the system, if we ignore the losses in the filter, the energy of the storage element-capacitor or inductance

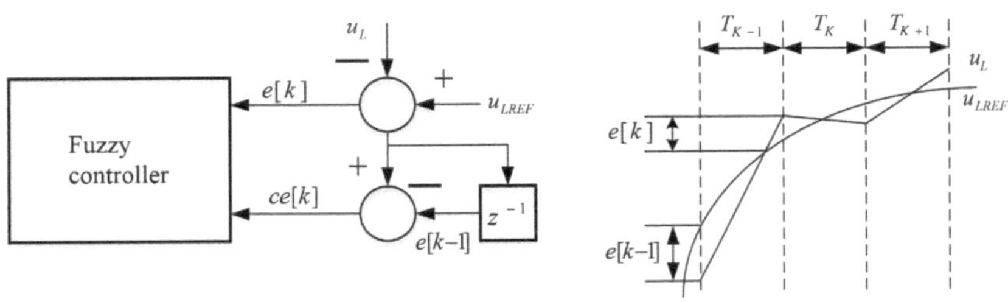

FIG. 33 Realization of fuzzy logic control.

should be constant. In that case, the used energy storage element is a capacitor, two factors influence the variation of its voltage. The first one is the balance of the active power in the transient modes, as in fact there are active power losses in the filter. The other one is the reactive power of the load, which should be compensated in each period this energy storage element dissipates or consumes reactive power, so that there should be variations of its voltage. There are different methods to keep this voltage constant, as it is not necessary that this regulation loop should have so quick action as the loop for tracking of the reference curve for the current. Often the difference between the reference value and the actual value of the voltage U_d is sent to the PI-regulator. The output signal of the regulator is multiplied by the defined reference curve for the current $i_{F_{REF}}$ of the active power filter or of the source current $i_{S_{REF}}$, as the resulting curve is followed by a control system using one of the presented methods. Usually in the control systems of active power filters there is internal loop for current control and external loop for control of the voltage of the capacitor U_d. The predefined value for single phase filters should be higher than the maximum value of the phase voltage, and for three phase filters—than the maximum value of the line to line voltage.

Control algorithms are presented also in Ref. [56]. In the practical use of the active power filters should be taken in consideration the technical and economic factors [57].

REFERENCES

[1] L. Gyugyi, E. Strycula, in: Active AC power filters, IEEE Industrial Applications Society Annual Meeting, 1976, pp. 529–535.

[2] N. Mohan, H.A. Peterson, W.F. Long, G.R. Dreifuerst, J.J. Vithaythil, Active filters for AC harmonic suppression, IEEE Winter Power Meeting, 1977.

[3] D. Detjen, S. Schroder, R.W. De Doncker, Embedding DSP control algorithms in Pspice, IEEE Trans. Power Electron. 18 (1) (2003) 294–299.

[4] S.-H. Li, C.-M. Liaw, Paralleled DSP-based soft switching-mode rectifiers with robust voltage regulation control, IEEE Trans. Power Electron. 19 (4) (2004) 937–946.

[5] W. Zhang, G. Feng, Y.-F. Liu, B. Wu, A digital power factor correction (PFC) control strategy optimized for DSP, IEEE Trans. Power Electron. 19 (6) (2004) 1474–1485.

[6] H. Akagi, Modern active filters and traditional passive filters, Bull. Polish Acad. Sci. Tech. Sci. 54 (3) (2006) 255–269.

[7] H. Akagi, New Trends in Active Filters for Improving Power Quality, Okayama University, Electrical Engineering Fields, 1996, pp. 417–425.

[8] T.P. Cheng, Z. Salam, A. Jusoh, Harmonics mitigation using active power filter: a technological review, Elektrika: J. Electr. Eng. 8 (2) (2006) 17–26.

[9] J.N. Chiasson, L.M. Tolbert, K.J. McKenzie, Z. Du, A complete solution to the harmonic elimination problem, IEEE Trans. Power Electron. 19 (2) (2004) 491–499.

[10] N.K. Poon, B.M.H. Pong, C.K. Tse, A constant-power battery charger with inherent soft switching and power factor correction, IEEE Trans. Power Electron. 18 (6) (2003) 1262–1269.

[11] A. Prasai, D. Divan, Zero-energy sag corrector with reduced device count, IEEE Trans. Power Electron. 24 (6) (2009) 1646–1653.
[12] H. Akagi, H. Fujita, A new power line conditioner for harmonic compensation in power systems, IEEE Trans. Power Deliv. 10 (3) (1995) 1570–1575.
[13] M. McGranaghan, S. Santoso, Challenges and trends in analyses of electric power quality measurement data, EURASIP J. Adv. Signal Process. 2007.
[14] Z. Pan, F.Z. Peng, S. Wang, Power factor correction using a series active filter, IEEE Trans. Power Electron. 20 (1) (2005) 148–153.
[15] E.A. Ribeiro, I. Barbi, Harmonic voltage reduction using a series active filter under different load conditions, IEEE Trans. Power Electron. 21 (5) (2006) 1394–1402.
[16] V.F. Corasantini, M.B. Barbieri, P.L. Arnera, M.I. Valla, Hybrid active filter for reactive and harmonics compensation in a distribution network, IEEE Trans. Ind. Electron. 56 (3) (2009) 670–677.
[17] M.K. Darwish, M. El-Habrouk, I. Kasikci, EMC compliant harmonic and reactive power compensation using passive filters cascaded with shunt active filter, EPE J. 12 (3) (2002) 43–50.
[18] G. Meckien, R. Strzelecki, in: Single phase power line conditioners – without transformers, Proc. BEC, 2002, pp. 385–388.
[19] G. Molleykutty, Artifical intelligence based three-phase unified power quality conditioner, J. Comput. Sci. 3 (7) (2007) 465–477.
[20] S. Rahmani, F. Fnaiech, A hybrid structure of series active and passive filters to achieving power quality criteria, in: IEEE International Conference on Systems, Man and Cybernetics, vol. 3, (2002).
[21] H. Akagi, Trends in active power line conditioners, IEEE Trans. Power Electron. 9 (3) (1994) 263–268.
[22] Z. Sun, X. Jiang, D. Zhu, G. Zhang, A novel active power quality compensator topology for electrified railway, IEEE Trans. Power Electron. 19 (4) (2004) 1036–1042.
[23] F. Barrero, S. Martinez, F. Yeves, P.M. Martinez, Universal and reconfigurable to UPS active power filter for line conditioning, IEEE Trans. Power Deliv. 18 (1) (2003) 283–290.
[24] T. Sousa, V. Monteiro, J. Pinto, J. Afonso, J.L. Afonso, Unified architecture of single-phase active power filter with battery interface for UPS operation, Proceedings of the Word Congress on Engineering, vol. 1, WCE, London, UK, 2017.
[25] K.J.P. Macken, K. Vanthournout, J.V. den Keybus, G. Deconinck, R.J.M. Belmans, Distributed control of renewable generation units with integrated active filter, IEEE Trans. Power Electron. 19 (5) (2004) 1353–1360.
[26] S. Srinath, S. Poongothai, T. Aruna, PV integrated shunt active filter for harmonic compensation, Energy Procedia 117 (2017) 1134–1144.
[27] J.R. Babu, R.S. Rao, P.L. Reddy, Shunt active power filter connected to MPPT based photovoltaic array for PQ enhancement, Int. J. Pure Appl. Math. 114 (9) (2017) 389–398.
[28] M.H. Antchev, M.P. Petkova, V.T. Gurgulitsov, H.M. Antchev, in: Study of single-phase bidirectional AC/DC converter with high power factor, EUROCON, St. Petersburg, Russia, 2009, p. 2009.
[29] A. Chaghi, A. Guettafi, A. Benoudjit, Four-legged active power filter compensation for a utility distribution system, J. Electr. Eng. 55 (1–2) (2004) 31–35.
[30] E.C. Secaran, P.N. Anbalagan, C. Palanisamy, Analysis and simulation of a new shunt active power filter using cascaded multilevel inverter, J. Electr. Eng. 58 (5) (2007) 241–249.
[31] J. Mossoba, P.W. Lehn, A controller architecture for bandwidth active power filters, IEEE Trans. Power Electron. 18 (1) (2003) 317–325.

[32] T. Lee, J. Li, P. Cheng, Discrete frequency tuning active filter for power system harmonics, IEEE Trans. Power Electron. 24 (5) (2009) 1209–1222.

[33] A. Roux, H. Mouton, H. Akagi, DFT-based repetitive control of a series active filter integrated with 12-pulse rectifier, IEEE Trans. Power Electron. 24 (6) (2009) 1515–1521.

[34] H. Akagi, New trends in active filters for power conditioning, IEEE Trans. Ind. Appl. 32 (6) (1996) 1312–1322.

[35] T. Skvarenina, The Power Electronics Handbook, CRC Press, Boca Raton, 2002.

[36] L.S. Czarnecki, On some misinterpretations of the instantaneous reactive power p-q theory, IEEE Trans. Power Electron. 3 (2004) 19.

[37] P. Zhu, X. Li, Y. Kang, J. Chen, Control scheme for a universal power quality manger in a two-phase synchronous rotating frame, IEE Proc. Gener. Transm. Distrib. 151 (5) (2004) 590–595.

[38] M.H. Antchev, M.P. Petkova, Operating characteristics of shunt active power filter with hyteresis – current control, 5-th IASTED Conf. Power and Energy Systems, EuroPES, Benalmadena, Spain, Proceedings, 2005, pp. 399–405.

[39] S. Bhattacharya, A. Veltman, D.M. Divan, R.D. Lorenz, Flux-based active filter controller, IEEE Trans. Ind. Appl. 32 (3) (1996) 491–502.

[40] M. Salo, H. Tuusa, A novel open-loop control method for a current-source active power filter, IEEE Trans. Ind. Electron. 50 (2) (2003) 960–971.

[41] D. Nedeljkovic, M. Nemec, V. Ambrozic, Application of direct current control in a three-phase parallel active power filter, Electrotech. Rev. (Slovenia) 76 (3) (2009) 85–91.

[42] R. Costa-Castello, R. Grino, E. Fossas, Odd-harmonic digital repetitive control of a single-phase current active filter, IEEE Trans. Power Electron. 19 (4) (July 2004) 1060–1068.

[43] B.-R. Lin, Y.-C. Lee, Three-phase power quality compensator under the unbalanced sources and nonlinear loads, IEEE Trans. Ind. Electron. 51 (5) (2004) 1009–1017.

[44] S. Seguf, F. Gimeno, S. Ords, R. Masot, M. Alcartiz, Control of shunt unbalanced active compensators for reactive and asymmetry elimination in four wire electrical systems using symmetrical components, 33rd Annual PESC, Aachen, Germany, 2004, pp. 3139–3145.

[45] K. Uyyuri, M.K. Mishra, A. Ghosh, An optimization-based algorithm for shunt active filter under distorted supply voltages, IEEE Trans. Power Electron. 24 (5) (2009) 1223–1249.

[46] H. Zhong, P. Chen, Z. Lu, Z. Quian, in: Novel control scheme based on per – phase current calculation for active series power filter with fundamental current bypass channel in unbalanced conditions, 35th Annual PESC, Aachen, Germany, 2004, pp. 999–1002.

[47] H. Fujita, A single-phase active filter using an H-Brihge PWM converter with a sampling frequency quadruple of the switching frequency, IEEE Trans. Power Electron. 24 (4) (2009) 934–941.

[48] M.H. Antchev, M.P. Petkova, M. Petkov, Single phase shunt active power filter with frequency limitation and hysteresis - current control, 5-th Power Conversion Conference PCC Nagoya, Japan, Proceedings, 2007, pp. 97–102.

[49] M.H. Antchev, M.P. Petkova, V.T. Gurgulitsov, in: Sliding mode control of a series active power filter, IEEE Conf. EUROCON, Warshaw, Poland, Proceedings, 2007, pp. 1344–1349.

[50] M. Odavic, P. Zanchetta, M. Summer, Z. Jacopovic, in: High performance predictive current control for active shunt filters, EPE-PEMC, Portoroz, Slovenia, 2006, pp. 1677–1681.

[51] R. Panigrahi, B. Subudhi, P. Panda, A comparative assessment of hysteresis and dead-beat controllers for performances of three phase shunt active power filtering, J. Power Technol. 94 (4) (2014) 286–295.

[52] P. Kiramanich, R.M. O'Connell, Fuzzy logic control of an active power line conditioner, IEEE Trans. Power Electron. 19 (6) (2004).

[53] A.J. Laxmi, G.T.R. Das, K.U. Rao, Role of PI and fuzzy controller in unified power quality conditioner, ARPN J. Eng. Appl. Sci. 2 (2) (2007) 1–10.
[54] A. Tomova, M. Antchev, M. Petkova, H. Antchev, Fuzzy logic hysteresis control of a single-phase on-grid inverter: computer investigation, Int. J. Power Electron. Drive Syst. 3 (2) (2013) 1–8. ISSN: 2088-8694.
[55] J.R. Vazquea, P. Salmeron, Active power filter control using neural network technologies, IEE Proc. Electr. Power Appl. 150 (2) (2003).
[56] Y. Hoon, M.A.M. Radzy, M. Hassan, N.F. Mailah, Control algorithms of shunt active power filter for harmonics mitigation: a review. Energies 10 (2017), https://doi.org/10.3390/en10122038.
[57] M.H. Antchev, A.F. Zobaa, M.P. Petkova, Economics reasons to use shunt active power filters, Proc. Tech. Univ. Sofia 56 (1) (2006) 177–183. ISSN: 1311-0829.

Chapter 10

Optimization-Based Power Capacitor Model Parameterization for Decision Support in Power Distribution Systems

Guilherme Spavieri*, Guilherme G. Lage*, Ricardo A.S. Fernandes*, Daniel Barbosa† and Mário Oleskovicz‡

**Department of Electrical Engineering, Federal University of São Carlos, São Carlos, SP, Brazil,* †*Polytechnic School, Federal University of Bahia, Salvador, BA, Brazil,* ‡*Department of Electrical and Computer Engineering, University of São Paulo, São Carlos, SP, Brazil*

CHAPTER POINTS

- This chapter presents the formulation of the capacitor model parameterization problem when the voltage across the capacitors' terminals has harmonic distortions.
- The optimization-based parameterization methodology determines the parameters of an electrical model by minimizing the difference between the reference current acquired in laboratory tests and the calculated current flowing through the terminals of the parameterized power capacitor model.
- To realize the actual parameterization, a methodology to acquire the necessary reference currents in laboratory tests is also presented.
- Three capacitor models, two metalized polypropylene power capacitors (one all-film and one impregnated), and four voltage waveforms (one sinusoidal and three with harmonic distortions) are considered to show the effectiveness of the optimization-based parameterization methodology.
- Such an optimization-based methodology is applicable as a prior tool to support capacitor allocation studies in power distribution systems.

Classical and Recent Aspects of Power System Optimization. https://doi.org/10.1016/B978-0-12-812441-3.00010-0
© 2018 Elsevier Inc. All rights reserved.

1 INTRODUCTION

Due to inductive and nonlinear loads in power distribution systems, preventive and corrective measures must be deployed to either correct power factors or mitigate harmonic distortions. Power capacitors are regarded as one of the most practical devices to be used for such purposes [1]. Nevertheless, special attention must be given to power capacitor allocation since these devices might resonate with a specific harmonic component, causing amplification of the voltage and current distortions resulting in equipment overheating. These excessive currents result in additional losses and adverse effects, such as fuses blowing and reduced service life and compromising the capacitor bank operation [2, 3].

Methodologies for optimally allocating power capacitors in distribution feeders have been proposed in the literature [1, 4–8] with the objective of improving voltage profile and reducing power losses. However, these papers disregard the capacitors' assembly aspect and their response for different frequencies (i.e., their proper modeling). Capacitor modeling by electrical circuits considering their assembly aspect and response for different frequencies may be found in Refs. [9–13]. However, their parameterization is based on characteristics that are not provided by capacitor manufacturers, for example, the distributed resistance of the metalized film and the capacitance by area unit [14].

It is crucial that capacitor models have their parameters estimated to be used in power system operation and planning studies. Nevertheless, their proper parameterization becomes difficult when the voltage across the capacitors' terminals has harmonic distortions. In this context, the formulation of the problem of power capacitor parameterization considering harmonic distortions is very important to evaluate the capacitor behavior in real applications with nonlinear loads and harmonic generators.

Thus, this chapter presents an optimization-based parameterization methodology to determine the parameters of a capacitor's electrical model by minimizing the difference between the reference current acquired in laboratory tests and the calculated current flowing through the terminals of the parameterized model. The motivation to formulate and solve such a problem is that manufacturers usually define capacitor parameters in terms of the network operation frequency, and these do not present reliable results when electrical power systems are under the influence of harmonic distortions. In fact, only a range of values is known for many of these parameters, as their values vary according to the type of capacitor and network operation frequency. Thus, the use of more detailed capacitor models becomes crucial to allow precise power systems simulation.

Therefore, the formulation of the capacitor parameterization problem is very important and it consists of minimizing the mean squared error between the reference current and the calculated current flowing through the capacitor's parameterized model, subject to the parameter's lower and upper bounds. Since such a calculated current is a function of the capacitor model itself and its parameters, for each capacitor model there is a different formulation of the parameterization problem. It is important to highlight at this point that the mean squared error between the reference and calculated currents is chosen as the objective function because it results in a differentiable function (differently than the mean absolute error, for example), allowing result comparisons among different metaheuristics and mathematical programming methods.

The development of each problem in terms of three capacitor models found in the literature [9, 12, 13] is presented and discussed. After formulating the capacitor parameterization problem, the optimization-based parameterization methodology for three capacitor models is presented, considering two types of metalized polypropylene power capacitors (one all-film [15] and one impregnated [16]), and four different voltage waveforms (one sinusoidal and three with harmonic distortions). As three capacitor models, two types of power capacitors, and four different voltage waveforms are considered, the resulting 24 problems of power capacitor parameterization are, then, solved by an optimization algorithm. Such a methodology is intended to be applicable as a prior tool to support capacitor allocation studies in power distribution systems.

The remainder of this chapter is organized as follows. In Section 2, the three power capacitor models whose parameters are to be estimated are presented. Section 3 presents the development and formulation of the capacitor parameterization problem. The laboratory setup and data acquisition methodology are presented in Section 4. At last, Section 5 presents a summary of the power capacitor optimization-based parameterization methodology and some numerical results obtained for the metalized propylene power capacitors considered in this chapter.

2 CAPACITOR MODELS

As previously mentioned, many electrical circuits have been proposed in the literature to model the actual response of capacitors. The three models in [9, 12, 13] have been chosen to allow such an analysis and compare their efficacy.

In [9, 12], the authors present compact models considering resistances associated with the capacitor's material, insulation and electrode contacts, an inductance and its characteristic capacitance. Despite their compactness, these models are well suited to model power capacitors, since these devices operate at low frequencies. In [13], the authors present a model that is suited to capacitors operating at both low and high frequencies. In the latter, a few considerations on the modeling must be made so that it may be used to model power capacitors.

2.1 Capacitor Model 1

The capacitor model presented in [9], namely *Model 1* and depicted in Fig. 1, has four parameters to be estimated: resistance (R_{ES}), conductance (G_{ES}), inductance

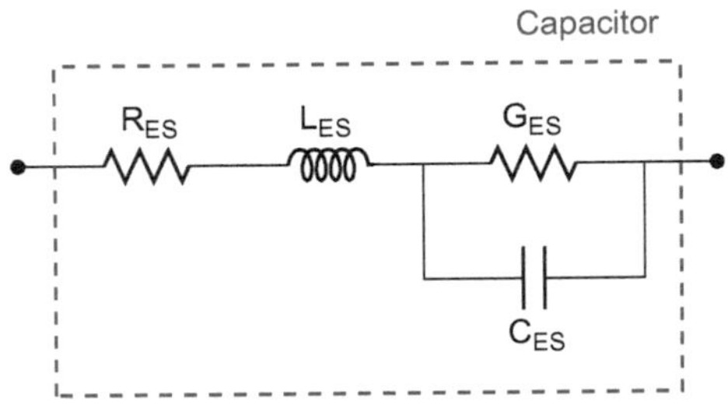

FIG. 1 Capacitor *Model 1*. Equivalent circuit for a capacitor proposed in [9].

(L_{ES}), and capacitance (C_{ES}). This model is considered adequate for many applications at low frequencies and it is commonly used in power system simulators.

In *Model 1*, the equivalent series resistance and capacitance are represented, respectively, by R_{ES} and C_{ES}. These parameters are derived from diffusion equations based on the distributed resistance of the metalized film and its capacitance by area unit. The parameter L_{ES} represents the series equivalent inductance associated with the capacitor's terminal connections and the length of the metalized film. At last, the parameter G_{ES} represents the capacitor's dielectric conductance; for metalized polypropylene power capacitors, the dielectric conductance is practically zero, being disregarded in the modeling [9].

The generalized impedance of the equivalent circuit in Fig. 1 in terms of the complex frequency s, namely $\mathbf{Z}_{M1}(s)$, is:

$$
\begin{aligned}
\mathbf{Z}_{M1}(s) &= R_{ES} + L_{ES}\,s + \left(\frac{\dfrac{1}{G_{ES}}\cdot\dfrac{1}{C_{ES}\,s}}{\dfrac{1}{G_{ES}}+\dfrac{1}{C_{ES}\,s}}\right)\\
&= R_{ES} + L_{ES}\,s + \left(\frac{\dfrac{1}{\cancel{G_{ES}C_{ES}\,s}}}{\dfrac{C_{ES}\,s+G_{ES}}{\cancel{G_{ES}C_{ES}\,s}}}\right)\\
&= R_{ES} + L_{ES}\,s + \frac{1}{C_{ES}\,s+G_{ES}}\\
&= \frac{(L_{ES}\,s+R_{ES})(C_{ES}\,s+G_{ES})+1}{C_{ES}\,s+G_{ES}}\\
&= \frac{L_{ES}C_{ES}\,s^2+L_{ES}G_{ES}\,s+R_{ES}C_{ES}\,s+R_{ES}G_{ES}+1}{C_{ES}\,s+G_{ES}}\\
&= \frac{L_{ES}C_{ES}\,s^2+(L_{ES}G_{ES}+R_{ES}C_{ES})\,s+R_{ES}G_{ES}+1}{C_{ES}\,s+G_{ES}}
\end{aligned}
$$

Since $s = j\omega$ for periodic sinusoidal signals, the impedance of the equivalent circuit in Fig. 1, in terms of the frequency ω, namely $\mathbf{Z}_{M1}(j\omega)$, is:

$$\mathbf{Z}_{\mathrm{M1}}(j\omega)=\frac{\mathrm{L_{ES}C_{ES}}\,(j\omega)^2+(\mathrm{L_{ES}G_{ES}}+\mathrm{R_{ES}C_{ES}})\,(j\omega)+\mathrm{R_{ES}G_{ES}}+1}{\mathrm{C_{ES}}\,(j\omega)+\mathrm{G_{ES}}}$$

$$=\frac{\left(-\omega^2\mathrm{L_{ES}C_{ES}}+\mathrm{R_{ES}G_{ES}}+1+j\omega\mathrm{L_{ES}G_{ES}}+j\omega\mathrm{R_{ES}C_{ES}}\right)}{\mathrm{G_{ES}}+j\omega\mathrm{C_{ES}}}\cdot\frac{(\mathrm{G_{ES}}-j\omega\mathrm{C_{ES}})}{(\mathrm{G_{ES}}-j\omega\mathrm{C_{ES}})}$$

$$=\frac{\cancel{-\omega^2\mathrm{L_{ES}G_{ES}C_{ES}}}+\mathrm{R_{ES}G_{ES}^2}+\mathrm{G_{ES}}+\cancel{\omega^2\mathrm{L_{ES}G_{ES}C_{ES}}}+\omega^2\mathrm{R_{ES}C_{ES}^2}}{\mathrm{G_{ES}^2}+\omega^2\mathrm{C_{ES}^2}}+$$

$$j\frac{\omega\mathrm{L_{ES}G_{ES}^2}+\cancel{\omega\mathrm{R_{ES}G_{ES}C_{ES}}}+\omega^3\mathrm{L_{ES}C_{ES}^2}-\cancel{\omega\mathrm{R_{ES}G_{ES}C_{ES}}}-\omega\mathrm{C_{ES}}}{\mathrm{G_{ES}^2}+\omega^2\mathrm{C_{ES}^2}}$$

$$=\frac{\mathrm{R_{ES}}\left(\mathrm{G_{ES}^2}+\omega^2\mathrm{C_{ES}^2}\right)+\mathrm{G_{ES}}}{\mathrm{G_{ES}^2}+\omega^2\mathrm{C_{ES}^2}}+j\frac{\omega\left[\mathrm{L_{ES}}\left(\mathrm{G_{ES}^2}+\omega^2\mathrm{C_{ES}^2}\right)-\mathrm{C_{ES}}\right]}{\mathrm{G_{ES}^2}+\omega^2\mathrm{C_{ES}^2}}$$

Thus, the impedance of the equivalent circuit in Fig. 1 in terms of the harmonic order n, namely $\mathbf{Z}_{\mathrm{M1}_n}(jn\omega)$, is:

$$\mathbf{Z}_{\mathrm{M1}_n}(jn\omega)=\frac{\mathrm{R_{ES}}\left(\mathrm{G_{ES}^2}+n^2\omega^2\mathrm{C_{ES}^2}\right)+\mathrm{G_{ES}}}{\mathrm{G_{ES}^2}+n^2\omega^2\mathrm{C_{ES}^2}}+j\frac{n\omega\left[\mathrm{L_{ES}}\left(\mathrm{G_{ES}^2}+n^2\omega^2\mathrm{C_{ES}^2}\right)-\mathrm{C_{ES}}\right]}{\mathrm{G_{ES}^2}+n^2\omega^2\mathrm{C_{ES}^2}}$$

In polar coordinates, the modulus and phase of the impedance of the equivalent circuit in Fig. 1, in terms of the harmonic order n are, respectively:

$$|\mathbf{Z}_{\mathrm{M1}_n}|=\frac{\sqrt{\left[\mathrm{R_{ES}}\left(\mathrm{G_{ES}^2}+n^2\omega^2\mathrm{C_{ES}^2}\right)+\mathrm{G_{ES}}\right]^2+n^2\omega^2\left[\mathrm{L_{ES}}\left(\mathrm{G_{ES}^2}+n^2\omega^2\mathrm{C_{ES}^2}\right)-\mathrm{C_{ES}}\right]^2}}{\mathrm{G_{ES}^2}+n^2\omega^2\mathrm{C_{ES}^2}} \tag{1}$$

$$\phi_{\mathrm{M1}_n}=\tan^{-1}\frac{n\omega\left[\mathrm{L_{ES}}\left(\mathrm{G_{ES}^2}+n^2\omega^2\mathrm{C_{ES}^2}\right)-\mathrm{C_{ES}}\right]}{\mathrm{R_{ES}}\left(\mathrm{G_{ES}^2}+n^2\omega^2\mathrm{C_{ES}^2}\right)+\mathrm{G_{ES}}} \tag{2}$$

2.2 Capacitor Model 2

The capacitor model presented in [12], namely *Model 2* and depicted in Fig. 2, has five parameters to be estimated. Its fundamental characteristic is the fact that it considers dielectric losses in the device, which become important in

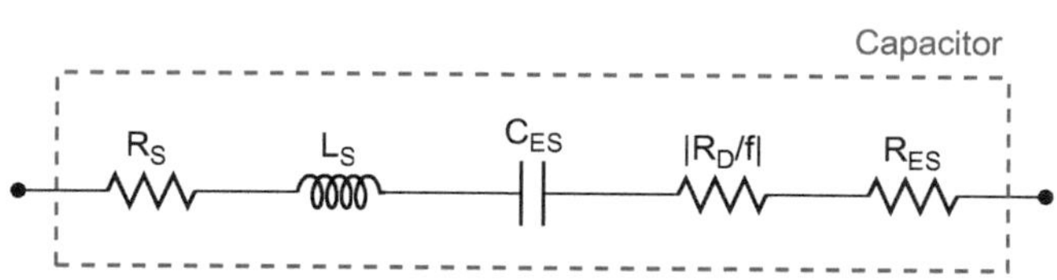

FIG. 2 Capacitor *Model 2*. Equivalent circuit for a capacitor proposed in [12].

analyses carried out at low frequencies. In addition, this model also considers parameters associated with external elements, such as the supply source and other devices connected to the capacitor.

In this model, the parameters R_S and L_S represent an external impedance. The parameters R_{ES} and C_{ES} have the same meaning as in [9]. The parameter with great importance in this model is the resistance $|R_D/f|$, which is inversely proportional to the frequency and represents the dielectric losses at low frequencies.

Despite having different physical origins, R_S and R_{ES} cannot be distinguished from one another from the viewpoint of parametric estimation and will, therefore, be represented by $(R_S + R_{ES})$. The generalized impedance of the equivalent circuit in Fig. 2 in terms of the complex frequency s, namely $\mathbf{Z}_{M2}(s)$, is:

$$\begin{aligned}
\mathbf{Z}_{M2}(s) &= (R_S + R_{ES}) + |R_D/f| + L_S s + \frac{1}{C_{ES} s} \\
&= \frac{[(R_S + R_{ES}) + |R_D/f| + L_S s] C_{ES} s + 1}{C_{ES} s} \\
&= \frac{L_S C_{ES} s^2 + [(R_S + R_{ES}) + |R_D/f|] C_{ES} s + 1}{C_{ES} s}
\end{aligned}$$

Because $s = j\omega$ for periodic sinusoidal signals, the impedance of the equivalent circuit in Fig. 2 in terms of the frequency ω, namely $\mathbf{Z}_{M2}(j\omega)$, is:

$$\begin{aligned}
\mathbf{Z}_{M2}(j\omega) &= \frac{L_S C_{ES}(j\omega)^2 + [(R_S + R_{ES}) + |R_D/f|] C_{ES}(j\omega) + 1}{C_{ES}(j\omega)} \\
&= \frac{\left[-\omega^2 L_S C_{ES} + 1 + j\omega C_{ES}(R_S + R_{ES}) + j\omega C_{ES}|R_D/f|\right]}{j\omega C_{ES}} \frac{(-j\omega C_{ES})}{(-j\omega C_{ES})} \\
&= \frac{\cancel{\omega^2 C_{ES}^2}(R_S + R_{ES}) + \cancel{\omega^2 C_{ES}^2}|R_D/f|}{\cancel{\omega^2 C_{ES}^2}} + j\frac{\omega^3 L_S C_{ES}^2 - \omega C_{ES}}{\omega^2 C_{ES}^2} \\
&= (R_S + R_{ES}) + |R_D/f| + j\frac{\cancel{(\omega C_{ES})}\left(\omega^2 L_S C_{ES} - 1\right)}{\cancel{(\omega C_{ES})}(\omega C_{ES})} \\
&= (R_S + R_{ES}) + |R_D/f| + j\frac{\omega^2 L_S C_{ES} - 1}{\omega C_{ES}}
\end{aligned}$$

Thus, the impedance of the equivalent circuit in Fig. 2 in terms of the harmonic order n, namely $\mathbf{Z}_{M2_n}(jn\omega)$, is:

$$\mathbf{Z}_{M2_n}(jn\omega) = (R_S + R_{ES}) + |R_D/f| + j\frac{n^2\omega^2 L_{ES} C_{ES} - 1}{n\omega C_{ES}}$$

In polar coordinates, the modulus and phase of the impedance of the equivalent circuit in Fig. 2 in terms of the harmonic order n are, respectively:

$$|\mathbf{Z}_{M2_n}| = \sqrt{[(R_S + R_{ES}) + |R_D/f|]^2 + \frac{(n^2\omega^2 L_S C_{ES} - 1)^2}{n^2\omega^2 C_{ES}^2}} \quad (3)$$

$$\phi_{M2_n} = \tan^{-1}\frac{n^2\omega^2 L_S C_{ES} - 1}{n\omega C_{ES}[(R_S + R_{ES}) + |R_D/f|]} \quad (4)$$

2.3 Capacitor Model 3

While the model presented in [12] considers dielectric losses at low frequencies, the model presented in [13] is suited to capacitors operating at both low and high frequencies. The authors considered two types of metalized capacitors to conduct their tests and find an equivalent impedance. Nevertheless, different parasite parameters arise depending on the device's operating frequency, which interfere significantly in the capacitor's modeling. Thus, to propose a model that would be suited to capacitors operating at both low and high frequencies, it was necessary to consider the dielectric absorption for low frequencies and a parasite impedance for high frequencies in the capacitor modeling. It is noteworthy to mention that this model also considers Foucault currents.

The model presented in [13], namely *Model 3* and depicted in Fig. 3, has 17 parameters to be estimated, in which not only intrinsic and external parameters, but also parasite impedances associated with operating frequencies different from the nominal are considered.

In this model, R_S represents the resistance of the electrode contacts and the capacitor's terminal connections; L_2 represents the series equivalent inductance associated with the capacitor's terminal connections; L_1 and R_{L1} represent, respectively, the parasite inductance in high frequencies and the skin effect; R_P represents dielectric losses; C is the capacitor's characteristic capacitance provided by the manufacturer; and R_{DA} and C_{DA}

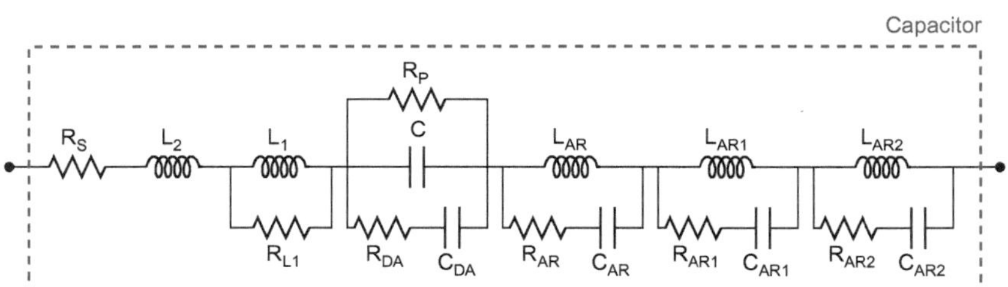

FIG. 3 Capacitor *Model 3*. Equivalent circuit for a capacitor proposed in [13].

represent the dielectric absorption at low frequencies. At last, L_{AR}, R_{AR}, C_{AR}, L_{AR1}, R_{AR1}, C_{AR1}, L_{AR2}, R_{AR2}, and C_{AR2} model parasite currents in capacitors operating close to their resonance frequency, which may generate localized heating in the device. Because the resonance frequency of power capacitors is higher than 1500 Hz, that is, the 25th harmonic for a network operation frequency of 60 Hz, such parameters will be disregarded in the modeling.

$$
\begin{aligned}
\mathbf{Z}_{M3}(s) &= R_S + L_2 s + \cfrac{1}{\cfrac{1}{L_1 s} + \cfrac{1}{R_{L1}}} + \cfrac{1}{\cfrac{1}{R_P} + \cfrac{1}{\cfrac{1}{Cs}} + \cfrac{1}{R_{DA} + \cfrac{1}{C_{DA} s}}} \\
&= R_S + L_2 s + \frac{L_1 R_{L1} s}{L_1 s + R_{L1}} + \cfrac{1}{\cfrac{R_P C s + 1}{R_P} + \cfrac{C_{DA} s}{R_{DA} C_{DA} s + 1}} \\
&= \frac{(L_2 s + R_S)(L_1 s + R_{L1}) + L_1 R_{L1} s}{L_1 s + R_{L1}} + \cfrac{1}{\cfrac{(R_P C s + 1)(R_{DA} C_{DA} s + 1) + R_P C_{DA} s}{R_P R_{DA} C_{DA} s + R_P}} \\
&= \frac{L_2 L_1 s^2 + (L_2 R_{L1} + R_S L_1 + L_1 R_{L1}) s + R_S R_{L1}}{L_1 s + R_{L1}} \\
&\quad + \frac{R_P R_{DA} C_{DA} s + R_P}{R_P C R_{DA} C_{DA} s^2 + (R_P C + R_P C_{DA} + R_{DA} C_{DA}) s + 1} \\
&= \frac{a_4 s^4 + a_3 s^3 + a_2 s^2 + a_1 s + a_0}{b_3 s^3 + b_2 s^2 + b_1 s + b_0}
\end{aligned}
$$

where

$$
\begin{aligned}
a_4 &= L_2 L_1 R_P C R_{DA} C_{DA}, \\
a_3 &= L_2 L_1 (R_P C + R_P C_{DA} + R_{DA} C_{DA}) + (L_2 R_{L1} + R_S L_1 + L_1 R_{L1}) R_P C R_{DA} C_{DA}, \\
a_2 &= L_2 L_1 + (L_2 R_{L1} + R_S L_1 + L_1 R_{L1})(R_P C + R_P C_{DA} + R_{DA} C_{DA}) \\
&\quad + R_S R_{L1} R_P C R_{DA} C_{DA} + L_1 R_P R_{DA} C_{DA}, \\
a_1 &= L_2 R_{L1} + R_S L_1 + L_1 R_{L1} + R_S R_{L1} (R_P C + R_P C_{DA} + R_{DA} C_{DA}) + L_1 R_P \\
&\quad + R_{L1} R_P R_{DA} C_{DA}, \\
a_0 &= R_S R_{L1} + R_{L1} R_P,
\end{aligned}
$$

and

$$
\begin{aligned}
b_3 &= L_1 R_P C R_{DA} C_{DA}, \\
b_2 &= L_1 (R_P C + R_P C_{DA} + R_{DA} C_{DA}) + R_{L1} R_P C R_{DA} C_{DA}, \\
b_1 &= L_1 + R_{L1} (R_P C + R_P C_{DA} + R_{DA} C_{DA}), \\
b_0 &= R_{L1}.
\end{aligned}
$$

Since $s = j\omega$ for periodic sinusoidal signals, the impedance of the equivalent circuit in Fig. 3 in terms of the frequency ω, namely $\mathbf{Z}_{M3}(j\omega)$, is:

$$\begin{aligned}
\mathbf{Z}_{M3}(j\omega) &= \frac{a_4(j\omega)^4 + a_3(j\omega)^3 + a_2(j\omega)^2 + a_1(j\omega) + a_0}{b_3(j\omega)^3 + b_2(j\omega)^2 + b_1(j\omega) + b_0} \\
&= \frac{\omega^4 a_4 - \omega^2 a_2 + a_0 - j\omega^3 a_3 + j\omega a_1}{-\omega^2 b_2 + b_0 - j\omega^3 b_3 + j\omega b_1} \\
&= \frac{\omega^4 a_4 - \omega^2 a_2 + a_0 + j(-\omega^3 a_3 + \omega a_1)}{-\omega^2 b_2 + b_0 + j(-\omega^3 b_3 + \omega b_1)} \cdot \frac{-\omega^2 b_2 + b_0 - j(-\omega^3 b_3 + \omega b_1)}{-\omega^2 b_2 + b_0 - j(-\omega^3 b_3 + \omega b_1)} \\
&= \frac{\begin{array}{c}\omega^6(-a_4 b_2 + a_3 b_3) + \omega^4(a_4 b_0 - a_3 b_1 + a_2 b_2 - a_1 b_3) + \\ \omega^2(-a_2 b_0 + a_1 b_1 - a_0 b_2) + a_0 b_0\end{array}}{\omega^6 b_3^2 + \omega^4(b_2^2 - 2b_3 b_1) + \omega^2(b_1^2 - 2b_2 b_0) + b_0^2} \\
&\quad + j\frac{\begin{array}{c}\omega^7 a_4 b_3 + \omega^5(-a_4 b_1 + a_3 b_2 - a_2 b_3) + \\ \omega^3(-a_3 b_0 + a_2 b_1 - a_1 b_2 + a_0 b_3) + \omega(a_1 b_0 - a_0 b_1)\end{array}}{\omega^6 b_3^2 + \omega^4(b_2^2 - 2b_3 b_1) + \omega^2(b_1^2 - 2b_2 b_0) + b_0^2}
\end{aligned}$$

The impedance of the equivalent circuit in Fig. 3 in terms of the harmonic order n, namely $\mathbf{Z}_{M3_n}(jn\omega)$, is:

$$\begin{aligned}
\mathbf{Z}_{M3_n}(jn\omega) &= \frac{\begin{array}{c}n^6\omega^6(-a_4 b_2 + a_3 b_3) + n^4\omega^4(a_4 b_0 - a_3 b_1 + a_2 b_2 - a_1 b_3) + \\ n^2\omega^2(-a_2 b_0 + a_1 b_1 - a_0 b_2) + a_0 b_0\end{array}}{n^6\omega^6 b_3^2 + n^4\omega^4(b_2^2 - 2b_3 b_1) + n^2\omega^2(b_1^2 - 2b_2 b_0) + b_0^2} \\
&\quad + j\frac{\begin{array}{c}n^7\omega^7 a_4 b_3 + n^5\omega^5(-a_4 b_1 + a_3 b_2 - a_2 b_3) + \\ n^3\omega^3(-a_3 b_0 + a_2 b_1 - a_1 b_2 + a_0 b_3) + n\omega(a_1 b_0 - a_0 b_1)\end{array}}{n^6\omega^6 b_3^2 + n^4\omega^4(b_2^2 - 2b_3 b_1) + n^2\omega^2(b_1^2 - 2b_2 b_0) + b_0^2}
\end{aligned}$$

In polar coordinates, the modulus and phase of the impedance of the equivalent circuit in Fig. 3 in terms of the harmonic order n are, respectively:

$$|\mathbf{Z}_{M3_n}| = \sqrt{\mathrm{Re}\{\mathbf{Z}_{M3_n}(jn\omega)\}^2 + \mathrm{Im}\{\mathbf{Z}_{M3_n}(jn\omega)\}^2} \tag{5}$$

$$\phi_{M3_n} = \tan^{-1}\frac{\mathrm{Im}\{\mathbf{Z}_{M3_n}(jn\omega)\}}{\mathrm{Re}\{\mathbf{Z}_{M3_n}(jn\omega)\}} \tag{6}$$

3 FORMULATION OF THE CAPACITOR MODEL PARAMETERIZATION PROBLEM

The formulation of the capacitor model parameterization problem consists in minimizing the mean squared error between a reference current acquired in laboratory tests and the calculated current flowing through the terminals of the parameterized model, subject to the parameters' lower and upper bounds. Thus, it may be generically stated as:

$$\begin{aligned} \min \quad & \frac{1}{N}\sum_{k=1}^{N}\left(I_{k_{\text{ref.}}} - I_{k_{\text{calc.}}}(p)\right)^2 \\ \text{subject to}: \quad & p^{\min} \leq p \leq p^{\max} \end{aligned}$$

where p is the vector of capacitor model parameters, with $p^{\min}$ and $p^{\max}$ representing, respectively, the lower and upper bounds of these parameters; N is the number of signal samples; $I_{k_{\text{ref.}}}$ is the punctual value of the reference current signal acquired in laboratory tests; and $I_{k_{\text{calc.}}}$ is the punctual value of the calculated current flowing through the terminals of the parameterized model.

Since only harmonic components of up to the 25th order are considered in the representation of harmonic distortions [17], the voltage across the terminals of the power capacitor may be expressed by:

$$v(t) = \sum_{n=1}^{25} v_n(t)$$

where

$$v_n(t) = \sqrt{2}\,V_n \cos(n\omega t + \theta_n)$$

is the harmonic component of order n, with V_n and θ_n corresponding to RMS harmonic voltage magnitude and phase angle, respectively. The voltage phasor of the harmonic component of order n is, then:

$$\mathbf{V}_n = V_n \underline{/\theta_n}$$

The equivalent impedance between the terminals of the parameterized model, in polar coordinates, is:

$$\mathbf{Z}_n(jn\omega) = |\mathbf{Z}_n(p)| \, \underline{/\phi_n(p)}$$

where $|\mathbf{Z}_n(p)|$ and $\phi_n(p)$ are, respectively, the modulus and phase of the impedance in terms of the harmonic order n and capacitor model parameters p.

Thus, the current phasor of the harmonic component of the calculated current flowing through the terminals of the parameterized model is:

$$\mathbf{I}_n = \frac{\mathbf{V}_n}{\mathbf{Z}_n(jn\omega)} = \frac{V_n}{|\mathbf{Z}_n(p)|} \underline{/\theta_n - \phi_n(p)}$$

In the time domain, the harmonic component of such a current is given by:

$$i_n(t) = \sqrt{2}\,\frac{V_n}{|\mathbf{Z}_n(p)|}\cos(n\omega t + \theta_n - \phi_n(p))$$

and the current flowing through the terminals of the parameterized model is:

$$i(t) = \sum_{n=1}^{25} i_n(t)$$

Therefore, for a sampling time T_s (s), the kth punctual value of the calculated current flowing through the terminals of the parameterized model is

$$I_{k_{\text{calc.}}}(p) = \sum_{n=1}^{25} \sqrt{2}\,\frac{V_n}{|\mathbf{Z}_n(p)|}\cos\left(n\omega(k-1)T_s + \theta_n - \phi_n(p)\right) \tag{7}$$

Thus, since the calculated current $I_{k_{\text{calc.}}}$ is a function of the voltage across the terminals of the capacitor, the capacitor model itself and its parameters, for each of the three capacitor models considered in this chapter there will be a different formulation of the problem of power capacitor model parameterization. Thus, the determination of the calculated current $I_{k_{\text{calc.}}}$ and the specific formulation of such a problem for each model depends on the impedance of the equivalent circuit of the power capacitor model.

3.1 Capacitor Model 1: Parameterization Problem

Substituting $|\mathbf{Z}_n(p)|$ and $\phi_n(p)$ in Eq. (7) by the modulus (1) and the phase (2) of the impedance of the equivalent circuit in Fig. 1 results in:

$$\begin{aligned} I_{k_{\text{calc.}}}(p) &= \sum_{n=1}^{25}\left[\sqrt{2}\frac{V_n\left(\mathrm{G}_{\mathrm{ES}}^2 + n^2\omega^2\mathrm{C}_{\mathrm{ES}}^2\right)}{\sqrt{\begin{array}{l}\left[\mathrm{R}_{\mathrm{ES}}\left(\mathrm{G}_{\mathrm{ES}}^2 + n^2\omega^2\mathrm{C}_{\mathrm{ES}}^2\right) + \mathrm{G}_{\mathrm{ES}}\right]^2 + \\ n^2\omega^2\left[\mathrm{L}_{\mathrm{ES}}\left(\mathrm{G}_{\mathrm{ES}}^2 + n^2\omega^2\mathrm{C}_{\mathrm{ES}}^2\right) - \mathrm{C}_{\mathrm{ES}}\right]^2\end{array}}}\right. \\ &\quad \left.\times\cos\left(n\omega(k-1)T_s + \theta_n - \tan^{-1}\frac{n\omega\left[\mathrm{L}_{\mathrm{ES}}\left(\mathrm{G}_{\mathrm{ES}}^2 + n^2\omega^2\mathrm{C}_{\mathrm{ES}}^2\right) - \mathrm{C}_{\mathrm{ES}}\right]}{\mathrm{R}_{\mathrm{ES}}\left(\mathrm{G}_{\mathrm{ES}}^2 + n^2\omega^2\mathrm{C}_{\mathrm{ES}}^2\right) + \mathrm{G}_{\mathrm{ES}}}\right)\right] \end{aligned} \tag{8}$$

Table 1 presents the lower and upper bounds for the parameters in the equivalent circuit of a capacitor in Fig. 1 [9, 14].

TABLE 1 Lower and Upper Bounds for Parameters in Model 1

Parameter	Lower Bound	Upper Bound
R_{ES} (Ω)	0.05	0.75
L_{ES} (mH)	0.05	0.5
G_{ES} (nS)	0.04	14.3
C_{ES} (μF)	127.0	147.0

Thus, the formulation of the problem of power capacitor parameterization for the model proposed in [9] is:

$$\begin{aligned} \min \quad & \frac{1}{N}\sum_{k=1}^{N}\left(I_{k_{\text{ref.}}} - I_{k_{\text{calc.}}}(p)\right)^2 \\ \text{subject to:} \quad & 0.05 \le \mathrm{R_{ES}} \le 0.75 \\ & 0.05\mathrm{E}-03 \le \mathrm{L_{ES}} \le 0.5\mathrm{E}-03 \\ & 0.04\mathrm{E}-09 \le \mathrm{G_{ES}} \le 14.3\mathrm{E}-09 \\ & 127.0\mathrm{E}-06 \le \mathrm{C_{ES}} \le 147.0\mathrm{E}-06 \end{aligned}$$

where the punctual value of the calculated current flowing through the terminals of the parameterized model $I_{k_{\text{calc.}}}(p)$ is given by Eq. (8).

3.2 Capacitor Model 2: Parameterization Problem

Substituting $|\mathbf{Z}_n(p)|$ and $\phi_n(p)$ in Eq. (7) by the modulus (3) and the phase (4) of the impedance of the equivalent circuit in Fig. 2 results in:

$$I_{k_{\text{calc.}}}(p) = \sum_{n=1}^{25}\left[\sqrt{2}\frac{V_n}{\sqrt{\left[(\mathrm{R_S}+\mathrm{R_{ES}})+|\mathrm{R_D}/\mathrm{f}|\right]^2+\dfrac{\left(n^2\omega^2\mathrm{L_S}\mathrm{C_{ES}}-1\right)^2}{n^2\omega^2\mathrm{C_{ES}^2}}}} \times \cos\left(n\omega(k-1)T_s+\theta_n-\tan^{-1}\frac{n^2\omega^2\mathrm{L_S}\mathrm{C_{ES}}-1}{n\omega\mathrm{C_{ES}}\left[(\mathrm{R_S}+\mathrm{R_{ES}})+|\mathrm{R_D}/\mathrm{f}|\right]}\right)\right] \tag{9}$$

Table 2 presents the lower and upper bounds for the parameters in the equivalent circuit of a capacitor in Fig. 2 [12, 14].

TABLE 2 Lower and Upper Bounds for Parameters in Model 2

Parameter	Lower Bound	Upper Bound
$(\mathrm{R_S}+\mathrm{R_{ES}})$ (Ω)	0.05	0.75
$\mathrm{L_S}$ (mH)	0.05	0.5
$\mathrm{C_{ES}}$ (μF)	127.0	147.0
$|\mathrm{R_D}/\mathrm{f}|$ (Ω)	0.01	0.1

Thus, the formulation of the problem of power capacitor parameterization for the model proposed in [12] is:

$$\begin{aligned} \min \quad & \frac{1}{N}\sum_{k=1}^{N}\left(I_{k_{\text{ref.}}}-I_{k_{\text{calc.}}}(p)\right)^2 \\ \text{subject to:} \quad & 0.05 \le (\mathrm{R_S}+\mathrm{R_{ES}}) \le 0.75 \\ & 0.05\mathrm{E}-03 \le \mathrm{L_S} \le 0.5\mathrm{E}-03 \\ & 127.0\mathrm{E}-06 \le \mathrm{C_{ES}} \le 147.0\mathrm{E}-06 \\ & 0.01 \le |\mathrm{R_D}/\mathrm{f}| \le 0.1 \end{aligned}$$

where the punctual value of the calculated current flowing through the terminals of the parameterized model $I_{k_{\text{calc.}}}(p)$ is given by Eq. (9).

3.3 Capacitor Model 3: Parameterization Problem

Substituting $|\mathbf{Z}_n(p)|$ and $\phi_n(p)$ in Eq. (7) by the modulus (5) and the phase (6) of the impedance of the equivalent circuit in Fig. 3 results in:

$$I_{k_{\text{calc.}}}(p) = \sum_{n=1}^{25}\left[\sqrt{2}\frac{V_n}{\sqrt{\mathrm{Re}\{\mathbf{Z}_{\mathrm{M3}_n}(jn\omega)\}^2+\mathrm{Im}\{\mathbf{Z}_{\mathrm{M3}_n}(jn\omega)\}^2}} \times\cos\left(n\omega(k-1)T_s+\theta_n-\tan^{-1}\frac{\mathrm{Im}\{\mathbf{Z}_{\mathrm{M3}_n}(jn\omega)\}}{\mathrm{Re}\{\mathbf{Z}_{\mathrm{M3}_n}(jn\omega)\}}\right)\right] \tag{10}$$

Table 3 presents the lower and upper bounds for the parameters in the equivalent circuit of a capacitor in Fig. 3 [13, 14].

TABLE 3 Lower and Upper Bounds for Parameters in Model 3

Parameter	Lower Bound	Upper Bound
R_S (Ω)	0.05	0.75
L_2 (mH)	0.05	0.5
L_1 (μH)	0.1	1.0
R_{L1} (Ω)	0.005	0.05
R_P (GΩ)	7.0	25.0
C (μF)	127.0	147.0
R_{DA} (kΩ)	50.0	500.0
C_{DA} (nF)	0.1	1.0

Thus, the formulation of the problem of power capacitor parameterization for the model proposed in [13] is:

$$
\begin{aligned}
\min \quad & \frac{1}{N}\sum_{k=1}^{N}\left(I_{k_{\mathrm{ref.}}} - I_{k_{\mathrm{calc.}}}(p)\right)^2 \\
\text{subject to:} \quad & 0.05 \le \mathrm{R_S} \le 0.75 \\
& 0.05\mathrm{E}-03 \le \mathrm{L_2} \le 0.5\mathrm{E}-03 \\
& 0.1\mathrm{E}-06 \le \mathrm{L_1} \le 1.0\mathrm{E}-06 \\
& 0.005 \le \mathrm{R_{L1}} \le 0.05 \\
& 7.0\mathrm{E}+09 \le \mathrm{R_P} \le 25.0\mathrm{E}+09 \\
& 127.0\mathrm{E}-06 \le \mathrm{C} \le 147.0\mathrm{E}-06 \\
& 50.0\mathrm{E}+03 \le \mathrm{R_{DA}} \le 500.0\mathrm{E}+03 \\
& 0.1\mathrm{E}-09 \le \mathrm{C_{DA}} \le 1.0\mathrm{E}-09
\end{aligned}
$$

where the punctual value of the calculated current flowing through the terminals of the parameterized model $I_{k_{\mathrm{calc.}}}(p)$ is given by Eq. (10).

4 LABORATORY SETUP AND DATA ACQUISITION

This section presents the laboratory setup to obtain the punctual reference currents $I_{k_{\mathrm{ref.}}}$ flowing through the terminals of power capacitors supplied with known harmonic voltages in the problem formulations presented earlier.

The data acquisition was carried out by means of laboratory tests with two different metalized polypropylene power capacitors (one all-film [15] and one impregnated [16]), both with nominal capacitance of 137 μF and 2.5 kVar. The laboratory setup is shown in Fig. 4.

The voltage across the terminals of these power capacitors was generated by a configurable source of 5 kVA from California Instruments, capable of generating voltage signals with harmonic components up to the 50th order. In the tests, only harmonic components up to the 25th order were considered.

The software CIGui controls the source, and it is responsible for sampling the current flowing through the capacitors' terminals. Since the network operation frequency is 60 Hz and only harmonic components up to the 25th order are considered, the sampling frequency must be, at least, 1500 Hz. The current signals are obtained for 1024 points with a sampling time of 25.6 μs; the number of samples is established according to [17]. Thus, a sampling time of 25.6 μs results in a sampling frequency of 39.06 kHz, guaranteeing the acquisition of 651 samples/cycle, which is enough to represent harmonic distortions.

The capacitors must be supplied, one at a time, with a pure sinusoidal waveform (220 V, 60 Hz) to allow the assessment of the approach in an ideal operating condition, namely SIN waveform, and with three different voltage

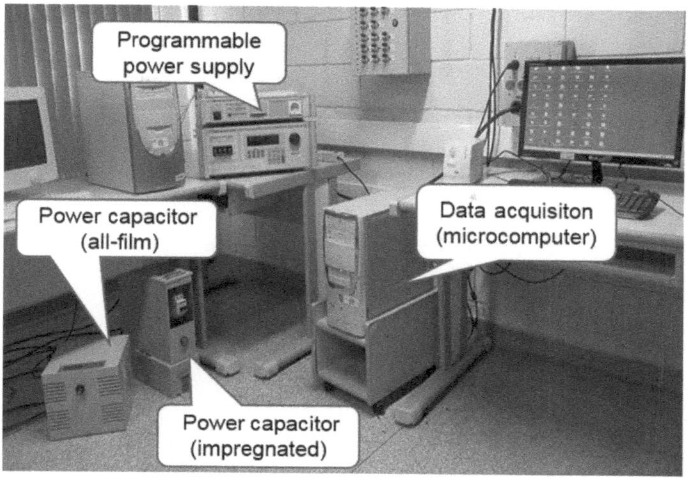

FIG. 4 Laboratory setup used for data acquisition. Configurable source (programmable power supply), all-film and impregnated polypropylene power capacitors, and microcomputer for controlling the source and sampling the reference currents flowing through the capacitors' terminals [14].

waveforms with harmonic distortions according to the following standards and/or recommendations: IEC [18], IEEE [19], and PRODIST [17].

The data for voltage phasors of the harmonic components for each voltage signal are presented in Table 4.

Fig. 5 shows the sinusoidal voltage waveform, and Figs. 6–8 show, respectively, the three voltage waveforms with harmonic distortions generated by the source in the laboratory according to IEC, IEEE, and PRODIST standards.

Figs. 9–12 show, respectively, the four reference currents flowing through the terminals of the all-film polypropylene power capacitor supplied with the voltage signals shown earlier.

5 OPTIMIZATION-BASED POWER CAPACITOR MODEL PARAMETERIZATION METHODOLOGY

The problem of power capacitor parameterization, as in Section 3, was implemented in the AMPL mathematical programming language [20], and the capacitor parameters were obtained by a mathematical programming method based on reduced gradient for nonlinear objectives (implemented in the solver MINOS [21]). The results presented in Tables 5–10 consider the three capacitor models, the two metalized polypropylene power capacitors (one all-film and one impregnated) and four voltage waveforms (one sinusoidal and three with harmonic distortions).

TABLE 4 Voltage Phasors of Harmonic Components

	SIN		IEC		IEEE		PRODIST	
Harmonic	V_n *(V)*	θ_n *(degrees)*	V_n *(V)*	θ_n *(degrees)*	V_n *(V)*	θ_n *(degrees)*	V_n *(V)*	θ_n *(degrees)*
1	220.0	−90.0	220.0	−90.0	220.0	−90.0	220.0	−90.0
2	0.0	0.0	0.242	−90.0	0.242	−90.0	5.236	−91.1
3	0.0	0.0	1.914	−74.7	4.532	−84.6	7.832	−87.4
4	0.0	0.0	0.088	−90.0	0.110	−90.0	3.234	−92.7
5	0.0	0.0	0.858	−87.5	3.300	−91.4	10.560	−92.6
6	0.0	0.0	0.000	0.0	0.044	−90.0	0.044	−90.0
7	0.0	0.0	0.352	−99.9	2.178	−94.2	9.790	−94.9
8	0.0	0.0	0.286	−90.0	0.022	−90.0	0.132	−90.0
9	0.0	0.0	0.682	−100.3	4.312	−96.8	0.132	−90.0
10	0.0	0.0	0.176	−90.0	0.044	−90.0	0.044	−90.0
11	0.0	0.0	0.594	−100.0	3.234	−97.6	3.190	−98.3
12	0.0	0.0	0.176	−90.0	0.044	−90.0	0.066	−90.0
13	0.0	0.0	0.374	−100.1	3.168	−99.7	2.068	−99.7
14	0.0	0.0	0.176	−90.0	0.088	−90.0	0.088	−90.0
15	0.0	0.0	0.176	−90.0	0.044	−90.0	0.044	−90.0
16	0.0	0.0	0.000	0.0	0.044	−90.0	0.044	−90.0

17	0.0	0.0	0.198	−90.0	0.044	−90.0	0.044	−90.0
18	0.0	0.0	0.066	−90.0	0.044	−90.0	0.044	−90.0
19	0.0	0.0	0.286	−90.0	0.044	−90.0	0.044	−90.0
20	0.0	0.0	0.286	−90.0	0.044	−90.0	0.044	−90.0
21	0.0	0.0	0.374	−104.3	0.044	−90.0	0.044	−90.0
22	0.0	0.0	0.176	−90.0	0.044	−90.0	0.044	−90.0
23	0.0	0.0	0.374	−104.6	0.044	−90.0	0.044	−90.0
24	0.0	0.0	0.154	−90.0	0.022	−90.0	0.044	−90.0
25	0.0	0.0	0.154	−90.0	0.022	−90.0	0.044	−90.0

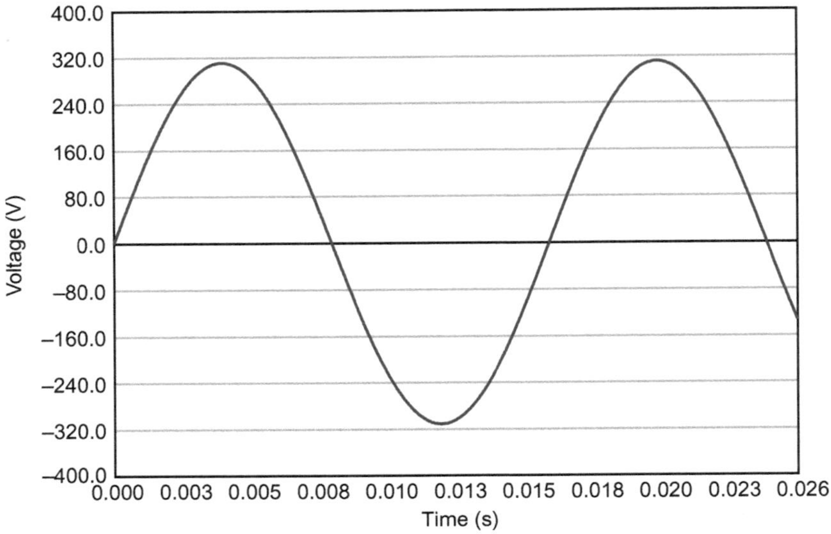

FIG. 5 SIN waveform. Sinusoidal voltage waveform, 220 V and 60 Hz.

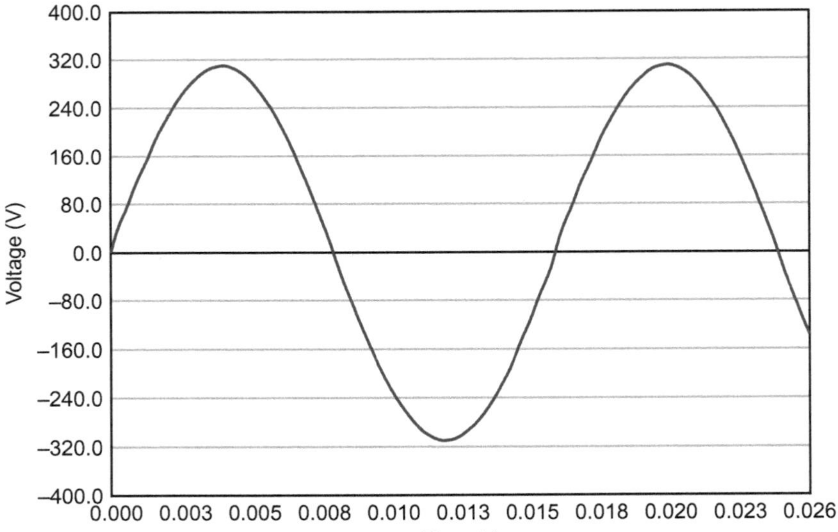

FIG. 6 IEC waveform. Distorted voltage waveform according to IEC standards.

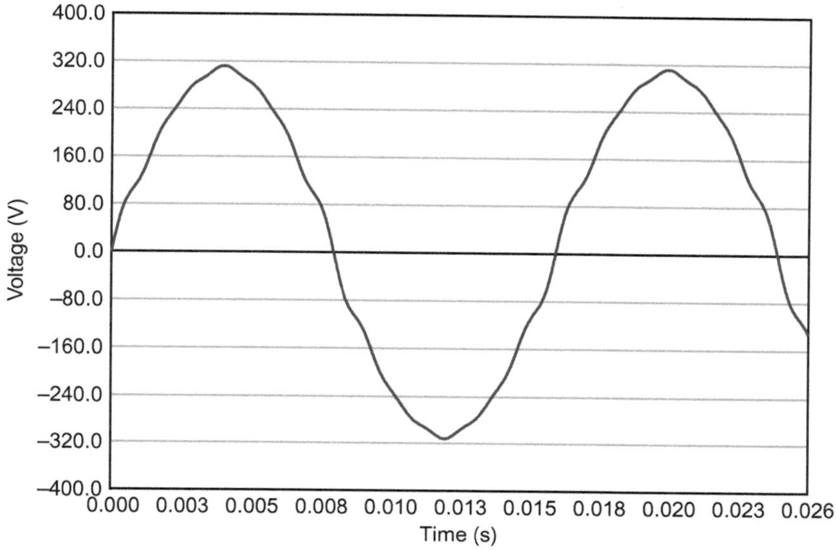

FIG. 7 IEEE waveform. Distorted voltage waveform according to IEEE standards.

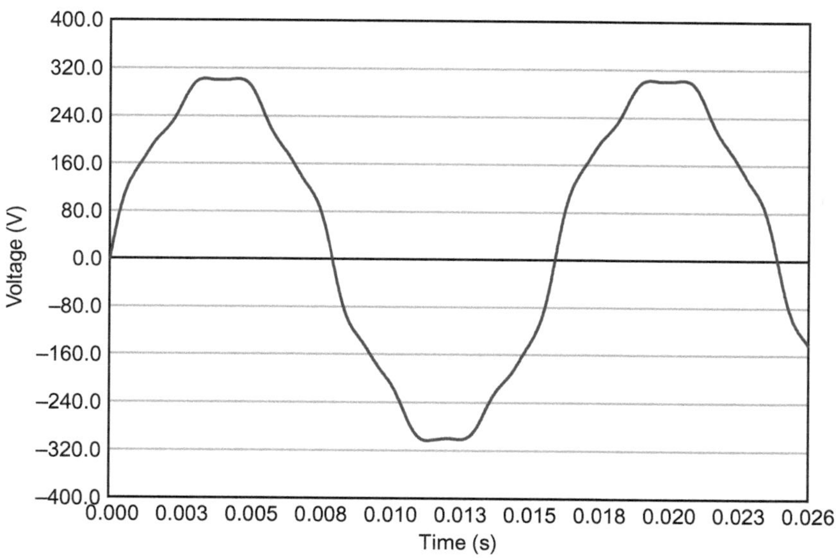

FIG. 8 PRODIST waveform. Distorted voltage waveform according to PRODIST standards.

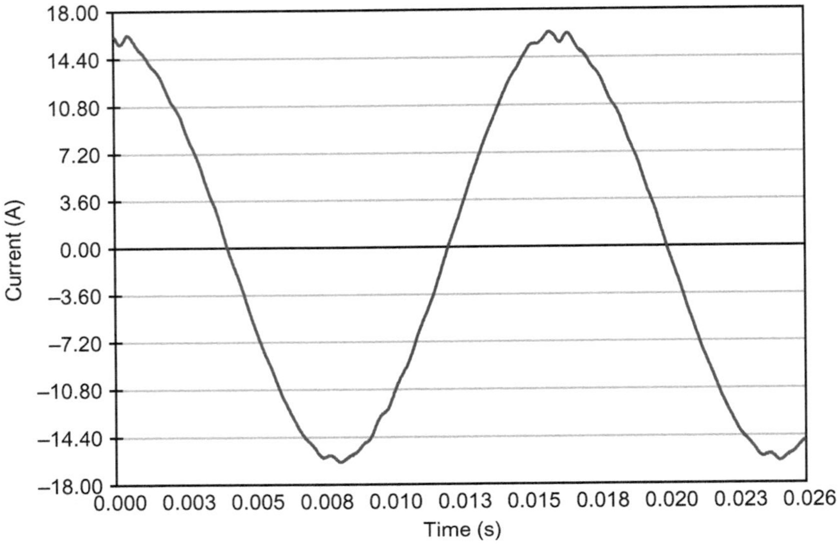

FIG. 9 Reference current for SIN waveform. Reference current flowing through the terminals of the all-film polypropylene power capacitor supplied with SIN voltage signal.

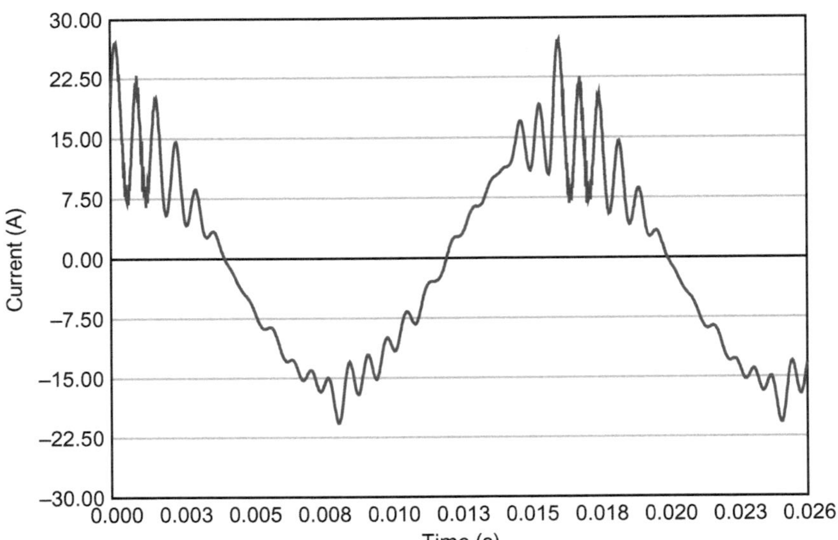

FIG. 10 Reference current for IEC waveform. Reference current flowing through the terminals of the all-film polypropylene power capacitor supplied with IEC voltage signal.

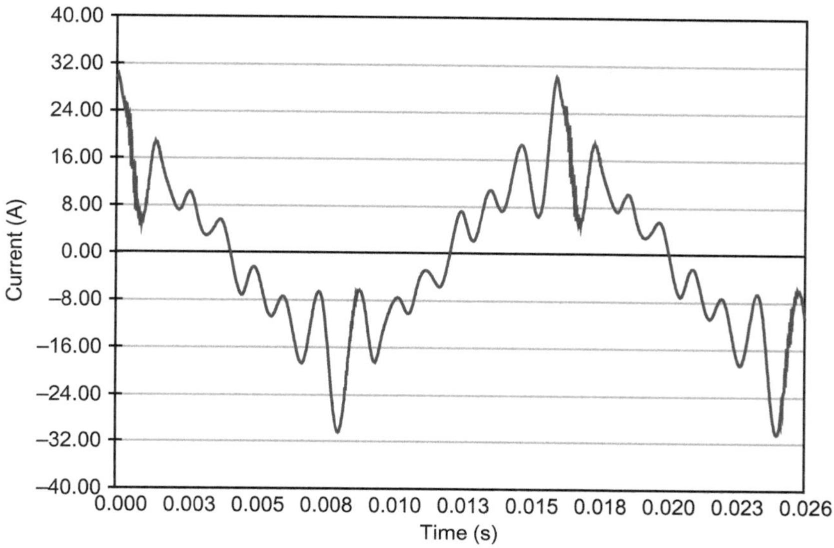

FIG. 11 Reference current for IEEE waveform. Reference current flowing through the terminals of the all-film polypropylene power capacitor supplied with IEEE voltage signal.

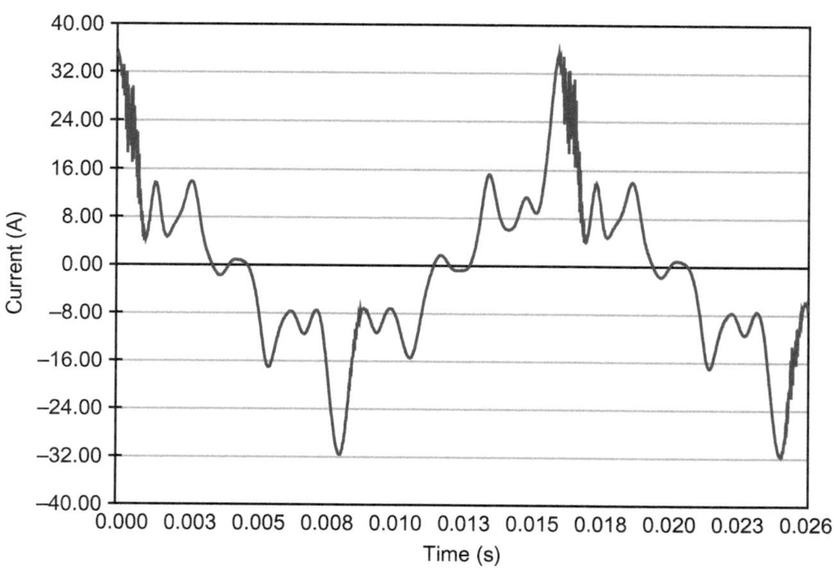

FIG. 12 Reference current for PRODIST waveform. Reference current flowing through the terminals of the all-film polypropylene power capacitor supplied with PRODIST voltage signal.

TABLE 5 Obtained Parameters of Model 1 for the All-Film Polypropylene Power Capacitor

Parameter	SIN	IEC	IEEE	PRODIST
R_{ES} (Ω)	0.75	0.27021	0.61544	0.65621
L_{ES} (mH)	0.5	0.10289	0.11403	0.076881
G_{ES} (nS)	14.3	14.3	14.3	14.3
C_{ES} (μF)	135.75	136.49	136.66	137.73

TABLE 6 Obtained Parameters of Model 1 for the Impregnated Polypropylene Power Capacitor

Parameter	SIN	IEC	IEEE	PRODIST
R_{ES} (Ω)	0.75	0.23602	0.62345	0.68811
L_{ES} (mH)	0.5	0.10285	0.11413	0.079566
G_{ES} (nS)	14.3	14.3	14.3	14.3
C_{ES} (μF)	130.77	131.39	131.82	132.67

TABLE 7 Obtained Parameters of Model 2 for the All-Film Polypropylene Power Capacitor

Parameter	SIN	IEC	IEEE	PRODIST
$(R_S + R_{ES})$ (Ω)	0.70372	0.21521	0.56044	0.60121
L_S (mH)	0.275	0.10289	0.11403	0.076881
C_{ES} (μF)	136.34	136.49	136.66	137.73
$\|R_D/f\|$ (Ω)	0.055	0.055	0.055	0.055

The comparison between the reference and calculated currents flowing through the power capacitors' terminals in terms of the capacitor model and voltage waveform with harmonic distortions are presented in Figs. 13–18.

At last, to assess the effectiveness of the model parameterizations, Figs. 19 and 20 show the mean squared error between the reference and calculated currents.

TABLE 8 Obtained Parameters of Model 2 for the Impregnated Polypropylene Power Capacitor

Parameter	SIN	IEC	IEEE	PRODIST
(R_S +R_{ES}) (Ω)	0.75	0.22602	0.56845	0.63311
L_S (mH)	0.5	0.10285	0.11413	0.079566
C_{ES} (μF)	130.77	131.39	131.82	132.67
$\|R_D/f\|$ (Ω)	0.066951	0.01	0.055	0.055

TABLE 9 Obtained Parameters of Model 3 for the All-Film Polypropylene Power Capacitor

Parameter	SIN	IEC	IEEE	PRODIST
R_S (Ω)	0.75	0.27021	0.61408	0.65621
L_2 (mH)	0.5	0.10289	0.1131	0.076882
L_1 (μH)	1.0	1.0	1.0	1.0
R_{L1} (Ω)	0.005	0.0275	0.014562	0.005
R_P (GΩ)	16.0	16.0	16.0	16.0
C (μF)	135.74	136.49	136.66	137.73
R_{DA} (kΩ)	275.0	275.0	275.0	275.0
C_{DA} (nF)	1.0	1.0	1.0	0.55

TABLE 10 Obtained Parameters of Model 3 for the Impregnated Polypropylene Power Capacitor

Parameter	SIN	IEC	IEEE	PRODIST
R_S (Ω)	0.75	0.23602	0.62206	0.6881
L_2 (mH)	0.5	0.10285	0.11319	0.079566
L_1 (μH)	1.0	1.0	1.0	1.0
R_{L1} (Ω)	0.0275	0.0275	0.014478	0.005
R_P (GΩ)	16.0	16.0	16.0	16.0
C (μF)	130.77	131.39	131.82	132.76
R_{DA} (kΩ)	275.0	275.0	275.0	275.0
C_{DA} (nF)	1.0	1.0	1.0	0.55

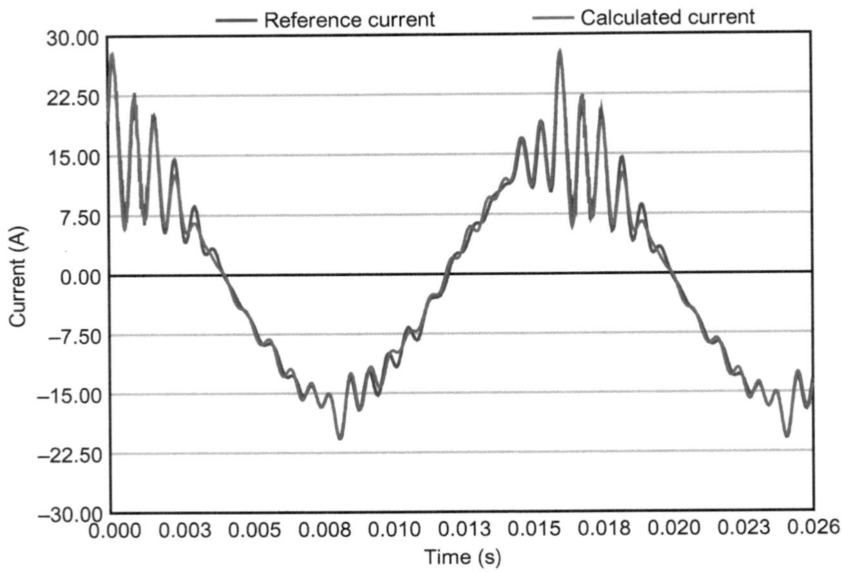

FIG. 13 Currents for IEC waveform across the terminals of the all-film polypropylene power capacitor. Comparison between reference and calculated currents considering IEC distorted voltage waveform and capacitor *Model 1* for the all-film polypropylene power capacitor.

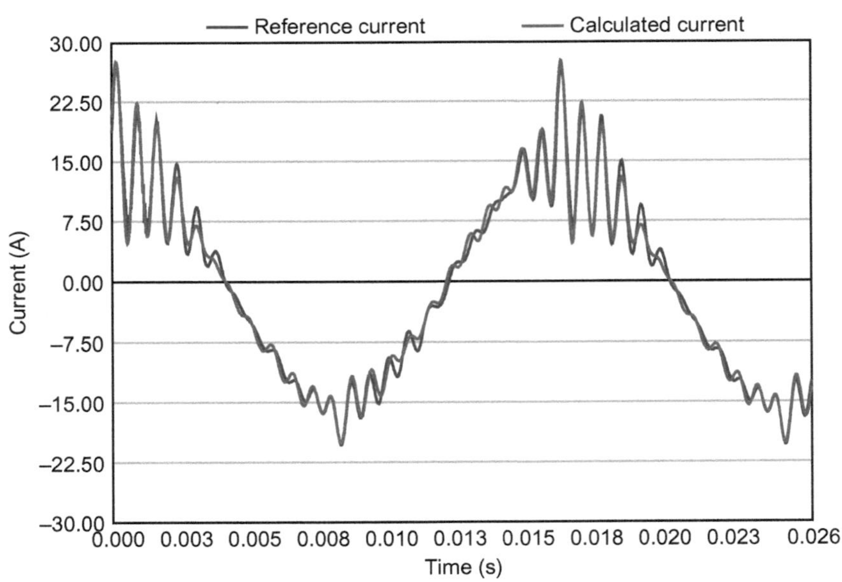

FIG. 14 Currents for IEC waveform across the terminals of the impregnated polypropylene power capacitor. Comparison between reference and calculated currents considering IEC distorted voltage waveform and capacitor *Model 1* for the impregnated polypropylene power capacitor.

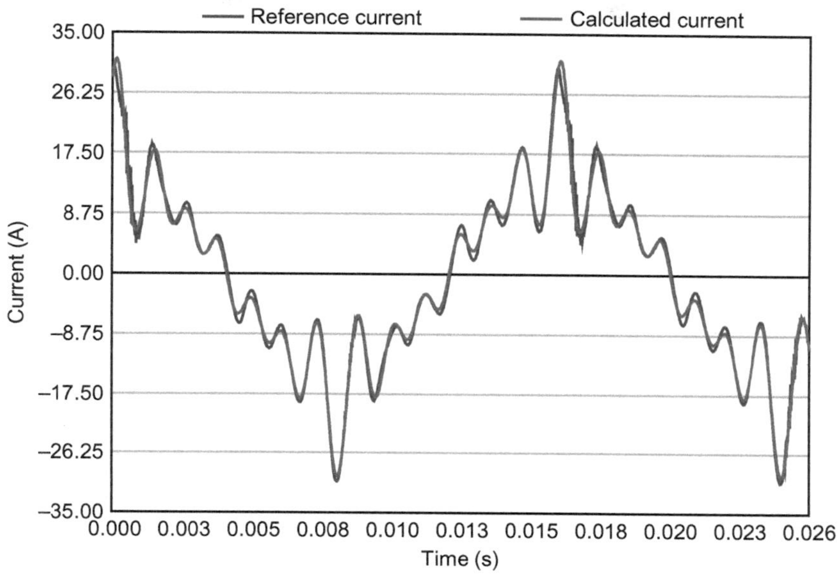

FIG. 15 Currents for IEEE waveform across the terminals of the all-film polypropylene power capacitor. Comparison between reference and calculated currents considering IEEE distorted voltage waveform and capacitor *Model 2* for the all-film polypropylene power capacitor.

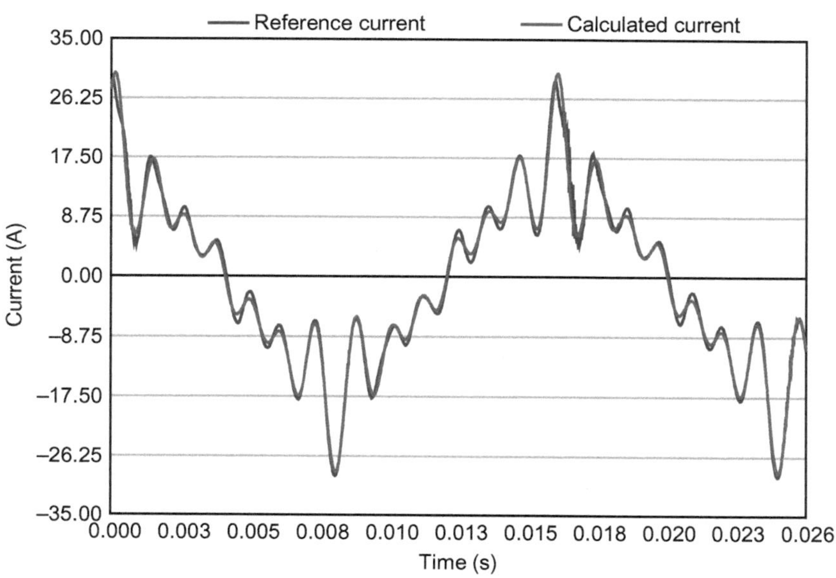

FIG. 16 Currents for IEEE waveform across the terminals of the impregnated polypropylene power capacitor. Comparison between reference and calculated currents considering IEEE distorted voltage waveform and capacitor *Model 2* for the impregnated polypropylene power capacitor.

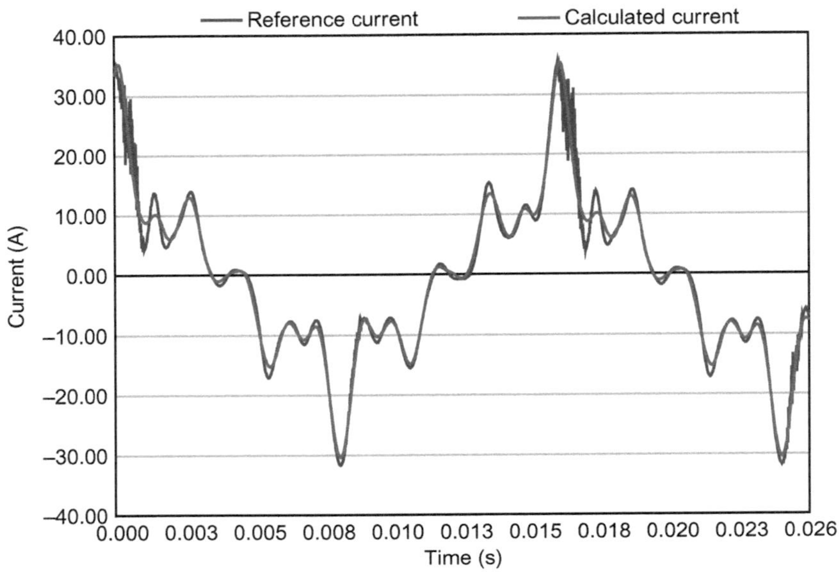

FIG. 17 Currents for PRODIST waveform across the terminals of the all-film polypropylene power capacitor. Comparison between reference and calculated currents considering PRODIST distorted voltage waveform and capacitor *Model 3* for the all-film polypropylene power capacitor.

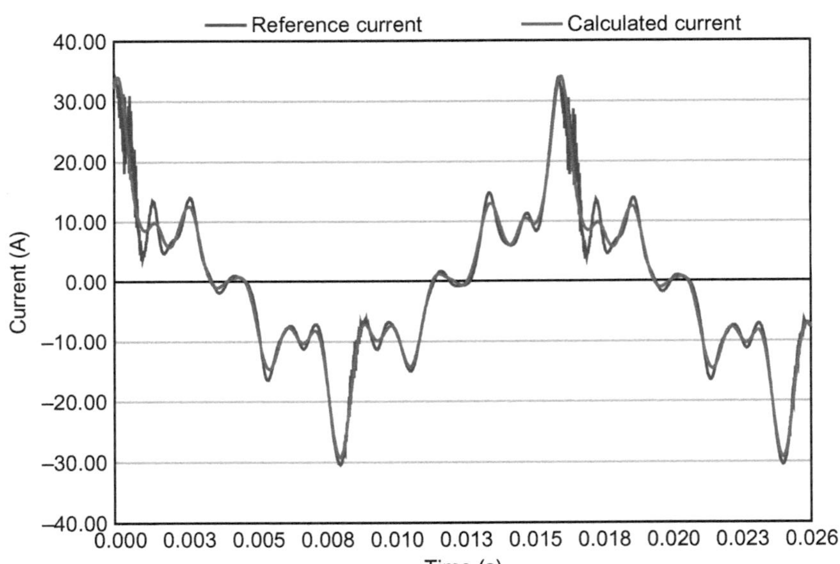

FIG. 18 Currents for PRODIST waveform across the terminals of the impregnated polypropylene power capacitor. Comparison between reference and calculated currents considering PRODIST distorted voltage waveform and capacitor *Model 3* for the impregnated polypropylene power capacitor.

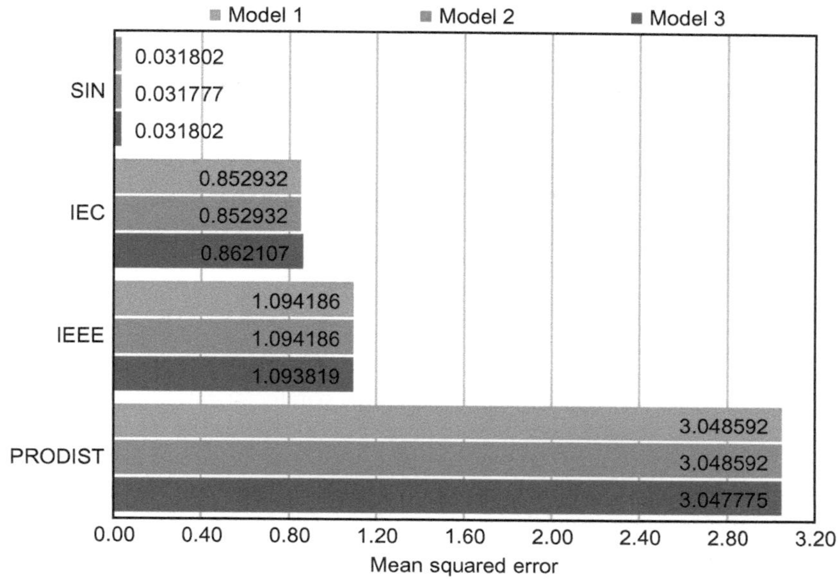

FIG. 19 Assessment of the all-film polypropylene power capacitor parameterization. Mean squared error between measured and calculated currents for the all-film polypropylene power capacitor.

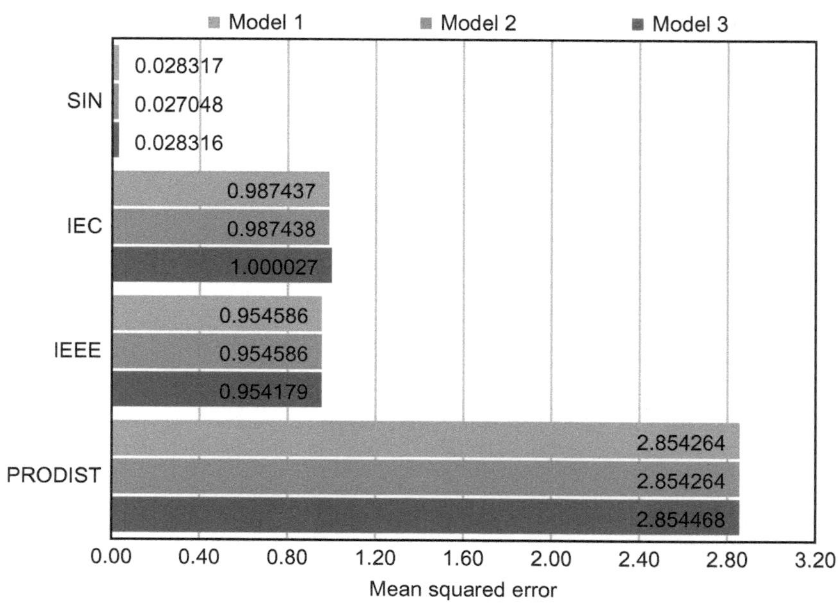

FIG. 20 Assessment of the impregnated polypropylene power capacitor parameterization. Mean squared error between measured and calculated currents for the impregnated polypropylene power capacitor.

It is important to emphasize that the focus of this chapter is not on an optimization method implementation or a specific capacitor model, but on the methodology to determine the capacitor model parameters to support applications in power distribution systems, avoiding unnecessary operative risks.

Additionally, the obtained parameters show that such capacitor models can behave differently depending on the voltage waveform across the capacitor terminals, that is, the knowledge of the harmonics present in power system is crucial to ensure the results. Thus, the capacitor model selection depends on the power distribution system application and the desired result accuracy.

6 SUMMARY

The optimization-based power capacitor model parameterization methodology presented in this chapter may be summarized in the following steps:

1. Choose a capacitor model that considers its assembly aspects and response for different frequencies by an equivalent electrical circuit.
2. Make necessary considerations and/or assumptions on the capacitor modeling so that it may be used to model capacitors operating at low frequencies, such as power capacitors, as for the capacitor model in [13] in Section 2.3.
3. Determine the modulus and phase of the impedance of the capacitor's equivalent electrical circuit, as in Section 2.
4. Establish the range of values for the capacitor parameters, that is, the parameters' lower and upper bounds.
5. Determine the expression for the punctual calculated current $I_{k_{\text{calc.}}}$ flowing through the terminals of the power capacitor, as in Sections 3.1–3.3.
6. Formulate the problem of power capacitor parameterization, as in Section 3.
7. Program the configurable source to generate voltage signals with harmonic components according to standards and/or recommendations.
8. Sample the current flowing through the capacitor's terminals, considering a sampling frequency of at least 1500 Hz.
9. Solve the problem of power capacitor parameterization by metaheuristics or nonlinear mathematical programming methods.
10. Assess the model parameterization by means of the mean squared error between the reference and calculated currents.

PROBLEMS

1. An all-film propylene power capacitor with nominal capacitance of 137 μF and 2.5 kVar is supplied with a pure sinusoidal waveform (220 V, 60 Hz). Considering that such a capacitor is modeled by the simple equivalent circuit depicted in Fig. 21 [10] estimate the parameters C (capacitance),

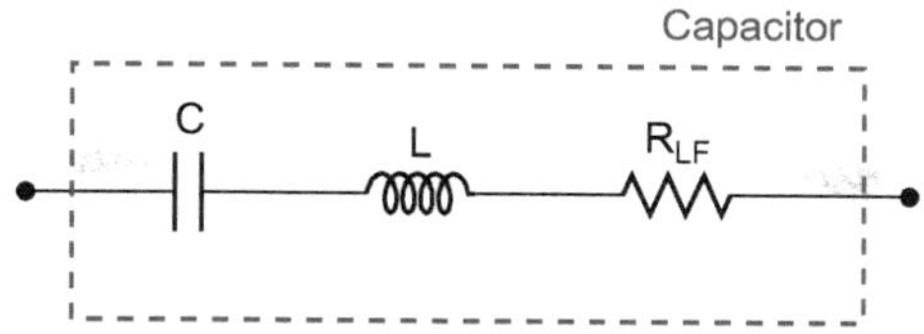

FIG. 21 Simple equivalent circuit of a capacitor.

TABLE 11 Lower and Upper Bounds for Parameters in the Capacitor Model in Fig. 21

Parameter	Lower Bound	Upper Bound
C_{ES} (μF)	127.0	147.0
L (mH)	0.05	0.5
R_{LF} (Ω)	0.05	0.75

L (inductance of connection wires), and R_{LF} (series resistance at low frequencies), whose lower and upper bounds are presented in Table 11.

Observation. The reference current data are available at http://sites.ieee.org/pes-iss/data-sets/, in the Power Quality section.

Answers. C = 135.748 μF; L = 0.5 mH; R_{LF} = 0.75 Ω; mean squared error = 0.0318023 A^2. (Results obtained with MINOS [21].)

2. An impregnated propylene power capacitor with nominal capacitance of 137 μF and 2.5 kVar is supplied with a distorted voltage waveform according to IEEE standards. Considering the equivalent circuit of a capacitor from Problem 1, whose lower and upper bounds are presented in Table 11, estimate the parameters C, L and R_{LF} according to the optimization-based power capacitor parameterization presented in this chapter.

Observation. The reference current data are available at http://sites.ieee.org/pes-iss/data-sets/, in the Power Quality section.

Answers. C = 131.816 μF; L = 0.114127 mH; R_{LF} = 0.623448 Ω; mean squared error = 0.954586 A^2. (Results obtained with MINOS [21].)

ACKNOWLEDGMENTS

This work has been supported by the Brazilian funding agencies: São Paulo Research Foundation (FAPESP), National Council for Scientific and Technological Development (CNPq), and Federal Agency for Support and Evaluation of Graduate Education (CAPES).

REFERENCES

[1] G.W. Chang, W.C. Chang, C.S. Chuang, D.Y. Shih, Fuzzy logic and immune-based algorithm for placement and sizing of shunt capacitor banks in a distorted power network, IEEE Trans. Power Delivery 26 (4) (2011) 2145–2153.

[2] G.J. Wakileh, Power Systems Harmonics: Fundamentals, Analysis and Filter Design, Springer, New York, NY, 2010.

[3] A. Baggini, Handbook of Power Quality, John Wiley & Sons, New York, NY, 2008.

[4] A.A. Eajal, M.E. El-Hawary, Optimal capacitor placement and sizing in unbalanced distribution systems with harmonics consideration using particle swarm optimization, IEEE Trans. Power Delivery 25 (3) (2010) 1734–1741.

[5] H. Mohkami, R. Hooshmand, A. Khodabakhshian, Fuzzy optimal placement of capacitors in the presence of nonlinear loads in unbalanced distribution network using BF-PSO algorithm, Appl. Soft Comput. 11 (2011) 3634–3642.

[6] S.M. Kannan, P. Renuga, S. Kalyani, E. Muthukumaran, Optimal capacitor placement and sizing using Fuzzy-DE and Fuzzy-MAPSO methods, Appl. Soft Comput. 11 (2011) 4997–5005.

[7] S.A. Taher, R. Bagherpour, A new approach for optimal capacitor placement and sizing in unbalanced distorted distribution systems using hybrid honey bee colony algorithm, Int. J. Electr. Power Energy Syst. 49 (2013) 430–448.

[8] M. Mohammadi, Bacterial foraging optimization and adaptive version for economically optimum sitting, sizing and harmonic tuning orders setting of LC harmonic passive power filters in radial distribution systems with linear and nonlinear loads, Appl. Soft Comput. 29 (2015) 345–356.

[9] W.J. Sarjeant, Capacitor fundamentals, in: Proceedings of the 19th Electrical Electronics Insulation Conference, 1989, pp. 1–51.

[10] S. Siami, C. Joubert, C. Glaize, High frequency model for power electronics capacitors, IEEE Trans. Power Electron. 16 (2) (2001) 157–166.

[11] R.W. Brown, Using distributed circuits to model power capacitor behavior, in: IEEE Region 10 TENCON, 2005, pp. 1–6.

[12] R.W. Brown, Empirically-derived capacitor characteristic formulas from distributed modelling, IET Circuits Devices Syst. 1 (2) (2007) 117–125.

[13] M. Makdessi, A. Sari, P. Venet, Improved model of metalized film capacitors, IEEE Trans. Dielectr. Electr. Insul. 21 (2) (2014) 583–593.

[14] G. Spavieri, R.T.M. Ferreira, R.A.S. Fernandes, G.G. Lage, D. Barbosa, M. Oleskovicz, Particle swarm optimization-based approach for parameterization of power capacitor models fed by harmonic voltages, Appl. Soft Comput. 56 (2017) 55–64.

[15] T. Umemura, K. Abe, K. Akiyama, Y. Tanaka, All-film power capacitor with folded electrode foil, IEEE Trans. Power Delivery PWRD-2 (1) (1987) 185–188.

[16] A. Tomago, T. Shimizu, Y. Iijima, I. Yamauchi, Development of oil-impregnated, all-polypropylene-film power capacitor, IEEE Trans. Electr. Insul. EI-12 (4) (1977) 293–301.

[17] ANEEL, PRODIST—Procedimentos de Distribuiç ao de Energia Elétrica no Sistema Elétrico Nacional Módulo 8 (Qualidade da Energia Elétrica)., ANEEL, 2013.

[18] IEC Std. 61000-3-2, Electromagnetic Compatibility (EMC)—Part 3-2 (Limits for harmonic current emissions). 1995.

[19] IEEE Std. 1159-2009, IEEE Recommended practice for monitoring electric power quality. 2009.

[20] R. Fourer, D.M. Gay, B.W. Kernighan, AMPL: A Modeling Language for Mathematical Programming, second ed., Cengage Learning, Duxbury, 2002. ISBN 0534388094.

[21] B.A. Murtagh, M.A. Saunders, Large-scale linearly-constrained optimization, Math. Program. 14 (1) (1978) 41–72.

Chapter 11

Two-Level Multidimensional Enhanced Melody Search Algorithm for Dynamic Planning of MV Open-Loop Distribution Networks

Mohammad Kiani-Moghaddam*, Mojtaba Shivaie† and Philip D. Weinsier‡

*Department of Electrical Engineering, Shahid Beheshti University, Tehran, Iran, †Faculty of Electrical Engineering and Robotic, Shahrood University of Technology, Shahrood, Iran, ‡Department of Applied Electrical Engineering, Bowling Green State University, Firelands, OH, United States

NOMENCLATURE

Indices and Sets

a*, *Na	Index and set of all substation types (expansion options)
b*, *Nb	Index and set of all branch types (expansion options)
cbr*, *Ncbr	Index and set of all candidate branches for reinforcement
clb*, *Nclb	Index and set of all candidate locations for branch installation
cns*, *Ncns	Index and set of all candidate nodes for substation installation
csr*, *Ncsr	Index and set of all candidate substations for reinforcement
es*, *Nes	Index and set of all existing substations
h*, *Nh	Index and set of all failure events
k*, *Nk	Index and set of all load sectors in each load node
ll*, *Nll	Index and set of all load levels
ln*, *Nln	Index and set of all load nodes
m*, *m′*, *Nm	Index and set of all optimization algorithms
p*, *Np	Index and set of all periods of the planning horizon
q*, *Nq	Index and set of all objective functions
y	Index of the expansion plans

Classical and Recent Aspects of Power System Optimization. https://doi.org/10.1016/B978-0-12-812441-3.00011-2
© 2018 Elsevier Inc. All rights reserved.

Parameters

B_p Budget at period p (M$)

C_{ll}^{energy} Cost of energy at load level ll ($/MWh)

D_{ll} Durations of load level ll (h)

$FR_{p,\, ll,\, h, ln,\, k}^{\text{aver}}$ Average failure rate at load sector k of affected load node ln in case of failure event h, at load level ll of period p (failure per year)

$h_{ln,\, ll,\, p}^{\max}$ Maximum nodal-load shedding in load node ln at load level ll of period p (MW)

Ir Interest rate (%)

$I_{es}^{\max}$ Maximum current of existing substation es (p.u)

L_{rb} Length of reinforced branch rb (km)

L_{ib} Length of installed branch ib. (km)

$P_{ll,\, p}^{\text{elec-mark}}$ Electricity market price at load level ll of period p ($/MWh)

$\mathbf{P}^{\text{max-dgr}}$ Maximum capacity of DGRs that can be installed in the MV open-loop distribution network (MW)

$P_{id}^{\max}$ Maximum capacity of installed DGR id (MW)

R_{es} Resistance of existing substation es (Ω)

TB Total budget (M$)

$V_{ln,\, ll,\, p}^{\min}$ Minimum voltage in load node ln at load level ll of period p (p.u)

$V_{ln,\, ll,\, p}^{\max}$ Maximum voltage in load node ln at load level ll of period p (p.u)

$\mathbf{Y}^{\text{e-ne}}$ Node-branch incidence matrix for the existing MV open-loop distribution network

$\mathbf{Y}_{\text{column}:eb}^{\text{e-ne}}$ Column eb of the node-branch incidence matrix of the existing MV open-loop distribution network

Z_{eb} Impedance of existing branch eb of the MV open-loop distribution network (Ω)

δ_s Economic lifetime of the substations (year)

Variables

$C_{y,\, p}^{\text{inv}}$ Investment costs for the expansion plan y at period p (M$)

$C_{rs,\, p}^{\text{inv}}(a)$ Investment costs for the reinforced substation rs, at period p, associated with substation type a (M$)

$C_{is,\, p}^{\text{inv, fix}}$ Fixed investment costs for the installed substation is, at period p, associated with substation type a (M$)

$C_{is,\, p}^{\text{inv, var}}(a)$ Variable investment costs for the installed substation is, at period p, associated with substation type a (M$)

$C_{ib,\, p}^{\text{inv}}(b)$ Investment costs for the installed branch ib., at period p, associated with branch type b (M$/km)

$C_{rb,\, p}^{\text{inv}}(b)$ Investment costs for the reinforced branch rb, at period p, associated with branch type b (M$/km)

$C^{inv}_{id,\,p}$ Investment costs for the installed DGR *id* at period *p* ($/MW)
$C^{maint}_{y,\,p}$ Total maintenance costs for expansion plan *y* at period *p* (M$)
$C^{oper}_{y,\,p,\,ll}$ Total operational costs for expansion plan *y* at load level *ll* of period *p* (M$)
$C^{maint}_{es,\,p}$ Maintenance costs of existing substation *es* at period *p* (M$)
$C^{oper}_{id,\,ll,\,p}$ Operational costs of installed DGR *id* at load level *ll* of period *p* ($/MW h)
$C^{cu\text{-}out}_{y,\,p}$ Customer outage costs of expansion plan *y* at period *p* (M$)
$C^{dam\text{-}fun}_{p,\,ll,\,h,ln,\,k}$ Outage costs, based on the outage duration time D^{out}_{h}, using the SCDF at load sector *k* of affected load node *ln* in case of failure event *h*, at load level *ll* of period *p* ($/kW)
$\mathbf{d}^{no\text{-}load}_{ll,\,p}$ Column vector of the nodal load at load level *ll* of period *p*
$D^{out}_{p,\,ll,\,h,ln,\,k}$ Outage duration time at load sector *k* of affected load node *ln* in case of failure event *h*, at load level *ll* of period *p* (h)
$\mathbf{f}^{e\text{-}b}_{ll,\,p}$ Row vector of the current for existing branches at load level *ll* of period *p*
$\mathbf{g}^{no\text{-}inj}_{ll,\,p}$ Column vector of the nodal injection at load level *ll* of period *p*
$h_{ln,\,ll,\,p}$ Nodal-load shedding in load node *ln* at load level *ll* of period *p* (MW)
$\mathbf{h}^{no\text{-}load\text{-}sh}_{ll,\,p}$ Column vector of nodal-load shedding at load level *ll* of period *p*
$I_{es,\,ll,\,p}$ Injected current of existing substation *es* at load level *ll* of period *p* (p.u)
$I_{eb,\,ll,\,p}$ Current of existing branch *eb* at load level *ll* of period *p* (p.u)
$N^{i\text{-}s}_{cns,\,p}$ Number of installed substations in candidate node *cns* at period *p*
$N^{r\text{-}s}_{csr,\,p}$ Number of reinforcements of candidate substation *csr* at period *p*
$N^{i\text{-}b}_{clb,\,p}$ Number of installed branches in candidate location *clb* at period *p*
$N^{r\text{-}b}_{cbr,\,t}$ Number of reinforcements of candidate branch *cbr* at period *p*
$P_{es,\,ll,\,p}$ Injected power of existing substation *es* at load level *ll* of period *t* (MW)
$P^{cap}_{id,\,p}$ Total operational capacity of installed DGR *id* at period *p* (MW)
$P^{res}_{id,\,p}$ Total reserve capacity of installed DGR *id* at period *p* (MW)
$P_{id,\,ll,\,p}$ Generated power of the installed DGR *dg* at load level *ll* of period *p* (MW)
$P^{dema}_{p,\,ll,\,h,ln,\,k}$ Amount of power demand at load sector *k* of affected load node *ln* in case of failure event *h*, at load level *ll* of period *p* (MW)
TIC_y Total investment costs for expansion plan *y* (M$)
TOC_y Total operational costs along with maintenance costs for expansion plan *y* (M$)
$TCOC_y$ Total customer outage costs of expansion plan *y* (k$/h)
$\mathbf{v}_{ll,\,p}$ Row vector of nodal voltages at load level *ll* of period *p*
$V_{ln,\,ll,\,p}$ Voltage in load node *ln* at load level *ll* of period *p* (p.u)
$\zeta_{u,\,3}$ Skewness of probabilistic variable *u*.

NOTE

- Corresponding indexes, set, parameters, and variables are also considered for branches and DGRs by changing index s to b or d, respectively.
- Corresponding indexes, set, parameters, and variables are also considered for reinforcement and installation by changing index e to r or i, respectively.
- Corresponding indexes, set, parameters, and variables are also considered for voltage conversion nodes and generation nodes by changing index ln to cn or gn, respectively.
- For example: The maintenance cost of existing substation es at period p ($C^{\text{main}}_{es,\ p}$) is converted to the maintenance cost of reinforcement branch rb at period p ($C^{\text{main}}_{rb,\ p}$) by changing e to r and s to b.

1 INTRODUCTION

1.1 Background and Motivation

In a very general sense, optimization is the process of obtaining one or more solutions from the all the feasible solutions by minimizing (or maximizing) one or more objective functions under specified circumstances—if any [1]. During the past few decades, this process has been organized into many techniques, such as linear conic optimization [2], convex or concave minimization [3,4], branch and bound optimization [5], constructive optimization [6], gradient descent optimization [7], Newton, Gauss-Newton, Quasi-Newton optimization [8], etc. Each of these classical optimization techniques has been constructed on a given theoretical principle along with a large linear/nonlinear collection of equations. In such a situation, each iteration of these classical optimization techniques requires tackling a wide range of computational complexities, which for real-world, large-scale engineering optimization problems can be prohibitively expensive. At the other end of the spectrum from these classical optimization techniques, modern metaheuristic algorithms have been well-suited for solving engineering optimization problems with a combination of discrete and continuous decision-making variables. This is true when the feasible region of the search space and/or dimensions of the optimization problem increases, as in cases where the gradient does not exist and when the objective and/or constraint functions are discontinuous [9,10]. A modern metaheuristic algorithm is defined as an iterative process to explore and exploit near-optimal/optimal solutions of the search space with respect to multiple goals and conditions [8–10]. For the reasons identified above, modern metaheuristic algorithms are widely employed in engineering applications; therefore, researchers and specialists must have an appropriate understanding of them. As a result, researchers and engineers in different areas of the sciences are enthusiastic about employing innovative alternatives in these modern metaheuristic algorithms to enhance the performance and efficiency of solving real-world, large-scale engineering optimization problems.

1.2 Literature Review and Contributions

There are many different classifications available for organizing modern metaheuristic algorithms. In general, these algorithms can be categorized based on four major perspectives: (i) certain behaviors of a swarm of fauna; (ii) given attributes of biological and neurobiological systems; (iii) physics/mathematics concepts; and, (iv) specific behaviors of human and man-made societies [9]. Swarm-based optimization algorithms have been inspired by the individual and collective behaviors of interacting fauna that pursue some ordinary rules; examples include fireflies [11], ants [12], bees [13], particles [14], bacteria [15], fish [16], cuckoo [17], monkeys [18], and bats [19]. Table 1 shows the most well-known swarm-based optimization algorithms. From an implementation standpoint, these algorithms usually bring about a premature convergence and get stuck in a local optimum point, because the final solution point mostly depends on the beginning point. In a general sense, biological/neurobiological-based optimization algorithms have a direct link to any biological/neurobiological behavior, and do not directly utilize the swarming patterns such as genetic algorithm [20], differential evolution [21], brainstorm algorithm [22], shuffled leap-frogging algorithm [23], dolphin echolocation algorithm [24], etc. Table 1 also presents the most popular biological/neurobiological-based optimization algorithms. In addition to the aforementioned disadvantages, these algorithms usually require a relatively large population and a lot of generations before obtaining the final solution point, and are not directly appropriate for solving constrained, real-world, large-scale optimization problems from an implementation point of view. The physics/mathematics-based optimization algorithms have been constructed by considering the rules of physics/mathematics, such as: simulated annealing [25], gravitational search [26], big bang/big crunch [27], electromagnetism [28], spiral optimization [29], harmony search algorithm [30], etc. The human-driven optimization algorithms have also been inspired by individual and social behaviors, immune systems, and human skills, examples of which include artificial immune system [31], key cutting algorithm [32], seeker optimization algorithm [33], and imperialist competitive algorithm [34]. Table 1 illustrates the most well-known physics/mathematics-based and human-driven optimization algorithms. These algorithms have a low convergence rate, high computational efforts, and also simply fall into precocity phenomena. Accordingly, the physics/mathematics-based and human-driven optimization algorithms may not be well-suited techniques for tackling complex real-world, large-scale optimization problems having a nonlinear, mixed-integer structure. Thus, from an investigative standpoint, existing modern metaheuristic algorithms can lead to inappropriate designs and result in implementation difficulties such as poor global and local search functionalities, explore a barely sufficient search space, converge comparatively slowly, require initial value settings, do not sweep all the existing stored solution vectors to produce a new solution vector, and so on. As a result, such drawbacks cause these techniques to collapse and have no chance in dealing with complex

TABLE 1 A Summary of Attributes and Power Systems Applications of the Most Popular Modern Metaheuristic Optimization Algorithms

Category	Refs.	Name	Inspired Idea	Power Systems Applications
Swarm-based optimization algorithms	[11]	Firefly algorithm	Fireflies shine	Flexible alternative current transmission system (FACTS) placement
	[12]	Ant colony algorithm	Swarm behavior of ants	Distribution system reconfiguration and capacitor placement
	[13]	Bee colony algorithm	Swarm behavior of bees	Tie switch allocation and distributed generation (DG) placement
	[14]	Particle swarm optimization algorithm	Swarm behavior of particles	Power system state estimation
	[15]	Bacterial foraging optimization algorithm	Bacterial foraging	Economic dispatch
	[16]	Artificial fish swarm algorithm	Swarm behavior of fish	Hydrothermal scheduling
	[17]	Cuckoo search algorithm	Behavior of cuckoo for flyblow	Hydrothermal scheduling
	[18]	Monkey search algorithm	Swarm behavior of monkeys	Allocation of capacitor banks in distribution systems
	[19]	Bat algorithm	Sound system of bats	Load frequency controller design

Biological/neurobiological-based optimization algorithms	[20]	Genetic algorithm	Natural evolution	Electric network design
	[21]	Differential evolution algorithm	Differential evolution	Power flow
	[22]	Brainstorm algorithm	Human brainstorming process	Energy optimization in grid systems
	[23]	Shuffled leap-frog algorithm	Natural memetics	Photovoltaic model identification
	[24]	Dolphin echolocation algorithm	Intelligent behaviors of dolphins	Reactive power dispatch
Physics/mathematics-based optimization algorithms	[25]	Simulated annealing algorithm	Heating and controlled cooling of a material	Energy resource management
	[26]	Gravitational search algorithm	Newton's famous law of gravity and the law of motion	Power flow
	[27]	Big bang/big crunch algorithm	Energy dissipation and randomly distributed particles	Placement and sizing of distributed generators
	[28]	Electromagnetism algorithm	Attraction-repulsion mechanism among charges to evolve the members of a population	Engineering application
	[29]	Spiral algorithm	Analogy of spiral phenomena in nature	Combined economic and emission dispatch
	[30]	Harmony search algorithm	Music phenomenon	Placement and sizing of active power filters

Continued

TABLE 1 A Summary of Attributes and Power Systems Applications of the Most Popular Modern Metaheuristic Optimization Algorithms—cont'd

Category	Refs.	Name	Inspired Idea	Power Systems Applications
Human-driven optimization algorithms	[31]	Artificial immune system algorithm	Principles and processes of the vertebrate immune system	Distribution system reconfiguration
	[32]	Key cutting algorithm	Work of locksmiths to defeat the lock	Power system state estimation
	[33]	Seeker algorithm	Concept of simulating human searching behaviors	Economic load dispatch
	[34]	Imperialist competitive algorithm	Imperialistic competition	Unit commitment

real-world, large-scale optimization problems with big data. However, in 2011, for the first time, a new computational technique called a two-level computational, multidimensional, continuous melody search algorithm (2D-cMSA), a physics/mathematics-based optimization algorithm, was proposed based on the main concepts of the single-level computational, one-dimensional harmony search algorithm (1D-HSA) [35]. Organizing the 2D-cMSA brought about an innovative direction in the architecture of modern metaheuristic optimization algorithms for solving complex real-world, large-scale optimization problems with big data. This technique, when compared with other modern metaheuristic optimization algorithms (see again Table 1) has several significant privileges.

- It is a two-level computational algorithm, whereas other modern metaheuristic optimization algorithms have but a single computational level.
- It has multiple solution memories so that each represents a single-level computational, one-dimensional optimization algorithm, whereas other modern metaheuristic optimization algorithms have only one solution memory.
- It considers interactive relationships between solution memories, whereas other modern metaheuristic optimization algorithms do not.

These features can increase the flexibility, robustness, and parallel capability of the 2D-cMSA and, therefore, can enhance the possibility of generating better solution vectors. Nevertheless, this algorithm was developed only for optimization problems with continuous decision-making variables.

The focus of this chapter, then, is on identifying two targets in the context of the modern metaheuristic optimization algorithms. Also introduced is a new enhanced version of the melody search algorithm with two-level computational, multidimensional features to improve the performance of the original 2D-cMSA in solving continuous optimization problems and offering a new discrete version of the two-level computational, multidimensional melody search algorithm for use in real-world, large-scale discrete optimization problems. The combination of the two new optimization algorithms is called a two-level computational, multidimensional, continuous/discrete, enhanced melody search algorithm (2D-cdEMSA). In addition, a midterm dynamic planning problem of the medium voltage (MV), open-loop distribution networks (DyMODN) in the presence of distributed generation resources (DGRs) has been employed to evaluate the performance of the 2D-cdEMSA, compared with other existing optimization algorithms. The DyMODN problem is a complicated, large-scale, nonconvex, optimization problem having a nonlinear, mixed-integer nature in which expansion options are a reinforcement and/or installation of substations, branches, and DGRs. The planning methodology considers the total investment cost (TIC), total operational cost (TOC), and total customer outage cost (TCOC) in the optimization problem as three problem objectives. Concurrently, Kirchhoff's current and voltage rules, operational limits, financial restrictions, and logical limits are considered as optimization problem constraints. To handle risks of planning arising from uncertainty of demand, a two-point estimation method

(2-PEM) is employed. To demonstrate the feasibility and capability of the new algorithm, the performance of the 2D-cdEMSA is compared with: the original 1D-HSA; the single-level computational, one-dimensional improved harmony search algorithm (1D-IHSA); the original 2D-cMSA; the nondominated sorting genetic algorithm-II (NSGA-II); and, the particle swarm optimization (PSO) algorithm. Simulation results illustrate the sufficiency and profitableness of the 2D-cdEMSA, when compared with other optimization algorithms.

1.3 Reading Map

The authors provide the following reading map to the readers. In Section 2, an overview of the concepts and the computational structures of the music-inspired optimization algorithms, focusing in particular on the original 1D-HSA and 2D-cMSA, is briefly presented. The new 2D-cdEMSA is presented in Section 3. In Section 4, the implementation of the new optimization algorithm for solving a midterm DyMODN problem is discussed in detail, including mathematical modeling of the DyMODN problem, numerical results, and case studies. The chapter ends with a brief summary and some concluding remarks in Section 5.

2 THE MUSIC-INSPIRED OPTIMIZATION ALGORITHMS: CONCEPTS AND COMPUTATIONAL STRUCTURES

2.1 Overview of the Concepts

The harmony search algorithm (HSA) is a new population-based, metaheuristic optimization algorithm, which can be conceptualized from a music performance process (i.e., improvisation by a clarinet player) involving searching for a fantastic harmony [30,36–38]. There are several technical publications in the area of improving performance of the original HSA. As further explanation, all improvements of the original HSA cover two distinct perspectives: (1) improvement of the original HSA in terms of parameter settings, and (2) improvement of the original HSA with respect to hybridizing its components with other metaheuristic optimization algorithms. It is relevant to note that the hybridization of the original HSA with other metaheuristic optimization algorithms covers two different approaches: (1) incorporating some components of other metaheuristic optimization algorithms into the original HSA structure, and (2) incorporating some components of the original HSA into the structure of other metaheuristic optimization algorithms. These perspectives are out of the scope of this chapter, but the interested reader may look to the work by Moh'd-Alia and Mandava [39] for a discussion of these aspects. The original HSA was organized by a musician with a single improvisation procedure. In other words, the original HSA is like a musician who plays a song with a musical instrument (e.g., guitar, piano, violin, contrabass, or trumpet). Therefore, the

HSA is labeled as a 1D-HSA. The melody search algorithm (MSA) is a new population-based, metaheuristic optimization algorithm, which was inspired by borrowing the main concepts of the HSA; however, the MSA is completely different in structure [40,41]. The HSA employs an individual harmony memory (HM), while the MSA uses multiple memories, called Player memories (PMs). Multiple PMs are organized into a melody memory (MM). Therefore, the MSA is an enhanced version of the HSA. The MSA has been established by multiple music players with both single- and group-level improvisation. In other words, the MSA is similar to a group of music players all playing the same musical instrument, for example a guitar. As a result, the MSA is labeled as a 2D-cMSA.

2.2 Computational Structure

2.2.1 Single-Level Computational, One-Dimensional Harmony Search Algorithm

The process of solving an optimization problem using the 1D-HSA can be performed in two principal phases.

Phase 0: Input required data. Table 2 shows how the preparation of the optimization problem and initialization of the 1D-HSA parameters are performed.

Phase 1: Single-level computational.

Subphase 1-1: Preparation of the initial HM. Multiple randomized harmony vectors (solution vectors), depending on the HMS, and relevant

TABLE 2 The 1D-HSA Parameters

No.	1D-HSA Parameters	Abbreviation	Parameter Range
1	Distance bandwidth	BW	$BW \in R > 0$
2	Harmony memory	HM	–
3	Harmony memory considering rate	HMCR	$0 \leq HMCR \leq 1$
4	Harmony memory size	HMS	$HMS \geq 1$
5	Maximum number of iterations	MNI	$MNI \geq 1$
6	Number of decision-making variables	D	$D \geq 1$
7	Pitch adjusting rate	PAR	$0 \leq PAR \leq 1$

objective function values are stored in the initial HM via Eqs. (1)–(3). It is worth noting that Eqs. (2), (3) are generally considered for discrete and continuous decision-making variables, respectively.

$$\mathrm{HM}=\begin{bmatrix} h_1^1 & \cdots & h_1^d & \cdots & h_1^{\mathrm{D}} & | & f(\mathbf{h}_1) \\ \vdots & & \vdots & & \vdots & & \vdots \\ h_g^1 & \cdots & h_g^d & \cdots & h_g^{\mathrm{D}} & | & f(\mathbf{h}_g) \\ \vdots & & \vdots & & \vdots & & \vdots \\ h_{\mathrm{HMS}}^1 & \cdots & h_{\mathrm{HMS}}^d & \cdots & h_{\mathrm{HMS}}^{\mathrm{D}} & | & f(\mathbf{h}_{\mathrm{HMS}}) \end{bmatrix};\forall\{d\in\mathrm{D}, g\in\mathrm{HMS}\} \quad (1)$$

$$h_g^d = h^d(z);\forall\{d\in\mathrm{D}, g\in\mathrm{HMS}, z\in rand(\{1,2,\ldots,Z\})\} \quad (2)$$

$$h_g^d = LL^d + rand\in(0,1)\times\left(UL^d - LL^d\right);\forall\{d\in\mathrm{D}, g\in\mathrm{HMS}\} \quad (3)$$

where, LL^d and UL^d are the lower and upper limits of continuous decision-making variable d, respectively, and Z is the number of values of discrete decision-making variables.

After filling the initial HM, all stored harmony vectors in the initial HM are arranged in the desired order, based on their objective function values.

Subphase 1-2: Generate a new harmony vector (improvisation). Here, the inherent nature of the 1D-HSA and the foundation that is the structure of this music-inspired algorithm is shown. In this subphase, improvising a new harmony vector is conducted by taking three main operators into account—the harmony memory consideration, the pitch adjustment rate, and the random selection. Table 3 provides the pseudo code of the improvisation procedure for discrete and continuous decision-making variables in the 1D-HSA.

Subphase 1-3: Update the HM. First, the newly improvised harmony vector from the previous subphase is evaluated in terms of its objective function value. Second, a comparison between this harmony vector and the worst harmony vector stored in the HM is performed. If the newly improvised harmony vector yields a better fitness than that of the worst harmony vector stored in the HM, it will be included in the HM and the existing worst harmony vector will be excluded from the HM. Otherwise, the newly improvised harmony vector will be ignored. Third, all stored harmony vectors in the HM are rearranged in the desired order, based on their objective function values.

Subphase 1-4: Check the stopping index. The process of solving the optimization problem continues by iterating subphases 1-2 and 1-3 until the given stopping index is satisfied. Eventually, the best harmony vector in the HM is selected as the optimal solution of the optimization problem under investigation. This was not intended to be an all-encompassing overview of discrete and continuous versions of the 1D-HSA.

TABLE 3 The Improvisation Procedure for Discrete and Continuous Decision-Making Variables in the 1D-HSA

Algorithm 1: Improvisation Procedure for the 1D-HSA

Input: BW, D, Z, HMCR, HMS, LL^d, PAR, UL^d

Output: h_{new} (new harmony vector)

start main body

for *each decision-making variable* $\{d \in \text{D}\}$ **do**

  if $rand \in (0,1) \leq$ HMCR **then**

    operator 1: *harmony memory consideration with probability* HMCR

  $h^d_{\text{new}} = h^d_g$; $\forall\, g = rand(\{1,2,\ldots,\text{HMS}\})$; for CD-DMV

  if $rand \in (0,1) \leq$ PAR **then**

    operator 2: *pitch adjustment with probability* HMCR*PAR

   $h^d_{\text{new}} = h^d_{\text{old}}(m)$; $\forall\, m = \text{rand}(\{\ldots,-2,-1,1,2,\ldots\})$; for D-DMV

   $h^d_{\text{new}} = h^d_{\text{old}} \pm rand \in (0,1) \times \text{BW}$; for C-DMV

  end if

  else if

    operator 3: *random selection with probability* 1-HMCR

  $h^d_{\text{new}} = h^d(z)$; $\forall\, z = rand(\{1,2,\ldots,\text{Z}\})$; for D-DMV

  $h^d_{\text{new}} = LL^d \pm rand \in (0,1) \times (UL^d - LL^d)$; for C-DMV

  end if

end for

end main body

Please refer to the work by Moh'd-Alia and Mandava [39] and Lee et al. [42] for a comprehensive discussion of these versions of the 1D-HSA.

2.2.2 *Two-Level Computational, Multidimensional, Continuous Melody Search Algorithm*

The optimization process of the 2D-cMSA can be classified into three main phases for every specific optimization problem.

Phase 0: Load required data. The definition and preparation of the optimization problem is conducted. In addition, Table 4 shows how the 2D-cMSA parameters, which can alter in any optimization problem, are determined.

Phase 1: Single improvisation level (SIL), or first-level computational. In the SIL, each music player improvises their melody separately without the influence of other music players, similar to the structure of the 1D-HSA.

Subphase 1-1: Preparation of the initial MM. All PMs are randomly filled by initial melodies and the MM structure is also organized by multiple PMs, according to Eqs. (4)–(6). It is worth noting that the original 2D-cMSA is presented only for continuous decision-making variables.

$$\text{MM} = [PM_1 \;\ldots\; PM_m \;\ldots\; PM_{\text{PMN}}]^{\text{T}}; \forall\{m \in \text{PMN}\} \tag{4}$$

TABLE 4 The 2D-cMSA Parameters

No.	2D-cMSA Parameters	Abbreviation	Parameter Range
1	Distance bandwidth	BW	BW∈R>0
2	Melody memory	MM	–
3	Maximum number of iterations for first-level computational	MNI	MNI≥1
4	Maximum number of iterations for second-level computational	MNII	MNII≥1
5	Number of decision-making variables	D	D≥1
6	Player memory number	PMN	PMN≥1
7	Player memory considering rate	PMCR	0≤PMCR≤1
8	Player memory size	PMS	PMS≥1
9	Pitch adjusting rate	PAR	0≤PAR≤1

$$PM_m = \begin{bmatrix} h^1_{m,1} & \cdots & h^d_{m,1} & \cdots & h^{\mathrm{D}}_{m,1} & | & f(\mathbf{h}_{m,1}) \\ \vdots & & \vdots & & \vdots & & \vdots \\ h^1_{m,g} & \cdots & h^d_{m,g} & \cdots & h^{\mathrm{D}}_{m,g} & | & f(\mathbf{h}_{m,g}) \\ \vdots & & \vdots & & \vdots & & \vdots \\ h^1_{m,\mathrm{PMS}} & \cdots & h^d_{m,\mathrm{PMS}} & \cdots & h^{\mathrm{D}}_{m,\mathrm{PMS}} & | & f(\mathbf{h}_{m,\mathrm{PMS}}) \end{bmatrix}; \quad \forall\{m \in \mathrm{PMN}, d \in \mathrm{D}, g \in \mathrm{PMS}\} \tag{5}$$

$$h^d_{m,g} = LL^d + rand \in (0,1) \times \left(UL^d - LL^d\right); \forall\{m \in \mathrm{PMN}, d \in \mathrm{D}, g \in \mathrm{PMS}\} \tag{6}$$

where, LL^d and UL^d are the lower and upper limits of the continuous decision-making variable d, respectively. These parameters cannot change in the SIL, though they can in the group improvisation level (GIL). After filling the initial MM, all stored melody vectors in each PM are arranged in the desired order, based on their objective function values.

Subphase 1-2: Generate a new melody vector for each PM. A new melody vector is improvised for each PM. In the original 2D-cMSA, an alternative improvisation procedure (AIP) was proposed for generating new melodies, which was inspired from the main concepts of the harmony improvisation.

Subphase 1-3: Update each PM. Initially, the newly improvised melody vector by the music player, m, is evaluated in terms of the objective function value. Then, this melody vector is compared with the state of the worst melody vector in the corresponding PM. In this case, if the newly improvised melody vector is better than the worst melody vector, it replaces the old one and the worst melody is discarded from the corresponding PM. Otherwise, the newly improvised melody vector will be discarded. Finally, all stored melody vectors in the corresponding PM are rearranged in the desired order, based on their objective function values.

Subphase 1-4: Check the stopping index of the SIL. Iterate subphases 1-2 and 1-3 until the stopping index of the SIL, the MNI, is satisfied.

Phase 2: Group improvisation level, or second-level computational. In the GIL, a search for the well-designed configurations of pitches is accomplished by an interactive-based process among the music players, where feasible areas of pitches (limits of the decision-making variables) are updated in each iteration of the GIL.

Subphase 2-1: Generate a new melody vector for each PM. The AIP is applied for improvising a new melody vector for each PM in accordance with the feasible areas of pitches.

Subphase 2-2: Update each PM. Each PM is updated, similar to the process performed in subphase 1-3.

Subphase 2-3: Find the feasible areas of pitches for subsequent improvisation. By storing well-designed melody variables of all PMs, new feasible areas of pitches can be determined for the subsequent improvisation (just for randomization). Table 5 addresses this procedure. It is noteworthy that the subscript "best" is used for well-designed melodies in each PM.

Subphase 2-4: Check the stopping index of the GIL. Iterate subphases 2-1, 2-2, and 2-3 until the stopping index of the GIL, the MNII, is satisfied.

Finally, the best melody vector in the MM is selected as the optimal solution of the optimization problem under investigation.

TABLE 5 The Procedure to Find the Feasible Areas of Pitches in the 2D-cMSA

Algorithm 3: Procedure to find the feasible areas of pitches

Input: $h^d_{m,\text{best}}$; $\forall\{m \in \text{PMN}, d \in \text{D}\}$

Output: LL^d, UL^d (lower and upper limits of the decision-making variable *d*, respectively)

```
start main body
```

1: **for** *each decision-making variable* $\{d \in \text{D}\}$ **do**

2: $LL^d = \min(h^d_{m,\text{best}})$; $\forall\{m \in \text{PMN}\}$; for C-DMV

3: $UL^d = \max(h^d_{m,\text{best}})$; $\forall\{m \in \text{PMN}\}$; for C-DMV

4: **end for**

```
end main body
```

TABLE 6 The Improvisation Procedure for Continuous Decision-Making Variables in the 2D-cMSA

Algorithm 2: The AIP for the 2D-cMSA

Input: $BW_{\min}$, $BW_{\max}$, LL^d, D, MNI, MNII, PMN, $PAR_{\min}$, $PAR_{\max}$, PMCR, PMS, UL^d.

Output: $h_{m,\text{new}}$ (new melody vector for all music players)

start main body

for *each music player* $\{m \in \text{PMN}\}$ **do**
  for *each decision-making variable* $\{d \in \text{D}\}$ **do**
    if $rand \in (0,1) \leq$ HMCR **then**
      operator 1: *player memory consideration with probability* PMCR
      if *iteration counter* $\{i \in \text{MNI} + \text{MNII}\}$ = odd **then**
        $h^d_{m,\text{new}} = h^d_{m,g} + rand \in (0,1) \times BW_i;\ \forall\ g = rand(\{1,2,\ldots,\text{PMS}\})$
      else if
        $h^d_{m,\text{new}} = h^n_{m,g} + rand \in (0,1) \times BW_i;\ \forall\ g = \text{rand}(\{1,2,\ldots,\text{PMS}\});\ \forall\ n = \text{rand}(\{1,2,\ldots,\text{D}\})$
      end if
      if $rand \in (0,1) \leq PAR_i$ **then**
        operator 2: *pitch adjustment with probability* HMCR*PAR
        $h^d_{m,\text{new}} = h^d_{m,\text{best}}$
      end if
    else if
      operator 3: *random selection with probability* 1-HMCR
      $h^d_{m,\text{new}} = LL^d \pm rand \in (0,1) \times (UL^d - LL^d)$
    end if
  end for
end for

end main body

Table 6 shows the pseudo code of the AIP for improvising new melody vectors in both SIL and GIL of the 2D-cMSA [40]. In order to improve performance of the 2D-cMSA and avoid getting trapped in local minimum, dynamic relationships for the PAR and BW parameters are chosen. The PAR and BW parameters are dynamically updated by increasing the generation number, according to Eqs. (7), (8).

$$PAR_i = \text{PAR}_{\min} + ((\text{PAR}_{\max} - \text{PAR}_{\min})/\text{MNI}) \times i \tag{7}$$

where, PAR_i is the pitch adjustment rate in generation i; and, $\text{PAR}_{\min}$ and $\text{PAR}_{\max}$ are the minimum and maximum pitch adjustment rates, respectively.

$$BW_i = \text{BW}_{\max} \times \exp^{((Ln(\text{BW}_{\min}/\text{BW}_{\max})/\text{MNI}) \times i)} \tag{8}$$

where, BW_i is the distance bandwidth in generation i; and, $\text{BW}_{\min}$ and $\text{BW}_{\max}$ are the minimum and maximum bandwidths, respectively.

3 TWO-LEVEL COMPUTATIONAL, MULTIDIMENSIONAL, CONTINUOUS/DISCRETE, ENHANCED MELODY SEARCH ALGORITHM (2D-cdEMSA): AN INNOVATIVE IMPROVISATION PROCEDURE

In many cases, and particularly in engineering applications, optimization problems have been modeled as continuous problems for simplifying; more recently, its decision-making variables have been coded as continuous quantities. However, real-world, large-scale engineering optimization problems are faced with a combination of continuous and discrete decision-making variables, denoted by mixed-integer variables. In these circumstances, the complexity of the optimization problems is significantly increased. Therefore, on the one hand, the continuous optimization algorithms (see Table 1 for the continuous version of the optimization algorithms and, in particular, the 2D-cMSA) could not be directly employed in the presence of discrete variables. On the other hand, the existing discrete optimization algorithms (see Table 1 for the discrete version of the optimization algorithms) were unable completely to handle the effects of mixed-integer search spaces. As a result, it was necessary to develop a new strategy to provide a well-designed discrete optimization algorithm for treating the mixed-integer nature of the real-world, large-scale optimization problems. To cope with the drawbacks of the original 2D-cMSA, a new enhanced version of the original 2D-cMSA and a new discrete version of the two-level computational, multidimensional melody search algorithm have been developed by the authors for tackling the real-world, large-scale continuous/discrete optimization problems. As previously mentioned, the combination of these two new optimization algorithms is called the 2D-cdEMSA. Similar to the original 2D-cMSA, the computational structure of the 2D-cdEMSA consists of three different phases; only the differences between the optimization algorithm and the original version are examined here. Unlike the original 2D-cMSA, in the new 2D-cdEMSA, the size of the memory of each music player, PMS, is considered to be different from other music players (PMS in the original 2D-cMSA vs PMS_m in the new 2D-cdEMSA). Hence, from subphase 1-1, Eqs. (5), (6) are replaced with Eqs. (9), (10), and Eq. (11) is added for discrete decision-making variables. It is worth noting that Eq. (4) remains unchanged. In the 2D-cdEMSA, a new improvisation procedure is employed to generate new melody vectors for both discrete and continuous decision-making variables (see Table 7).

$$\mathrm{PM}_m = \begin{bmatrix} h^1_{m,1} & \cdots & h^d_{m,1} & \cdots & h^{\mathrm{D}}_{m,1} & | & f(\mathbf{h}_{m,1}) \\ \vdots & & \vdots & & \vdots & & \vdots \\ h^1_{m,g} & \cdots & h^d_{m,g} & \cdots & h^{\mathrm{D}}_{m,g} & | & f(\mathbf{h}_{m,g}) \\ \vdots & & \vdots & & \vdots & & \vdots \\ h^1_{m,PMS_m} & \cdots & h^d_{m,PMS_m} & \cdots & h^{\mathrm{D}}_{m,PMS_m} & | & f(\mathbf{h}_{m,PMS_m}) \end{bmatrix}; \quad \forall\{m \in \mathrm{PMN}, d \in \mathrm{D}, g \in PMS_m\} \tag{9}$$

$$h^d_{m,g} = LL^d + rand \in (0,1) \times \left(UL^d - LL^d\right); \forall\{m \in \mathrm{PMN}, d \in \mathrm{D}, g \in PMS_m\} \tag{10}$$

$$h^d_{m,g} = h^d(z); \forall\{m \in \mathrm{PMN}, d \in \mathrm{D}, g \in PMS_m, z \in rand(\{1,2,\ldots,Z\})\} \tag{11}$$

This improvisation procedure was established based on three principal operators: (1) player memory consideration with probability PMCR; (2) pitch

TABLE 7 The Improvisation Procedure for Continuous/Discrete Decision-Making Variables in the 2D-cdEMSA

Algorithm 2: Improvisation Procedure for the 2D-cdEMSA

Input: BW_{min}, BW_{max}, D, LL^d, MNI, MNII, PMN, PAR_{min}, PAR_{max}, PMCR, PMS_m, UL^d, Z.

Output: $h_{m,\text{new}}$ (new melody vector for all music players)

`start main body`

for *each music player* $\{m \in \text{PMN}\}$ **do**
  for *each decision-making variable* $\{d \in \text{D}\}$ **do**
    if $rand \in (0,1) \le \text{HMCR}$ **then**
      operator 1: *player memory consideration with probability* PMCR
      if *iteration counter* $\{i \in \text{MNI} + \text{MNII}\} = \text{odd}$ **then**
        $h^d_{m,\text{new}} = h^d_{m,g}; \forall g = rand(\{1,2,\ldots,PMS_m\})$; for D-DMV
        $h^d_{m,\text{new}} = h^d_{m,g} + rand \in (0,1) \times BW_i; \forall g = rand(\{1,2,\ldots,PMS_m\})$; for C-DMV
      else if
        $h^d_{m,\text{new}} = h^n_{m,g}; \forall g = rand(\{1,2,\ldots,PMS_m\}); \forall n = rand(\{1,2,\ldots,\text{D}\})$;for D-DMV
        $h^d_{m,\text{new}} = h^n_{m,g} + rand \in (0,1) \times BW_i; \forall g = rand(\{1,2,\ldots,PMS_m\}); \forall n = rand(\{1,2,\ldots,\text{D}\})$;for C-DMV
      end if
      if $rand \in (0,1) \le PAR_i$ **then**
        operator 2: *pitch adjustment with probability* HMCR*PAR
        switch 1
          case *iteration counter* $\{i \in \text{MNI} + \text{MNII}\} \le \text{MNI}$ **then**
          $h^d_{m,\text{new}} = h^d_{m,\text{best}}$; for CD-DMV
          case *iteration counter* $\{i \in \text{MNI} + \text{MNII}\} \ge \text{MNI}$ *and iteration counter* $\{i \in \text{MNI} + \text{MNII}\} \le$ MNII **then**
          $h^d_{m,\text{new}} = h^d_{\text{best,best}}$; for CD-DMV
        end switch
      end if
    else if
      operator 3: *random selection with probability* 1-HMCR
      $h^d_{m,\text{new}} = h^d_m(z); \forall z = rand(\{1,2,\ldots,\text{Z}\})$; for D-DMV
      $h^d_{m,\text{new}} = LL^d \pm rand \in (0,1) \times (UL^d - LL^d)$; for C-DMV
    end if
  end for
end for

`end main body`

adjustment with probability PMCR $\times$ PAR; and, (3) random selection with probability 1-PMCR. If a uniform random number in the range (0,1) [i.e., $rand \in (0,1)$] is less than PMCR, the value of each new continuous/discrete decision-making variable of the new melody vector for music player m is generated by operator 1; otherwise, this value is obtained by operator 3. After applying operator 1, if the newly generated $rand \in (0,1)$ is less than PAR, the value of each new continuous/discrete decision-making variable of the new melody vector for music player m is updated by operator 2.

In the first operator, two conflicting conditions are frequently applied to generate new continuous/discrete decision-making variables. For discrete decision-making variables, each condition includes a chosen decision-making variable from the current PM. For continuous decision-making variables, however, each condition includes a linear combination of a selected decision-making variable from the current PM and a proportion of the bandwidth. If the iteration counter is odd, the first condition, the value of each new discrete/continuous decision-making variable of the new melody vector for music player m is determined by using a corresponding decision-making variable of the current PM; otherwise, this value is determined by using a random decision-making variable from among any of the decision-making variables in the current PM, according to the second condition. Another important difference between the new 2D-cdMSA and the original 2D-cMSA makes its appearance in the second operator. In the original 2D-cMSA, the second operator is similar to the SIL and GIL. In the 2D-cdEMSA, however, this operator is considered to be different in the SIL and GIL. If the iteration counter is smaller and equal to the MNI, namely the SIL, the value of each new discrete/continuous decision-making variable of the new melody vector for music player *m* is determined by using a corresponding decision-making variable of the best melody vector stored in the current PM (reflective of the best melody vector stored in its memory); otherwise, for the GIL, this value is determined by using a corresponding decision-making variable from the best melody vector stored in the memory of the best music player (reflective of the best melody vector stored in the memory of the best music player). In operator 3, the value of each new discrete/continuous decision-making variable of the new melody vector for music player *m* is determined by a random reinitialization between the lower and upper limits (feasible area of the solution space). Table 5 shows how these limits for the continuous decision-making variables are updated again in the GIL. As was the case with the original 2D-cMSA, improving performance for the new 2D-cdEMSA, and to avoid getting trapped in local minimum, dynamic relationships of the PAR and the BW parameters (see Eqs. 7, 8) were chosen here.

4 IMPLEMENTATION OF THE 2D-cdEMSA TO SOLVE DYNAMIC PLANNING OF AN MV OPEN-LOOP DISTRIBUTION NETWORK

Here, a large-scale distribution network expansion planning problem, the DyMODN problem, in the presence of the DGRs was considered in the comparison of the performance of the 2D-cdEMSA and other metaheuristic optimization algorithms. For a comprehensive overview of distribution network expansion planning, please refer to the work by Khator and Leung [43], Willis [44], and Ault et al. [45].

4.1 Mathematical Modeling of the DyMODN Problem

The proposed midterm, large-scale DyMODN problem is formulated from the distribution network operator's (DNO) perspective. The DNO expands the MV open-loop distribution network using expansion options so that the supply of energy for current and future customers and facilities is served at minimum initial investment costs, maximum electrical efficiency (minimum operating costs), minimum maintenance costs, and maximum service continuity (minimum customer outage costs). The main decisions of the DNO are to determine which substations, branches, and DGRs should be reinforced and/or installed in the existing MV open-loop distribution network; where and when the action should occur; how many should be reinforced or installed; and, which different objective functions of the DyMODN problem subject to techno-economic limitations are satisfied. The DyMODN problem is modeled by taking the following assumptions into account:

- The daily load variations during the period, p, of the planning horizon are modeled using a load duration curve (LDC). Due to the unavailability of the continuous LDC, it is approximated by the discrete curve. The discrete LDC is divided into Nll load levels.
- For load level ll of the LDC, there is a related electricity market price.
- The DNO purchases the required power of the customers and facilities from the electricity market (i.e., upstream network), and/or generates it by the DGRs.
- For the MV open-loop distribution network expansion, the DNO is authorized to choose among the expansion options in an integrated framework.

In this regard, the DNO simultaneously optimizes three objective functions containing the TIC, the TOC, and the TCOC, as follows:

- Total investment costs. The DNO's required budget for purchase and installation of the MV open-loop distribution network equipment is considered as an important factor in distribution network expansion [46–48]. Hence, the first target of the DyMODN problem is considered to minimize the TIC index, as determined by Eqs. (12), (13):

$$\mathrm{OF}_1 = TIC_y = \sum_{p=1}^{Np} \left(\frac{1}{(1+\mathrm{Ir})^p} \right) \cdot C_{y,p}^{\mathrm{inv}}; \forall \{p \in Np\} \tag{12}$$

$$\begin{aligned} C_{y,p}^{\mathrm{inv}} = & \left(\frac{(1+\mathrm{Ir})^{\delta_s}}{(1+\mathrm{Ir})^{\delta_s} - 1} \right) \cdot \left(\sum_{rs=1}^{Nrs} C_{rs,p}^{\mathrm{inv}}(a) + \sum_{is=1}^{Nis} C_{is,p}^{\mathrm{inv,\,fix}} + C_{is,p}^{\mathrm{inv,var}}(a) \right) \\ & + \left(\frac{(1+\mathrm{Ir})^{\delta_b}}{(1+\mathrm{Ir})^{\delta_b} - 1} \right) \cdot \left(\sum_{rb=1}^{Nrb} C_{rb,p}^{\mathrm{inv}}(b) \cdot (L_{rb}) + \sum_{ib=1}^{Nib} C_{ib,p}^{\mathrm{inv}}(b) \cdot (L_{ib}) \right) \\ & + \left(\frac{(1+\mathrm{Ir})^{\delta_d}}{(1+\mathrm{Ir})^{\delta_d} - 1} \right) \cdot \left(\sum_{id=1}^{Nid} C_{id,p}^{\mathrm{inv}} \cdot \left(P_{id,p}^{\mathrm{cap}} + P_{id,p}^{\mathrm{res}} \right) \right); \\ & \forall \{rs \in Nrs, is \in Nis, rb \in Nrb, ib \in Nib, id \in Nid, a \in Na, b \in Nb, p \in Np\} \end{aligned} \tag{13}$$

Eq. (12) represents the net present value of the TIC index over the planning horizon. The investment costs at period p has three main parts: the investment cost for the reinforced and installed substations, the investment cost for the reinforced and installed branches, and the investment cost for the installed DGRs (see Eq. 13).

- Total operation costs. On the one hand, the maintenance costs and the possibility of operator error increases because the MV open-loop distribution network and related equipment is more complicated in its ability to supply growing current and future demands with maximum continuity and flexibility [46]. On the other hand, electrical efficiency of the MV open-loop distribution network can generally be affected by increasing the losses in transformers, branches, and utilization equipment [46,49,50]. Therefore, the second target of the DyMODN problem is formulated with the aim of minimizing the maintenance and operational costs, namely the TOC index, as shown in Eqs. (14)–(16):

$$\mathrm{OF}_2 = TOC_y = \sum_{p=1}^{Np}\left(\left((1+\mathrm{Ir})^{-p} + \frac{(1+\mathrm{Ir})^{-Np}}{\mathrm{Ir}}\right)\cdot\right.$$

$$\left.\left(C_{y,p}^{\text{maint}} + \sum_{ll=1}^{Nll}\left(8760\cdot C_{ll}^{\text{energy}}\cdot\frac{D_{ll}}{24}\right)\cdot C_{y,p,ll}^{\text{oper}}\right)\right);\ \forall\{p\in Np, \forall ll\in Nll\} \tag{14}$$

$$C_{y,p}^{\text{maint}} = \left(\sum_{es=1}^{Nes} C_{es,p}^{\text{maint}} + \sum_{rs=1}^{Nrs} C_{rs,p}^{\text{maint}} + \sum_{is=1}^{Nis} C_{is,p}^{\text{maint}}\right) + \left(\sum_{eb=1}^{Neb} C_{eb,p}^{\text{maint}} + \sum_{rb=1}^{Nrb} C_{rb,p}^{\text{maint}} + \sum_{ib=1}^{Nib} C_{ib,p}^{\text{maint}}\right);$$

$$\forall\{es\in Nes, rs\in Nrs, is\in Nis, eb\in Neb, rb\in Nrb, ib\in Nib\} \tag{15}$$

$$C_{y,p,ll}^{\text{oper}} = P_{ll,p}^{\text{elec-mark}}\cdot\left(\sum_{es=1}^{Nes} R_{es}\cdot\left(I_{es,ll,p}\right)^2 + \sum_{rs=1}^{Nrs} R_{rs}\cdot\left(I_{rs,ll,p}\right)^2 + \sum_{is=1}^{Nis} R_{is}\cdot\left(I_{is,ll,p}\right)^2\right)$$

$$+ P_{ll,p}^{\text{elec-mark}}\cdot\left(\sum_{eb=1}^{Neb} R_{eb}\cdot\left(I_{eb,ll,p}\right)^2 + \sum_{rb=1}^{Nrb} R_{rb}\cdot\left(I_{rb,ll,p}\right)^2\right.$$

$$\left.+\sum_{ib=1}^{Nib} R_{ib}\cdot\left(I_{ib,ll,p}\right)^2\right) + \left(\sum_{id=1}^{Nid} C_{id,ll,p}^{\text{oper}}\cdot P_{id,ll,p}\right) + \left(\sum_{es=1}^{Nes} P_{ll,p}^{\text{elec-mark}}\cdot P_{es,ll,p}\right.$$

$$\left.+\sum_{rs=1}^{Nrs} P_{ll,p}^{\text{elec-mark}}\cdot P_{rs,ll,p} + \sum_{is=1}^{Nis} P_{ll,p}^{\text{elec-mark}}\cdot P_{is,ll,p}\right);$$

$$\forall\{es\in Nes, rs\in Nrs, is\in Nis, eb\in Neb, rb\in Nrb, ib\in Nib, id\in Nid\} \tag{16}$$

Eq. (14) shows the net present value of the TOC index over the planning horizon, which has two major parts: (1) the maintenance cost and (2) the operational cost. The maintenance cost at period p is divided into the maintenance cost of the substations and the branches. The first term of

Eq. (15) represents the maintenance cost of existing, reinforced, and installed substations in the MV open-loop distribution network. The second term of Eq. (15) demonstrates the maintenance cost of existing, reinforced, and installed branches in the MV open-loop distribution network. Eq. (16) gives the operational cost at load level ll of period p, which has four principal parts: (1) the loss cost of existing, reinforced, and installed substations (the first term); (2) the loss cost of existing, reinforced, and installed branches (the second term); (3) the operational costs of installed DGRs (the third term); and, (4) the cost of power purchased from the electricity market (upstream network) via existing, reinforced, and installed substations (the fourth term).

- Total customer outage costs. Reliability is one of the main indices that must be considered in the DyMODN problem. Ignoring this index in this problem makes it virtually impossible for the obtained optimal plan for expansion of the open-loop distribution network to satisfy the predetermined reliability levels [46,51]. Therefore, the third target of the DyMODN problem is minimization of the total customer outage costs, namely the TCOC index, as given by Eqs. (17), (18):

$$\mathrm{OF}_3 = TCOC_y = \sum_{p=1}^{Np}\left((1+\mathrm{Ir})^{-p} + \frac{(1+\mathrm{Ir})^{-Np}}{\mathrm{Ir}}\right)\cdot C_{y,p}^{\text{cu-out}};\forall\{p \in Np\} \tag{17}$$

$$C_{y,p}^{\text{cu-out}} = \sum_{ll=1}^{Nll}\left(8760\cdot\frac{D_{ll}}{24}\cdot\sum_{h=1}^{Nh}\sum_{ln=1}^{Nln}\sum_{k=1}^{Nk} P_{p,ll,h,ln,k}^{\text{dema}}\cdot C_{p,ll,h,ln,k}^{\text{dam-fun}}\left(D_{p,ll,h,ln,k}^{\text{out}}\right)\cdot FR_{p,ll,h,ln,k}^{\text{aver}}\right);$$
$$\forall\{p \in Np, ll \in Nll, h \in Nh, ln \in Nln, k \in Nk\} \tag{18}$$

Eq. (17) explains the net present value of the TCOC index over the planning horizon. Eq. (18) represents the customer outage costs at period p. The step-by-step process of reliability worth evaluation is as follows:

(1) Set the number of periods equal to Np.
(2) Set the period counter ($p=1$).
(3) Set the number of load levels equal to Nll.
(4) Set the load levels counter ($ll=1$).
(5) Set the number of failure events equal to Nh.
(6) Set the failure events counter ($h=1$).
(7) Set the number of affected load nodes equal to Nln.
(8) Set the affected load nodes counter ($ln=1$).
(9) Set the number of load sectors equal to Nk.
(10) Set the load sector counter ($k=1$).
(11) Find the average failure rate ($FR_{p,\ ll,\ h,ln,\ k}$) and outage duration time ($D^{\text{out}}_{p,\ ll,\ h,ln,\ k}$) at load sector k of affected load node ln in failure

event h at load level ll of period p, according to the MV open-loop distribution network configuration and transfer capability of alternate supplies (reserve feeder and/or DGRs).

(12) Evaluate the cost of the outage, based on the outage duration time at load sector k of affected load node ln in failure event h at load level ll of period p ($D^{\text{out}}_{p,\,ll,\,h,ln,\,k}$), using the sector customer damage function (SCDF) at load sector k of affected load node ln in failure event h at load level ll of period p ($C^{\text{dam-fun}}_{p,\,ll,\,h,ln,\,k}(D^{\text{out}}_{p,\,ll,\,h,ln,\,k})$).

(13) Calculate the expected customer outage costs at load sector k of load node ln in failure event h at load level ll of period p, according to ($FR^{\text{aver}}_{p,\,ll,\,h,ln,\,k}\cdot C^{\text{dam-fun}}_{p,\,ll,\,h,ln,\,k}(D^{\text{out}}_{p,\,ll,\,h,ln,\,k})\cdot P^{\text{dema}}_{p,\,ll,\,h,ln,\,k}$).

(14) $k=k+1$; if $k \leq Nk$ go to step (11); otherwise go to the next step.

(15) $ln=ln+1$; if $ln \leq Nln$ go to step (9); otherwise go to the next step.

(16) $h=h+1$; if $h \leq Nh$ go to step (7); otherwise go to the next step.

(17) $ll=ll+1$; if $ll \leq Nll$ go to step (5); otherwise go to the next step.

(18) $p=p+1$; if $p \leq Np$ go to step (3); otherwise go to the next step.

(19) Stop.

We will not dwell on the reasons for choosing this type of reliability index, due to noncompliance with the targets of the chapter. For a detailed description of the different types of reliability indices, please refer to the work by Lotero and Contrera [46].

The constraints of the midterm, large-scale DyMODN problem are divided into four distinct classes [46–50,52,53]: (1) Kirchhoff's current and voltage rules; (2) the operational restrictions on network equipment; (3) the availability of financial resources; and, (4) the following logical restrictions:

- Kirchhoff's current rule at any node in any existing, reinforced, and installed MV open-loop distribution network must be met at each load level of period p, as determined by Eq. (19):

$$\left[\mathrm{Y}^{\text{e-ne}}\right]\cdot\left[\mathrm{f}^{\text{e-b}}_{ll,p}\right]^{\mathrm{Tr}}+\left[\mathrm{Y}^{\text{r-ne}}\right]\cdot\left[\mathrm{f}^{\text{r-b}}_{ll,p}\right]^{\mathrm{Tr}}+\left[\mathrm{Y}^{\text{i-ne}}\right]\cdot\left[\mathrm{f}^{\text{i-b}}_{ll,p}\right]^{\mathrm{Tr}}$$
$$+\left[\mathrm{g}^{\text{no-inj}}_{ll,p}\right]+\left[\mathrm{h}^{\text{no-load-sh}}_{ll,p}\right]=\left[\mathrm{d}^{\text{no-load}}_{ll,p}\right];\ \forall\{p\in Np, ll\in Nll\} \tag{19}$$

where, the superscript Tr demonstrates the vector transposition.

- Kirchhoff's voltage rule must be satisfied at each load level of period p. For the branches of any existing, reinforced, and installed MV open-loop distribution network, refer to Eqs. (20)–(22):

$$Z_{eb}\cdot I_{eb,ll,p}+\left[\mathrm{v}_{ll,p}\right]\cdot\left[\mathrm{Y}^{\text{e-ne}}_{\text{column}:eb}\right]=0;\forall\{eb\in Neb, ll\in Nll, p\in Np\} \tag{20}$$

$$Z_{rb}\cdot I_{rb,ll,p}+\left[\mathrm{v}_{ll,p}\right]\cdot\left[\mathrm{Y}^{\text{r-ne}}_{\text{column}:rb}\right]=0;\forall\{rb\in Nrb, ll\in Nll, p\in Np\} \tag{21}$$

$$Z_{ib}\cdot I_{ib,ll,p}+\left[\mathrm{v}_{ll,p}\right]\cdot\left[\mathrm{Y}^{\text{i-ne}}_{\text{column}:ib}\right]=0;\forall\{ib\in Nib, ll\in Nll, p\in Np\} \tag{22}$$

- Operational constraints. For the existing, reinforced, and installed substations, the limits to current injections at load level ll of period p are given by Eqs. (23)–(25):

$$0 \leq I_{es,ll,p} \leq I_{es}^{\max}; \forall \{es \in Nes, ll \in Nll, p \in Np\} \tag{23}$$

$$0 \leq I_{rs,ll,p} \leq I_{rs}^{\max}; \forall \{rs \in Nrs, ll \in Nll, p \in Np\} \tag{24}$$

$$0 \leq I_{is,ll,p} \leq I_{is}^{\max}; \forall \{is \in Nis, ll \in Nll, p \in Np\} \tag{25}$$

Limits on existing, reinforced, and installed branch currents at load level ll of period p are defined using Eqs. (26)–(28):

$$0 \leq I_{eb,ll,p} \leq I_{eb}^{\max}; \forall \{eb \in Neb, ll \in Nll, p \in Np\} \tag{26}$$

$$0 \leq I_{rb,ll,p} \leq I_{rb}^{\max}; \forall \{rb \in Nrb, ll \in Nll, p \in Np\} \tag{27}$$

$$0 \leq I_{ib,ll,p} \leq I_{ib}^{\max}; \forall \{ib \in Nib, ll \in Nll, p \in Np\} \tag{28}$$

The range of allowable values for the voltage at voltage conversion, load, and generation nodes can be determined using Eqs. (29)–(31):

$$V_{cn}^{\min} \leq V_{cn,ll,p} \leq V_{cn}^{\max}; \forall \{cn \in Ncn, ll \in Nll, p \in Np\} \tag{29}$$

$$V_{ln}^{\min} \leq V_{ln,ll,p} \leq V_{ln}^{\max}; \forall \{ln \in Nln, ll \in Nll, p \in Np\} \tag{30}$$

$$V_{gn}^{\min} \leq V_{gn,ll,p} \leq V_{gn}^{\max}; \forall \{gn \in Ngn, ll \in Nll, p \in Np\} \tag{31}$$

The range of acceptable values for the installed DGRs power generation at load level ll of period p is presented by Eq. (32):

$$0 \leq P_{id,ll,p} \leq P_{id}^{\max}; \forall \{id \in Nid, ll \in Nll, p \in Np\} \tag{32}$$

The limit on the total capacity of the DGRs that can be installed in an MV open-loop distribution network is given by Eq. (33):

$$\sum_{p=1}^{Np} \sum_{id=1}^{Nid} P_{id,p}^{\text{cap}} + P_{id,p}^{\text{res}} \leq \text{P}^{\max\text{-dgr}}; \forall \{p \in Np, id \in Nid\} \tag{33}$$

The permissible range of load-shedding in load node ln at load level ll of period p is given by Eq. (34):

$$0 \leq h_{ln,ll,p} \leq h_{ln,ll,p}^{\max}; \forall \{ln \in N\ ln, ll \in Nll, p \in Np\} \tag{34}$$

- Financial constraints. The restrictions related to the investment costs in period p and in all periods are depicted by Eqs. (35), (36):

$$\left(\sum_{rs=1}^{Nrs} C_{rs,p}^{\text{inv}} + \sum_{is=1}^{Nis} C_{is,p}^{\text{inv, fix}}(a) + C_{is,p}^{\text{inv,var}}(a) \right)$$
$$+ \left(\sum_{rb=1}^{Nrb} C_{rb,p}^{\text{inv}}(b) \cdot (L_{rb}) + \sum_{ib=1}^{Nib} C_{ib,p}^{\text{inv}}(b) \cdot (L_{ib}) \right) + \left(\sum_{id=1}^{Nid} C_{id,p}^{\text{inv}} \cdot \left(P_{id}^{\text{cap}} + P_{id}^{\text{res}}\right) \right)$$

$$\leq B_p; \forall\{rs \in Nrs, is \in Nis, rb \in Nrb, ib \in Nib, id \in Nid, a \in Na, b \in Nb, p \in Np\} \tag{35}$$

$$\sum_{p=1}^{Np}\left(\frac{1}{(1+\text{Ir})^p}\right)\cdot\left(\left(\sum_{rs=1}^{Nrs} C_{rs,p}^{\text{inv}} + \sum_{is=1}^{Nis} C_{is,p}^{\text{inv, fix}}(a) + C_{is,p}^{\text{inv, var}}(a)\right)\right.$$
$$\left.+\left(\sum_{rb=1}^{Nrb} C_{rb,p}^{\text{inv}}(b)\cdot(L_{rb}) + \sum_{ib=1}^{Nib} C_{ib,p}^{\text{inv}}(b)\cdot(L_{ib})\right) + \left(\sum_{id=1}^{Nid} C_{id,p}^{\text{inv}}\cdot\left(P_{id}^{\text{cap}} + P_{id}^{\text{res}}\right)\right)\right) \leq TB;$$
$$\forall\{rs \in Nrs, is \in Nis, rb \in Nrb, ib \in Nib, id \in Nid, a \in Na, b \in Nb, p \in Np\} \tag{36}$$

- Logical constraints. The number of installed substations in each candidate node at period p and at all periods must meet the conditions specified by Eqs. (37), (38):

$$N_{cns,p}^{\text{i-s}} \leq 1; \forall\{cns \in Ncns, p \in Np\} \tag{37}$$

$$\sum_{p=1}^{Np} N_{cns,p}^{\text{i-s}} \leq 1; \forall\{cns \in Ncns, p \in Np\} \tag{38}$$

The candidate substation for reinforcement can only be reinforced once in period p and in all periods, in accordance with Eqs. (39), (40):

$$N_{csr,p}^{\text{r-s}} \leq 1; \forall\{csr \in Ncsr, p \in Np\} \tag{39}$$

$$\sum_{p=1}^{Np} N_{csr,p}^{\text{r-s}} \leq 1; \forall\{csr \in Ncsr, p \in Np\} \tag{40}$$

The number of installed branches in each candidate location at period p and at all periods must satisfy Eqs. (41), (42):

$$N_{clb,p}^{\text{i-b}} \leq 1; \forall\{clb \in Nclb, p \in Np\} \tag{41}$$

$$\sum_{p=1}^{Np} N_{clb,p}^{\text{i-b}} \leq 1; \forall\{clb \in Nclb, p \in Np\} \tag{42}$$

The candidate branch for reinforcement can only be reinforced once in period p and in all periods, according to Eqs. (43), (44):

$$N_{cbr,p}^{\text{r-b}} \leq 1; \forall\{cbr \in Ncbr, p \in Np\} \tag{43}$$

$$\sum_{p=1}^{Np} N_{cbr,p}^{\text{r-b}} \leq 1; \forall\{cbr \in Ncbr, p \in Np\} \tag{44}$$

The implementation of the open-loop configuration of the distribution network is one of the most complex limitations in the process of solving the DyMODN problem. In this study, however, a new graph-based scheme was employed to meet the restriction spelled out by Eq. (45):

$$\text{Open-loop configuration of the MV distribution network} \tag{45}$$

The MV distribution network configuration can be demonstrated by a simple linear graph: $G=(V, E)$. G consists of two different sets: a set of objects $V=\{v_1, v_2, \ldots\}$ called vertices (corresponding with the MV distribution network load nodes) and a set of arcs $E=\{e_1, e_2, \ldots\}$ called edges (corresponding with the MV distribution network branches). Also, in the graph theory, a tree is an undirected graph in which any two vertices, or load nodes, are connected by exactly one path, or branch. In simple terms, a tree must have two conditions: no loop and connectivity. In addition, a forest is a disjoint union of trees. Hence, in order to meet the restriction expressed in Eq. (45), the obtained trees in any expansion plan (solution vector) must satisfy these two conditions. For a thorough discussion on the approaches to evaluate these conditions, please see the work by Deo [54].

4.2 Incorporating Probabilistic Characteristics of Power Demand in the DyMODN Problem

With the continuous growth in population and economic development, power demand on distribution networks will also grow. For forecasting future power demand, previous studies have employed prediction methodologies such as time series, regression, and artificial neural networks. The forecasts prepared by these methodologies contained errors and were not accurate, in that they would impose uncertainties on future distribution network studies. Hence, due to the weakness of these prediction methodologies in accurately predicting power demand growth, it is necessary to develop a new strategy that can provide a well-designed model that includes uncertainty effects of power demand in distribution network expansion planning studies. In the proposed DyMODN problem, stochastic behavior (uncertainty) of the estimated power demand is modeled using the Gaussian probability distribution function (normal PDF).

In the context of power systems planning, various approaches are employed to tackle the risks arising from severe uncertainties, such as: the Taylor series expansion method [55]; the first-order, second-moment method [56]; Gram Charlier [57]; and, discretization methods [58]. For more information about the key benefits of each approach, and comparisons to other existing ones, please refer to the discussion by Verbic and Canizares [59] on this topic. Here, however, the stochastic behavior of power demand is applied in the optimal power flow (OPF) calculation by using 2-PEM [59,60]. In the proposed DyMODN problem, the amount of power demand of load node ln, at load level ll, of period p, is considered as uncertain input variables, as given by Eq. (46):

$$X = \left[d_{ln,ll,p}^{\text{no-dem}}\right]; \forall \{ln \in Nln, ll \in Nll, p \in Np\} \tag{46}$$

Thus, the DyMODN problem is a function of uncertain input variables; for example, power demand that can be represented by Eq. (47):

$$Y = h(X) \tag{47}$$

The uncertain output variables, which directly depend on the uncertain input variables, can be represented by Eq. (48):

$$Y_1 = \left[C_{y,p}^{\text{maint}}, C_{y,p,ll}^{\text{oper}}, C_{y,p}^{\text{cu-out}}\right]; \forall \{p \in Np\} \tag{48}$$

Also, the uncertain output variables, which indirectly depend on the uncertain input variables, can be represented by Eq. (49):

$$Y_2 = \left[C_{y,p}^{\text{inv}}\right]; \forall \{p \in Np\} \tag{49}$$

In this regard, the proposed probabilistic OPF for computing the moments of the uncertain output variables can be applied using the following step-by-step process:

(1) Set the number of uncertain input variables equal to Nu.
(2) Set $E(Y)=0$ and $E(Y^2)=0$.
(3) Set the uncertain input variable counter ($u=1$).
(4) Compute the locations of two concentrations and the probabilities of concentrations by using Eqs. (50), (51):

$$\vartheta_{u,r} = (\zeta_{u,3}/2) + (-1)^{(r+1)} \cdot \sqrt{Nu + (\zeta_{u,3}/2)^2}; \forall \{r \in \{1,2\}\} \tag{50}$$

$$\rho_{u,r} = \left(\left((-1)^{(r)} \cdot (\vartheta_{u,r})\right) \Big/ \left(2 \cdot Nu \cdot \sqrt{Nu + (\zeta_{u,3}/2)^2}\right)\right); \forall \{r \in \{1,2\}\} \tag{51}$$

(5) Compute two concentration points using Eq. (52):

$$x_{u,r} = \mu_{X,u} + \vartheta_{u,r} \cdot \sigma_{X,u}; \forall \{r \in \{1,2\}\} \tag{52}$$

(6) Run the deterministic OPF for two concentration points $x_{u,r}; \forall \{r \in \{1,2\}\}$ in accordance with vector X. This vector can be determined using Eq. (53):

$$X = \left[\mu_{X,1}, \mu_{X,2}, \ldots, x_{u,r}, \ldots, \mu_{X,U}\right]; \forall \{r \in \{1,2\}\} \tag{53}$$

(7) Update $E(Y)$ and $E(Y^2)$ using Eqs. (54), (55):

$$E(Y) \cong \sum_{u=1}^{Nu} \sum_{r=1}^{2} \left(\rho_{u,r} \cdot h\left(\left[\mu_{X,1}, , \mu_{X,2}, \ldots, x_{u,r}, \ldots, \mu_{X,U}\right]\right)\right); \forall \{r \in \{1,2\}\} \tag{54}$$

$$E(Y^2) \cong \sum_{u=1}^{Nu} \sum_{r=1}^{2} \left(\rho_{u,r} \cdot h\left(\left[\mu_{X,1}, , \mu_{X,2}, \ldots, x_{u,r}, \ldots, \mu_{X,U} \right] \right)^2 \right); \forall \{ r \in \{1,2\} \} \tag{55}$$

(8) Compute the mean and standard deviation of Y using Eq. (56):

$$\mu_Y = E(Y) \tag{56}$$

$$\sigma_Y = \sqrt{E(Y^2) - \mu_Y^2} \tag{57}$$

(9) $u = u + 1$; if $u \leq Nu$ go to step (4).
(10) Stop.

4.3 Test the MV Distribution Network Description

A modified three-phase MV open-loop distribution network is considered to calculate the optimal plan for the DyMODN problem over a 3-year planning horizon divided into 1-year periods. The base apparent power and voltage for the entire MV distribution network is equal to 1 MVA and 13.8 kV, respectively. This MV distribution network consists of 24 load nodes, three voltage conversion (substation) nodes, and 39 branches. Fig. 1 shows the primary single-line diagram of the modified MV open-loop distribution network.

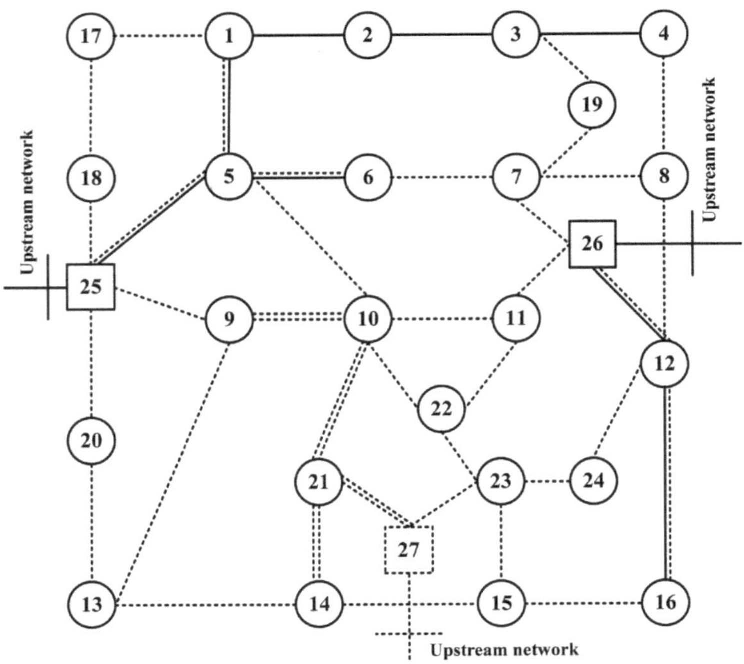

FIG. 1 Primary single-line diagram of the modified MV open-loop distribution network.

In Fig. 1, circles denote the load nodes and squares represent substations (those with solid lines are existing substations in the initial MV open-loop distribution network that can be reinforced in period p; those with dotted lines are candidates for installation in period p). Also, solid lines represent the existing branches in the initial MV open-loop distribution network that cannot be reinforced; double lines (one solid line plus one dotted line) denote the existing branches in the initial MV open-loop distribution network that can be reinforced in period p. In addition, dotted lines are candidate branches for installation in period p; double dotted lines are candidate branches for installation in period p and reinforcement in period $p+1$. To describe the daily load duration curve in the proposed DyMODN problem, period p considers three different load levels: (1) load level $ll1$, also called the on-peak load level; (2) load level $ll2$, or mid-peak load level; and, (3) load level $ll3$, or off-peak load level. Then, each load is categorized into three different sectors: residential, commercial, and industrial. Table 8 summarizes the estimated power demand in each load level of period p and its dispersion across the number of customers in period p. Table 9 gives the technical and economic features of the substations and branches used in the DyMODN problem.

Table 10 shows the SCDF and its duration at each load sector, the economic lifetime of the equipment of the MV open-loop distribution network, and the data associated with the load levels and market prices. The interest rate is set at 10%. For simplicity, the failure rate and repair times are considered similar to all branches. These values are set at 0.4 failures per year and 10 h, respectively. The considered budget at period p and the total budget for MV open-loop distribution network expansion are \$20 M and \$45 M, respectively. The lower and upper voltage limits in generation and load nodes are set at 0.95 and 1.05 per unit, respectively. The voltage at the voltage conversion nodes is fixed at 1.05 per unit. In the proposed DyMODN problem, each node of the MV open-loop distribution network could be a candidate node for installing DGRs. The candidate DGR has size multiples of 1 MW. The maximum number of the DGRs at each candidate node of the MV distribution network is considered to be four DGRs along with one DGR in reserve. In addition to this, total capacity of the DGRs that can be installed in the MV open-loop distribution network is considered to be 15 MW. The investment and operation costs of installed DGRs are considered to be \$0.75 M/MW and \$50/MW h, respectively. It is worth noting that the power demand estimation error in each load node of the MV open-loop distribution network can be varied between −20% and +20%. Additional required data can be found in the study by Lotero and Contreras [46].

4.4 Proposed Solution Method

As previously mentioned, the newly developed DyMODN problem is a complicated, large-scale, nonlinear mixed-integer optimization problem.

TABLE 8 Estimated Power Demand and Customer Data

Node No.	Loads (MVA) and Customer Data																				
	Period 1							*Period 2*							*Period 3*						
	Load Level			*Load Sector*				*Load Level*			*Load Sector*				*Load Level*			*Load Sector*			
	1	2	3	*No.*	*Res.*	*Com.*	*Ind.*	*1*	2	3	*No.*	*Res.*	*Com.*	*Ind.*	*1*	2	3	*No.*	*Res.*	*Com.*	*Ind.*
1	1.2	0.72	0.24	100	0.5	0.2	0.3	1.2	0.72	0.24	100	0.5	0.2	0.3	1.2	0.72	0.24	100	0.5	0.2	0.3
2	0	0	0	–	–	–	–	1.2	0.72	0.24	10	0	0.1	0.9	1.2	0.72	0.24	10	0	0.1	0.9
3	0	0	0	–	–	–	–	0	0	0	–	–	–	–	1.2	0.72	0.24	50	0.2	0.2	0.6
4	1.2	0.72	0.24	10	0	0.2	0.8	1.2	0.72	0.24	10	0	0.2	0.8	1.2	0.72	0.24	10	0	0.2	0.8
5	1.2	0.72	0.24	50	0.2	0.2	0.6	1.2	0.72	0.24	50	0.2	0.2	0.6	1.2	0.72	0.24	50	0.2	0.2	0.6
6	1.2	0.72	0.24	30	0.15	0.2	0.65	1.2	0.72	0.24	30	0.15	0.2	0.65	1.2	0.72	0.24	30	0.15	0.2	0.65
7	0	0	0	–	–	–	–	1.2	0.72	0.24	10	0	0	1	1.2	0.72	0.24	10	0	0	1
8	1.2	0.72	0.24	50	0.1	0.2	0.7	1.2	0.72	0.24	50	0.1	0.2	0.7	1.2	0.72	0.24	50	0.1	0.2	0.7
9	0	0	0	–	–	–	–	1.2	0.72	0.24	50	0.1	0.2	0.7	1.2	0.72	0.24	100	0.1	0.2	0.7
10	0	0	0	–	–	–	–	0	0	0	–	–	–	–	2.4	3.6	1.2	10	0	0	1
11	1.2	0.72	0.24	50	0.5	0.4	0.1	1.2	2.4	0.48	50	0.5	0.4	0.1	2.4	3.6	1.2	50	0.5	0.4	0.1
12	0	0	0	–	–	–	–	1.2	0.72	0.24	50	0.4	0.4	0.2	1.2	0.72	0.24	50	0.4	0.4	0.2
13	1.2	0.72	0.24	10	0	0.15	0.85	1.2	2.4	0.48	10	0	0.15	0.85	2.4	3.6	1.2	10	0	0.15	0.85
14	0	0	0	–	–	–	–	0	0	0	–	–	–	–	2.4	1.2	0.48	15	1	0	0

15	0	0	0	–	–	–	–	0	0	0	–	–	–	–	2.4	1.2	0.48	15	1	0	0
16	1.2	0.72	0.24	50	0.3	0.2	0.5	1.2	0.72	0.24	50	0.3	0.2	0.5	1.2	0.72	0.24	50	0.3	0.2	0.5
17	0	0	0	–	–	–	–	1.2	0.72	0.24	10	0	0.25	0.75	2.4	1.2	0.48	15	0	0.25	0.75
18	0	0	0	–	–	–	–	0	0	0	–	–	–	–	1.2	2.4	0.48	20	0.8	0.2	0
19	1.2	0.72	0.24	10	0	0	1	1.2	2.4	0.48	10	0	0	1	1.2	2.4	0.48	10	0	0	1
20	0	0	0	–	–	–	–	1.2	0.72	0.24	50	0	0	1	1.2	0.72	0.24	50	0	0	1
21	1.2	0.72	0.24	100	0.6	0.25	0.15	1.2	2.4	0.48	100	0.6	0.25	0.15	1.2	2.4	0.48	100	0.6	0.25	0.15
22	0	0	0	–	–	–	–	0	0	0	–	–	–	–	1.2	0.72	0.24	10	0	1	0
23	0	0	0	–	–	–	–	0	0	0	–	–	–	–	1.2	0.72	0.24	10	1	0	0
24	1.2	0.72	0.24	50	0.5	0.2	0.3	1.2	0.72	0.24	50	0.5	0.2	0.3	1.2	0.72	0.24	50	0.5	0.2	0.3

TABLE 9 Technical and Economic Features of the Substations and Branches (13.8 kV and 1 MVA)

Substations												
	Existing Substations in the Initial MV Distribution Network				*Candidate New Substations*							
					Substation Option 1				*Substation Option 2*			
Node No.	*R* *(Ω)*	I_{es}^{max} *(p. u)*	C_{es}^{main} *(M$)*	$C_{is}^{inv,\ fix}$ *(M$)*	*R (Ω)*	I_{rs}^{max}, I_{is}^{max} *(p. u)*	C_{rs}^{main} *(M$)*, C_{is}^{main} (M$)	C_{rs}^{inv}, $C_{is}^{inv,\ var}$ *(M$)*	*R (Ω)*	I_{rs}^{max}, I_{is}^{max} *(p. u)*	C_{rs}^{main} *(M$)*, C_{is}^{main} (M$)	C_{rs}^{inv}, $C_{is}^{inv,\ var}$ *(M$)*
25	0.13	15	0.3	0.5	0.25	12	0.1	4.875	0.16	15	0.2	5.100
26	0.13	15	0.3	0.5	0.25	12	0.1	4.875	0.16	15	0.2	7.100
27	–	–	–	0.5	0.25	12	0.1	4.875	0.16	15	0.2	7.100

Branches												
	Existing Branches in the Initial MV Distribution Network				*Candidate New Branches*							
					Branch Option 1				*Branch option 2*			
Branch Status	*R* *(Ω/ km)*	*Z* *(Ω/ km)*	I_{eb}^{max}, I_{rb}^{max}*(p. u)*	C_{eb}^{main}, C_{rb}^{main}, C_{ib}^{main}*(M$)*	*R* *(Ω/ km)*	*Z* *(Ω/ km)*	I_{rb}^{max}, I_{ib}^{max} *(p. u)*	C_{rb}^{inv}, C_{ib}^{inv} *(M$/ km)*	*R* *(Ω/ km)*	*Z* *(Ω/ km)*	I_{rb}^{max}, I_{ib}^{max} *(p. u)*	C_{rb}^{inv}, C_{ib}^{inv} *(M$/ km)*
Existing	0.76	1	12	0.048	–	–	–	–	–	–	–	–
Reinforcement	0.76	1	12	0.033	0.44	0.7	10	0.568	0.25	0.5	12	1.079
Installation	–	–	–	0.043	0.76	1	11	1.136	0.44	0.7	15	1.42

TABLE 10 SCDF Data, Economic Lifetime Data of the Equipment, Load Levels, and Market Price Data

No.	Customer Sectors Data						No.	Load Levels Data			No.	Economic Lifetime Data		
		[Duration (min) and Outage Costs ($/kW)]												
No.	*Customer sector*	*1 min*	*20 min*	*60 min*	*240 min*	*480 min*	No.	*Load Level*	*Duration (h/day)*	*Market Price ($/MWh)*	No.	*Equipment*	*Index*	*Value (Year)*
1	Residential	0.001	0.093	0.482	4.914	15.69	2	1	3	70	3	Transformer	δ_s	15
	Commercial	0.381	2.969	8.552	31.32	83.01		2	13	49		Branch	δ_b	25
	Industrial	1.625	3.868	9.085	25.16	55.81		3	8	35		DGR	δ_d	25

Hence, the 2D-cdEMSA is widely used for calculating optimal solutions in multiobjective problems. After calculation of the optimal solutions, it is favorable to determine a flexible solution that indicates a compromise between different objective functions. Although many approaches have been reported for choosing a trade-off solution among the optimal solutions, a fuzzy satisfying approach (FSA) is of high interest, due to its simplicity and resemblance to human ratiocination. Here, The FSA is applied based on a conservative method to choose the final solution. A more detailed description of this approach can be found in studies by the current authors [30,36]. The decision-making variables of the melody vector of the 2D-cdEMSA are: (1) the time of the installed/reinforced equipment of the MV open-loop distribution network over the planning horizon; (2) the site, size and generated power of the candidate installed DGRs; (3) the site and size of the candidate installed substations/branches; and, (4) the size of the candidate reinforced substations/branches.

4.5 Simulation Results

The software code is written in MATLAB. The OPF formulation for the MV open-loop distribution network was applied in the proposed DyMODN problem. For further discussion on the OPF formulation, please refer to the work by Sheng et al. [61] To analyze the sufficiency and effectiveness of the 2D-cdEMSA in the DyMODN problem, the following two cases are investigated:

- Case 1The DyMODN problem in the absence of the DGRs.
- Case 2The DyMODN problem in the presence of the DGRs.

Table 11 gives the parameter adjustments of the 2D-cdEMSA. These parameters are the best among several implementations of the offered DyMODN problem.

4.5.1 Analysis of the Simulation Results

Figs. 2 and 3 show optimal expansion plans for Cases 1 and 2, respectively.

In these figures, letters *I* and *R* are associated with installed and reinforced distribution network equipment, respectively. Also, the number following these letters represents the size of the distribution network equipment, either 1 or 2. Letter *p* and the following number are related to the period in which the installation or reinforcement of the distribution network equipment is to be performed.

As shown in these two figures, the obtained configurations of the MV open-loop distribution network in each period, for Cases 1 and 2, are different. This is due to the fact that the different expansion options are available to the DNO. When the DGRs are not utilized (Case 1) in the DyMODN problem, candidate

TABLE 11 Parameter Adjustments of the 2D-cdEMSA

No.	2D-cdEMSA Parameters	Value	No.	2D-cdEMSA Parameters	Value	No.	2D-cdEMSA Parameters	Value
1	BW_{min}	0.4	5	PAR_{min}	0.2	9	PMS_m	250 $\{m\in [1,2,3]\}$
2	BW_{max}	0.9	6	PAR_{max}	2	10	PMS_m	200 $\{m\in [4,5]\}$
3	MNI	400	7	PMN	10	11	PMS_m	300 $\{m\in [6,7,8]\}$
4	MNII	600	8	PMCR	0.85	12	PMS_m	280 $\{m\in [9,10]\}$

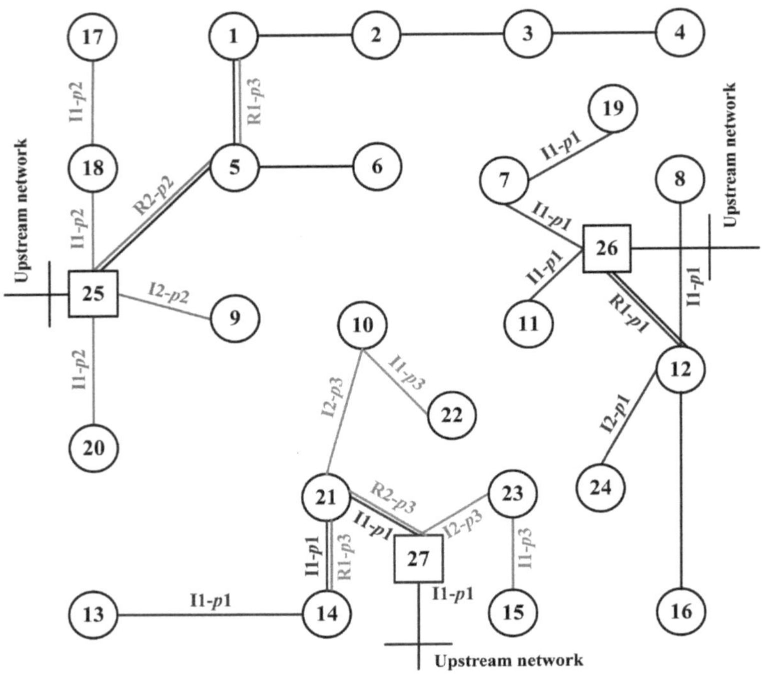

FIG. 2 Optimal expansion plans of the modified MV open-loop distribution network for Case 1: in the absence of the DGRs.

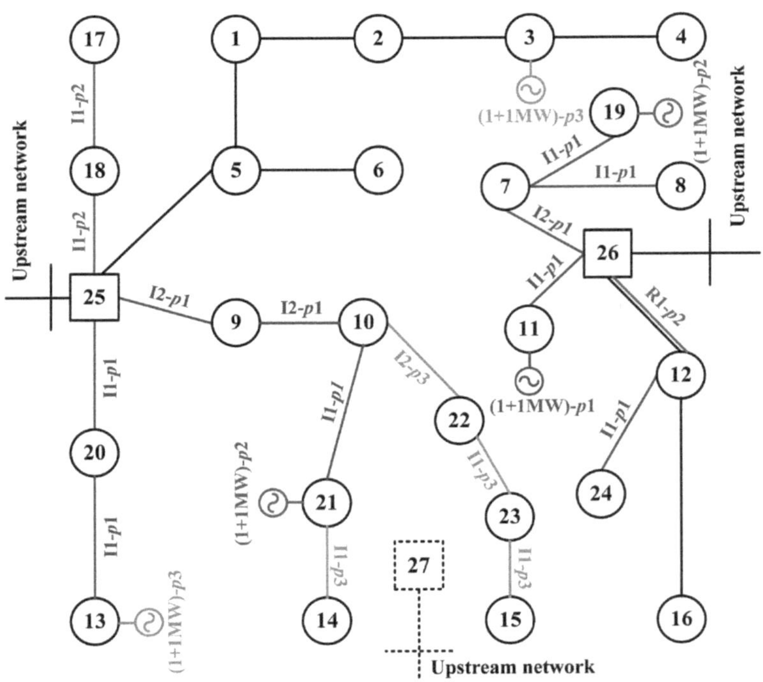

FIG. 3 Optimal expansion plans of the modified MV open-loop distribution network for Case 2: in the presence of the DGRs.

node 27 for substation installation has been expanded. On the contrary, when the DGRs are incorporated (Case 2) in the DyMODN problem, there is no need to install and/or reinforce any substation. It is clear that incorporating DGRs in the DyMODN problem can result in a well-designed model in distribution network expansion planning studies so that technical/economic limitations associated with expensive equipment of the distribution networks, especially substations, are partially alleviated. Another important point that can be seen from the obtained topologies for the two cases is that some special branches are always available. These branches, so-called robust branches, are installations and reinforcements of branches between nodes 25 and 18, 18 and 17, 26 and 11, 26 and 12, 26 and 7, 7 and 19, 12 and 24, 25 and 20, 21 and 14, 21 and 10, and 23 and 15. Table 12 shows the value of the DyMODN problem indices over the planning periods for the two cases.

By considering DGRs, these indices are lower than those without taking DGRs into account. It is clear that regarding the DGRs as an alternative expansion option in the DyMODN problem can provide optimal plans with lower investment, operation, and customer outage costs. Fig. 4 compares the changes of voltage profiles at load nodes in period 3 of the planning horizon for the two cases. It is obvious that the presence of the DGRs in the DyMODN problem provides better voltage values in the majority of the load nodes. The maximum voltage drops are for Cases 1 and 2 are 4.55% and 3.33%, respectively.

One thing that must be highlighted in the values obtained in Table 12 is the relationship between the values of the DyMODN problem indices. It is well recognized that with an increase in the TIC index, the TOC index is also increased. This is because with the reinforcement of existing equipment and or installation of new equipment in the primary distribution network to meet load growth, the maintenance and operation costs increase. Also, in the mesh or multifeed networks, increasing the TIC index will inversely influence the TCOC index. Typically, in the mesh networks, the TCOC index can be decreased by supplying multiple utility power sources/services and or multiple connection paths (increasing the investment cost) to the loads served. For further discussion on the relationship between the investment costs and customer outage costs in the mesh networks, please review the work by Shivaie et al. [62]. In open-loop, single-feed distribution networks, however, increasing the investment costs to meet the growing demand for electricity throughout the network creates an increase in costumer outage cost. In such networks, any failure in any given branch will lead to an interruption of service continuity to all downstream customers, which are supplied from the respective branch.

As a result, there is no alternate route to supply the disconnected loads, which leads to an increase in the TCOC index.

TABLE 12 Index Values of the Final Expansion Plan for Cases 1 and 2

No.	Index	Period 1		Period 2		Period 3		Total	
		Case 1	*Case 2*	*Case 1*	*Case 2*	*Case 1*	*Case 2*	*Case 1*	*Case 2*
1	TIC (M$)	16.379474	14.013378	5.478204	5.927071	8.064624	8.242823	29.922301	28.183272
2	TOC (M$)	1.276165	1.119260	2.745894	2.675075	4.593978	4.281952	8.616037	8.076287
3	TCOC (M$)	0.735634	0.693765	1.196897	1.021358	2.024121	1.881854	3.956652	3.596977

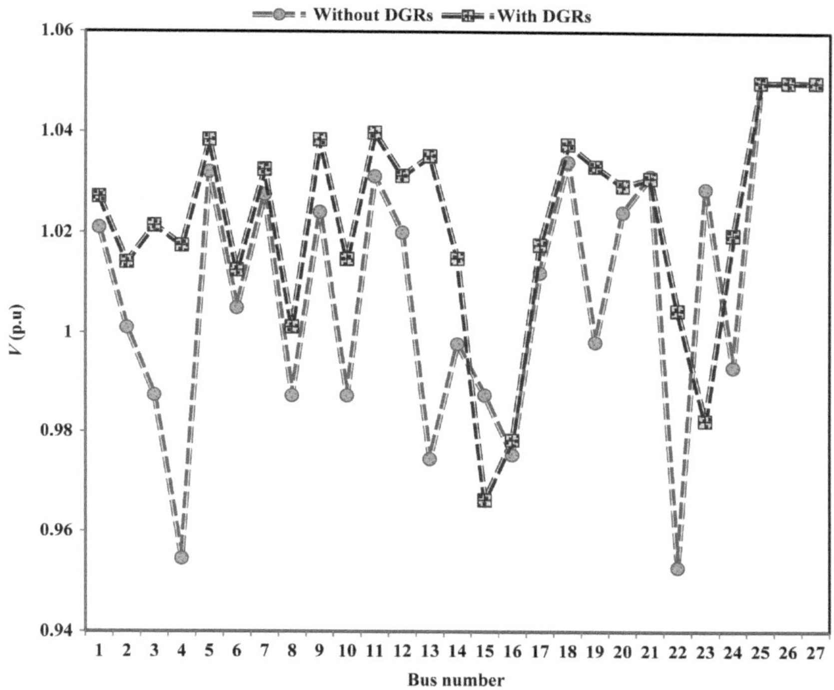

FIG. 4 Changes of voltage profiles at load nodes in period 3 for the two cases studied.

4.5.2 Performance Evaluation of the 2D-cdEMSA

In this section, performance of the 2D-cdEMSA is compared with the 2D-cMSA, 1D-HSA, 1D-IHSA, NSGA-II, and PSO under Case 2. For a detailed description of the 1D-HSA, 1D-IHSA, and NSGA-II please look at a previous publication by the authors [30]. Additional information on the PSO can be found in a study by Tungadio et al. [14]. Table 13 shows the calculated values of the DyMODN problem indices by the proposed optimization algorithms. In accordance with the results in Table 13, the multiobjective DyMODN problem in the presence of the DGRs (i.e., Case 2) by the 2D-cdEMSA leads to more efficient results than other optimization algorithms. In addition, an index of cost saving (ICS) is defined and implemented to more clearly assess the performance of these optimization algorithms, as given in Eq. (58) [41]:

$$ICS_q^{m,m'} = \begin{cases} 1-\left(F_q^m/F_{q'}^{m'}\right); \text{ for minimization} \\ \left(F_q^m/F_{q'}^{m'}\right)-1; \text{ for maximization} \end{cases} ;\forall\{m,m' \in Nm, q \in Nq\} \quad (58)$$

This index represents the amount of the superiority, or weakness, of optimization algorithm m compared with optimization algorithm m', from the

TABLE 13 The DyMODN Problem Indices, Based on the Optimization Algorithms

			Calculated Values of the DyMODN Problem Indices					
No.	Index	Period	*2D-cdEMSA*	*2D-cMSA*	*1D-HSA*	*1D-IHSA*	*NSGA-II*	*PSO*
1	TIC (M$)	1	14.013378	15.522863	17.524526	16.235620	16.326520	18.625360
		2	5.927071	6.510393	7.710268	7.581125	7.573125	8.385672
		3	8.242823	7.986342	8.420573	7.472355	8.069899	9.147401
		Total	28.183272	30.019598	33.655367	31.289100	31.969544	36.158433
2	TOC (M$)	1	1.119260	1.274320	1.684521	1.402680	1.524120	1.825420
		2	2.675075	2.762850	3.076350	2.875241	2.915674	3.232523
		3	4.281952	4.582431	4.924885	4.686202	4.784669	5.256350
		Total	8.076287	8.619601	9.685756	8.964123	9.224463	10.314293
3	TCOC (M$)	1	0.693765	0.738412	0.873893	0.804783	0.829433	0.918654
		2	1.021358	1.117336	1.314709	1.183907	1.238357	1.452437
		3	1.881854	1.945434	2.125819	2.012709	2.060456	2.243547
Total	Total	Total	3.596977	3.801182	4.314421	4.001399	4.128246	4.614638

perspective of the obtained value of objective function q by optimization algorithm m. Table 14 give the ICS for the optimization algorithms. As the results show, the results of the 2D-cdEMSA are more favorable than comparable results from the other optimization algorithms. For example, the ICS from the perspective of the obtained TIC index by the 2D-cdEMSA is 6.117090%, 16.259204%, 9.926229%, 11.843371%, and 22.056157% better (positive sign) than that by the 2D-cMSA, 1D-HSA, 1D-IHSA, NSGA-II, and PSO, respectively. Also, the ICS from the perspective of the obtained TIC index by the 2D-cMSA is 6.515659% worse (negative sign) than that by the 2D-cdEMSA, and is 10.802939%, 4.057329%, 6.099386%, and 16.977602% better than 1D-HSA, 1D-IHSA, NSGA-II, and PSO, respectively. Fig. 5 shows a comparison of the convergence characteristic of the 2D-cdEMSA and the convergence characteristics of the 2D-cMSA, 1D-HSA, 1D-IHSA, NSGA-II, and PSO. According to this figure, the convergence characteristic of the 2D-cdEMSA is optimized better, faster, and with more precision than that of the other optimization algorithms.

5 CONCLUSIONS

In this chapter, the authors presented the 2D-cdEMSA, a two-level computational, multidimensional, continuous/discrete, enhanced melody search algorithm designed to improve performance of the music-inspired optimization algorithms in dealing with complex, nonconvex, large-scale, nonlinear mixed-integer optimization problems. The newly proposed 2D-cdEMSA is organized by multiple music players with a two-level improvisation. Single and group improvisation levels are the two different levels used to improvise new melodies. Using multiple music players with different tastes, ideas, and experiences under interactive relationships among music players can conduct players to choose better pitches.

The 2D-cdEMSA was successfully implemented on the midterm, large-scale DyMODN problem. The proposed DyMODN problem, with a new point of view, modeled and scrutinized the probabilistic characteristics of the power demand needed to provide a well-designed framework of distribution network expansion planning. Investment costs—operational cost along with maintenance cost—and costumer outage costs were considered in the optimization problem as three objective functions. The DyMODN problem takes into account Kirchhoff's current and voltage rules, operational constraints on the capacities of the distribution network equipment, constraints related to the financial resources, and the logical constraints.

In this study, the DyMODN satisfied the challenges of distribution network expansion planning, and accurately modeled the probabilistic characteristics of the power demand. In addition, the authors performed a thorough analysis in which the 2D-cdEMSA was compared with state-of-the-art optimization algorithms. Additionally, the performance of the optimization algorithms,

TABLE 14 The ICS for Different Indices, Based on Various Optimization Algorithms

No.	The DyMODN Indices	Optimization Algorithms	Optimization Algorithms					
			2D-cdEMSA	*2D-cMSA*	*1D-HSA*	*1D-IHSA*	*NSGA-II*	*PSO*
1	ICS (%) for TIC	2D-cdEMSA	0	6.117090	16.259204	9.926229	11.843371	22.056157
		2D-cMSA	−6.515659	0	10.802939	4.057329	6.099386	16.977602
		1D-HSA	−19.416109	−12.111318	0	−7.562592	−5.273215	6.922495
		1D-IHSA	−11.020112	−4.228910	7.030875	0	2.128413	13.466659
		NSGA-II	−13.434466	−6.495576	5.009076	−2.174699	0	11.584818
		PSO	−28.297499	−20.449424	−7.437345	−15.562394	−13.102749	0
2	ICS (%) for TOC	2D-cdEMSA	0	6.303238	16.616865	9.904326	12.447076	21.698103
		2D-cMSA	−6.727274	0	11.007452	3.843343	6.557151	16.430520
		1D-HSA	−19.928328	−12.368960	0	−8.050235	−5.000757	6.093844
		1D-IHSA	−10.993121	−3.996959	7.450456	0	2.822278	13.090281
		NSGA-II	−14.216632	−7.017285	4.762591	−2.904244	0	10.566211
		PSO	−27.710828	−19.660910	−6.489292	−15.061931	−11.814563	0
3	ICS (%) for TCOC	2D-cdEMSA	0	5.372144	16.628975	10.107015	12.869121	22.052889
		2D-cMSA	−5.677128	0	11.895895	5.003674	7.922589	17.627731
		1D-HSA	−19.945749	−13.502089	0	−7.822814	−4.509784	6.505754
		1D-IHSA	−11.243386	−5.267230	7.255249	0	3.072660	13.288994
		NSGA-II	−14.769875	−8.604270	4.315179	−3.170066	0	10.540198
		PSO	−28.292119	−21.400080	−6.958454	−15.325615	−11.782049	0

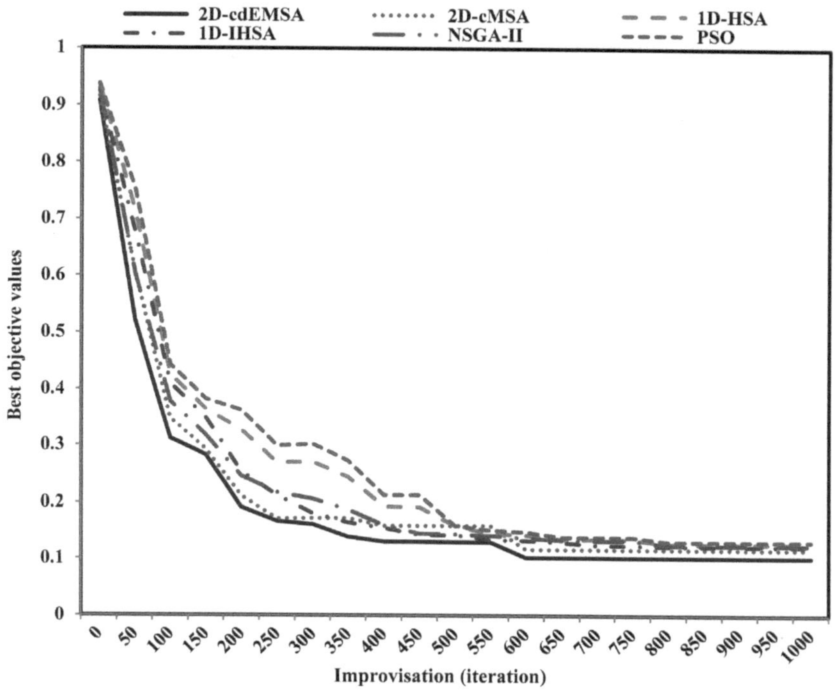

FIG. 5 Convergence characteristics of the proposed optimization algorithms.

especially the 2D-cdEMSA, was thoroughly assessed through an index of cost saving. The simulation results show that the newly proposed 2D-cdEMSA outperformed the other optimization algorithms in all indices of the DyMODN problem. Hence, the 2D-cdEMSA seems to surface as an effective optimization algorithm for future large-scale engineering optimization problems with big data.

In future studies, the authors plan to evaluate a new class of the music-inspired optimization algorithms and investigate whether they can outperform the current 2D-cdEMSA implementation, by focusing on solving complex and multilevel optimization problems of power systems.

REFERENCES

[1] A.D. Belegundu, T.R. Chandrupatla, Optimization Concepts and Applications in Engineering, Prentice Hall, Upper Saddle River, NJ, 1999.

[2] D.G. Luenberger, Y. Ye, Linear and Nonlinear Programming, Springer, New York, NY, 1984.

[3] S. Boyd, L. Vandenberghe, Convex Optimization, Cambridge University Press, Cambridge, 2004.

[4] P.M. Pardalos, J. Rosen, Methods for global concave minimization: a bibliographic survey, SIAM Rev. 28 (3) (1986) 367–379.

[5] A.H. Land, A. Doig, An automatic method of solving discrete programming problems, Econometrica 28 (1960) 497–520.
[6] A.I. Tyatyushkin, O.V. Morzhin, Constructive methods of control optimization in nonlinear systems, Autom. Remote Control 70 (5) (2009) 772–786.
[7] D.F. Shanno, Recent advances in gradient based unconstrained optimization techniques for large problems, ASME J. Mech. Transm. Autom. Des. 105 (1983) 155–159.
[8] S.S. Rao, Engineering Optimization: Theory and Practice, fourth ed., Wiley, Hoboken, NJ, 2009.
[9] E.G. Talbi, Metaheuristics: From Design to Implementation, Wiley, Hoboken, NJ, 2009.
[10] K.Y. Lee, M.A. El-Sharkawi, Modern Heuristic Optimization Techniques: Theory and Applications to Power Systems, IEEE Press Series on Power Engineering, Wiley Interscience, Hoboken, 2008.
[11] S. Ranganathan, M.S. Kalavathi, C.A. Rajan-C, Self-adaptive firefly algorithm based multi-objectives for multi-type FACTS placement, IET Gener. Transm. Distrib. 10 (11) (2016) 2576–2584.
[12] C.F. Chang, Reconfiguration and capacitor placement for loss reduction of distribution systems by ant colony search algorithm, IEEE Trans. Power Syst. 23 (4) (2008) 1747–1755.
[13] M.M. Aman, G.B. Jasmon, H. Mokhlis, A.H. Abu-Bakar, Optimum tie switches allocation and DG placement based on maximisation of system loadability using discrete artificial bee colony algorithm, IET Gener. Transm. Distrib. 10 (10) (2016) 2277–2284.
[14] D.H. Tungadio, B.P. Numbi, M.W. Siti, A.A. Jimoh, Particle swarm optimization for power system state estimation, Neurocomputing 148 (2015) 175–180.
[15] I.A. Farhat, M.E. El-Hawary, Dynamic adaptive bacterial foraging algorithm for optimum economic dispatch with valve-point effects and wind power, IET Gener. Transm. Distrib. 4 (9) (2010) 989–999.
[16] N. Fang, J. Zhou, R. Zhang, Y. Liu, Y. Zhang, A hybrid of real coded genetic algorithm and artificial fish swarm algorithm for short-term optimal hydrothermal scheduling, Int. J. Electr. Power Energy Syst. 62 (2014) 617–629.
[17] T.T. Nguyen, D.N. Vo, A.V. Truong, Cuckoo search algorithm for short-term hydrothermal scheduling, Appl. Energy 132 (2014) 276–287.
[18] F.G. Duque, L.W. de-Oliveira, E.J. de-Oliveira, A.L.M. Marcato, I.C. Silva, Allocation of capacitor banks in distribution systems through a modified monkey search optimization technique, Int. J. Electr. Power Energy Syst. 73 (2015) 420–432.
[19] S.M. Abd-Elazim, E.S. Ali, Load frequency controller design via BAT algorithm for nonlinear interconnected power system, Int. J. Electr. Power Energy Syst. 77 (2016) 166–177.
[20] F.M. Gonzalez-Longatt, P. Wall, P. Regulski, V. Terzija, Optimal electric network design for a large offshore wind farm based on a modified genetic algorithm approach, IEEE Syst. J. 6 (1) (2012) 164–172.
[21] A.A. Abou-El-Ela, M.A. Abido, S.R. Spea, Optimal power flow using differential evolution algorithm, Electr. Power Syst. Res. 80 (8) (2010) 878–885.
[22] M. Arsuaga-Rios, M.A. Vega-Rodriguez, Multi-objective energy optimization in grid systems from a brain storming strategy, Soft. Comput. 19 (11) (2015) 3159–3172.
[23] H.M. Hasanien, Shuffled frog leaping algorithm for photovoltaic model identification, IEEE Trans. Sustain. Energy 6 (2) (2015) 509–515.
[24] K. Lenin, B.R. Reddy, M.S. Kalavathi, Dolphin echolocation algorithm for solving optimal reactive power dispatch problem, Int. J. Comput. 12 (1) (2014).

[25] T. Sousa, H. Morais, Z. Vale, P. Faria, J. Soares, Intelligent energy resource management considering vehicle-to-grid: a simulated annealing approach, IEEE Trans. Smart Grid 3 (1) (2012) 535–542.

[26] A. Bhattacharya, P.K. Roy, Solution of multi-objective optimal power flow using gravitational search algorithm, IET Gener. Transm. Distrib. 6 (8) (2012) 751–763.

[27] M.M. Othman, W. El-Khattam, Y.G. Hegazy, A.Y. Abdelaziz, Optimal placement and sizing of distributed generators in unbalanced distribution systems using supervised big bang-big crunch method, IEEE Trans. Power Syst. 30 (2) (2015) 911–919.

[28] C. Lee, F. Chang, Fractional-order PID controller optimization via improved electromagnetism-like algorithm, Expert Syst. Appl. 37 (12) (2010) 8871–8878.

[29] L. Benasla, A. Belmadani, M. Rahli, Spiral optimization algorithm for solving combined economic and emission dispatch, Int. J. Electr. Power Energy Syst. 62 (2014) 163–174.

[30] M. Shivaie, A. Salemnia, M.T. Ameli, A multi-objective approach to optimal placement and sizing of multiple active power filters using a music-inspired algorithm, Appl. Soft Comput. 22 (2014) 189–204.

[31] F.R. Alonso, D.Q. Oliveira, A.C. Zambroni-de-Souza, Artificial immune systems optimization approach for multiobjective distribution system reconfiguration, IEEE Trans. Power Syst. 30 (2) (2015) 840–847.

[32] Y. Kongjeen, P. Inrawong, K. Buayai, T. Kerdchuen, Key cutting algorithm application to measurement placement for power system state estimation, Energy Procedia 34 (2013) 142–147.

[33] B. Shaw, V. Mukherjee, S.P. Ghoshal, Seeker optimisation algorithm: application to the solution of economic load dispatch problems, IET Gener. Transm. Distrib. 5 (1) (2011) 81–91.

[34] M. Moghimi-Hadji, B. Vahidi, A solution to the unit commitment problem using imperialistic competition algorithm, IEEE Trans. Power Syst. 27 (1) (2012) 117–124.

[35] S.M. Ashrafi, A.B. Dariane, A novel and effective algorithm for numerical optimization: melody search (MS), 11th International Conferences on Hybrid Intelligent Systems (HIS), Malacca, Malaysia, 5–8 December, 2011, , pp. 109–114.

[36] M. Shivaie, A. Salemnia, M.T. Ameli, Optimal multi-objective placement and sizing of passive and active power filters by a fuzzy-improved harmony search algorithm, Int. Trans. Electr. Energy Syst. 25 (3) (2013) 520–546.

[37] M. Shivaie, M.T. Ameli, An implementation of improved harmony search algorithm for scenario-based transmission expansion planning, Soft Comput. 18 (8) (2014) 1615–1630.

[38] M. Kiani-Moghaddam, M. Shivaie, An Innovative Multi-Stage Multi-Dimensional Multiple-Inhomogeneous Melody Search Algorithm: Symphony Orchestra Search Algorithm (SOSA), IGI Global, PA, 2017, pp. 1–40.

[39] O. Moh'd-Alia, R. Mandava, The variants of the harmony search algorithm: an overview, Artif. Intell. Rev. 36 (1) (2011) 49–68.

[40] S.M. Ashrafi, A.B. Dariane, Performance evaluation of an improved harmony search algorithm for numerical optimization: melody search (MS), Eng. Appl. Artif. Intell. 26 (2013) 1301–1321.

[41] M. Shivaie, M.T. Ameli, Strategic multiyear transmission expansion planning under severe uncertainties by a combination of melody search algorithm and Powell heuristic method, Energy 115 (2016) 338–352.

[42] K.S. Lee, Z.W. Geem, S. Lee, K. Bae, The harmony search heuristic algorithm for discrete structural optimization, Eng. Optim. 37 (7) (2005) 663–684.

[43] S.K. Khator, L.C. Leung, Power distribution planning: a review of models and issues, IEEE Trans. Power Syst. 12 (3) (1997) 1151–1159.

[44] H.L. Willis, Power Distribution Planning Reference Book, second ed., Marcel Dekker, New York, 2004.
[45] G.W. Ault, C.E.T. Foote, J.R. McDonald, Distribution system planning in focus, IEEE Power Eng. Rev. 22 (1) (2002) 60–62.
[46] R.C. Lotero, J. Contreras, Distribution system planning with reliability, IEEE Trans. Power Deliv. 26 (4) (2011) 2552–2562.
[47] S. Wong, K. Bhattacharya, J.D. Fuller, Electric power distribution system design and planning in a deregulated environment, IET Gener. Transm. Distrib. 3 (12) (2009) 1061–1078.
[48] E.G. Carrano, R.T.N. Cardoso, R.H.C. Takahashi, C.M. Fonseca, O.M. Neto, Power distribution network expansion scheduling using dynamic programming genetic algorithm, IET Gener. Transm. Distrib. 2 (3) (2008) 444–455.
[49] Z.N. Popovic, D.S. Popovic, Graph theory based formulation of multi-period distribution expansion problems, Electr. Power Syst. Res. 80 (10) (2010) 1256–1266.
[50] H. Khan, M.A. Choudhry, Implementation of distributed generation (IDG) algorithm for performance enhancement of distribution feeder under extreme load growth, Int. J. Electr. Power Energy Syst. 32 (9) (2010) 985–997.
[51] R. Billinton, W. Zhang, Cost related reliability evaluation of bulk power systems, Int. J. Electr. Power Energy Syst. 23 (2) (2001) 99–112.
[52] T. Gönen, B.L. Foote, Mathematical dynamic optimization model for electrical distribution system planning, Electr. Power Energy Syst. 4 (2) (1982) 129–136.
[53] T. Gönen, B.L. Foote, Distribution-system planning using mixed-integer programming, IEE Proc. C Gener. Transm. Distrib. 128 (2) (1981) 70–79.
[54] N. Deo, Graph Theory With Applications to Engineering and Computers Science, Prentice-Hall, Englewood Cliffs, NJ, 1974.
[55] G.H. Hahn, S.S. Shapiro, Statistical Models in Engineering, Wiley, New York, 1967.
[56] M. Madrigal, K. Ponnambalam, V.H. Quintana, in: Probabilistic Optimal Power Flow, Proc. IEEE Can. Conf. Electrical Computer Engineering, Waterloo, ON, Canada, 1998, , pp. 385–388.
[57] P. Zhang, S.T. Lee, Probabilistic load flow computation using the method of combined and Gram-Charlier expansion, IEEE Trans. Power Syst. 19 (1) (2004) 676–682.
[58] S. Kaplan, On the method of discrete probability distributions in risk and reliability calculation-application to seismic risk assessment, Risk Anal. 1 (3) (1981) 189–196.
[59] G. Verbic, C.A. Canizares, Probabilistic optimal power flow in electricity markets based on a two-point estimate method, IEEE Trans. Power Syst. 21 (4) (2006) 1883–1893.
[60] H.P. Hong, An efficient point estimate method for probabilistic analysis, Reliab. Eng. Syst. Saf. 59 (3) (1998) 261–267.
[61] W. Sheng, K.-y. Liu, S. Cheng, Optimal power flow algorithm and analysis in distribution system considering distributed generation, IET Gener. Transm. Distrib. 8 (2) (2014) 261–272.
[62] M. Shivaie, M.S. Sepasian, M.K. Sheikh-El-Eslami, Multi-objective transmission expansion planning based on reliability and market considering phase shifter transformers by fuzzy-genetic algorithm, Int. Trans. Electr. Energy Syst. 23 (8) (2013) 1468–1489.

Chapter 12

Demand-Side Management—Recent Aspects and Challenges of Optimization for an Efficient and Robust Demand-Side Management

Michael Diekerhof*,†, Antonello Monti*,† and Sebastian Schwarz*,†
*E.ON ERC, RWTH Aachen University, Aachen, Germany, †Institute for Automation of Complex Power Systems, Faculty of Electrical Engineering and Information Technology, RWTH Aachen University, Aachen, Germany

1 INTRODUCTION

1.1 Flexibility for Demand Side Management and Demand Response

The electrical distribution network is currently going through significant changes. In the past, most electrical energy was generated by large fossil fuel power plants and some smaller combined heat and power (CHP) units on transmission (system) level [1].

The operation planning of the energy generation has been demand-driven. This means that the operation of the individual power plants is scheduled, considering the local constraints of each power plant, to meet a certain demand and minimize the total generation costs. This is traditionally known as the optimal dispatch problem [1].

Although, the optimal dispatch problem is still a highly relevant tool for scheduling the operation of power plants on transmission level, the increasing amount of highly volatile renewable energy generation units, on the distribution level, introduces a paradigm shift: the distribution network changes from demand to supply driven. Efficient and robust Demand Side Management (DSM) that exploits local flexibility is thus, a promising solution for balancing electrical supply and demand in line with an increasing amount of distributed energy resources (DER), such as photovoltaic (PV) units and wind turbines [2].

Classical and Recent Aspects of Power System Optimization. https://doi.org/10.1016/B978-0-12-812441-3.00012-4
© 2018 Elsevier Inc. All rights reserved.

For the sake of an optimal balance of demand and supply, flexibility is used to shift the operation of flexible loads, such as electric vehicles or electro-thermal heating units equipped with thermal storage, to times of renewable energy supply for an optimal usage of the available renewable energy.

This requires the optimization and coordination of supply and demand over a specific time horizon, which is usually day-ahead or intraday. In other words, given the volatile generation, the supply side rules, and the demand must adapt to the volatile generation from PV units and wind turbines.

This chapter adopts the definition of flexibility from the Expert Group 3 of the Smart Grids Task Force of the European Commission [3]. "On an individual level, flexibility is the modification of generation injection and/or consumption patterns in reaction to an external (signal or activation) to provide services within the energy system." The electrical flexibility is characterized by multiple parameters, such as the amount of power modulation, rate of change, response time, duration, etc. [3]. Whereby flexibility can, in general, be provided both by the supply and the demand side, electrical flexibility on distribution level comes mainly from the demand side. However, CHP units equipped with a thermal energy storage (TES) can provide demand side flexibility on the supply side in the distribution network as well. Customers that provide flexibility can be residential, nonresidential, or industrial. Dall'Anese et al. [4] extend the flexibility definition to the "system's ability to provide secure and economical supply-demand balance across spatial and temporal scales by leveraging and seamlessly coordinating various controllable assets." In particular, Dall'Anese et al. [4] identify local neighborhoods and districts as the main driving factors for the integration of the increasing amount of DER and also link the efficient deployment of multi-energy systems to socioeconomic, environmental, and operational efficiency benefits [4].

Built upon the definition of flexibility and because Demand Response (DR) makes use of flexibility on a larger scale, DSM and DR strategies and methods are discussed next. In general, DR is well discussed in literature including relevant surveys and summaries for the deployment of DR, such as in [5,6].

One of the most cited and relevant survey on DSM and DR is authored by Palensky et al. [5]. According to [5], DSM is every action that changes the energy system on the demand side and integrates DR in the context of DSM and shows that DSM can be categorized according to the impact on process quality, its potential for economic or ecologic benefits, and timing. Four main categories are established for classifying DSM, that is, (1) energy efficiency, (2) time of use, (3) demand response and (4) spinning reserve. This is shown in Fig. 1, which is an extension on the work in [5], where the focus on market DR and physical DR of this chapter is highlighted by the dashed box. Note, the faster the impact decisions are performed, for example, spinning reserve in the timing of seconds, the more unwanted effects can occur on customer level, such as deterioration of customer comfort on building level or manufacturing output. Energy efficiency belongs to the lower layer of DSM, which encourages

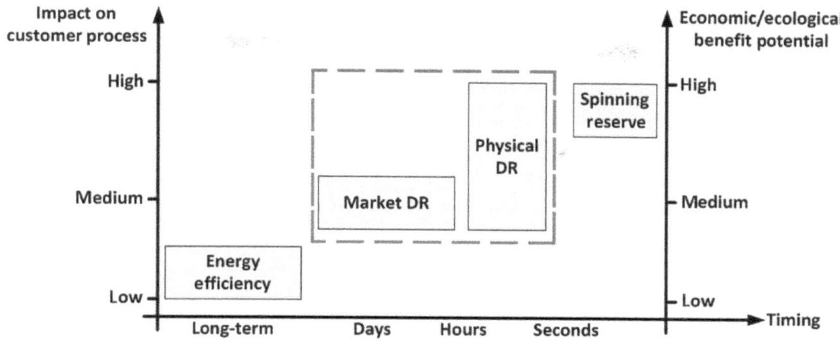

FIG. 1 DSM categories according to the impact on process quality and timing. *(Based on P. Palensky, D. Dietrich, demand side management: demand response, intelligent energy systems, and smart loads, IEEE Trans. Industr. Inform. 7(3) (2011) 381–388.)*

customers in the long term to, for example, invest in a more efficient electro-thermal heating unit. Although energy efficiency initially seems to have the lowest economic and/or ecological benefit potential from a global system level perspective, the investment in energy efficiency may accumulatively lead to high benefits considering its long-term operation time scale. For this reason, the DSM category should not be neglected.

Moreover, Palensky et al. [5] distinguish DR between market-based DR and physical DR. On the one hand, market based DR is based upon pricing mechanism of relevant market places, for example, EEX—European Energy Exchange [7] in Leipzig, Germany, and uses real-time pricing or other incentives, such as discussed in [6]. However, time-of-use tariffs, which are more stable over time and do not change as frequently as compared to real-time pricing, are also said to belong to time-based DR rates within market DR. Physical DR, on the other hand, refers to grid management based upon emergency signals. Binding requests, such as load shedding, ensure a stable grid operation and prevent issues such as grid congestion. Because market-based DR can lead to critical grid situations, a good mixture between market-based and physical DR is suggested by Palensky et al. [5] to run a grid in stable conditions.

Vardakas et al. [6] provide a comprehensive overview on DR programs in smart grids based upon different pricing techniques and optimization algorithms. This survey shows which optimization techniques are the most appropriate per type of objective function for residential, commercial, and industrial customers. Relevant pricing schemes, such as time-of-use prices or real-time pricing, are discussed according to their effectiveness for reducing the peak demand. Albadi et al. [8] provide a good summary on DR for deregulated energy markets, which highlights the most common metrics for DR measurement applied to a realistic use case.

As mentioned, [6] provides an overview on optimization techniques applied to DR in the electrical energy sector. However, Dall'Anese et al. [4] extend the idea of DR and balancing of demand and supply to multienergy systems, such as heat, electricity and gas. Relevant challenges in the optimization for balancing demand and supply across multienergy systems and across spatial and temporal scales are discussed, whereby the focus is in the motivation for distributed optimization (DO) strategies instead of central optimization for DR purposes. However, the intelligent coordination of DER, for example, in the offline day-ahead time frame, is intended to balance supply and demand, and is also heavily affected by uncertainty. Typical sources for uncertainty include wind and PV generation due to weather uncertainty, the electrical/thermal energy demand due to user behavior and weather uncertainty, as well as fluctuations in the electrical energy price. Therefore, relevant distributed optimization techniques that address the impact of uncertainty within the DR portfolio optimization are discussed within this chapter as well.

The rest of the chapter is organized as follows: Section 2 introduces the portfolio coordination problem for DR including relevant flexibility sources. In Section 3, relevant challenges and requirements for an efficient flexibility exploitation on district level are discussed, and Section 4 introduces an architecture for DR that addresses the predefined requirements. Section 5 introduces DO as major tool for flexibility exploitation, while Section 6 links the optimization techniques to uncertainty in DSM. Numerical results for a use case that links flexibility provider and flexibility user via a robust DO are carried out in Section 7. Finally, conclusions are drawn in Section 8. The Appendix provides a simple example for the application of the distributed optimization algorithm to energy management.

2 DEMAND RESPONSE AND FLEXIBILITY SOURCES

2.1 The Portfolio Coordination Problem

The formulation of the DR problem is, in principle, given by the portfolio coordination problem. This problem serves as reference for the description of flexibility deployment in DR. Whereby the classical optimal dispatch problem schedules the operation of (large) electrical power plants in order to meet a pre-estimated electrical demand, the portfolio coordination problem extends the dispatch problem to the scheduling of flexible devices on the electrical demand side to balance demand and supply across spatial and temporal scales [1,4].

The portfolio coordination problem is well defined, for example, in [1] and reads, with the following provided data information, as follows:

- A coordination and optimization horizon $\tau = [0, T]$ with $t \in \tau$ discrete time slots
- A portfolio of $N = \{1, \ldots, M\}$ customers with M production and consumption units in total
- Set C_i of local operating constraints for each customer $i \in N$

- A global objective Function F, which describes the system (high) level coordination objective, such as minimizing the total operation costs of the DR aggregator

The general portfolio coordination problem aims at determining the consumption and generation power references $\boldsymbol{P}_i \in \mathbb{R}^T$ for each customer $i \in N$ that solve the following optimization problem:

$$\underset{P_i}{\text{minimize}} \sum_{i \in N} \sum_{t \in \tau} F\left(P_i^t\right) \tag{1}$$

Subject to:

$$\boldsymbol{P}_i \in C_i \ \ \forall i \in N \tag{2}$$

This is a very high-level view of the portfolio coordination problem which only includes the system level objective function F. The flexibility provided is integrated via temporal power shifts in time of power consumption and generation devices over the optimization horizon τ according to the local and private set of constraints C_i per customer. The global objective function F could also include a set of local and private objective functions f_i for each customer i. It is important to note that in the traditional optimal dispatch problem, all devices, that is, power plants, that are coordinated by a single entity, such as an aggregator, are assumed to fully collaborate. This is not the case for the portfolio coordination problem in DR. It might be the case that the DR aggregator coordinates a fleet of customers in which each customer has an individual, not strictly monetary, objective. This private objective might not be in line with the high level objective function of the DR aggregator. Therefore, the coordination of the DR aggregator aims to find a trade-off between customer and system level objectives while satisfying the global equality constraints, such as the balancing of electrical power and demand in order to avoid undesired power imbalances [1].

2.2 Sources for Flexibility

The exploitation of energy storage is necessary for the deployment of flexibility and for balancing supply and demand. The energy storage can be in the form of electrical energy, heat, or gas storage [4].

On the distribution level, a promising flexibility source is the integration of electro-thermal heating units equipped with a TES. Because a TES will probably be available at much lower cost than electrical storages according to [4], a TES is a key factor for flexibility and energy shifting on local district level. A TES decouples the thermal energy demand of the end-user from the thermal energy generation, which in case of electrical heaters (EHs) and electrical heat pumps (EHPs) consumes electrical energy or in case of CHP units generates electricity. In particular, as evaluated in [4,9] a share of both CHP and EH/EHP units on local district level might lead to system level benefits, such as the integration of volatile renewable energy generation. Next to energy storages, the following other sources for demand side flexibility exist and are well described, for instance, in [10]:

- Curtailable Loads
  A curtailable load provides flexibility by, under some constraints, reducing the load demand of a physical process within a specific time interval. In other words, a curtailable load supports and, in some cases, can penalize the deviation from a desired operation (load) profile. Examples are specific processes that can run in partial load, such as electrical charging processes.
- Deferrable Loads
  A deferrable load is in general not characterized by a certain cost (objective) function but needs to consume a certain minimum amount of energy E within a given time interval. Nonetheless, the operation is restricted by the minimum and maximum energy consumption. Examples are classical white goods like washing machines or dish washers.
- Electric Vehicles (EV)
  EVs are flexible loads that combine characteristics of storage, deferrable loads, and curtailable loads. The end-user might decide charging time intervals and thus provide flexibility to the energy system. Deviations from a certain required charging profile are penalized, whereby a specific amount of energy must be charged within a certain time horizon. Notice that EVs can also provide flexibility in terms of electrical power generation for example, via vehicle-to-grid applications.

3 CHALLENGES AND REQUIREMENTS ON THE ARCHITECTURE AND OPTIMIZATION ALGORITHM

The requirements derived in this section are based upon the work in [2,11]. To maximize the aggregated flexibility potential of heterogeneous customers on district level, the following requirements must be followed:

(1) The optimization procedure for the portfolio balancing problem should be integrated in an architecture that provides the option to consider customer (lower level) and system (upper level) objectives. For example, a two- or multilevel hierarchy should enable the integration and interconnection of flexibility providers (customers, aggregation service provider) and flexibility users, such as the distribution system operator (DSO), transmission system operator (TSO), and balance responsible party (BRP).
(2) The architecture and optimization algorithm should be scalable to cater many participants.
(3) The optimization algorithm should enable each customer to choose and consider its own local and private objective function, among a predefined

set, such as cost- or CO_2-emission reduction, and still be able to optimize toward a system level (aggregator) objective.

(4) The algorithm needs to account for uncertainty in relevant sources of uncertainty, such as renewable generation or electrical and thermal power demand.

(5) The algorithm needs to converge to optimal or near-optimal results in a suitable time to stay within the operation time window of the actions of the flexibility provider and flexibility user.

(6) The optimization algorithm needs to ensure that sensitive data, such as private customer information, is treated carefully and data privacy is always respected.

4 AN ARCHITECTURE FOR THE DEPLOYMENT OF FLEXIBILITY IN CITY DISTRICTS

This section explains the most relevant architecture schematics for flexibility deployment in energy management and provides a detailed explanation to a state-of-the-art architecture, which is recommend by the Expert Group 3 of the Smart Grids Task Force of the European Commission [3] and which is also adopted for the use case example at the end of this chapter.

4.1 Distributed and Decentralized Architecture Schematics

For the descriptions of relevant architectures, this chapter sticks to the explanations of Kargarian et al. [12], which distinguish between decentralized and distributed control architectures. Both distributed and decentralized have many similarities, but the major difference is that a distributed architecture always contains a central coordinating entity, whereby a decentralized architecture does not have a central coordinator, as visualized in Fig. 2. This means that a distributed architecture always contains (at least) two levels. The lower level consists of the controllable assets (customers) and the upper level of the coordinating entity (aggregator) [12].

Note, that both distributed and decentralized approaches do in general satisfy Requirements (3)–(5). Distributed architectures imply that each entity on the lower level solves its own local and private objective function and only certain information is exchanged with the coordinator on the upper level. In other words, Requirements (1)–(2) and Requirement (6) are also satisfied. However, in hierarchical architectures like that, the optimization problem could also be solved using central computation methods. Though it implies that the coordinator contains all private local information from (each customer) in the lower level. This immediately violates Requirement (6) as well as Requirements (1) and (2).

Decentralized approaches do not contain any central coordinator (aggregator) and all information is only exchanged between local neighboring agents.

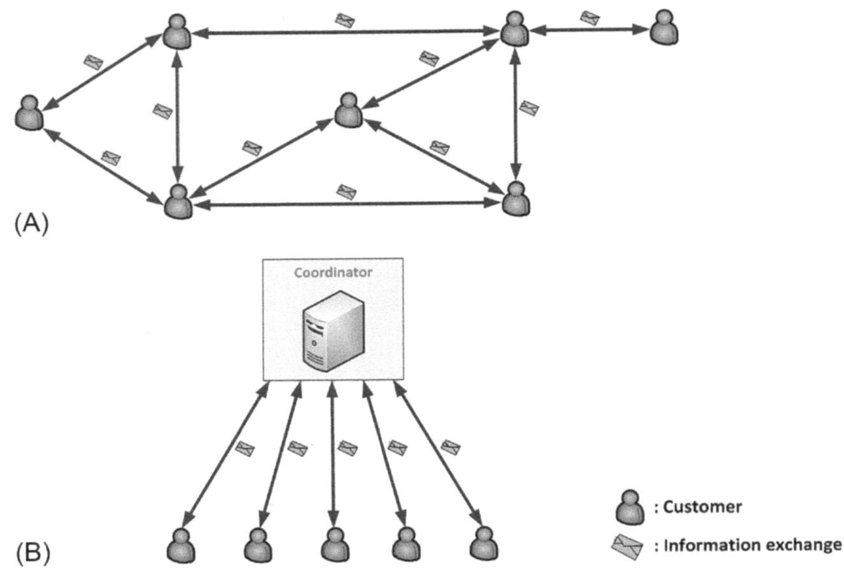

FIG. 2 Overview on different architectures for the deployment of flexibility. (A) Decentralized control structure and (B) distributed control structure. *(Based on A. Kargarian et al., Toward distributed/decentralized DC optimal power flow implementation in future electric power systems, IEEE Trans. Smart Grid, (2016) 1.)*

Although Requirement (1) is not ensured in the decentralized approach, the other requirements are addressed. The reader should note that a distributed architecture, despite its advantages, still contains one single-point-of-failure, which is the coordinator. Kargarian et al. [12] relate this, for example, to risks of cyber-attacks. In comparison, the total decentralized structure does not have that single-point-of-failure. Both approaches are studied and applied to both dynamic energy management and DR problems. Total decentralized approaches for DR problems are well described and solved in [10] or [13], whereby the hierarchical distributed architecture for example is applied to DR problems and analyzed in [2,11,14].

4.2 Reference Architecture for the Deployment of Flexibility

This book chapter, similar to recent research such as [2,11], adopts and adapts the architecture recommended by [3] for the deployment of flexibility. The architecture satisfies all predefined major requirements and is shown in Fig. 3. It links flexibility users with a flexibility provider. The aggregator terminates a flexibility purchase contract with its customers and very likely, such as claimed in [3], the aggregator will also act as supplier to its customers. It is important to stress that the customers can be residential, commercial, or industrial and that each customer can commit to the aggregator of its choice. In fact,

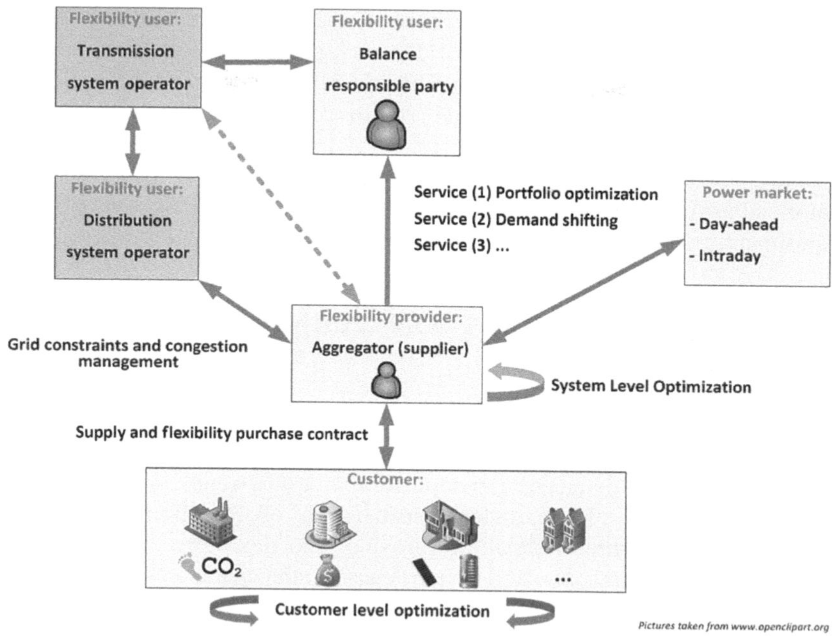

FIG. 3 Reference architecture for the deployment of flexibility. *(Based on Smart Grids Task Force, Regulatory recommendations for the deployment of flexibility: Sgtf-Eg3 Report, Http://Ec.Europa.Eu/Energy/Sites/Ener/Files/Documents/Eg3%20final%20-%20january%202015.Pdf, 2015.)*

given heterogeneous customers, such as residential and small and medium enterprises, the aggregator might identify flexibility synergies. Dall'Anese et al. [4] identify and address relevant synergies for flexibility deployment across multienergy carriers, such as electrical, heat, and gas. To address the defined requirements, it is important that the aggregator takes care that each customer still pursues its own local private objective according to its own local set of private constraints.

In this architecture, and in the recommendations from [3], the aggregator is the flexibility provider, whereby the flexibility user can vary. The aggregator might provide technical services such as a portfolio optimization, generation or demand adjustment, or peak shaving to the DSO or TSO. One of the most promising services, according to [3] is the portfolio optimization required by a BRP. The TSO takes care of and covers occurring imbalances, but the occurring costs are relocated to the BRP. In other words, the BRP oversees the imbalances within its own portfolio and in case the energy balance is achieved, the BRP gets monetary rewarded, but also gets penalized if there is a mismatch and balance is not achieved. Thus, it is a suitable choice for the BRP to outsource the portfolio optimization and purchase this service directly from the aggregator. This also requires that the aggregator should be able to choose its own mathematical objective function according to the required service and service user, while the portfolio balancing of demand and supply is ensured due to a global equality constraint. The timeline of the flexibility service provision can be integrated in the day-ahead and real-time, that is, intraday horizon operation of the BRP [3].

For a comprehensive overview on a set of possible services that the aggregator can provide, further information about the role of the BRP and financial adjustment mechanism, the reader should refer to [3].

5 DISTRIBUTED OPTIMIZATION TECHNIQUES FOR DEMAND SIDE MANAGEMENT

5.1 Architecture Choice

The portfolio optimization in the hierarchical architecture of the previous section could be solved in a central computation, but this would violate Requirements (2) and (6). Thus, under consideration of Requirement (1), the recommended structure of flexibility provider and flexibility user connected in a hierarchical architecture and Requirement (2) the characteristics of a hierarchical coordination schematic, see Figs. 2 and 3, a reasonable design choice for the coordination to the predefined requirements is in DO.

Molzahn et al. [15] therefore describe the relevant advantages of DO over centralized approaches in the following way: First, DO algorithms allow parallel computing and hence, in particular for a larger sets of entities participating in the optimization process, might overwhelm the performance of a central computation in terms of solution speed and problem size respectively scalability. Second, in distributed algorithms participants only share a limited amount of private information. Whereby in a central computation the aggregator needs to possess full information of private and sensitive information, such as cost functions, charging cycles etc., distributed algorithms can respect data privacy on the customer level. Third, decentralized and distributed algorithms do have a higher robustness against failure of individual participants in the communication than within a central control algorithm. However, in a hierarchical architecture, the coordinating central entity on the upper level (in this chapter it is the DR aggregator) is still a vulnerable single point of failure. This is not the case in a fully decentralized control architecture in which each agent only exchanges information with its neighboring agent. In that case, Molzahn et al. [15] also list the limited amount of communication infrastructure and the advantages in cybersecurity, such as no central point for cyber-attacks, as major advantages for decentralized approaches.

As soon as the decision to apply DO is made, the next step is to identify what kind of DO is the most suitable for the specific optimization problem and that satisfies all predefined requirements. Molzahn et al. [15] provide a comprehensive survey on offline and online DO algorithms for electrical power systems. Although the focus is on DO applied to offline and online optimal power flow problems, the DO techniques are applicable in a broader field, such as to DR and EV charging concepts under the hut of DSM and portfolio coordination.

5.2 Relevant Metrics for the Evaluation of Demand Response Flexibility Services

The evaluation of the DR service can be done according to different metrics and should depend on the chosen objective. Very often, for example in [11] or [16], market-based DR algorithms are evaluated in terms of the total energy costs for the individual customer and the aggregator. In case the optimization is affected by uncertainty, the amount of energy imbalances in comparison to the precalculated deterministic optimization is considered, such as in [2]. For stochastic optimization, the so-called value of perfect information (EVPI) and value of stochastic solution (VSS) describe the main benefits for using stochastic optimization in contrast to purely deterministic analysis.

The effect on reduction of the peak load due to DR is often analyzed, such as in [13] or [17]. Typical metrics are the peak to average, peak to valley, and peak reduction ratio. These metrics are explained in more detail in [2,18].

Further, for portfolio balancing it is important to evaluate the number of imbalances and the necessary adjustments made to the original (day-ahead) planning schedule. These adjustments do occur due to the impact of uncertainty.

Whereby these metrics describe the impact of DR explicitly, also the pure performance of the optimization algorithms toward convergence, such as algorithm runtime, number of iterations, and optimality gap are necessary to consider. All metrics derived from DO techniques can be compared to a central benchmark, which solves the optimization problem using central computation methods.

5.3 The Alternating Direction Method of Multipliers Algorithm

For use cases related to DR, EV scheduling and optimal power flow, Dual Decomposition (DD), the Alternating Direction Method of Multipliers (ADMM) as well as Dantzig-Wolfe Decomposition are frequently mentioned and applied. One widely-used optimization algorithm for optimal portfolio balancing is the DO algorithm DD. In literature, several research works demonstrate a balancing of demand and supply by DD, such as in [11,19]. However, DD only works with many mathematical assumptions, such as strict convexity of the system and local objective functions and is known to be slow in convergence. For this reason, the combination of DD and the Augmented Lagrangian Method leads to the ADMM, which is applied in this chapter. This method, similarly to DD, is an iterative DO algorithm where the global system level objective is distributed into smaller subproblems, solved independently and separately by each customer of the architecture. Despite DD, ADMM only requires mathematical objective functions that are close, proper, and convex, but not necessarily strict convex [20].

In summary, Boyd et al. [20] describe the ADMM algorithm and its benefits as "an algorithm that is intended to blend the decomposability of dual ascent with the superior convergence properties of the method of multiplier." The usage of ADMM for problems related to DSM is already introduced in recent literature. For instance, the authors of [14] use the ADMM algorithm for coordinating EV charging in a distributed manner, whereas [1] schedules heat pumps within a group of buildings for contributing to a system level objective, which is the minimization of grid losses in that case. ADMM is also successfully applied to optimal power flow problems, such as in [21]. For a more detailed description on other DO techniques and further variants, such as on Proximal Message Passing ADMM, the reader is kindly referred to [15,20], which provide a good overview on the different ADMM methods.

In this chapter, the focus is on the Exchange ADMM problem formulation, which is known as a special case for economic problem statements and successfully applied to dynamic energy management in literature. According to [20], the Exchange ADMM problem formulation can be viewed as a form of optimal exchange between goods or commodities via tâtonnement or price adjustment process. In other words, a competitive market seeks solutions in the direction of a market equilibrium. Transferred to the optimal portfolio balancing problem, this means that a (shadow) price vector for the electrical power demand and generation is increased or decreased based upon the increase or decrease of electrical energy demand or supply. Mathematically, the Exchange ADMM problem is

$$\underset{x_i}{\text{minimize}} \sum_{i=1}^{N} f_i(x_i) \tag{3}$$

Subject to:

$$\sum_{i=1}^{N} x_i = 0 \tag{4}$$

where N entities, that is, customers, exchange a common good $\boldsymbol{x} = [x_1, x_2, \ldots, x_N]^T$, for example, electrical power, under fixed objectives f_i per customer. Eq. (4) is called the equilibrium constraint, which forces the sum over all goods x_i to be zero. By introducing a new function g with

$$g(z) = \begin{cases} 0, & \text{if } z = 0 \\ \infty, & \text{otherwise} \end{cases} \tag{5}$$

and a new variable $\boldsymbol{z} = [z_1, z_2, \ldots, z_N]^T$, the optimization problem in Eqs. (3), (4) can equivalently be written in a more standard ADMM form:

$$\underset{x_i, z_i}{\text{minimize}} \sum_{i=1}^{N} f_i(x_i) + g\left(\sum_{i=1}^{N} z_i\right) \tag{6}$$

Subject to:

$$x_i - z_i = 0,\ i = 1, \ldots, N \tag{7}$$

Forming the augmented Lagrangian function L_ρ for Eqs. (6), (7) yields:

$$L_\rho(\boldsymbol{x}, \boldsymbol{z}, \boldsymbol{\lambda}) = \sum_{i=1}^{N} f_i(x_i) + g\left(\sum_{i=1}^{N} z_i\right) + \boldsymbol{\lambda}^T \cdot (\boldsymbol{x} - \boldsymbol{z}) + \frac{\rho}{2} \cdot \|\boldsymbol{x} - \boldsymbol{z}\|_2^2 \tag{8}$$

In Eq. (8) variable ρ introduces the augmented Lagrangian penalty term and represents a step size parameter for the ADMM algorithm. Variable $\boldsymbol{\lambda} = [\lambda_1, \lambda_2, \ldots, \lambda_N]^T$ is a vector of Lagrange variables, that is, the dual variable and represents the shadow prices [14]. The ADMM algorithm now solves the Exchange ADMM problem formulation in Eqs. (3), (4) by iteratively solving the set of Eqs. (9)–(11), where k is the current iteration number:

$$\boldsymbol{x}^{k+1} = \underset{\boldsymbol{x}}{\operatorname{argmin}}\ L_\rho\left(\boldsymbol{x}, \boldsymbol{z}^k, \boldsymbol{\lambda}^k\right) = \underset{\boldsymbol{x}}{\operatorname{argmin}} \sum_{i=1}^{N} f_i(x_i) + \boldsymbol{\lambda}^{k^T} \cdot \left(\boldsymbol{x} - \boldsymbol{z}^k\right) + \frac{\rho}{2} \cdot \left\|\boldsymbol{x} - \boldsymbol{z}^k\right\|_2^2 \tag{9}$$

$$\begin{aligned}\boldsymbol{z}^{k+1} &= \underset{\boldsymbol{z}}{\operatorname{argmin}}\ L_\rho\left(\boldsymbol{x}^{k+1}, \boldsymbol{z}, \boldsymbol{\lambda}^k\right) \\ &= \underset{\boldsymbol{z}}{\operatorname{argmin}}\ g\left(\sum_{i=1}^{N} z_i\right) + \boldsymbol{\lambda}^{k^T} \cdot \left(\boldsymbol{x}^{k+1} - \boldsymbol{z}\right) + \frac{\rho}{2} \cdot \left\|\boldsymbol{x}^{k+1} - \boldsymbol{z}\right\|_2^2\end{aligned} \tag{10}$$

$$\boldsymbol{\lambda}^{k+1} = \underset{\boldsymbol{\lambda}}{\operatorname{argmin}}\ L_\rho\left(\boldsymbol{x}^{k+1}, \boldsymbol{z}^{k+1}, \boldsymbol{\lambda}\right) = \underset{\boldsymbol{\lambda}}{\operatorname{argmin}}\ \boldsymbol{\lambda}^{k^T} \cdot \left(\boldsymbol{x}^{k+1} - \boldsymbol{z}^{k+1}\right) \tag{11}$$

The Exchange ADMM solution can be further simplified using the KKT conditions as demonstrated in an application in Rivera et al. [14] and more generally in [20]. Doing so, the solution for the Exchange ADMM problem formulation in Eqs. (3), (4) becomes:

$$x_i^{k+1} = \underset{x_i}{\operatorname{argmin}}\ f_i(x_i) + \lambda_i^k \cdot x_i + \frac{\rho}{2} \cdot \left\|x_i - \left(x_i^k - \overline{x}^k\right)\right\|_2^2 \tag{12}$$

$$\overline{x}^{k+1} = \frac{1}{N} \cdot \sum_{i=1}^{N} x_i^{k+1} \tag{13}$$

$$\boldsymbol{\lambda}^{k+1} = \boldsymbol{\lambda}^k + \rho \cdot \overline{x}^{k+1} \tag{14}$$

Regarding the optimal portfolio balancing problem, the different variables in Eqs. (12)–(14) have a comprehensible interpretation. Variable x_i^k represents the current power schedule per customer i, whereas the average value $\overline{x}^k$ gives the average power profile for all N participating customers within each iteration step k. Thus, it is desired to let $\overline{x}^k$ move to zero to balance the overall power supply and power demand $\sum_{i=1}^{N} x_i^k$. The dual variable $\boldsymbol{\lambda}^k$ can be understood as the shadow price vector, which serves as an incentive for the participating customers. Within Eq. (12), there is no penalty against the mean.

This means that the term $x_i^k - \overline{x}^k$ would be the contribution that one participant would have, if the exchange might fail to clear. Hence, x_i implies to move the power consumption or generation in the necessitated direction, which would lead the ADMM augmented Lagrangian penalty term to be zero [20,22].

Eqs. (13), (14) must be computed by the coordinating DR aggregator, whereas the optimization problem in Eq. (12) can be solved by each customer i independently and in parallel, that is, without violating Requirement (6). Moreover, it is permitted to restrict each variable x_i by a set of constraints C_i, which could be device-specific constraints for example, physical limitations and/or user specifications. Thus, in iteration step k each participant is required to solve its individual optimization problem under its desired objective function f_i and its set of local constraints C_i according to Requirement (3) in Eqs. (15), (16).

$$\underset{x_i}{\operatorname{argmin}}\ f_i(x_i) + \lambda_i^k \cdot x_i + \frac{\rho}{2} \cdot \left\| x_i - \left(x_i^k - \overline{x}^k \right) \right\|_2^2 \tag{15}$$

Subject to:

$$x_i \in C_i \ \ \forall i \in N \tag{16}$$

In order to fulfill Requirement (5), after each iteration step k the stopping criterion, adopted from [20], defined in Eqs. (17), (18) should be evaluated by the DR aggregator. Expression (17) represents the primal residual r^k and expression (18) the dual residual $\boldsymbol{s}^k = [s_1^k, s_2^k, \ldots, s_N^k]^T$, where $\varepsilon^{\text{primal}} > 0$ and $\varepsilon^{dual} > 0$ is the primal respectively dual tolerance. Only if both conditions (17) and (18) are fulfilled at the same time is the stopping criterion is reached, which means the ADMM algorithm converged to the optimal or at least to the near-optimal solution of Eqs. (3), (4) for sufficient small ε^{primal} and ε^{dual}. Furthermore, it holds that the optimal solution is obtained for $k \rightarrow \infty$ in any case. [20]

$$\left\| r^k \right\|_2 = \left\| \overline{x}^{k+1} \right\|_2 \leq \varepsilon^{primal} \tag{17}$$

$$\left\| \boldsymbol{s}^k \right\|_2 = \left\| -\rho \cdot N \cdot \left(\boldsymbol{x}^{k+1} - \boldsymbol{x}^k + \left(\overline{x}^k - \overline{x}^{k+1} \right) \cdot \mathbf{1} \right) \right\|_2 \leq \varepsilon^{dual} \tag{18}$$

For further information on the general Exchange ADMM working principle and to achieve a better understanding of the iterative algorithm design, the reader is kindly referred to the simple use case example provided in the Appendix of this chapter.

6 UNCERTAINTY IN DSM

The efficient coordination of electrical power and consumption devices for balancing demand and supply across temporal scales is highly affected by uncertainty. Therefore, this section introduces relevant sources of uncertainty and techniques to address the challenges of uncertainty in DSM and is primarily based upon the work published in [2]. The impact of uncertainty is explained according to the standard form of a linear problem, given in Eqs. (19)–(21) with $\boldsymbol{x} \in \mathbb{R}^n$, $A \in \mathbb{R}^{m \times n}$, $m \leq n$, $\boldsymbol{b} \in \mathbb{R}^m$ and $\boldsymbol{c} \in \mathbb{R}^n$:

$$\underset{\boldsymbol{x}}{\text{minimize}}\ \boldsymbol{c}^T \boldsymbol{x} \tag{19}$$

Subject to:

$$A\boldsymbol{x} = \boldsymbol{b} \tag{20}$$

$$\boldsymbol{x} \geq \boldsymbol{0} \tag{21}$$

Due to difficulties in weather forecasting, such as wind speed and solar irradiation, the electrical power generation output of renewable energy sources, such as PV units or wind turbines, is highly uncertain and influences the coefficients in the matrix A. Further, the energy demand (electrical/thermal) depends on certain random factors, such as outdoor temperature, user behavior etc. and hence, affects coefficients in the vector $\boldsymbol{b}$. Additionally, electrical prices, such as day-ahead, intraday prices or fuel costs, are uncertain and influence the electrical cost vector $\boldsymbol{c}$ within the objective function [23].

The variety of options that exist to consider uncertainty in the scheduling of flexible devices for DR can be separated into three main categories. These are Model Predictive Control (MPC), Stochastic Optimization (SO) and Robust Optimization (RO). There are overlaps and these techniques are often combined, such as a SO with MPC, as done, for instance, in [24]. The following section gives a very brief introduction in these three concepts.

6.1 Model Predictive Control

MPC is a control algorithm based upon a moving or so-called rolling horizon approach, which is well described in Agachi et al. [25]. The main idea of MPC is in the robustification of physical control processes under uncertainty carried-out in an online manner. For example, such a physical control process could represent a DR power scheduling application: An optimal DR power schedule is determined for a desired day-ahead optimization scenario with optimization horizon $\tau = [0, T]$ and based upon required parameter estimations, for example, weather forecast, at an initial point in time $t_0 < 0$. Therefore, the power schedule can be seen as an optimal trajectory for this specific DR power scheduling application. However, because all parameter estimations are subject to uncertainty, at time $t_k \geq 0$ the real power schedule might differ from the optimal trajectory. To capture this deviation in terms of MPC, at time t_k another optimization is performed taking into account the next p future timeslots, that is, the MPC optimization horizon is set to $\tau_{MPC} = [k, k+p]$, where p denotes the MPC step sizes. Consequently, the deviation is likely to get compensated due to the new optimization but with identical objective, so that the real DR power schedule follows the optimal trajectory again at its best in near future time steps. However, because the MPC procedure must be repeated after each p time slots, the tradeoff for applying MPC might result in high computational effort [26].

An efficient MPC application to DSM and DR is for instance provided in [27].

6.2 Stochastic Optimization

Conejo et al. [28] provide a comprehensive overview on techniques for decision-making under uncertainty in electricity markets, whereby the focus is on SO, including risk management. The portfolio coordination explained in this chapter cannot be solved under perfect information as a deterministic problem, given the sources of uncertainty that were explained before. Often, these uncertain input data are given by probability distribution functions. This motivates the usage of SO, in which the uncertain input data are modeled as random variables. These random variables are represented by a finite set of scenarios, as for example well explained in [28]. Typically, a random variable σ can be characterized by $\sigma(w)$, $w = 1, \ldots, N_\Omega$, whereby N_Ω represents the total number of scenarios, with the scenario index w and the set of scenarios Ω. Thus, σ_Ω is the set of possible scenarios with the random variable σ, which leads to $\sigma_\Omega = \{\sigma(1), \ldots, \sigma(N_\Omega)\}$. In SO, the uncertainty is addressed because each realization of the random variable $\sigma(w)$ is given with a probability of occurrence $\pi(w)$ [28]:

$$\pi(w) = P(w | \sigma = \sigma(w)), \text{ where} \sum_{w \in \Omega} \pi(w) = 1 \tag{22}$$

For most problems related to energy management, the random variables are treated as stochastic processes. For example, if σ represents the outdoor temperature for the next day, $\sigma(w)$ represents a 24×1 vector (hourly resolution) with one possible realization of the outdoor temperature for the next day. For further information, and in particular to detailed tutorials in scenario generation and reduction techniques, the reader is referred to [28]. A two-stage SO for DR under uncertainty using the flexibility of electro-thermal heating systems is for example given in [16], whereby the work of Conejo et al. [28] or [29] provide relevant examples that compare and link two- or multistage optimization to market clearing problems. The major benefit of SO is in the realistic consideration of uncertain parameters, treated as random variables, though the computational effort increases significantly.

This becomes even more challenging if the SO is linked with MPC. In such cases, decomposition methods are promising solutions to improve the computational performance. The problem is decomposed toward scenarios, so that many scenarios can be solved in parallel. Zhu et al. [24] provides a concept for decomposing scenarios in stochastic MPC for the optimal scheduling of energy storages and generation units.

6.3 Robust Optimization

The application of RO limits the ability to achieve the best theoretical optimum but decreases the probability to be affected by uncertainty. However, RO techniques are generally seen as more conservative than SO algorithms. Therefore, this chapter refers to the methodology of Bertsimas et al. [30] known as "price of robustness," which allows adjustment of the optimization problem between the deterministic, that is, nonrobust, and the most conservative approach of Soyster et al. [31].

In the following section, this is applied to a DR use case. The concept of the "price of robustness" is explained before based upon the information in [30] and the application for DR in [2]. The "price of robustness" is described, based upon the linear optimization problem formulation in Eqs. (23)–(25) with $\boldsymbol{x} \in \mathbb{R}^n$, $A \in \mathbb{R}^{m \times n}$, $m \leq n$, $\boldsymbol{b} \in \mathbb{R}^m$ and $\boldsymbol{c} \in \mathbb{R}^n$:

$$\underset{x}{\text{minimize}}\ \boldsymbol{c}^T \boldsymbol{x} \tag{23}$$

Subject to:

$$\sum_{j \epsilon J_i} a_{ij} x_j + \beta_i(x, \Gamma_i) \leq b_i \ \ \forall i \tag{24}$$

$$\beta_i(x, \Gamma_i) = \max_{\{S_i \subseteq J_i \,:\, |S_i| = \lfloor \Gamma_i \rfloor\}} \sum_{j \in S_i} \hat{a}_{ij} x_j \ \ \forall i \tag{25}$$

The uncertainty in the coefficient $\tilde{a}_{ij}$ in the matrix A that contains m rows and n columns can deviate within the interval $[a_{ij} - \hat{a}_{ij}, a_{ij} + \hat{a}_{ij}]$. The set of uncertain coefficients is denoted by $\tilde{a}_{ij}, j \epsilon J_i$. The main idea behind this formulation for the "price of robustness" is that it is very unlikely that all uncertain coefficients will switch to their worst-case value at the exact same time step. Therefore, the introduced parameter Γ_i allows to define the level of robustness. The values of Γ_i can vary between $[0, |J_i|]$ for every i and, in case Γ_i is chosen as an integer value, allow to protect the ith constraint of matrix A against uncertainty using the protection parameter β_i. In other words, the cardinality $\lfloor \Gamma_i \rfloor$ defines the level of considered robustness and conservativeness in the optimization. If $\lfloor \Gamma_i \rfloor = |J_i|$, all uncertain parameters in each row of the matrix are protected against uncertainty, which is the most conservative RO approach of Soyster et al. [31]. However, if $\lfloor \Gamma_i \rfloor = 0$, the optimization would result in the deterministic case without considering any robustness. In other words, the adjustment of Γ_i allows a trade-off between risk and conservativeness in the optimization [2,30].

This method has been used for DR scheduling purposes successfully in literature, such as in [2,32]. For a very good overview on state-of-the-art RO techniques, the reader is referred to the dissertation of Degel [33].

7 DISTRIBUTED OPTIMIZATION FOR THE DEPLOYMENT OF FLEXIBILITY UNDER UNCERTAINTY: A USE CASE

7.1 Use Case Conditions

This section demonstrates the benefits of DO for the deployment of flexibility under uncertainty using the Exchange ADMM algorithm and is based upon the work published in [2,22]. Therefore, a day-ahead DR scheduling use case simulation is applied to a small city district, which comprises a distributed architecture as described in Section 4. Fig. 4 shows the overall Exchange ADMM algorithm working sequence applied to the hierarchical architecture, as already presented in Fig. 3.

The city district consists, on the one hand, of one DR aggregator and, on the other hand, of 50 residential customers each represented by a single-family house (SFH) containing an electro-thermal heating unit plus TES and a PV unit. In order to assess a broad and heterogeneous choice of SFHs, the SFHs in this use case simulation are modeled based upon the data provided by the EU project TABULA [34]. TABULA represents national building typologies and characterizes typical, in this case, German SFH including their specific energy need for thermal heating. Using historical weather data for the calendar year 2010 from [35] and standard load profiles from [36], the thermal load curve per SFH is calculated. Furthermore, PV generation profiles are estimated based upon the method provided in [37], which is again using the weather data from [35], and taking into account the applicable roof area per SFH for installing PV units given in [34]. The electro-thermal heating units, which are air-to-air EHPs

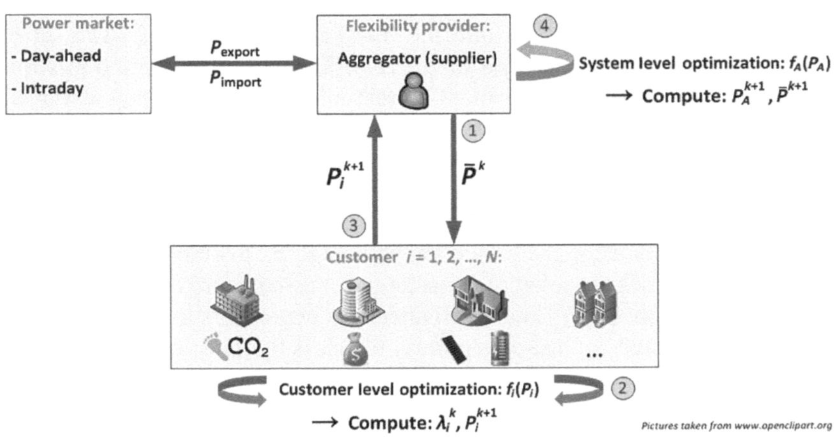

FIG. 4 Exchange ADMM applied in the hierarchical architecture. *(Based on M. Diekerhof, F. Peterssen, A. Monti, Hierarchical distributed robust optimization for demand response services. IEEE Trans. Smart Grid, (2017) 1; Smart Grid Task Force, Regulatory recommendations for the deployment of flexibility: Sgtf-Eg3 Report, Http://Ec.Europa.Eu/Energy/Sites/Ener/Files/Documents/Eg3%20final%20-%20january%202015.Pdf, 2015.)*

for this use case are modeled based upon the data from [38] and are dimensioned as monovalent heating systems. This means that their nominal thermal power generation is capable of covering the corresponding SFH's maximum thermal heat demand at any time, meaning thermal comfort is always guaranteed to the user. Then, corresponding TES is sized twice the EHP's thermal energy generation per hour. This allows the EHP to be turned off for at least 2 h while the stored thermal energy within the associated TES will cover the SFH's heating demand. Thus, electro-thermal heating systems equipped with TES are the flexibility resources within this example.

For applying the ADMM algorithm to this specific use case scenario, the ADMM Exchange problem formulation in Eqs. (3), (4) is adapted. Both the local objectives of all customers and the global objective for the DR aggregator are in the minimization of energy costs, meaning a pure cost optimization is applied. However, the difference in the costs per timeslot for the customers is in the usage of time-of-use tariffs from [39], which are scaled/normed to match the German mean energy price, compared to the DR aggregator, who wants to minimize the costs for power import and export based on historical day-ahead price data from [7] instead. Other individual convex customer objectives, such as CO_2-minimization are realizable as well, but not further considered in this example. This yields the use case specific Exchange ADMM optimization problem formulation in Eqs. (26)–(28):

$$\underset{\boldsymbol{P}_i,\boldsymbol{P}_a}{\text{minimize}} \sum_{i=1}^{N} [f_i(\boldsymbol{P}_i)] + f_a(\boldsymbol{P}_a) \tag{26}$$

Subject to:

$$\sum_{i=1}^{N} \boldsymbol{P}_i + \boldsymbol{P}_a = \boldsymbol{0} \tag{27}$$

$$\boldsymbol{P}_i \in C_i \ \forall i \tag{28}$$

The dimension of all vectors in Eqs. (26)–(28) corresponds to the total amount of discrete time steps t within the desired optimization horizon τ. Because a 24 h day-ahead scheduling is applied in this use case, $T=96$ is set, that is, there exist 96 discrete time slots t of length $\Delta t = 15$ min each.

In Eq. (26), $f_i(\boldsymbol{P}_i) = \boldsymbol{c}_{tou}^T \cdot \boldsymbol{P}_i$ is the objective function per customer i and $f_a(\boldsymbol{P}_a) = \boldsymbol{c}_{da}^T \cdot \boldsymbol{P}_a$ the objective function of the DR aggregator. In this context, variable $\boldsymbol{c}_{tou}$ is a vector of time-of-use prices, whereas variable $\boldsymbol{c}_{da}$ is a vector of day-ahead prices. Constraint (27) represents the coupling constraint and ensures a balance between power demand and power supply on system level. In this regard, vector $\boldsymbol{P}_a$ represents the power profile of the DR aggregator, meaning power imports or exports, on the energy market that become necessary to avoid power imbalance situations. In Eq. (28), the set of constraints C_i described through Eqs. (29)–(33) that restricts the power profile $\boldsymbol{P}_i$ per customer

i comprises both the physical constraints of the EHP plus TES model as well as the physical constraints of the PV unit model.

$$\boldsymbol{P}_i = \boldsymbol{P}_{El.EHP,i} + \boldsymbol{P}_{El.PV,i} \quad \forall i \tag{29}$$

$$\boldsymbol{P}_{El.PV,i} = -\boldsymbol{P}_{El.PV.Forecast,i} \quad \forall i \tag{30}$$

$$\boldsymbol{0} \leq \boldsymbol{P}_{El.EHP,i} \cdot \boldsymbol{COP}_i \leq \boldsymbol{P}_{Th.EHP.Max,i} \quad \forall i \tag{31}$$

$$\boldsymbol{0} \leq \boldsymbol{E}_{Th.Storage,i} \leq \boldsymbol{E}_{Th.Storage.Max,i} \quad \forall i \tag{32}$$

$$E^t_{Th.Storage,i} = E^{t-1}_{Th.Storage,i} + \left[P^t_{El.EHP,i} \cdot COP^t_i - P^t_{Th.SFH,i}\right] \cdot \Delta t - E^t_{Loss,i} \quad \forall t, i \tag{33}$$

In Eq. (29) the optimization variable $\boldsymbol{P}_i$ is defined as the sum of the electrical power demand from the EHP unit $\boldsymbol{P}_{El.\ EHP,\ i}$ and the PV generation unit $\boldsymbol{P}_{El.PV,i}$ per customer/SFH i. Because curtailment of renewable energy is not considered in this example, a PV unit is assumed to be an inflexible device. For the entire optimization horizon τ the forecast PV generation profile $\boldsymbol{P}_{El.PV.Forecast,i}$ is assigned to variable $\boldsymbol{P}_{El.PV,i}$ in Eq. (30). The negative sign in Eq. (30) represents power generation at this point. A surplus of renewable energy is assumed to be exported by the DR aggregator to the energy market. Conversely, any shortage of power supply is assumed to be imported by the DR aggregator from the energy market. Eq. (31) defines the physical operating range of the EHP. The electrical power consumption of the EHP is bounded by zero and by the maximum thermal heating power generation $\boldsymbol{P}_{Th.EHP.Max,i}$. In this connection, variable COP_i defines a coefficient of performance for the conversion between thermal and electrical power, which is dynamically calculated based upon the work in [40]. In the same manner, Eq. (32) defines the physical limits of the TES, which is bounded by zero and by a maximum energy capacity $\boldsymbol{E}_{Th.Storage.Max,i}$ comparably. Eventually Eq. (33) defines the energy balance within each TES i and for all discrete time slots t, where variable $P^t_{Th.SFH,i}$ represents the required thermal heating demand of SFH i and variable $E^t_{Loss,i}$ the amount of total energy losses for the entire electro-thermal heating system of SFH i.

For the sake of exemplification only, the required thermal heating demand $\boldsymbol{P}_{Th.SFH,i}$ per SFH i is considered uncertain due to weather forecast deviations in this use case. This means that the uncertain thermal heating demand $\boldsymbol{P}_{Th.SFH,i}$ per SFH i is assumed to be lognormal-distributed $L(\mu, \sigma^2)$ with mean $\mu = 0$ and standard deviation $\sigma = 0.25$ in relation to the estimated thermal heating demand forecast. Furthermore, the uncertainty modeling assumes that the thermal demand uncertainty is autocorrelated over the entire optimization horizon. This means that the "real" thermal demand is either higher (because of a lower outside temperature) or lower (because of a higher outside temperature) than expected beforehand for every single time step t.

In this work, the Exchange ADMM algorithm is robustified by applying RO combined with an MPC strategy as described above. For the RO algorithm $\hat{a}_{lj} = 0.2$ is set and $\Gamma_l \in [0, 5, 10, 15, 20, 25, 30]$ is varied for evaluation purposes.

Moreover, for the MPC algorithm different step sizes are chosen in order to analyze their impact on the degree of robustness. In this connection, all results and metrics are determined through the application of a Monte-Carlo analysis. This means that n different day-ahead scheduling simulations are performed, whereupon for each simulation run and for each SFH i a "real" uncertain thermal heating demand profile $\boldsymbol{P}_{Th.SFH,i}$, other than the forecasted one, is generated via the underlying lognormal distribution $L(\mu=0,\sigma^2=0.25^2)$. Then, the power adjustments $\boldsymbol{P}_{El.Adj,i}$ per SFH i after being affected with the thermal heating demand uncertainty are evaluated per simulation and finally averaged over all simulation runs. Therefore, the Monte-Carlo analysis requires the additional constraint (34) besides Eqs. (29)–(33), which only holds for simulation and evaluation purposes to ensure that physical limitations such as the maximum TES capacity are not violated.

$$\boldsymbol{P}_{El.EHP,i}=\boldsymbol{P}^{opt}_{El.EHP,i}+\boldsymbol{P}_{El.Adj,i}\ \forall i \tag{34}$$

In constraint (34), variable $\boldsymbol{P}^{opt}_{El.EHP,i}$ is the optimal power schedule of the EHP obtained by the optimization for the estimated thermal heating power demand. Thus, for a positive power value of $P^t_{El.Adj,i}$, the EHP consumed a higher amount of power compared to $P^{opt,t}_{El.EHP,i}$, whereas for a negative power value of $P^t_{El.Adj,i}$, the EHP consumed a lower amount of power compared to $P^{opt,t}_{El.EHP,i}$ in time step t.

7.2 Use Case Results

The day-ahead scheduling applied in this use case focuses on a typical winter day within the annual heating period according to the German Guideline of [41] in order to access the flexibility potential provided through the TES of the SFHs within the city district.

First, to analyze the performance of the distributed Exchange ADMM algorithm for this winter day, it is focused on the deterministic, that is, nonrobust, case $\Gamma_l=0$, without application of MPC. Findings in previous work [2,22] successfully demonstrated that ADMM yields near-optimal results in comparison to different (central) benchmark algorithms with good convergence results. However, the ADMM convergence rate strongly depends on the value of the step size parameter ρ as well as on the chosen primal and dual tolerances. This is shown in Fig. 5, where the day-ahead scheduling for the considered winter day is performed via Exchange ADMM for different step sizes ρ and small tolerances $\varepsilon^{primal}=0.1$ and $\varepsilon^{dual}=1.0$. In Fig. 5, the required amount of iterations, before the Exchange ADMM algorithm terminates, strongly depends on the value of the step size parameter ρ and therefore varies in-between 100s to 1000s. For this use case, it becomes evident that a step size of $\rho=0.1$ is beneficial. However, the concrete choice of the step size parameter ρ highly depends on the analyzed problem/scenario and must be determined empirically in most

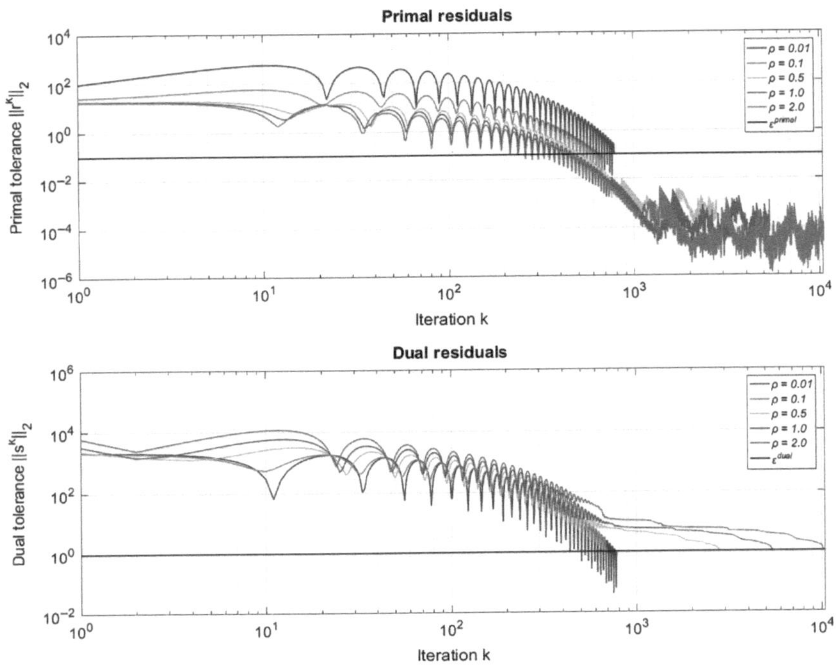

FIG. 5 Exchange ADMM convergence rate for different step sizes $\boldsymbol{\rho}$.

cases. This analysis is limited to constant step sizes $\rho^k = \rho$ for all iterations k. However, the literature, such as the work in [20], also provide methods that allow a dynamical adjustment of the step size parameter ρ^k per iteration k in order to improve the overall convergence speed.

The application of RO is also evaluated. Therefore, Fig. 6 shows the adjustment energy of the city district for the considered winter day, that is, the total electrical energy difference compared to the expected energy schedule due to the uncertainty in the thermal heating demand forecast $\boldsymbol{E}_{El.Adj,l} = \Delta t \cdot \left(\sum_{i \in N} \sum_{t \in \tau} P^t_{El.Adj,i}\right)$, for $\Gamma_l \in [0, 5, 10, 15, 20, 25, 30]$ and MPC step sizes 8, 16, 24, 48, and 96. At this point, it should be remarked that robustness values $\Gamma_l \geq 30$ are usually not advantageous, because such a high degree of robustness cannot be provided by the finite TES capacity. Instead, the TES state of charge would be kept on its half to capture both maximum positive respectively maximum negative adjustments for all time slots t. In Fig. 6, for small Γ_l and long MPC step sizes the adjustment energy of the city district increases. This behavior is counterintuitive, because of the following two reasons. First, the more robust the day-ahead schedule of the Exchange ADMM is, the less adjustment energy is required, because slight changes in the thermal heating demand forecasts are captured by the robust TES schedule.

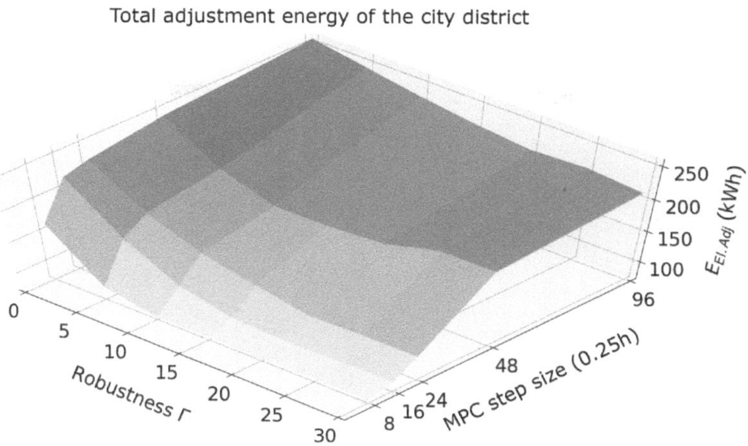

FIG. 6 Total adjustment energy of the city district.

Especially for the deterministic case, that is, for the nonrobust case $\Gamma_l=0$, it is evident that the highest adjustment energy is obtained. In comparison, a robustness value of Γ_l=30 would decrease the adjustment energy significantly. Second, the smaller the MPC step size is chosen, the less is the deviation propagation from the expected day-ahead schedule $\sum_{i \in N} \boldsymbol{P}^{opt}_{El.EHP,i}$ with ongoing time such that changes in the thermal heating demand forecasts are captured successively.

However, the application of RO also holds a clear trade-off, which is pictured in Fig. 7. This figure shows the energy costs of the city district including all customers plus the DR aggregator but without the costs for the adjustment energy. In contrast to Fig. 6, for high Γ_l-values and long MPC step sizes the total energy costs of the city district increase. The higher the robustness Γ_l is chosen, the higher are the total energy costs of the city district. Looking at the deterministic case again, that is, $\Gamma_l=0$, it is evident that the total energy costs are minimal, whereas for $\Gamma_l=30$ the costs are maximal. The main reason for this is because the increasing robustness decreases the option to import and export energy during rather attractive, that is, cheap time steps t because the objective function of cost minimization is bounded by the robustness.

Nonetheless, short MPC step sizes decrease the energy costs in turn. For this reason and based upon the previous result of low-energy adjustments for small MPC step sizes, the application of MPC with sufficient small MPC step sizes becomes reasonable in the majority of applications. Despite the trade-off described above, this usually holds for the application of RO as well. This is motivated by the fact that the DR aggregator must compensate for the adjustment energy in practice, which would usually (depending on the concrete tariff for adjustment energy compensation), lead to higher a-posteriori operation costs

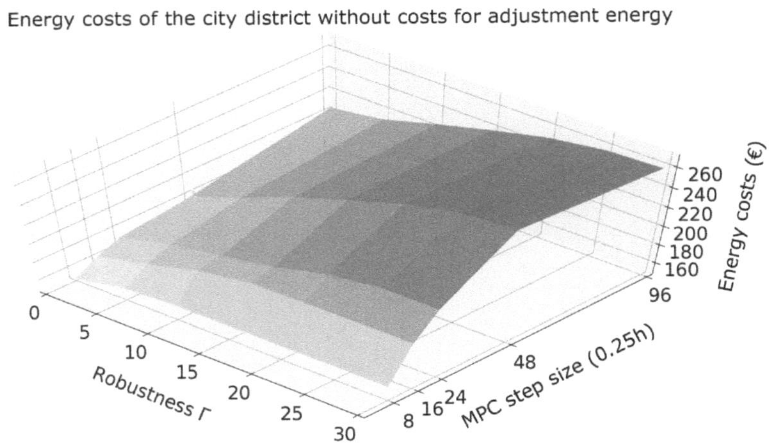

FIG. 7 Energy costs of the city district without costs for adjustment energy.

of the whole city district compared to the application of RO, involving its necessitated additional operation costs already in advance. In a conclusion, the application of both RO combined with a MPC strategy could form a fundamental pillar for an efficient and robust DSM in future.

8 CONCLUSION

This chapter provides an overview on recent challenges and applications in demand side management for city districts. Relevant requirements for the efficient deployment of flexibility are discussed and lead to the application of a state-of-the-art architecture and distributed optimization techniques. The work links relevant flexibility user, such as the balance responsible party or distribution system operator, to flexibility provider in an architecture which is recommended by the Smart Grids Task Force of the European Commission. This chapter introduces and discusses the standard portfolio coordination optimization problem, which is one of the main demand side flexibility use cases for demand response. Although multiple flexibility sources are introduced, the application focuses on flexibility from electro-thermal heating systems equipped with thermal storage due to their growing value in residential sectors. The first design choice from the hierarchical architecture and other requirements is distributed optimization. Therefore, an example application of the alternating direction method of multipliers proves how an aggregation service provider can successfully balance demand and supply across temporal scales.

Furthermore, derived from the requirements, uncertainty must be considered within demand response applications. In this work, model predictive control and robust optimization yield a reduction in energy deviations from the offline (day-ahead) and beforehand scheduled planning of flexible devices compared to the by uncertainty affected intraday operation. However, the application also shows that consideration of uncertainty always comes with a certain cost, such as computational intensive calculations and increasing energy costs. The latter, though, depends on the value of charged imbalance/adjustment cost. The additional adjustment costs could overwhelm the total energy costs of robust optimization and make the robust optimization with model predictive control an appealing strategy to reduce the total energy costs including adjustments.

APPENDIX EXCHANGE ADMM USE CASE EXAMPLE

To demonstrate the general working principal of the Exchange ADMM algorithm in more detail, in the following a simple use-case example is provided. Consider a basic distributed architecture as described in Section 4, consisting of one DR aggregator and two customers A and B. For sake of exemplification, assume that customer A possess a curtailable load and customer B a PV unit solely. A PV unit is assumed to be an inflexible power generation unit because curtailment of renewable generation units is not considered in this application. A surplus of renewable energy would have to be exported by the DR aggregator. A curtailable load offers flexibility potential as already explained in Section 2. Based on those assumptions, Eqs. (35)–(38) show fundamental mathematical optimization model realizations for the curtailable load respectively the PV unit.

$$\underset{P_A}{\text{minimize}}\, f_A(P_A) = P_{Nom} - P_A \tag{35}$$

$$\underset{P_B}{\text{minimize}}\, f_B(P_B) = P_B \tag{36}$$

Subject to:

$$P_{Min} \leq P_A \leq P_{Nom} \tag{37}$$

$$P_B = -P_{PV} \tag{38}$$

The objective function f_A for customer A is in the avoidance of curtailment from its nominal power consumption P_{Nom}. The optimization variable P_A represents the real power the curtailable load consumes. Because variable P_A is bounded by $[P_{Min}, P_{Nom}]$, the minimum power consumption of the curtailable load is P_{Min} and the maximum power consumption P_{Nom}. The objective function f_B for customer B is in the maximization of its PV power generation P_B. Because power generation is usually represented by a negative sign, this maximization intention is completely equivalent to the minimization of a negative value for P_B in a mathematical context. However, because the PV unit is assumed to be inflexible, the real power generation P_B is fixed to the constant PV power generation P_{PV} (for one arbitrary point in time) anyway. The objective of the DR aggregator is in the balancing of power demand and supply on system level to avoid power imbalances.

For this example, assume $P_{Min}=4$ kW, $P_{Nom}=6$ kW and $P_{PV}=5$ kW. Then, by ignoring the physical units due to readability, the optimization problems in Eqs. (35)–(38) become:

$$\underset{P_A}{\text{minimize}}\, f_A(P_A)=6-P_A \tag{39}$$

$$\underset{P_B}{\text{minimize}}\, f_B(P_B)=P_B \tag{40}$$

Subject to:

$$4\le P_A\le 6 \tag{41}$$

$$P_B=-5 \tag{42}$$

It is obvious that the optimal solutions for these optimization problems are $P_A=P_{Nom}=6$ kW and $P_B=-P_{PV}=-5$ kW, respectively. However, on system level, the power imbalance would result in $P_A+P_B=6$ kW -5 kW $=1$ kW then, which is not intended by the DR aggregator. Instead an operation in curtailment of $P_A=5$ kW for the curtailable load of customer A would lead to zero power imbalances on system level, which is the optimal result for this example. From a system level perspective, this optimal result is obtained by applying the Exchange ADMM algorithm. In this case, the Exchange ADMM problem formulation is:

$$\underset{P_A,P_B}{\text{minimize}}\, f_A(P_A)+f_B(P_B) \tag{43}$$

Subject to:

$$P_A+P_B=0 \tag{44}$$

Using Eqs. (15), (16) the optimization problems to solve for customers A and B in iteration step k are:

$$P_A^{k+1}=\underset{P_A}{\text{argmin}}\ f_A(P_A)+\lambda_A^k\cdot P_A+\frac{\rho}{2}\cdot\left\|P_A-\left(P_A^k-\overline{P}^k\right)\right\|_2^2 \tag{45}$$

$$P_B^{k+1}=\underset{P_B}{\text{argmin}}\ f_B(P_B)+\lambda_B^k\cdot P_B+\frac{\rho}{2}\cdot\left\|P_B-\left(P_B^k-\overline{P}^k\right)\right\|_2^2 \tag{46}$$

Subject to:

$$4\le P_A\le 6 \tag{47}$$

$$P_B=-5 \tag{48}$$

For sake of exemplification $\varepsilon^{primal}=0$ and $\varepsilon^{dual}=1$ are chosen and the penalty parameter is set to $\rho=1$. At this point it should be pointed out that the concrete choice of the penalty parameter ρ is in general highly problem-related. Nonetheless, literature offers efficient methods on how to determine parameter ρ dynamically in order to improve the convergence speed of the ADMM algorithm as depicted for instance in [14,20]. Additionally, based upon Eqs. (13), (14), all parameters are initialized with zero before the first iteration, that is, $P_A^1 = P_B^1 = \lambda_A^1 = \lambda_B^1 = \overline{P}^1 = 0$, albeit different values might be reasonable as well. This would represent a kind of a-priori adjustment.

Now for the first ADMM iteration, that is, $k=1$, customers A and B compute P_A^2 and P_B^2 in Eqs. (49), (50) by considering their individual objective functions and constraints in Eqs. (45)–(48):

$$\begin{aligned} P_A^2 &= \underset{P_A}{\text{argmin}}\ f_A(P_A) + \lambda_A^1 \cdot P_A + \frac{\rho}{2} \cdot \left\| P_A - \left(P_A^1 - \overline{P}^1\right) \right\|_2^2 \\ &= \underset{P_A}{\text{argmin}}\ 6 - P_A + \frac{1}{2} \cdot (P_A)^2 = 4 \end{aligned} \tag{49}$$

$$\begin{aligned} P_B^2 &= \underset{P_B}{\text{argmin}}\ f_B(P_B) + \lambda_B^1 \cdot P_B + \frac{\rho}{2} \cdot \left\| P_B - \left(P_B^1 - \overline{P}^1\right) \right\|_2^2 \\ &= \underset{P_A}{\text{argmin}} - P_B + \frac{1}{2} \cdot (P_B)^2 = -5 \end{aligned} \tag{50}$$

Then, customers A and B communicate P_A^2 and P_B^2 to the aggregator, which computes $\overline{P}^2$ in Eq. (51) and checks the stopping criterion in Eqs. (52), (53) afterwards:

$$\overline{P}^2 = \frac{1}{2} \cdot \left(P_A^2 + P_B^2\right) = \frac{1}{2} \cdot (4-5) = -\frac{1}{2} \tag{51}$$

$$\left\| r^1 \right\|_2 = \left\| \overline{P}^2 \right\|_2 = \frac{1}{2} \nleq \varepsilon^{primal} = 0 \tag{52}$$

$$\begin{aligned} \left\| s^1 \right\|_2 &= \left\| -\rho \cdot N \cdot \left(\begin{pmatrix} P_A^2 \\ P_B^2 \end{pmatrix} - \begin{pmatrix} P_A^1 \\ P_B^1 \end{pmatrix} + \left(\overline{P}^1 - \overline{P}^2\right) \cdot \begin{pmatrix} 1 \\ 1 \end{pmatrix} \right) \right\|_2 \\ &= \sqrt{162} \approx 12.728 \nleq \varepsilon^{dual} = 1 \end{aligned} \tag{53}$$

Because the stopping criterion is not fulfilled yet, the second ADMM iteration, that is, $k=2$, is performed in the same manner. Additionally, in Eq. (54) customers A and B must calculate variables λ_A^2 respectively λ_B^2 first based upon the value of $\overline{P}^2$ sent-out by the aggregator to all customers after iteration step $k=1$. However, because λ_A^1 and λ_B^1 possess the same initial value, it should be noticed that all λ_A^k and λ_B^k will be equal for this example.

$$\lambda_A^2 = \lambda_B^2 = \lambda_A^1 + \rho \cdot \overline{P}^2 = \lambda_B^1 + \rho \cdot \overline{P}^2 = 0 + 1 \cdot \left(-\frac{1}{2}\right) = -\frac{1}{2} \tag{54}$$

$$\begin{aligned} P_A^3 &= \underset{P_A}{\text{argmin}}\ f_A(P_A) + \lambda_A^2 \cdot P_A + \frac{\rho}{2} \cdot \left\| P_A - \left(P_A^2 - \overline{P}^2\right) \right\|_2^2 \\ &= \underset{P_A}{\text{argmin}}\ \frac{1}{2} \cdot (P_A)^2 - 6P_A + \frac{129}{8} = 6 \end{aligned} \tag{55}$$

$$P_B^3 = \underset{P_B}{\text{argmin}} f_B\ (P_B) + \lambda_B^2 \cdot P_B + \frac{\rho}{2} \cdot \left\| P_B - \left(P_B^2 - \overline{P}^2\right) \right\|_2^2$$
$$= \underset{P_B}{\text{argmin}}\ \frac{1}{2} \cdot (P_B)^2 + 3P_A + \frac{81}{8} = -5 \tag{56}$$

$$\overline{P}^3 = \frac{1}{2} \cdot \left(P_A^3 + P_B^3\right) = \frac{1}{2} \cdot (6-5) = \frac{1}{2} \tag{57}$$

$$\left\|r^2\right\|_2 = \left\|\overline{P}^3\right\|_2 = \frac{1}{2} \nleq \varepsilon^{primal} = 0 \tag{58}$$

$$\left\|s^2\right\|_2 = \left\| -\rho \cdot N \cdot \left(\begin{pmatrix} P_A^3 \\ P_B^3 \end{pmatrix} - \begin{pmatrix} P_A^2 \\ P_B^2 \end{pmatrix} + \left(\overline{P}^2 - \overline{P}^3\right) \cdot \begin{pmatrix} 1 \\ 1 \end{pmatrix} \right) \right\|_2$$
$$= \sqrt{8} \approx 2.828 \nleq \varepsilon^{dual} = 1 \tag{59}$$

Again, the stopping criterion is not fulfilled so that another ADMM iteration is needed. Therefore, the numerical solutions of further ADMM iteration steps are given in Table 1. The ADMM algorithm converges to the optimal solution after $k=6$ iteration steps. In other words, the optimal values for P_A and P_B are finally obtained. However, because $\varepsilon^{dual} \neq 0$ was chosen, it should be remarked that the optimal value **1** for the dual variable $\boldsymbol{\lambda}$ is still not obtained. To achieve optimality for variable $\boldsymbol{\lambda}$ as well, hundreds of additional iteration steps would be necessary. This behavior of the algorithm can be seen as one major drawback of ADMM and is analyzed in Boyd et al. [20] more precisely. In this connection, it is clearly stated by Boyd et al. [20] that the ADMM algorithm might have a very slow convergence for high accuracy solutions. However, for modest accuracy solutions, which are practically acceptable for most applications, the ADMM algorithm usually converges within a few tens of iterations. As demonstrated in this simple example, that particular behavior makes the ADMM algorithm useful for a large set of DO applications.

TABLE 1 Numerical Solutions for Different ADMM Iteration Steps *k* for the Simple Exchange ADMM Example

Iteration Step *k*	P_A^{k+1}	P_B^{k+1}	$\overline{P}^{k+1}$	$\lambda_A^{k+1}/\lambda_B^{k+1}$	$\|r^k\|_2$	$\|s^k\|_2$
1	4	−5	$-\frac{1}{2}$	$-\frac{1}{2}$	$\frac{1}{2}$	12.728
2	6	−5	$\frac{1}{2}$	$\frac{1}{2}$	$\frac{1}{2}$	2.828
3	6	−5	$\frac{1}{2}$	$\frac{1}{2}$	$\frac{1}{2}$	0
4	6	−5	$\frac{1}{2}$	1	$\frac{1}{2}$	0
5	5.5	−5	$\frac{1}{4}$	$\frac{5}{4}$	$\frac{1}{4}$	0.707
6	5	−5	0	$\frac{5}{4}$	0	0.707

REFERENCES

[1] M. Juelsgaard, Utilizing Distributed Resources in Smart Grids a Coordination Approach: A Coordination Approach (dissertation), Department of Electronic Systems and Automatin Control, Aalborg University, Denmark, 2014.

[2] M. Diekerhof, F. Peterssen, A. Monti, Hierarchical distributed robust optimization for demand response services, *IEEE Trans*. Smart Grid (2017) 1.

[3] Smart Grid Task Force, Regulatory Recommendations For The Deployment Of Flexibility: Sgtf-Eg3 Report, Http://Ec.Europa.Eu/Energy/Sites/Ener/Files/Documents/Eg3%20final%20-%20january%202015.Pdf, 2015.

[4] E. Dall'anese, P. Mancarella, A. Monti, Unlocking flexibility: integrated optimization and control of multienergy systems, IEEE Power Energy Mag. 15 (1) (2017) 43–52.

[5] P. Palensky, D. Dietrich, Demand side management: demand response, intelligent energy systems, and smart loads, IEEE Trans. Industr. Inform. 7 (3) (2011) 381–388.

[6] J.S. Vardakas, N. Zorba, C.V. Verikoukis, A survey on demand response programs in smart grids: pricing methods and optimization algorithms, IEEE Commun. Surv. Tutor. 17 (1) (2015) 152–178.

[7] *EEX—European Energy Exchange* Available from: http://www.eex.com. (Accessed 25 August 2017).

[8] M.H. Albadi, E.F. El-Saadany, A summary of demand response in electricity markets, Electr. Power Syst. Res. 78 (11) (2008) 1989–1996.

[9] C. Molitor, Residential City Districts as Flexibility Resource: Analysis, Simulation, and Decentralized Coordination Algorithms (dissertation), Insitute For Automation Of Complex Power Systems, Rwth Aachen, Aachen, 2015.

[10] M. Kraning, E. Chu, J. Lavaei, S. Boyd, Message passing for dynamic network energy management, Found. Trends Optimiz 1 (2012).

[11] M. Diekerhof, S. Schwarz, F. Martin, A. Monti, Distributed optimization for scheduling electrical demand in complex city districts, IEEE Syst. J. (2017) 1–12.

[12] A. Kargarian, et al., Toward distributed/decentralized DC optimal power flow implementation in future electric power systems, IEEE Trans. Smart Grid (2016) 1.

[13] S. Kolen, C. Molitor, L. Wagner, A. Monti, Two-level agent-based scheduling for a cluster of heating systems, Sustain. Cities Soc. 30 (2017) 273–281.

[14] J. Rivera, C. Goebel, H.A. Jacobsen, Distributed convex optimization for electric vehicle aggregators, IEEE Trans. Smart Grid 8 (4) (2017) 1852–1863.

[15] D.K. Molzahn, et al., A survey of distributed optimization and control algorithms for electric power systems, IEEE Trans. Smart Grid 8 (6) (2017) 2941–2962.

[16] N. Good, E. Karangelos, A. Navarro-Espinosa, P. Mancarella, Optimization under uncertainty of thermal storage-based flexible demand response with quantification of residential users' discomfort, IEEE Trans. Smart Grid 6 (5) (2015) 2333–2342.

[17] P. Mcnamara, S. Mcloone, Hierarchical demand response for peak minimization using Dantzig–Wolfe decomposition, IEEE Trans. Smart Grid 6 (6) (2015) 2807–2815.

[18] F. Salazar, F. Martin, M. Hormigo, Ideal Grid for All—Deliverable D7.1: KPI Definition, 2014.

[19] B. Moradzadeh, K. Tomsovic, Two-stage residential energy management considering network operational constraints, IEEE Trans. Smart Grid 4 (4) (2013) 2339–2346.

[20] S. Boyd, Distributed optimization and statistical learning via the alternating direction method of multipliers, Fnt Mach. Learn. 3 (1) (2010) 1–122.

[21] M. Bazrafshan, N. Gatsis, Decentralized stochastic optimal power flow in radial networks with distributed generation, IEEE Trans. Smart Grid 8 (2) (2017) 787–801.

[22] M. Diekerhof, S. Schwarz, A. Monti, Distributed optimization for electro-thermal heating units, IEEE PES Innovative Smart Grid Technologies, Europe, October, 9–12, Ljubljana: *Proceedings*, Ljubljana, Slovenia, 2016, pp. 1–6.

[23] N. Heitmann, Modellierung Von Investitionsentscheidungen Und Kraftwerkseinsatzplanung Unter Unsicherheit Mittels Stochastischer Optimierung Und Multi-Agenten-Ansatz Am Beispiel Des Deutschen Strommarktes (dissertation), Lehrstuhl Für Diskrete Mathematik, Optimierung Und Operations Research, Universität Augsburg, 2010.

[24] D. Zhu, G. Hug, Decomposed stochastic model predictive control for optimal dispatch of storage and generation, IEEE Trans. Smart Grid 5 (4) (2014) 2044–2053.

[25] P.Ş. Agachi, Model Based Control: Case Studies in Process Engineering, Wiley-Vch Verlag & Co, Weinheim, 2010.

[26] E.F. Camacho, C. Bordons, Model Predictive Control, Springer, London, 2007.

[27] S.H. Kim, Building demand-side control using thermal energy storage under uncertainty: an adaptive multiple model-based predictive control (Mmpc) approach, Build. Environ. 67 (2013) 111–128.

[28] A.J. Conejo, M. Carrión, J.M. Morales, Decision Making Under Uncertainty In Electricity Markets, Springer Science+Business Media LLC, Boston, MA, 2010.

[29] F. Abbaspourtorbati, A. Conejo, J. Wang, R. Cherkaoui, Three- or two-stage stochastic market-clearing algorithm? IEEE Trans. Power Syst. (2016) 99.

[30] D. Bertsimas, M. Sim, The price of robustness, Oper. Res. 52 (1) (2004) 35–53.

[31] A.L. Soyster, Technical note—convex programming with set-inclusive constraints and applications to inexact linear programming, Oper. Res. 21 (5) (1973) 1154–1157.

[32] E. Kuznetsova, C. Ruiz, Y.-F. Li, E. Zio, Analysis of robust optimization for decentralized microgrid energy management under uncertainty, Int. J. Electr. Power Energy Syst. 64 (2015) 815–832.

[33] D. Degel, Innovative Location Optimization for Rescue and Emergency Medical Services Adapting to a ... Dynamic Environment (dissertation), Logos Verlag Berlin, Berlin, 2015. Bochum University.

[34] *National Building Typologies: IEE projects episcope and Tabula.* Available from: http://episcope.eu/building-typology/. (Accessed 18 August 2015).

[35] *Bundesinstitut für Bau-, Stadt- und Raumforschung—Download Testreferenzjahr-Datensatz 2011.* Available from: http://www.bbsr-energieeinsparung.de/EnEVPortal/DE/Regelungen/Testreferenzjahre/Testreferenzjahre/TRY2011_Datensatz1_1.zip. (Accessed 3 April 2018).

[36] M. Hellwig, Entwicklung Und Anwendung Parametrisierter Standard-Lastprofile, Dissertation, Fakultät Für Elektrotechnik Und Informationstechnik, Lehrstuhl Am Institut Für Energietechnik, Tu München, München, 2003.

[37] W. Zhou, H. Yang, Z. Fang, A novel model for photovoltaic array performance prediction, Appl. Energy 84 (12) (2007) 1187–1198.

[38] *Dimplex—Heat Pumps.* Available from: Http://Www.Dimplex.De/En/Downloads/Product-Publications/Heat-Pumps.Html. (Accessed 25 August 2017).

[39] *Ontario Ministry of Energy—Smart Meters and Time-of-Use Prices.* Available from: Http://Www.Energy.Gov.On.Ca/En/Smart-Meters-And-Tou-Prices/. (Accessed 25 August 2017).

[40] A. Schafer, P. Baumanns, and A. Moser, "Modeling heat pumps as combined heat and power plants in energy generation planning," In *2012 Ieee Energytech*, Cleveland, Oh, USA, pp. 1–6.

[41] Vdi (The Association of German Engineers), Richtlinie Vdi 6007 Blatt 1: Berechnung Des Instationären Thermischen Verhaltens Von Räumen Und Gebäuden; Raummodell, Bauphysik 29 (5) (2007) 384.

Chapter 13

Demand-Side Management in Micro-Grids and Distribution Systems: Handling System Uncertainties and Scalabilities

Yubo Wang* and Hamidreza Nazaripouya†

**Department of Mechanical Engineering, University of California, Los Angeles, USA,* †*Electrical and Computer Engineering, Winston Chung Global Energy Center, University of California, Riverside, USA*

1 INTRODUCTION

1.1 Motivations

Coupled with the need for better demand-supply balance, future smart grid represents a vision to incorporate an increasing number of storage devices in distribution grids [1]. Facilitated by recent advances in demand side management (DSM), chemical [2], thermal [3] and mechanical [4] storage systems are optimally managed in response to electricity price incentives. Combined the with the target of cutting overall operation costs, DSM collaboratively addresses peak shaving and power quality improvements with cyber physical constraints [5,6].

The increasing penetration of distributed energy resource (DER), such as battery energy storage system (BESS) [7], electric vehicle (EV) [8], solar generation [9] etc., integrated to distribution grids introduces new challenges. On one hand, stochastic nature of DER (for example, EV) introduces uncertainties that affects the performance of the system; on the other hand, the distributed nature of DER calls for new energy management techniques that account for system scalability. Furthermore, DER integrations are studied in two typical scenarios, in microgrids [10,11], using lumped models, and distribution grids [12,13], using networked models. The two distinct models correspond to two groups of management roles in smart grids, that is, building manager and distribution system operator (DSO). Therefore, DSM in microgrids and distribution grids that considers DER integration has been a subject of significant ongoing research.

Classical and Recent Aspects of Power System Optimization. https://doi.org/10.1016/B978-0-12-812441-3.00013-6
© 2018 Elsevier Inc. All rights reserved.

1.2 Literature Review

Early works have primarily assessed DSM for DER integration using deterministic optimizations (DO) [14,15]. Two studies of DSM on a household, with one EV and solar generation, reported peak shifting with price incentives [14]. Mixed integer linear programming is used in both papers to achieve minimum operation cost. Chabaud et al. [15] modeled a grid-connected residential building with renewables and battery storage, which confirmed the importance of energy storage devices and diversity of renewable generation in cost saving. These researchers assumed perfect knowledge of uncertainties in DER and used deterministic optimization methods for DSM, without considering DER's stochastic nature.

Much of the research in DSM has examined uncertainties caused by DERs with robust optimization [16,17]. Malysz et al. [16] considered uncertainties in loads and solar generation of a microgrid and formulated the optimal control of a microgrid as a robust mixed integer linear programming problem. Bai et al. [17] evaluated a robust mixed integer quadratic programming method for large scale Vehicle-to-Grid (V2G) for EV aggregator considering EV instant power demand uncertainties. These methods addressed the uncertainties in the systems by estimating the worst case of the uncertainty sets, which might be conservative and resulted in high operational cost.

In Refs. [18,19], the wind generation uncertainties have been modeled in a microgrid with probabilistic constrained stochastic programming. The operation cost minimizations have been achieved with constraints of utilizing certain percentage of wind generation to meet minimum renewable utilization regulations. These papers considered the load and renewable uncertainties in the power grid and addressed the stochastic behavior with uncertainty-aware stochastic optimization. The proposed stochastic method works well with passive components (such as solar and wind) that cannot be controlled, however, active components such as EVs that could be controlled are also subject to probability. These researchers have not modeled how the randomness of EVs' energy demands, arrival/departure time would interact with the intermittent renewable generation, which if managed improperly, will increase the burden of grids.

Turning attention to the distribution grids (networked model), buses are connected by power line and interact with each other under power flow constraints. The isolated lumped model may turn out to either violate the bus voltage limits or result in excessive power losses in the power lines. Consequently, voltage regulation (VR) and optimal power flow (OPF) are the two important topics capturing researchers' interests in distribution grids. Several centralized methods have been proposed, including particle swarm optimization [20], genetic algorithm [21], mixed integer nonlinear programming [22], neural networks [23] and fuzzy logic [24]. These methods require expensive computational power especially when the system grows large. The scalability of these methods remains in question.

When dealing with fast growing controllable distributed generations and storages in distribution grids, it is natural to apply decentralized algorithms. Many recent studies have explored the applications of decentralized algorithms in DSM [25–27]. Zhang et al. [25] formulated the problem based on dual decomposition of a convex problem formulation. However, they have linearized the nonconvex OPF problem into DC power flow model which are typically used in transmission systems. Researchers in Refs. [26,27] have considered the OPF problem in a decentralized way, however, did not capture the uncertainties in the system.

1.3 Scopes and Organization

In this context, this chapter presents two typical scenarios of DER integration, that is in microgrids and distribution grids. The expected audiences of the former one is microgrids managers, for example, building managers; and the latter one is distribution grids managing authorities, for example, DSO. In the first scenario, we develop a two-stage stochastic programming framework that addresses the stochastic nature of the DERs. The two-stage formulation addresses the optimization problem of determining the first-stage variables under the uncertainties in the second stage. In this chapter, the second stage refers to the stochastic variables for EVs, where the first stage refers to the other decision variables. To make the optimization problem with stochastic variables tractable, a numerical method, sample average approximation (SAA), is applied to the original problem. Furthermore, a model-free distribution modeling method, Kernel density estimation (KDE), is applied to generate samples needed for SAA. After the formulation of DSM in Microgrids, we extend the work to distribution grids, in which power flows between different nodes needs to be considered. On top of the stochastic programming framework used in the microgrids scenario, the DSM is first formulated into a nonconvex optimization. The DSM is tightly relaxed to convex problem with restrictions. Moreover, a decentralized framework using alternating direction method of multipliers (ADMM) is proposed for the benefit of system scalability. Because the SAA-based method is computationally intensive, the decentralized method can dispatch the centralized computational load to decentralized nodes, fully utilizing the computational power in the system. Real-life data are used for validation of the proposed model followed by extensive results analysis.

The rest of this chapter is organized as follows: Section 2 studies the DSM in microgrids setting. A two-stage stochastic optimization framework is proposed to address the uncertainties of DER integration to microgrids. In Section 3, the DSM problem is reexamined in distribution systems. On top of the stochastic optimization framework developed in Section 2, this section further proposed a decentralized optimization framework. In Section 4, numerical simulations are carried out, followed by extensive analysis of the results. Finally, conclusions are drawn in Section 5.

2 DSM IN MICROGRIDS

In this section, we study DER integration in microgrids setting. The model we used is *lumped model*, that is, to ignore the power flow between different buses and treat all the loads and generation under a same bus. The *lumped model* is often used by building managers for managing the energy consumption in commercial buildings. We first introduce the system architecture and show the challenges of DSM under uncertainties. It is followed by modeling the uncertainties and formulating the DSM into a two-stage stochastic programming problem.

2.1 System Structure and Modeling

This chapter studies a grid-connected commercial building Microgrid consist of a variety of DERs: many BESSs, several EVs with V2G functionality [6] are integrated to the building. The system architecture of the grid-connected commercial building microgrid is shown in Fig. 1. It must be admitted that other forms of stationary storage exist in commercial buildings such as ice storage unit and heat storage unit [28]. However, BESSs allow bidirectional power flow, which adds more flexibility to the DSM design. We evaluate the BESSs' impact on DSM, but same modeling technique can be applied for analysis of other stationary storages.

2.1.1 DSM Model

For notation: In this chapter, bold variables stand for vectors while the unbolded variables stands for scalar. $\overline{(\cdot)}$ represents the upper limit of a scaler and similarly $\underline{(\cdot)}$ represents the lower limit. The expectation is denoted by $\mathbb{E}[\cdot]$.

For the tariff model, we use time-of-use price. We use a *slotted model* in this chapter, assuming variables stay constant between decision intervals and there are H slots in a day's operation. We assume that the price is perfectly

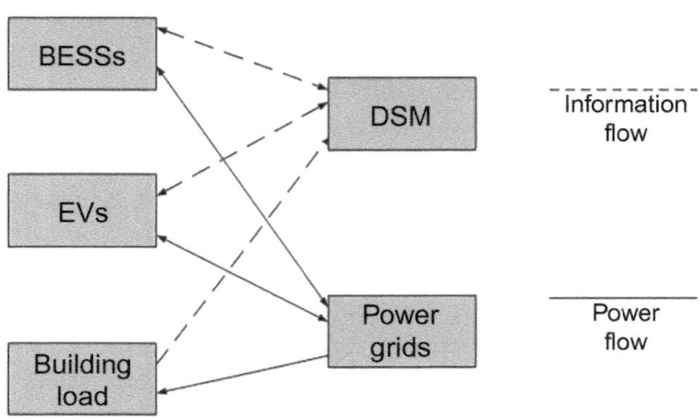

FIG. 1 Architecture of a grid-connected commercial building microgrid.

known, and the size of the generation and consumption of the commercial building Microgrid does not change the price. The economical compensations to EV users for V2G is a very complicated problem including proper business model [29] and battery degrading cost [30], which are not in the scope of this chapter and will be covered in future research. As building load is uncontrollable, and the grid interactions of EVs are subject to change until it is observed, the functionality of DSM is to help the building manager making day-ahead decisions on the operation of the BESSs. Without loss of generality, we assume there are m BESSs and n EVs in the commercial building microgrids. The optimal DSM cost function is formulated as follows:

$$\min \boldsymbol{C}\left[\sum_i \boldsymbol{P}_i^B + \boldsymbol{P}^L + \mathbb{E}\left\{\sum_i \boldsymbol{P}_i^E(\boldsymbol{\delta}_i, \boldsymbol{\nu}_i, \boldsymbol{\chi}_i)\right]\right. \tag{1}$$

where $\boldsymbol{C} \in \mathfrak{R}^H$ is the day-ahead price, $\boldsymbol{P}_i^B \in \mathfrak{R}^H$ denotes the power of the ith BESS, $\boldsymbol{P}^L \in \mathfrak{R}^H$ denotes the power of the building load, and $\boldsymbol{P}_i^E \in \mathfrak{R}^H$ corresponds to the power of the EV i. $\boldsymbol{\delta}_i, \boldsymbol{\nu}_i, \boldsymbol{\chi}_i \in \mathfrak{R}^H$ are stochastic variables representing the probabilities of EV i in arrival time, departure time, and energy demands.

Note that we ignore the time constant in this chapter when calculating energy. Moreover, we also ignore the transpose of vectors in equations. The DSM is modeled using two-stage stochastic optimization [31], that is, to decide the power output of BESSs in the first stage, given the observation of stochastic EVs in the second stage.

2.1.2 BESS Model

For each BESS i, the BESS can be modeled using the following equations:

$$\underline{P_i^B}\mathbf{1} \leq \boldsymbol{P}_i^B \leq \overline{P_i^B}\mathbf{1} \tag{2}$$

$$\underline{E_i^B} \leq E_{i,0}^B + \sum_{t \in T} P_i^B(t) \leq \overline{E_i^B} \tag{3}$$

$$E_{i,0}^B + \sum_{t=H} P_i^B(t) = E_{i,H}^B \tag{4}$$

where $\underline{P_i^B}$ and $\overline{P_i^B}$ stand for the lower and upper bound of the ith BESS output. Similarly, $\underline{E_i^B}$ and $\overline{E_i^B}$ are the lower and upper bounds of the State-of-Energy (SoE). $E_{i,0}^B$ and $E_{i,H}^B$ stands for the start and end of the SoE of BESS. $\boldsymbol{P}_i^B$ is the vector of $P_i^B(t)$ and represents the power output of BESS i. Eq. (2) reflects the power constraint on the inverter, Eq. (3) limits the maximum and minimum SoE of the BESS. Eq. (4) ensures that after a full day's operation, the BESS returns to a desired SoE.

2.1.3 EV Model

One major difference between EV and BESS is the uncertainties in EVs' availability and energy demand. An EV owner may arrive at a parking facility at a random time. Similarly, it is very likely the owner will not be committed to any schedule without additional incentives, therefore may leave at random time. In addition to the uncertain in availability, the energy demand of an EV highly depends on the travel distance of the owner. All these factors add up to the difficulty of DSM with EV integrations. Following the method used for modeling BESS, similarly, EV can be modeled as follows:

$$\underline{P_i^E} A_i^E(\delta_i, \nu_i, t) \leq P_i^E(t) \leq \overline{P_i^E} A_i^E(\delta_i, \nu_i, t) \tag{5}$$

$$A_i^E(\delta_i, \nu_i, t) = \begin{cases} 1, & EV\, i \,\text{available at}\, t \\ 0, & \text{otherwise} \end{cases} \tag{6}$$

$$\underline{E_i^E} \leq E_{i,t_s}^E(\chi_i) + \sum_{t \in T_i} A_i^E(\delta_i, \nu_i, t) P_i^E(t) \leq \overline{E_i^E} \tag{7}$$

$$E_{i,t_s}^E(\chi_i) + \sum_{t = t_{i,e}} A_i^E(\delta_i, \nu_i, t) P_i^E(t) = E_{i,t_e}^E(\chi_i) \tag{8}$$

where $\underline{P_i^E}$ and $\overline{P_i^E}$ are the lower and upper bounds for EV i's power. $\underline{E_i^E}$ and $\overline{E_i^E}$ are the upper and lower bounds on EV i's SoE. A_i^E stands for the availability vector of EV i. E_{i,t_s}^E and E_{i,t_e}^E are the start and end SoE of EV i. Eq. (5) constrains the power of the EV. The upper and lower bounds of the power are multiplied by the availability scaler A_i^E defined in Eq. (6). Therefore, when EV is not present, the power is constrained to zero. Eq. (7) ensures that the EV battery is not over discharged nor charged. And Eq. (8) represents that by the leave time, the EV is charged to a desired SoE. The modeling of EVs follows the same idea on modeling BESS. However, as shown in the equations, they are subject to stochastic variables in arrival time, departure time, and energy demands. The problem is hard to solve even knowing the distributions of stochastic variables. Therefore, a tractable method needs to be applied to make the problem computationally feasible.

2.2 Sample Average Approximation

SAA is a Monte Carlo simulation technique that serves as a statistical inference of the original problem. The idea behind it is to generate a large pool of samples of uncertainties based on their empirical distributions where each sample is called a *scenario*. Then solve the second-stage problems with each scenario before averaging them out to approximate the original problem. For a detailed discussion, see Ref. [31].

Without loss of generality, it is assumed that we generate N independent scenarios. Each scenario is one possible realization of the uncertainties, which includes EV availabilities and energy demands. Define N *i.i.d* samples forming the sample set: $\psi = \{\delta_i, \nu_i, \chi_i\}_{s=1}^{N}$. Hence, the EV constraints for each $s \in \psi$, are updated as follows:

$$\underline{P_i^E} A_i^E(t) \le P_{i,s}^E(t) \le \overline{P_i^E} A_{i,s}^E(t) \tag{9}$$

$$A_{i,s}^E(t) = \begin{cases} 1, & EV\, i \text{ available at } t \text{ in scenario } s \\ 0, & \text{otherwise} \end{cases} \tag{10}$$

$$\underline{E_i^E} \le E_{i,t_s,s}^E + \sum_{t \in T_{i,s}} A_{i,s}^E(t) P_{i,s}^E(t) \le \overline{E_i^E} \tag{11}$$

$$E_{i,t_s,s}^E + \sum_{t=t_{i,e,s}} A_{i,s}^E(t) P_{i,s}^E(t) = E_{i,t_e,s}^E \tag{12}$$

the subscript s represents the constraints needs to be held in *scenario s*. The EV constraints are now eliminated from stochastic variables, they are now deterministic in each *scenario*. When adding the scenario together, the averaged scenarios approximate the original problem with the increase of scenario size. The DSM could be also updated accordingly as follows:

$$\min_{P_i^B, P_{i,s}^E} C\left[\sum_i P_i^B + P^L + \frac{1}{N}\left\{\sum_s \sum_i P_{i,s}^E\right\}\right] \tag{13}$$
$$\text{subject to. } (2)-(4), (9)-(12)$$

The formulated DSM is a convex optimization problem and can be solved effectively. As SAA still relies on generated scenarios, we now present the method to generate scenarios based on the distribution of stochastic variables.

2.3 Kernel Density Estimation

To study the uncertainty of the system, data is collected to show the distribution of each stochastic variable which is called an *empirical* realization. Given an empirical realization of certain stochastic variables, here we apply KDE to capture the empirical probability density. The mathematical representation of KDE is described as follows:

$$\rho(x) = \frac{1}{N} \sum_{i=1}^{N} K\left(\frac{x - x_i}{b}\right) \tag{14}$$

where N is the number of kernels used, x_i is the center of one kernel, b denotes the bandwidth of the kernel, $K(\cdot)$ is the kernel function, and $\rho(\cdot)$ is the probability density function.

As empirical probability density is usually given in histogram, the center of each kernel naturally falls into the center of each bin. The number of kernels is equal to the number of bins in the histogram. The bandwidth then becomes the only free parameter to be determined. A common way to determine the bandwidth is to use the rule-of-thumb principle described in [32]. After determining each parameter, the discrete histogram can be modeled by continuous KDE. The modeled distribution can then be used to generate scenarios used in SAA.

To summarize, we tabulated the steps of formulating and solving the DSM in Microgrids in Algorithm 1.

3 DSM IN DISTRIBUTION SYSTEMS

In this section, we study DER integration in distribution grids setting. We use a *networked model* to consider the power flow between different buses in distribution grids. The *networked model* is often used by DSOs for achieving economic operation of distribution grids. We first introduce the system architecture and show the challenges in system scalability. We utilize some of the stochastic programming framework developed in Section 2. We first formulate the DSM into nonconvex optimization problem. It is followed by a tight convex relaxation. Finally, a decentralized framework is proposed to account for the system scalability.

3.1 System Structure and Modeling

As shown in Fig. 1, the studied system is in distribution grids connected to a substation at the root node. Without loss of generality, under each bus, it connects a set of BESSs, many EVs and uncontrollable loads. There are two major players in the distribution grids, namely DSO and sub-distribution system operator (SDSO). The DSO oversees managing the whole distribution grids and the SDSO is managing the BESSs, EVs, and loads under each bus. There exists two-way communication links between DSO and SDSO. In this section, again we consider a *slotted model* and slice the timeframe of a day into H equal time slots. The time is represented as t, and $t \in \{1, 2, \ldots, H\}$. The set of $\{t \mid t \in \{1, 2, \ldots, H\}\}$ is denoted as $\boldsymbol{T}$. As the target of this chapter is to manage the distribution grids under steady-state, we assume the power flow stays the same between each time slot.

Algorithm 1 Stochastic programming for DSM in Microgrids

1. **Data collection.** In this studied case, collect data on EV arrival time, departure time, energy demand.
2. **Model empirical distribution.** Using KDE in Eq. (14) to model the empirical distribution of each stochastic variables.
3. **Generate samples.** Generate N samples for the SAA
4. **Solve DSM.** Solve the DSM described in Eq. (13).

3.1.1 DSM Model

For notation: In this chapter, $\overline{(\cdot)}$ represents the upper limit of a scaler and similarly $\underline{(\cdot)}$ represents the lower limit. $(\cdot)^*$ stands for the conjugate of a scaler. For the same variable, capital letter is used to denote real variables, small letter is used to denote complex variables. The imaginary unit is denoted as $\boldsymbol{j}$. Other bold symbols represent sets. Finally, the expectation is denoted by $\mathbb{E}[\cdot]$.

The objective of the DSM proposed in this chapter is to minimize the overall nodal operation cost and power losses in distribution line while keeping the bus voltages regulated in the acceptable range. The distribution grids are represented as a connected graph $G=(\boldsymbol{N},\boldsymbol{\xi})$. Let $i\in\boldsymbol{N}=\{1,2,\ldots,N\}$ denote the node (bus) of the distribution grids, and $\boldsymbol{\xi}$ denote the lines. We use bus 1 to represent the root node shown in Fig. 2. If there is a line $(i,j)\in\boldsymbol{\xi}$ connecting bus i and bus j, let $z_{ij}=R_{ij}+\boldsymbol{j}X_{ij}$ be the line impedance, and $i_{ij}(t)$ be the line current from i to j at time t, $P_i^A(t)$ and $Q_i^A(t)$ are the aggregated active and reactive power drawn at bus i at time t, and the apparent power drawn at bus $i\in\boldsymbol{N}$ is represented as $s_i^A(t)=P_i^A(t)+\boldsymbol{j}Q_i^A(t)$. The positive sign for $s_i^A(t)$ represents power drawn from bus i. For each bus i, there exists a set of BESSs, denoted $\boldsymbol{M}_i^B$, and a set of EVs, denoted $\boldsymbol{M}_i^E$. Let $\Pi(t)\in\Re^{1+}$ be the energy price for a studied day, the DSM objective function is then presented as follows:

$$\min \sum_{t}\Pi(t)\sum_{i\in N\setminus\{1\}}P_i^A(t)+\gamma\sum_{t=H}\sum_{(i,j)\in\boldsymbol{\xi}}R_{ij}\left|i_{ij}(t)\right|^2 \tag{15}$$

The DSM objective is the summation of two parts: the first part minimizes the nodal operational cost, and the second part minimizes the power losses over distribution line. The two parts are connected by a weighting factor γ. The two terms in Eq. (15) are using different units. To avoid running into numerical issues, we normalized the Megawatt power loss with per-unit system. The weighting factor γ characterizes how an optimal decision leverages the different

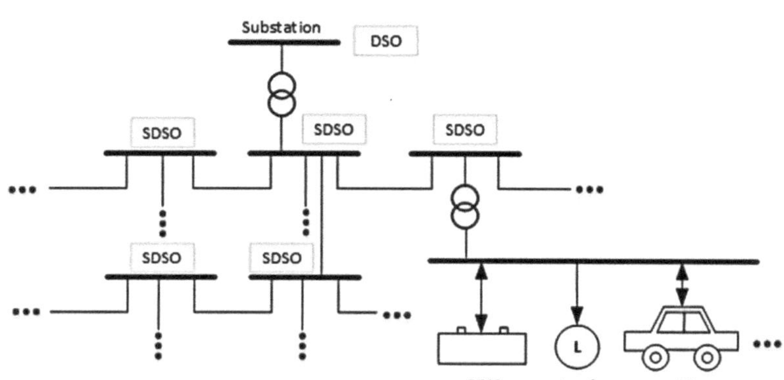

FIG. 2 Studied distribution grids architecture.

objectives. For simplicity, the time constant for each time slot is ignored, which does not affect the optimum of the system.

3.1.2 BESS Model

Let $P^B_{i,m}(t)$ and $Q^B_{i,m}(t)$ denote the active power and the reactive power output of the BESS m under bus i in time slot t. And $S^B_{i,m}(t)$ denote the apparent power limit of a physical inverter interfaced with the BESS. The constraint of the BESS for $\forall i \in \mathbf{N} \backslash \{1\}, \forall m \in \mathbf{M}^B_i$ can be represented as follows:

$$P^B_{i,m}2(t) + Q^B_{i,m}2(t) \leq S^B_{i,m}2(t) \quad (16a)$$

$$\underline{E^B_{i,m}} \leq ES^B_{i,m} + \sum_{t \in \mathbf{T}} P^B_{i,m}(t) \leq \overline{E^B_{i,m}} \quad (16b)$$

$$ES^B_{i,m} + \sum_{t=H} P^B_{i,m}(t) = EF^B_{i,m} \quad (16c)$$

where $ES^B_{i,m}$ represents the initial SoE of the BESS m under bus i at time $t=0$. $\underline{E^B_{i,m}}$ and $\overline{E^B_{i,m}}$ represent the upper and lower limits of the BESS, respectively. And $EF^B_{i,m}$ is the final SoE of the BESS after a full day's operation. Similarly, Eq. (16a) accounts for maximum power rating of an inverter. Without loss of generality, we also consider reactive power in this case, because reactive plays a major role in voltage regulation in distribution grids. Eq. (16b) constraints the BESS SoE in a desired range. And finally, Eq. (16c) accounts for final status of BESS after a full day's operation.

3.1.3 EV Model

Following the same philosophy in modeling uncertainties, this section uses the same method in modeling EV as Section 2. For a detailed modeling of uncertainties in EVs, please refer to Section 2.1.3. Let $\underline{P^E_{i,m}}$ and $\overline{P^E_{i,m}}$ be the maximum discharge power and charging power of EV i under bus m, with $\underline{P^E_{i,m}} \in \mathfrak{R}^{1-}$ representing the effect of V2G. Generating $Ni.i.d.$ samples denoted $\boldsymbol{\psi}$, for each $\forall i \in \mathbf{N} \backslash \{1\}, \forall m \in \mathbf{M}^E_i$, here we present the EV constraints in each scenario $s \in \boldsymbol{\psi}$ without detailed deduction:

$$\underline{P^E_{i,m}} A^E_{i,m,s}(t) \leq P^E_{i,m,s}(t) \leq \overline{P^E_{i,m}} A^E_{i,m,s}(t) \quad (17a)$$

$$A^E_{i,m,s}(t) = \begin{cases} 1, & EV\, i \,\text{available at}\, t \,\text{in scenario}\, s \\ 0, & \text{otherwise} \end{cases} \quad (17b)$$

$$\underline{E^E_{i,m}} \leq ES^E_{i,m,s} + \sum_{t \in \mathbf{T}^s_{i,m}} A^E_{i,m,s}(t) P^E_{i,m,s}(t) \leq \overline{E^E_{i,m}} \quad (17c)$$

$$ES^E_{i,m,s} + \sum_{t=t^{i,m}_{f,s}} A^E_{i,m,s}(t) P^E_{i,m,s}(t) = EF^E_{i,m,s} \quad (17d)$$

where $\underline{P^E_{i,m}}$ and $\overline{P^E_{i,m}}$ are the lower and upper power limits of EV m under bus i, $\underline{E^E_{i,m}}$ and $\overline{E^E_{i,m}}$ are the lower and upper SoE limit of the EV, $A^E_{i,m,s}$ is the availability vector for EV m under bus i in *scenario s*. Similarly, $P^E_{i,m,s}(t)$ is the power output of the EV, $ES^E_{i,m,s}$ and $EF^E_{i,m,s}$ are the initial and final SoE status of EV m under bus i in *scenario s*. We use $\boldsymbol{T}^s_{i,\ m}$ to denote the set of time when EV i under bus m in *scenario s* is available, and $t^{i,m}_{f,s}$ to denote the time instance when EV leaves in this scenario. On the constraints side, Eq. (17a) limits the power output of EV, and Eq. Eq. (17b) is an ancillary vector. Eq. (17c) represents the SoE constraint and Eq. (17d) represents the final SoE status constraint by the time EV leaves. The reactive power of EV bidirectional charging is not considered. The reason behind this modeling is, in practice, EV bidirectional chargers usually has much smaller power ratings compared to BESS, making the converter more economic at the price of two-quadrant controllability [6]. These constraints hold for each generated scenario, and within each scenario, the realization of stochastic variables reduce the problem to a deterministic problem.

3.1.4 Circuit Model

Following the notations in this section, for $\forall(i,j)\in\boldsymbol{\xi}, j\neq 1$ we can derive the power flow of the distribution grids using the following equations:

$$v_i(t) - v_j(t) = z_{ij} i_{ij}(t) \tag{18a}$$

$$s_{ij}(t) = v_i(t) i^*_{ij}(t) \tag{18b}$$

$$s_{ij}(t) - z_{ij}\left|i_{ij}(t)\right|^2 - \sum_{(j,k)\in\boldsymbol{\xi}} s_{jk}(t) = s^A_j(t) \tag{18c}$$

where the bus i voltage is denoted as v_i, and $s_{ij}(t) = P_{ij}(t) + jQ_{ij}(t)$ denote the apparent power flowing from bus i to bus j. The positive sign for s_{ij} is defined as power flowing from i to j. Eq. (18a) is the Ohm's law, Eqs. (18b), (18c) are the definitions for apparent power and power flow, respectively. Substituting Eqs. (18a), (18b) into Eq. (18c) under a radial network, we will have for $\forall(i,j)\in\boldsymbol{\xi}, j\neq 1$:

$$P^A_j(t) = P_{ij}(t) - R_{ij}\left|i_{ij}(t)\right|^2 - \sum_{(j,k)\in\boldsymbol{\xi}} P_{jk}(t) \tag{19a}$$

$$Q^A_j(t) = Q_{ij}(t) - X_{ij}\left|i_{ij}(t)\right|^2 - \sum_{(j,k)\in\boldsymbol{\xi}} Q_{jk}(t) \tag{19b}$$

$$|v_i(t)|^2 - \left|v_j(t)\right|^2 = 2\left(R_{ij}P_{ij}(t) + X_{ij}Q_{ij}(t)\right) - \left(R^2_{ij} + X^2_{ij}\right)\left|i_{ij}(t)\right|^2 \tag{19c}$$

$$|v_i(t)|^2\left|i_{ij}(t)\right|^2 = P^2_{ij}(t) + Q^2_{ij}(t) \tag{19d}$$

Eqs. (19a)–(19d) define the power flow of the distribution grids. Define the aggregated active and reactive load under bus i as P^A_i and Q^A_i. The aggregated power under each bus for $\forall i\in\mathcal{N}\backslash\{1\}$ is defined as follows:

$$P_i^A(t) = P_i^L(t) + \sum_{m\in \boldsymbol{M}_i^B} P_{i,m}^B(t) + \frac{1}{N}\sum_s \sum_{m\in \boldsymbol{M}_i^E} P_{i,m,s}^E(t) \tag{20a}$$

$$Q_i^A(t) = Q_i^L(t) + \sum_{m\in \boldsymbol{M}_i^B} Q_{i,m}^B(t) \tag{20b}$$

where $P_i^L(t)$ and $Q_i^L(t)$ are the uncontrollable load under bus i. For the root node, we have:

$$P_1^A(t) = \sum_{j:(1,j)\in\boldsymbol{\xi}} P_{1j}(t) \tag{20c}$$

$$Q_1^A(t) = \sum_{j:(1,j)\in\boldsymbol{\xi}} Q_{1j}(t) \tag{20d}$$

We put additional constraints to circuit model, which models the physical constraints on distribution grids:

$$\underline{V} \le |v_i| \le \overline{V}, \forall i \in \boldsymbol{N}\backslash\{1\}; |v_1| = V_c \tag{21a}$$

$$P_i(t)^2 + Q_i(t)^2 \le \overline{S_i}^2 \tag{21b}$$

$\underline{V}$ and $\overline{V}$ are the lower and upper voltage magnitude limits. The bus voltage at root node 1 is assumed to be constant V_c, which results from the fact that it is connected to the transmission system and be modeled as an infinite bus. $\overline{S_i}$ is the maximum transformer output at each bus. In other words, Eq. (21a) models the voltage regulation at each bus, Eq. (21b) the maximum power, the active and reactive power drawn at each bus. We define the decision variable set to be $\boldsymbol{\Omega} = \{P_i^A(t), Q_i^A(t), P_{i,m}^B(t), Q_{i,m}^B(t), P_{i,m,s}^E(t)\}$. The DSM problem now can be presented as:

$$\min_{\boldsymbol{\Omega}} \sum_t \Pi(t) \sum_{i\in \boldsymbol{N}\backslash\{1\}} P_i^A(t) + \gamma \sum_t \sum_{(i,j)\in\boldsymbol{\xi}} R_{ij}|i_{ij}(t)|^2$$
$$\text{subject to. } (16), (17), (19)-(21)$$

However, Eq. (19d) is nonconvex, making the problem impossible to solve with convex optimization solvers. Nonconvex solvers require much stronger computational power and the solutions are not guaranteed to be global optimum. As a result, we seek to convexify the DSM problem and solve it in a decentralized manner in the following section.

3.2 Convex Relaxation

According to [33,34], the nonconvexity constraint in Eq. (19d) can be relaxed to a convex second-order cone as follows:

$$|v_i(t)|^2 |i_{ij}(t)|^2 \ge P_{ij}^2(t) + Q_{ij}^2(t), \forall (i,j) \in \boldsymbol{\xi}, j \ne 1 \tag{22}$$

For a tight relaxation, there are three sufficient conditions, namely (a) the network must be radial; (b) the power injection to each bus cannot be too large;

and (c) bus voltages are kept around nominal values. The method to check the exactness of a relaxation is to solve for Eq. (22) and compare the left-hand-side in the equality to the right-hand-side. We define the decision variable set to be $\mathbf{\Omega}=\{P_i^A(t), Q_i^A(t), P_{i,m}^B(t), Q_{i,m}^B(t), P_{i,m,s}^E(t)\}$. With the convex relaxation, the relaxed DSM can be described as follows:

$$\min_{\mathbf{\Omega}} \sum_t \Pi(t) \sum_{i\in \mathcal{N}\setminus\{1\}} P_i^A(t) + \gamma \sum_t \sum_{(i,j)\in\xi} R_{ij}\left|i_{ij}(t)\right|^2$$
$$\text{subject to. } (16),(17),(19\text{a})-(19\text{c}),(20)-(22) \tag{23}$$

Given a tight convex relaxation, the DSM is in a centralized manner and therefore the DSO still must collect the information on each bus, raising privacy concerns to EV owners. Furthermore, with a growing size of BESSs, EVs and perhaps other controllable devices not included in this chapter, the DSO has a heavy computational burden for the system to scale up. These motivates us to develop a decentralized DSM that is scalable and privacy-preserving.

3.3 Decentralized Method

We develop the decentralized DSM using the ADMM method, see Ref. [35] for details. The key for using ADMM is to decompose a (convex) optimization problem into global and local subproblems. The ADMM method turns out to be a natural fit to DSM problem in this chapter. We decompose the original DSM problem into a global problem that solves for the OPF and VR at DSO, and local problems that solve for energy cost minimization at SDSOs.

Before start of the ADMM iterations, we first initialize the problem with random numbers: for the first iteration, DSO sets random values to P^i and Q^i, for $\forall i \in \mathcal{N}\setminus\{1\}$. For simplicity of notation, we neglect *time t* in each variable. We introduce two sets of dual variables for each bus $i\in\mathcal{N}\setminus\{1\}$, namely $\alpha_i(t)\in\mathfrak{R}^1$ and $\beta_i(t)\in\mathfrak{R}^1$.And SDSO randomly choose $P_{i,m}^B(t)$, $Q_{i,m}^B(t)$ for $\forall i\in\mathcal{N}\setminus\{1\}$, $\forall m\in \boldsymbol{M}_i^B, \forall t\in \boldsymbol{T}$, and $P_{i,m,s}^E(t)$ for $\forall s\in\boldsymbol{\psi}, \forall i\in\mathcal{N}\setminus\{1\}, \forall m\in\boldsymbol{M}_i^E, \forall t\in\boldsymbol{T}$.

Define $\mathbf{\Omega}_1=\{P_{i,m}^B(t), Q_{i,m}^B(t), P_{i,m,s}^E(t)\}$ for each iteration $k+1$, the SDSO at bus $i\in\mathcal{N}\setminus\{1\}$, get the iteration k update $(P_i^A)_k$, $(Q_i^A)_k$, $(\alpha_i)_k$ and $(\beta_i)_k$ from SDO and first solve a local DSM problem for $t\in\boldsymbol{T}$ as follows:

$$\min_{\mathbf{\Omega}_1} \sum_t \Pi(t) P_i^A(t) + \rho_1 \sum_t \left(-\left(P_i^A(t)\right)_k + P_i^L(t) + \sum_{m\in \boldsymbol{M}_i^B} P_{i,m}^B(t) \right.$$
$$\left. + \frac{1}{N}\sum_s \sum_{m\in\boldsymbol{M}_i^E} P_{i,m,s}^E(t) + (\alpha_i(t))_k \right)^2$$
$$+\rho_2 \sum_t \left(-\left(Q_i^A(t)\right)_k + Q_i^L(t) + \sum_{m\in\boldsymbol{M}_i^B} Q_{i,m}^B(t) + (\beta_i(t))_k \right)^2$$
$$\text{subject to.} (16),(17) \tag{24a}$$

where ρ_1, ρ_2 are the step size. The local DSM at each bus solves for the optimal schedule for BESSs and EVs in iteration $k+1$, denoted as $(P^B_{i,m}(t))_{k+1}$, $(P^E_{i,\,m,s}(t))_{k+1}$ and $(Q^B_{i,m}(t))_{k+1}$ for $t \in \boldsymbol{T}$. Then, the SDSO communicate the aggregated bus load $(P^a_i(t))_{k+1}$ and $(Q^a_i(t))_{k+1}$ (note: use "a" to denote the difference between P^A_i $(Q^A_i)_k$)for $t \in \boldsymbol{T}$ defined as follows to DSO:

$$\begin{aligned}
\left(P^a_i\right)_{k+1} &= P^L_i(t) + \sum_{m \in \boldsymbol{M}^B_i} \left(P^B_{i,m}(t)\right)_{k+1} + \frac{1}{N}\sum_s \sum_{m \in \boldsymbol{M}^E_i} \left(P^E_{i,m,s}(t)\right)_{k+1} \\
\left(Q^A_i\right)_{k+1} &= Q^L_l(t) + \sum_{m \in \boldsymbol{M}^B_i} \left(Q^B_i(t)\right)_{k+1}
\end{aligned}$$

After receiving all the information for iteration $k+1$ from all SDSOs (namely, $(P^a_i(t))_{k+1}$, $(Q^a_i(t))_{k+1}$), DSO solve a global DSM problem by every time step for all $t \in \boldsymbol{T}$ as follows:

$$\begin{aligned}
&\min_{P^A_i(t), Q^A_i(t)} \gamma \sum_{(ij)\in\xi} R_{ij}\left|i_{ij}(t)\right|^2 + \rho_1 \sum_{i \in \mathcal{N}\setminus\{1\}} \left(-P^A_i(t) + \left(P^a_i(t)\right)_{k+1} + (\alpha_i(t))_k\right)^2 \\
&\quad + \rho_2 \sum_{i \in \mathcal{N}\setminus\{1\}} \left(-Q^A_i(t) + \left(Q^a_i(t)\right)_{k+1} + (\beta_i(t))_k\right)^2 \\
&\text{subject to. } (19\text{a}) - (19\text{c}), (20) - (22)
\end{aligned} \tag{24b}$$

Note the aggregated active and reactive load solved by DSO are denoted with $P^A_i(t)$, $Q^A_i(t)$. After solving for each time step of a day, the DSO updates the two sets of dual variables for each bus $i \in \mathcal{N}\{1\}$ as follows:

$$\begin{aligned}
(\alpha_i(t))_{k+1} &= (\alpha_i(t))_k - \left(P^A_i(t)\right)_{k+1} + \left(P^a_i(t)\right)_{k+1} \\
(\beta_i(t))_{k+1} &= (\beta_i(t)\,)_k - \left(Q^A_i(t)\right)_{k+1} + \left(Q^a_i(t)\right)_{k+1}
\end{aligned} \tag{24c}$$

The updates of the dual variables signal the end of iteration $k+1$. The updated $(P^A_i(t))_{k+1}$, $(Q^A_i(t))_{k+1}$, $(\alpha_i)_{k+1}$ and $(\beta_i)_{k+1}$ are broadcasted to each SDSO for the next iteration. The iteration ends when the problem converges.

For convergence, define the primal residuals of the two dual variables at iteration k for each bus i to be:

$$\begin{aligned}
r^P_i &= \sum_{t=H} \left(-\left(P^A_i(t)\right)_k + \left(P^a_i(t)\right)_k\right)^2 \\
r^Q_i &= \sum_{t=H} \left(-\left(Q^A_i(t)\right)_k + \left(Q^a_i(t)\right)_k\right)^2
\end{aligned} \tag{25a}$$

The stopping criteria for the decentralized DSM is to make sure the two primal residuals to be sufficiently small:

$$r^i_p \le \epsilon, r^i_q \le \epsilon, \forall i \in \boldsymbol{N}\setminus\{1\} \tag{25b}$$

where ϵ is a predefined error boundary.

Algorithm 2 ADMM-based decentralized DSM

1. **Initialize** the problem. Set $k=1$. DSO randomly choose P_i^A, Q_i^A, α_i and β_i^*. SDSO at bus i randomly choose $P_{m,i}^B$, $Q_{m,i}^B$, and $P_{i,m,s}^E$.
2. **Repeat** $k=2, 3, \ldots$
3. **Solve the local DSM** in Eq. (24a) at each SDSO using $(P_i^A)_k$, $(Q_i^A)_k$, $(\alpha_i)_k$ and $(\beta_i)_k$ broadcasted by DSO, broadcast $(P_i^a)_{k+1}$ and $(Q_i^a)_{k+1}$
4. **Solve the global DSM** in Eq. (24b) at DSO by collecting $(P_i^a)_{k+1}$ and $(Q_i^a)_{k+1}$ from each SDSO
5. **Update dual variables** in Eq. (24c) and broadcast $(P_i^A)_{k+1}$, $(Q_i^A)_{k+1}$, $(\alpha_i)_{k+1}$ and $(\beta_i)_{k+1}$ to SDSO
6. **Until** convergence in Eqs. (25a), (25b)

*For simplicity of notation, time t is neglected for each variable.

3.4 Over-Relaxation

To accelerate the convergence speed, an over relaxation technique is proposed in [37]. Basically, it is to substitute the $(P_i^a)_{k+1}$ and $(Q_i^a)_{k+1}$ update in each iteration $k+1$ as follows:

$$\begin{aligned}\left(P_i^a(t)\right)_{k+1} &= \theta\left(P_i^a(t)\right)_{k+1} - (1-\theta)\left(-\left(P_i^A(t)\right)_k\right)\\ \left(Q_i^a(t)\right)_{k+1} &= \theta\left(Q_i^a(t)\right)_{k+1} - (1-\theta)\left(-\left(Q_i^A(t)\right)_k\right)\end{aligned} \tag{26}$$

where θ is a relaxation parameter. [36] suggested that $\theta \in [1.5,1.8]$ can improve the convergence speed. The summary of the ADMM based decentralized DSM is tabulated in Algorithm 2. The ADMM-based algorithm can distribute the computational burden at the DSO to SDSO. At the same time, for each iteration, the DSO only must know the aggregated load at each bus, preserving the privacy by hiding the individual information in the aggregated information.

4 RESULTS AND ANALYSIS

This section validates the proposed DSM and shows its performance. We first introduce the simulation setup. It is followed by case studies for validation, and extensive discussions on understanding the numerical results.

4.1 Experimental Setup

4.1.1 Microgrids Scenario

Price and loads: The DSM studied in this chapter uses 1 hour as time step. We use the hourly day-ahead price from ERCOT [37]. To demonstrate the correctness of the proposed work, we have randomly picked a day. Prices from 12 a.m.–11 p.m.

are as follows: [0.111,0.103,0.103,0.099,0.106,0.131,0.117,0.124,0.110,0.095, 0.091,0.084,0.080,0.077,0.082,0.093,0.137,0.159,0.159,0.144,0.139,0.132, 0.140,0.111]. The building load data is sourced from [38]. We have plotted out the building load in Fig. 3. For the microgrids scenario, we only account for the active power. And we have aggregated all the buses plotted in a lumped model.

EV and battery: We have collected 19 EVs' records for a whole year on the UCLA campus to accomplish this study. The recorded EVs fall into three popular models in the market, that is, Nissan Leaf, Tesla Model S, and Mitsubishi MiEV. Their arrival time, leaving time, and energy demands are recorded for generating a close to real-life study. Apart from filtering out data with communication failures, we made the following assumptions to the original data to verify the proposed DSM: (1) *V2G capability*. Very few charging stations support V2G nowadays, partly resulting from different interfacing protocols. The records we have collected are EV charging data, however, it reveals the V2G potential. An EV is usually fully charged before it is unplugged and therefore can perform V2G in spare time. Following this reasoning, we assume EVs are V2G capable once connected to the grid, however they need to meet their energy demands before leave. (2) *SoE acquisition*. The SoE data is also hard to obtain, and therefore, it is impossible to tell the starting SoE after EV connects. As we have data in EV energy demands, we will assume EVs become fully charged every time. In this way, it will be tractable to calculate the starting SoE using the collected energy demands data. Table 1 summarizes the key parameters used for three different EV models and BESS models in this simulation. All EV related parameters comes from real-life EV manufacturers.

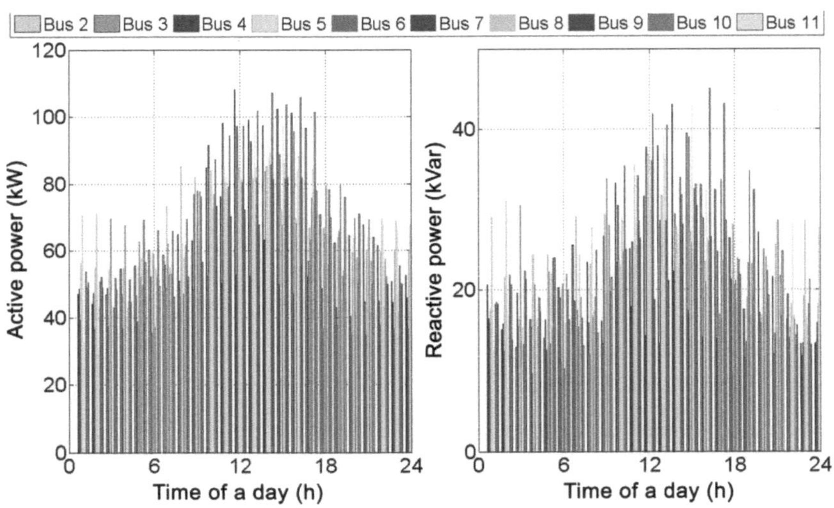

FIG. 3 A typical day load pattern.

TABLE 1 Key Parameters Used for BESS and EV

Device	Power Rating (Min, Max)	Capacity
Type A BESS	[−20,20] kVA	25 kWh
Type B BESS	[−15,15] kVA	20 kWh
Type C BESS	[−10,10] kVA	15 kWh
Tesla Model S	[−3.3,6.6] kW	65 kWh
Nissan Leaf	[−3.3,6.6] kW	24 kWh
Mitsubishi MiEV	[−3.3,6.6] kW	16 kWh

BESS-related parameters are chosen as these are the typical size of the BESS in distribution systems.

In observance of the large number of EV owners, we only show three EV owner profiles in Fig. 4 as the representatives of the 19 EV owners. The three owners drive a Tesla Model S, Nissan Leaf, and Mitsubishi MiEV, respectively. We capture the distribution of the real-life data using KDE described in Section 2.3. Fig. 4 compares the real-life data and distribution generated by KDE, showing a close match between the two. The KDE is then applied to generate the scenarios in SAA. To use the same data set as the distribution grid scenario, we duplicate 1 EV owner profile to make it 20. Also 11 BESSs that would be placed under each bus in the distribution grids case, have been put in the same microgrids.

4.1.2 Distribution Grids Scenario

Test architecture: We use a modified IEEE 13-bus radial test feeder [39] for demonstration of the proposed DSM. As the primary focus of this chapter is not on dynamics or electromagnetic transient, we made some minor modification to the original IEEE 13-bus radial test feeder to study the steady-state DSM. These modifications included: the in-line transformer between Bus 633 and 634 is removed; the breaker between Bus 671 and 692 is closed. Because of the extreme short line length between Bus 633 and 634, the two buses are treated as one single bus, that is, Bus 5 in Fig. 5. For similar reason, Bus 692 and 675 are treated as Bus 9 in Fig. 5. The root node is a 115-kV substation and it is connected with a 115 kV/4.16 kV transformer. We do not consider the three-phase imbalance in the distribution system, the three-phase diagram is therefore drawn as a one-line diagram. At each distribution bus, it is connected to several 4.16 kV/0.12 kV distribution transformers. The transformers are directly connected to the load, BESSs and EVs.

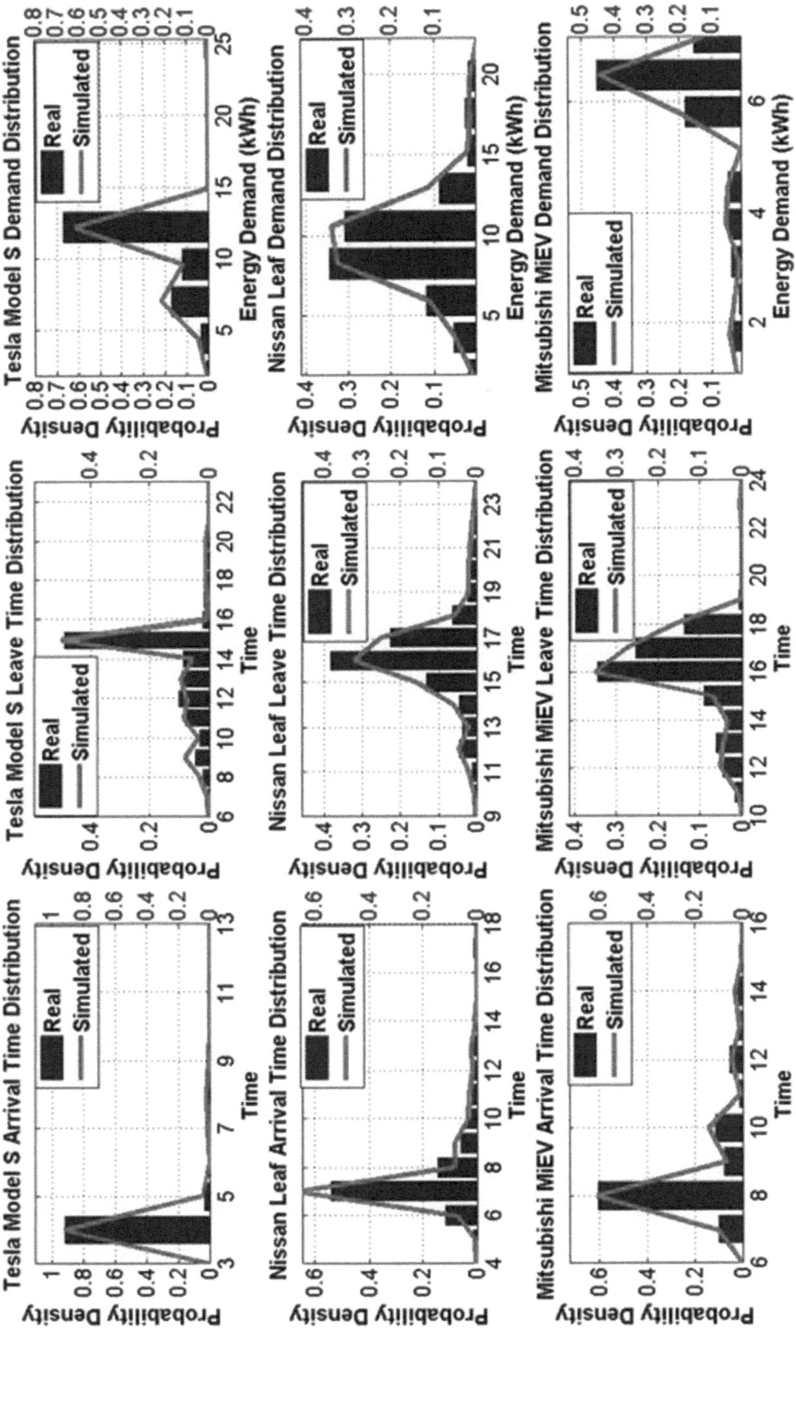

FIG. 4 Typical patterns of EV users.

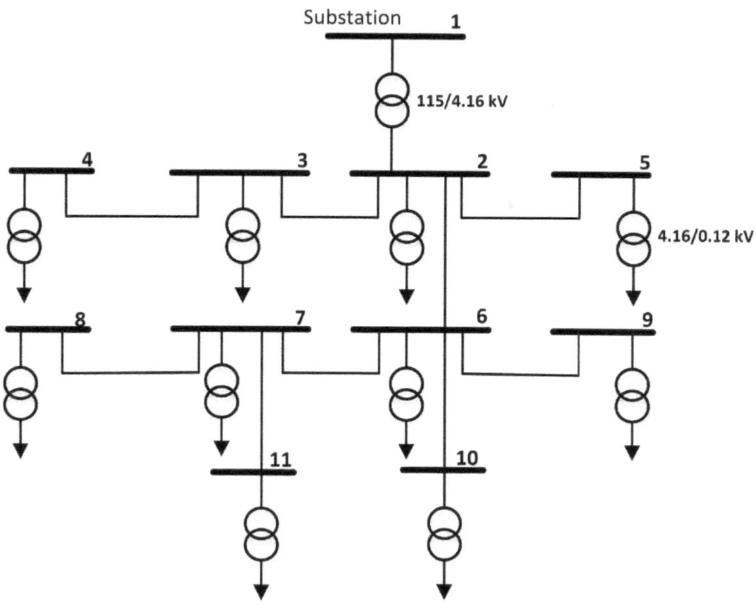

FIG. 5 Modified IEEE 13-bus radial test feeders.

Price and load: We use the same price model as the microgrids scenario. Fixed load profile at each bus is obtained from [38] and plotted in Fig. 3.

EVs and BESSs: We place one BESS at each bus in Fig. 5 except for the root bus. On EV side, we duplicate 1 out of the 19 EVs to make the total EV number 20, and we evenly distribute the two EVs at each distribution bus except for the root node. The EV data are collected the same way as described in the earlier section.

Network related settings: Table 2 summarizes the key parameters used for distribution network simulation. And finally, distribution network-related parameters are from IEEE benchmark test system and real-life regulations in US power systems. For the decentralized DSM, we set weighting factor

TABLE 2 Key Parameters Used for Distribution Networks

Parameters	Rating
Voltage magnitude limit	[0.95 1.05] p.u.
Substation power limit	5 MVA
Substation secondary voltage	4.16 kV

$\gamma = 1$, step size $\rho_1 = 1$, $\rho_2 = 2$ and predefined error boundary $\epsilon = 10^{-3}$. The over relaxation parameter is set to $\theta = 1.75$.

Both proposed DSMs are solved with Gurobi solver on a PC with 2.67 GHz CPU and 8 GB RAM.

4.2 Case Study

4.2.1 Microgrids Scenario

We generate 200 scenarios for the numerical simulation based on the KDE of EVs. To demonstrate the correctness of the proposed model, we run the numerical simulation, and its results are plotted in Fig. 6. In the figure, we show the energy profile of 1 of the 11 BESSs and 1 EV out of the 20 simulated EVs. Note that given we generate 200 scenarios, we are only plotting out 1 out of 200 scenarios for validation purposes. It is shown in the figure that after a full day's operation, the BESS returns to its original state-of-charge. On the EV side, the EV arrives at 4 a.m. and leaves at 3 p.m. The EV charge and discharge based on the price signal, however, reaching 100% SoC by the time it leaves. The simulated case demonstrates the correctness of the proposed model.

4.2.2 Distribution Grids Scenario

We generate 200 scenarios using the SAA and the convergence of the proposed DSM is presented in Fig. 7. Fig. 7A shows the convergence of objective function for both over-relaxed ADMM (o-ADMM) DSM and regular ADMM DSM. Both DSMs converges to the optimum solved in a centralized manner. After the DSMs converge, the performance of both DSMs is stable. Furthermore, we pick the slowest converging bus, Bus 11, to take a closer look at the convergence defined by Eq. (14) in Fig. 7B. It shows the primal dual residuals of the o-ADMM DSM and ADMM DSM at Bus 11 for both active (P) and reactive power (Q) corresponding to Eq. (14). It is observed that the over relaxation

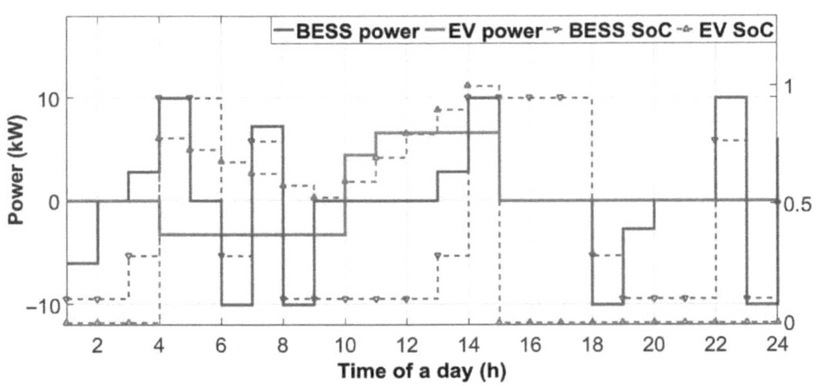

FIG. 6 One BESS and EV profile in microgrids.

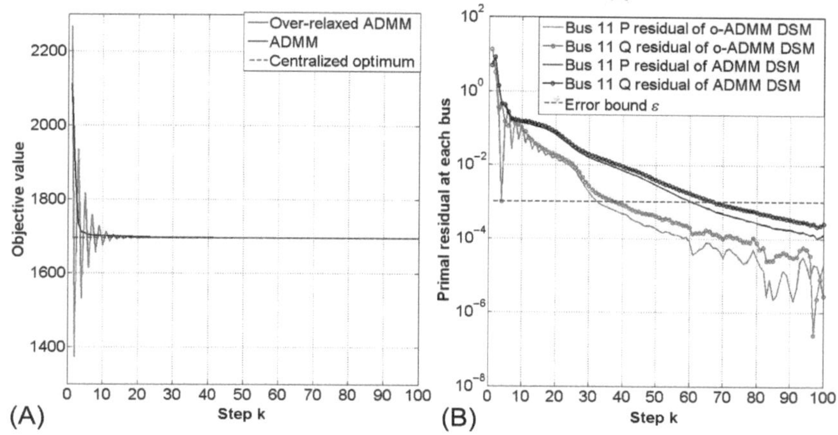

FIG. 7 (A) Convergence of the decentralized DSMs and (B) primal residuals of Bus 11.

speeds up the convergence: the o-ADMM DSM converges in 36 iterations while the ADMM DSM takes 70. After 80 iterations, the o-ADMM DSM hits its performance limitation; the residuals stop to decline and stay in an acceptable range. Moreover, we keep tracking the convex relaxation in Eq. (10) and found it to be tight for all iterations. In a word, the o-ADMM based DSM converges relatively fast.

To demonstrate the accuracy of the numerical experiments, we also present load profile under Bus 11. Bus 11 is picked for analysis as it is farthest from the root node. As voltage drop is most significant for the buses far away from the root node, we would like to make sure all the constraints are met at Bus 11. First, we examined the load profile under Bus 11, the battery's active and reactive power together with two EV profiles are shown in Fig. 8A. Note that the EV profile is one generated scenario out of 200 total scenarios. It is shown that the battery is most of the time supporting reactive power, which is expected for compensate for the voltage drop in distribution system. Furthermore, in Fig. 8B, we align the time axis and present the SoCs for BESS and EVs in the same scenario. For easy representation, we normalized the SoE used in Section 3 to SoC in figures. By comparing Fig. 8A and B, it is clearly shown that the EV1 arrives at 5 a.m. and leaves at 1 p.m., while EV2 arrives at 8 a.m. and leaves at 9 p.m. There are some V2G operations for both EVs. In Fig. 8B, we can also find out prevention of deep discharge of the battery when it hits the SoC lower limit. It needs to be pointed out that the info of battery, EV1, and EV2 in Fig. 8 are only accessible by SDSO using the proposed o-ADMM DSM, therefore persevering individual privacy from DSO.

Besides checking the load profiles and SoCs, we further present the bus aggregated load and voltage profile from a DSO's point of view. The aggregated active load, reactive load, and voltage profile for the same bus (Bus 11) is shown in Fig. 9. We notice that the voltage regulation is performed during

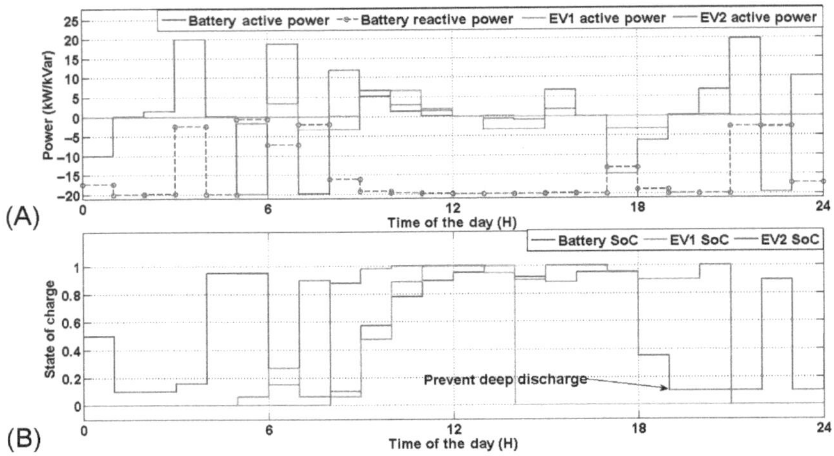

FIG. 8 SDSO accessible: (A) Load profile under Bus 11 and (B) SoC of batteries and EVs.

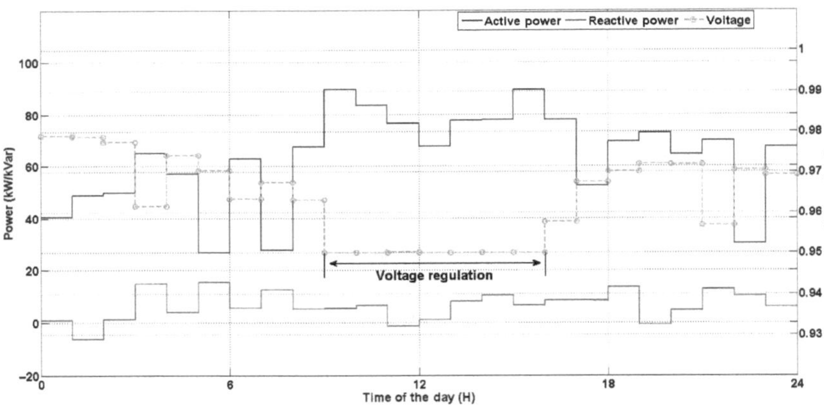

FIG. 9 DSO accessible: Bus 11 aggregated load profile and bus voltage.

10 a.m. and 5 p.m., a time window when system has heavy active load and voltage profile decays accordingly. By comparing Fig. 9 with Fig. 8A, we find out that battery is under reactive power compensation mode in the same time. Figs. 8 and 9 mutually verify the correctness of the proposed DSM.

Note that one specific stochastic energy management decision does not necessarily outperform the other candidates (because the realization of uncertainties could be an extreme case), however, the stochastic optimization gives the optimal decision for an "averaged" case. In other words, if the energy management experiment is repeated over a long period of time, the optimality reached by this paper will outperform the others.

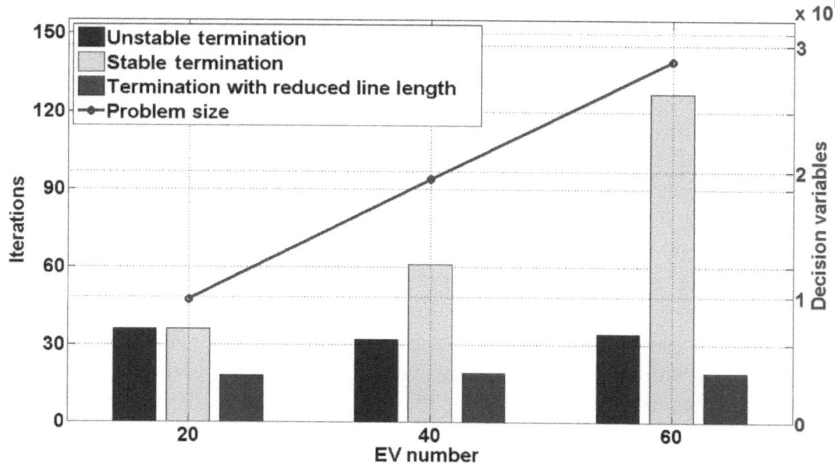

FIG. 10 DSM scalability and convergence vs. the size of the problem.

4.3 Discussions

4.3.1 Scalability

Fig. 10 shows the scalability of the proposed DSM under different size of the EV number. We change the number of EVs in the distribution network, from 20 EVs in the original setting to 40 EVs and 60 EVs. Given the limitation of the actual EV user profile we have collected, we duplicate the number of EVs under each bus to increase the EV number. We run the numerical experiment by keeping all the other settings the same. In Fig. 10, it can be clearly seen that as the EV number increases, the problem size scales up linearly with the EV number. However, as the stochastic optimization needs to generate many scenarios for each EV (in this case 200), the problem size grows very fast. By integrating 60 EVs, it is already extremely slow to solve the DSM in a centralized way. On the other hand, if we solve it in a decentralized manner, the computational burdens are decentralized to SDSO. In Fig. 10, we show three cases of numerical results, namely unstable termination, stable termination, and termination with reduced line length. The unstable termination refers to the case whose termination criteria is met according to Eq. (14) in certain iterations. However, if the algorithm continues to run for one more iteration, the termination criteria is violated. A good example of this case is shown in Fig. 7B: at Iteration 5, the oscillation introduced by the relaxed term in Eq. (15) has reduced the error below the preset error bound. However, such oscillation has enlarged the error above the error bound in the next iteration. The stable termination stands for termination where error continues to stay below error bound. An illustrative example is Fig. 7B at Iteration 36. After this iteration, the error continuously stays below error bound. We also show the scalability of the same problem with a reduced distribution line length. The line length in this case is reduced to one fifth of the original length.

There are three observations for this numerical experiment: (a) by comparing the unstable termination under different size of EV integration, we see a good chance that the o-ADMM based DSM terminates in 10s of iterations. The number of iterations does not grow much as the problem size grows. (b) By comparing the stable terminations, we find it scales as the problem scales up. By fitting a first order polynomial between the number of stable iterations and the problem size, it is found out that the decentralized method needs to increase one iteration with every increment of 2000 decision variables. There is a chance where the introduced oscillation terminates the algorithm in early stage, however, such termination is not guaranteed. However, to obtain guaranteed (stable) termination, the iteration scales up with the problem size. In other words, the decentralized DSM reduce the computational burden for the central controller at the expense of longer computational time. (c) By comparing the termination with reduced line length case, we further find out that the increase in stable termination results from a harder problem: the increase in EV numbers (and hence in energy demand) in the system has made the voltage regulation constraints hard meet. However, if we reduce the line length of the distribution system, the constraints are more easily met and hence iterations do not increase with the problem size.

If the EV size continuous to grow, it can be projected that the problem is harder to solve. With the introduction of more EVs, it not only increases the number of decision variables, at the same time, the EV charging demands also make the voltage and power flow constraints harder to be met. If the EV size is over several hundred, the model used in this chapter may need further adjustments: instead of modeling the stochastic behavior of each individual EV, stochastic modeling should be performed for aggregated EV load under each bus.

4.3.2 Lumped Model vs. Networked Model

We further study the traditional DSM which only accounts for the local effects (LDSM), that is, minimizing the operational cost under each bus. We compare the traditional LDSM with the DSM developed in this chapter in Fig. 11. The two cases are created under same simulation setup with the only difference being LDSM does not account for the network and power flow constraint. As shown in Fig. 11, we plot the aggregated power losses for the whole network at each time step for both DSMs. Moreover, the maximum and minimum voltage in the whole network at each time step for the two DSMs are also potted for comparisons. It is observed that the LDSM creates a much larger power loss over the distribution line. The total distribution loss is 353.9 kWh for the studied distribution grid by using the proposed DSM for operation. On the other hand, if the traditional lumped model-based DSM is applied, the total distribution losses

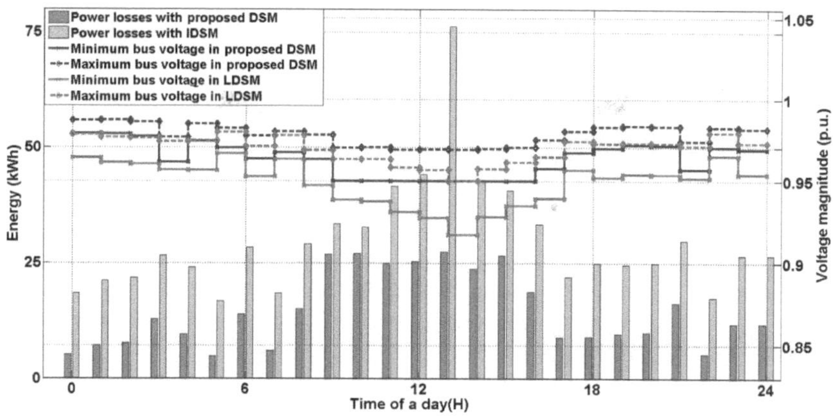

FIG. 11 Comparisons of power losses and voltage regulation for DSM and LDSM.

for the load profile will be 719.8 kWh, a loss twice the size of the proposed method. Furthermore, it is also observed that the minimum voltage in the system drops to 0.92 p.u. by using lumped model-based DSM, while the minimum voltage in the proposed networked based model is regulated at 0.95 p.u. The numerical results demonstrate the importance of switching from traditional LDSM to the proposed DSM which accounts for the network topology.

5 CONCLUSIONS

In this chapter, we have studied two DSMs. One is for the microgrids, which treat all loads and generations under a same bus using lumped models, while the other is for distribution systems which considers the power flow between different buses using the networked model. In both models, to account for the stochastic nature of DERs, a SAA-based stochastic optimization framework is proposed. For the distribution grids case, a decentralized framework is studied to alleviate the centralized computational burden. Numerical results show on one hand, and decentralized framework and SAA-based stochastic optimization are complementary. The decentralized method makes the computational demanding numerical approximation much more scalable. On the other hand, the networked model does show some benefits over the lumped model in terms of better managing the network voltage and economic operations, however, at the cost of more computational complexity. For DSM designers, there is no simple answer on which DSM outperforms the other: the selection is highly dependent on the application scenario. For future researchers, asynchronous decentralized methods will be studied to account for communication delay between different nodes.

REFERENCES

[1] L.T. Berger, K. Iniewski, Smart Grid: Applications, Communications, Security, Wiley, Hoboken, NJ, 2012.

[2] L. Barelli, G. Bidini, F. Bonucci, A micro-grid operation analysis for cost-effective battery energy storage and RES plants integration, Energy 113 (2016) 831–844.

[3] Q. Li, F. Bai, B. Yang, Z. Wang, B. El Hefni, S. Liu, S. Kubo, H. Kiriki, M. Han, Dynamic simulation and experimental validation of an open air receiver and a thermal energy storage system for solar thermal power plant, Appl. Energy 178 (2016) 281–293.

[4] S. Mukoyama, T. Matsuoka, H. Hatakeyama, H. Kasahara, M. Furukawa, K. Nagashima, M. Ogata, T. Yamashita, H. Hasegawa, K. Yoshizawa, Y. Arai, Test of REBCO HTS magnet of magnetic bearing for flywheel storage system in solar power system, IEEE Trans. Appl. Supercond. 25 (3) (2015) 1–4.

[5] S. Mocci, N. Natale, F. Pilo, S. Ruggeri, Demand side integration in LV smart grids with multi-agent control system, Electric Power Syst. Res. 125 (2015) 23–33.

[6] Y. Wang, O. Sheikh, B. Hu, C.C. Chu, R. Gadh, in: Integration of V2H/V2G hybrid system for demand response in distribution network, Smart Grid Communications (SmartGridComm), 2014 IEEE International Conference on, IEEE, 2014, pp. 812–817.

[7] H. Nazaripouya, Y. Wang, P. Chu, H.R. Pota, R. Gadh, in: Optimal sizing and placement of battery energy storage in distribution system based on solar size for voltage regulation, 2015 IEEE Power and Energy Society General Meeting, IEEE, 2015, pp. 1–5.

[8] Y. Wang, H. Nazaripouya, C.C. Chu, R. Gadh, H.R. Pota, in: Vehicle-to-grid automatic load sharing with driver preference in micro-grids, Innovative Smart Grid Technologies Conference Europe (ISGT-Europe), 2014 IEEE PES, IEEE, 2014, pp. 1–6.

[9] Y. Wang, B. Wang, R. Huang, C.C. Chu, H.R. Pota, R. Gadh, in: Two-tier prediction of solar power generation with limited sensing resource, Transmission and Distribution Conference and Exposition (T&D), 2016 IEEE/PES, IEEE, 2016, pp. 1–5.

[10] Y. Wang, B. Wang, C.C. Chu, H. Pota, R. Gadh, Energy management for a commercial building microgrid with stationary and mobile battery storage, Energy Buildings 116 (2016) 141–150.

[11] O. Erdinc, N.G. Paterakis, T.D.P. Mendes, A.G. Bakirtzis, J.P.S. Catalao, Smart household operation considering Bi-directional EV and ESS utilization by real-time pricing-based DR, IEEE Trans. Smart Grid 6 (3) (2015) 1281–1291.

[12] Y. Wang, W. Shi, B. Wang, C.C. Chu, R. Gadh, Optimal operation of stationary and mobile batteries in distribution grids, Appl. Energy 190 (2017) 1289–1301.

[13] P. Kayal, C.K. Chanda, Placement of wind and solar based DGs in distribution system for power loss minimization and voltage stability improvement, Int. J. Electric. Power Energy Syst. 53 (2013) 795–809.

[14] P. Mesarić, S. Krajcar, Home demand side management integrated with electric vehicles and renewable energy sources, Energy Buildings (2015). Available online 3 September.

[15] A. Chabaud, J. Eynard, S. Grieu, A new approach to energy resources management in a grid-connected building equipped with energy production and storage systems: a case study in the south of France, Energy Buildings 99 (15) (2015) 9–31.

[16] P. Malysz, S. Sirouspour, A. Emadi, An optimal energy storage control strategy for grid-connected microgrids, IEEE Trans. Smart Grid 5 (4) (2014) 1785–1796.

[17] X. Bai, W. Qiao, Robust optimization for bidirectional dispatch coordination of large-scale V2G, IEEE Trans. Smart Grid 6 (4) (2015) 1944–1954.

[18] Q. Wang, J. Wang, Y. Guan, Price-based unit commitment with wind power utilization constraints, IEEE Trans. Power Syst. 28 (3) (2013) 2718–2726.

[19] H.T. Nguyen, L.B. Le, Optimal energy management for building microgrid with constrained renewable energy utilization, IEEE Int. Conf. Smart Grid Communications 2014 (2014) 133–138.

[20] J. Yang, L. He, S. Fu, An improved PSO-based charging strategy of electric vehicles in electrical distribution grid, Appl. Energy 128 (2014) 82–92.

[21] A.F. Crossland, D. Jones, N.S. Wade, Planning the location and rating of distributed energy storage in LV networks using a genetic algorithm with simulated annealing, Int. J. Electric. Power Energy Syst. 59 (2014) 103–110.

[22] T. Sousa, H. Morais, J. Soares, Z. Vale, Day-ahead resource scheduling in smart grids considering vehicle-to-grid and network constraints, Appl. Energy 96 (2012) 183–193.

[23] P. Siano, C. Cecati, H. Yu, J. Kolbusz, Real time operation of smart grids via FCN networks and optimal power flow, IEEE Trans. Ind. Inform. 8 (4) (2012) 944–952.

[24] N. Yadaiah, A.G.D. Kumar, J.L. Bhattacharya, Fuzzy based coordinated controller for power system stability and voltage regulation, Electric Power Syst. Res. 69 (2) (2004) 169–177.

[25] Y. Zhang, G.B. Giannakis, Distributed stochastic market clearing with high-penetration wind power, IEEE Trans. Power Syst. 31 (2) (2016) 895–906.

[26] W. Shi, X. Xie, C.C. Chu, R. Gadh, Distributed optimal energy management in microgrids, IEEE Trans. Smart Grid 6 (3) (2015) 1137–1146.

[27] L. Wang, S. Sharkh, A. Chipperfield, Optimal coordination of vehicle-to-grid batteries and renewable generators in a distribution system, Energy 113 (2016) 1250–1264.

[28] Z. Xu, X. Guan, Q.-S. Jia, J. Wu, D. Wang, S. Chen, Performance analysis and comparison on energy storage devices for smart building energy management, IEEE Trans. Smart Grid 3 (4) (2012) 2136–2147.

[29] J. Lin, K.-C. Leung, V.O.K. Li, Optimal Scheduling With Vehicle-to-Grid Regulation Service, IEEE Internet Things J. 1 (6) (2014) 556–569.

[30] E. Sortomme, M.A. El-Sharkawi, Optimal Scheduling of Vehicle-to-Grid Energy and Ancillary Services, IEEE Trans. Smart Grid 3 (1) (2012) 351–359.

[31] A. Shapiro, D. Dentcheva, A. Ruszczyński, Lectures on Stochastic Programming: Modeling and Theory, second ed., Society for Industrial and Applied Mathematics, (2009).

[32] B.W. Silverman, Density Estimation for Statistics and Data Analysis, CRC Press, 1986.

[33] S.H. Low, Convex relaxation of optimal power flow—part I: formulations and equivalence, IEEE Trans. Control Netw. Syst. 1 (1) (2014) 15–27.

[34] S.H. Low, Convex relaxation of optimal power flow-part II: exactness, IEEE Trans. Control Netw. Syst. 1 (2) (2014) 177–189.

[35] S. Boyd, N. Parikh, E. Chu, B. Peleato, J. Eckstein, Distributed optimization and statistical learning via the alternating direction method of multipliers, Found. Trends Mach. Learn. 3 (1) (2010) 1–122.

[36] J. Eckstein, Parallel alternating direction multiplier decomposition of convex programs, J. Optim. Theory Appl. 80 (1) (1994) 39–62.

[37] ERCOT, Day-ahead Energy Price, [Online]. Available from: http://www.ercot.com/.

[38] Cornell University Facilities Service, Real Time Building Utility Use Data, [Online]. Available from: http://portal.emcs.cornell.edu/.

[39] W. Kersting, in: Radial distribution test feeders, Proc. IEEE PES Winter Meet., Columbus, OH, USA, January, 2001, pp. 908–912.

Chapter 14

Impact of Integrated Optimization of Independent Energy Carriers on Power Systems

M.J. Vahid-Pakdel, M. Pesaran Hajiabbas and B. Mohammadi-Ivatloo
Faculty of Electrical and Computer Engineering, University of Tabriz, Tabriz, Iran

1 CHAPTER ORGANIZATION

This chapter includes seven sections, which present an explanation about multicarrier energy systems in addition to their formulations and simulations. Section 2 introduces the topic of multicarrier energy system and discusses the basics of integrated energy systems formation. This section also gives explanations about fundamentals of an energy hub and reviews some literature in this regard. Section 3 explores the different sizes of multicarrier energy systems problems. Diverse scales of these problems from a mono energy hub optimization to wide spread integrated systems investigation have been mentioned in this section. The more detailed discussion about the simultaneous optimization of electrical and natural gas networks considering their technical constraints has been provided in Section 4. Section 5 presents a comprehensive mathematical formulation of optimal energy flow of multicarrier energy systems (electricity-gas) which was explained in Section 4. The simulation of proposed formulation in Section 5 has been done in Section 6 and its results have been demonstrated. A brief discussion about the impact of multicarrier energy system concept on power systems was served in Section 7 and, finally, Section 8 is the conclusion of the chapter.

2 MULTICARRIER ENERGY SYSTEMS

The world's energy crisis has received a great deal of attention in recent years. The limitation of fossil fuels as the major source of world's energy, environmental problems of power plants, and the incremental condition of energy

Classical and Recent Aspects of Power System Optimization. https://doi.org/10.1016/B978-0-12-812441-3.00014-8
© 2018 Elsevier Inc. All rights reserved.

demands beside the high investment cost of the energy networks have highlighted the importance of the energy management issue. Nowadays, energy management strategies are suggested in different scales with various targets. Energy management may be proposed for economical or environmental targets. The main goal of energy management is to suggest the optimal operational strategies according to the optimization objective concerns. Energy management issues are faced in all energy carriers including electricity, natural gas, and district heating. The main purpose of energy management studies is to meet the consumption demand with lower cost, lower greenhouse gases, and higher reliability within technical constraints considerations. For instance, the economic dispatch or optimal power flow problems in the electricity network or the optimal energy flow in the natural gas network are the samples of energy management optimization problems which search for the optimal operation points of a system. In other words, each energy system can provide the energy demands with different arrangements of its components; however, dissimilar arrangements will result in various operation cost and greenhouse gas emissions. Hence, energy management optimization is employed to find the best arrangement with respect to problem targets.

With the entrance of progressive energy convertors technologies, new challenges have been introduced to the energy management studies [1]. The capability of new devices for converting, to convert different kinds of energy to each other, makes the comprehensive energy management analysis possible. Devices such as combined heat and power (CHP) technologies, with the ability to convert natural gas to electricity and heat, propose new ways for providing the electrical demand other than conventional energy import from the main electrical grid. Now, new optimization problems can be defined in this integrated condition. Investigating the energy management strategies in the integrated situation of energy carriers, which is called multicarrier energy systems or energy hub, can introduce new optimization potentials which were not accessible in single-carrier system studies. Here, an example is given to provide a better understanding of the multicarrier energy systems.

A typical energy hub schematic has been shown in Fig. 1. This energy hub contains two boilers with different input energy carrier for providing heat demand, one electrical transformer, and one CHP unit which is utilized for meeting the electrical and heat demand, simultaneously.

As shown in the figure, there are two ways to meet the electrical demand and three ways to meet the heat demand. The electrical demand is supported by importing the electricity from grid and using the gas fueled CHP. Similarly, the heat demand is supplied by both gas and electrical boilers. The integrated devices' efficiencies are assumed as illustrated in Table 1. So it is clear that optimal operation point of the internal devices in this condition will depends on the electricity and natural gas prices. If the electricity price is more than natural gas price for each energy unit, all electrical and heat demand will be provided by gas-powered devices. In this example with the higher price of electricity,

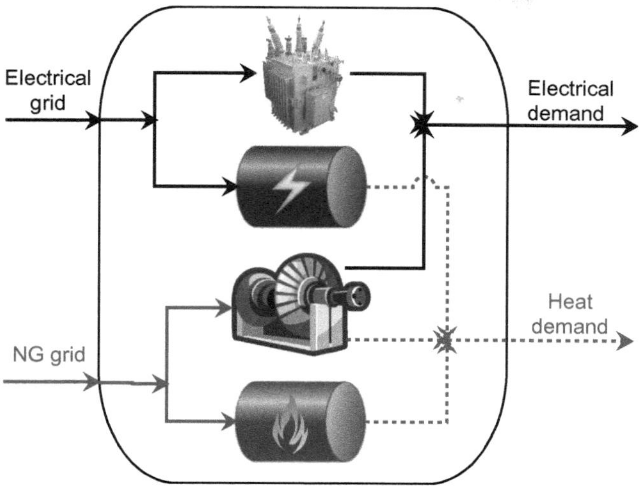

FIG. 1 Energy hub schematic.

TABLE 1 Devices' Efficiencies

Transformer	CHP (Gas to Electricity)	CHP (Gas to Heat)	Electrical Boiler	Gas-Powered Boiler
0.98	0.4	0.5	0.9	0.9

CHP will consume 2.5 p.u. natural gas energy to produce whole 1 p.u. electrical demand. On the other side, CHP heat production which will be 1.25 p.u. is sufficient for whole heat demand, and there is no necessity for utilization of a gas-powered boiler. Each hub should satisfy both relationships. First, the produced electrical energy should be equal to the consumed amount, to obtain the electrical energy balance, and second, the produced heat energy should not be lower than heat demand [2]. Now, consider the situation where natural gas is more costly than electricity. Energy hub manager's decision will be to use electrical-based devices to minimize the energy cost. So, in this condition, the imported power from grid will be 1.02 p.u. and electrical boiler input energy will be 1.11 p.u.

This simple example can explain the basic law of the multicarrier energy system. According to this law, the cheaper carrier will be used to support the demand as much as possible and this possibility will be specified by the technical constraints of devices or system.

Several optimization studies have been carried out on energy hubs in presence of new devices and technologies. A recent review of energy hub models and concepts is provided in [3]. Various energy storage systems, heat pumps, power to gas systems, and any other multienergy-related system can add new complexity as well as flexibility to the energy hub which makes its optimization more challenging and complicated. In addition, some researchers focused on operation optimization of energy hubs in the market environment [4]. In market-based studies the main issue is to find suitable coordination for the energy hub to optimize its objective with market price variations. Renewable energies will be one of the main components of future energy hubs. Hence, operation of renewable energy integrated energy hubs are widely studied [5]. The impact of uncertainty of wind power and energy demands on operation of smart energy hubs is studied in [6].

An important parameter in multicarrier energy system studies is the scale of optimization problem. This parameter can vary from a building size to a whole network scale comprising both electricity and natural gas energy carriers. A system comprising the natural gas network and electrical network, simultaneously, can be an instance of a large-scale multicarrier system for investigation [7]. This issue, which will be analyzed in greater detail in the following sections, can be summarized as the optimization of the coupled energy systems operation (gas and electricity) with satisfaction of each network's technical constraints. The coupling locations of the electricity network and natural gas system can be identified as the energy hubs or gas-powered power plants.

3 OPTIMIZATION PROBLEMS IN MULTICARRIER ENERGY SYSTEMS: RESIDENTIAL SCALE TO TRANSMISSION SCALE

Different optimization problems arise in the multicarrier energy system environment. These problems can be categorized by various criterion. Test system size is one criteria which may significantly influence the research. Multicarrier systems are investigated from a mono hub [8] to integrated electricity-gas girds [9]. Typically, mono hub is explored in the residential scale considering its user types. A considerable number of works in literature, which have focused on the mono hub optimization problems, have proposed energy management strategies for these hubs without taking into account the system details like voltage or lines capacity. The mono hubs optimization studies include different objective functions. The given energy management strategies for the control variables in the hubs optimization are based on determined targets. Whereas, most of studies in the mono hub subject have been done with economical target, other targets such as environmental issues [10] or technical attitudes [11] are considered, as well. Beside the mono hub studies classification according to their study targets, they can be categorized with respect to their optimization time period as well. Based on the optimization time period, the mono hubs investigation may be divided into two groups: (1) operation of an energy hub, (2) planning and designing of an energy hub. In the operational studies, the aim is to find the optimum set of hub's control variables like as devices setting while the determination of the type of devices or location of energy hub are classified as the energy hub planning and design issues. All mentioned subjects and study areas have been covered in literatures widely. Both operational and design studies have attempted to analyze the optimization problem with the different conditions. A group of operational studies have considered different devices in the energy hub and their impact on the hub's management [12]. Furthermore, the assessment of energy market impact on energy hub optimal operation [4] or the energy hub performance optimization inside a smart grid [13] are the examples of mono energy hub investigation. Furthermore, the studies concentrated on hub design and planning carried out their optimization in the presence of the various devices or conditions to make their study result more practical and reliable. The energy hub planning in presence of wind energy sources [8] is an example in this scope.

The probabilistic optimization of energy hubs have been provided in literature, as well. Different methods of uncertainty consideration in decision making are available in recent studies. The methods such as scenario-based algorithms [4] or robust-based optimization studies [14] have been proposed in the literature.

Mono energy hub studies have been developed by considering multiple hubs connection to each other [15]. In the network energy hubs, the possibility of exchanging between hubs gives more degrees of freedom for optimization issues. The mentioned electrical or thermal energy exchanging capability serves a type of locality and independency for hubs' network management.

Typically, the energy networks modeling resolution in the hubs' energy management optimizations is not so high. In spite of the sufficiency of the resolution for these optimization problems, adding the energy systems constraints can improve the practicality of studies. Recently, the concept of the energy hub or investigating the different energy carriers simultaneously is motivated the researchers toward multicarrier energy systems. Reference [2] introduced the mentioned novel aspect into the multicarrier energy systems studies. Reference [1] presented the optimal energy flow optimization problem in integrated energy systems. As matter of fact, in the current world, energy systems are not independent from each other and each system operation relies on the other system. For an instance, electrical and natural gas networks have different linkage locations which causes the coupled operation of these networks. The aforementioned linkage may be found in natural gas-powered power plants, electrical gas compressors, power to gas units, and distributed energy hubs. The progressive condition of energy convertor technologies reinforces this coupling and interdependency. Therefore, in this interdependent situation of energy systems, the optimization of each carrier system separately will offer tangible results. Consequently, the optimization study of multienergy systems considering all technical limitations of systems is one of the main interests of researchers.

4 OPTIMAL ENERGY FLOW PROBLEM IN MULTICARRIER ENERGY SYSTEM

As mentioned earlier, the energy networks in our current world are linked to each other. One of the wide spread example of these coupled energy networks is the electricity and natural gas coupling. The mentioned coupling means that a change or failure in each system affects the other network. For example, loss of a natural gas pipe which supplies a gas-powered power plant will result in power plant trip. In addition, the economical operation of a coupled energy network without considering the mentioned coupling and mutual impacts will be imperfect. The optimal operation of electrical network or natural gas network has been discussed separately in articles; however, when a network is modeled solitarily without considering the coupled network model, the model will not be comprehensive. In other words, when in a single network operational optimization, the enforced cost of the other network for determination of the optimal network setting objective is unknown. The whole system cost may be increased by \$1 to earn only \$0.5 from our optimization procedure. To present a comprehensive optimization, considering all systems in operation, the optimal energy flow problem of multicarrier energy systems have been discussed in literature. In this optimization, it may looks like some networks operation cost is increased in comparison to their single optimization, but it can be guaranteed that the whole system operation cost will be lower than the total sum of the single networks optimization results.

It is worth mentioning that the coupled energy systems should not be limited only to the electricity and natural gas networks. There are other networks which can be included in this simultaneous investigation. In recent studies, district heating network impact on the integrated operation of electricity-gas system have been explored [16, 17]. The coupling points of district heating, electricity, and natural gas networks, which make the simultaneous optimization possible and necessary, are the distributed energy hubs. In energy hubs, because of the possibility of supporting the heating load by electricity (electrical boiler), natural gas (gas-powered boiler), or district heating network, these three systems are joined to each other and can affect each other. Furthermore, there are other energy networks and inclusion of them in this procedure will result in challenging optimization problems. Liquid fossil fuels transportation networks, which join to the electrical network in fossil fuel powered power plants, is one example which shows that multicarrier energy system studies can be extended furthermore.

Studies, which have focused on the optimal energy flow problem in multicarrier energy systems, attempt to introduce novel optimization approaches for handling this problem. Different optimization concepts and ideas are available in literature. Reference [7] has proposed decomposition technique which utilizes the dummy variables concept for breaking down the problem into ordinary optimal power flow problems. Multiagent genetic algorithm as an evolutionary method has been employed to solve the optimization problem at the end. Linearization of problem have been done in [9] by using multidimension piecewise linear approximation technique. In addition, evolutionary algorithms-based frameworks for carrying out the optimization problem can be found in literature [18], as well.

In continuation, mathematical formulation of optimal energy flow in multi-carrier energy systems is presented and a discussion based on the results will be presented.

5 MULTICARRIER ENERGY SYSTEM OPTIMAL ENERGY FLOW: PROBLEM FORMULATION

In this section an optimal energy flow problem has been presented for coupled electricity and natural gas networks. The coupling locations are distributed energy hubs which schematic presentation were shown in Fig. 1. This optimization problem can be formulated as follows.

5.1 Objective Function

The optimization target is economical and the objective function includes the energy networks operation costs. Therefore, the problem's objective is definable as follows:

$$\text{OF} = \text{Cost}_\text{e} + \text{Cost}_\text{g} \tag{1}$$

$$\text{Cost}_\text{e} = \sum_{i_\text{e}=1}^{M_\text{e}} A_{i_\text{e}} P_{i_\text{e}}^2 + B_{i_\text{e}} P_{i_\text{e}} + C_{i_\text{e}} \tag{2}$$

$$\text{Cost}_\text{g} = \sum_{i_\text{g}=1}^{M_\text{g}} A_{i_\text{g}} P_{i_\text{g}}^2 + B_{i_\text{g}} P_{i_\text{g}} + C_{i_\text{g}} \tag{3}$$

where Cost_e and Cost_g are the operation cost of the electrical network and natural gas network, respectively. M_e and M_g show the number of power plants and natural gas supplier units, respectively. Power plants and gas supplier units operation cost function coefficients have been demonstrated by $A_{i_\text{e}}, B_{i_\text{e}}, C_{i_\text{e}}$ and $A_{i_\text{g}}, B_{i_\text{g}}, C_{i_\text{g}}$, respectively. P_{i_e} and P_{i_g} are responsible for the generated power of i_eth and i_gth generation units. It is worth mentioning that this formulation is presented based on the assumption that there is no private ownership in the networks. In other words, this optimization is from the perspective of a system operator who wants to minimize the whole system cost. In the following, the constraints of optimization problem are described.

5.2 Constraints

5.2.1 Electrical Network: AC Power Flow

Implementation of electrical power flow relationship is necessary for the modeling of the electrical network. The AC power flow equations of electrical network which imply the power balance have been demonstrated in Eqs. (4), (5).

$$P_i = d_i^p + \sum_{j=1}^{N_e} V_i V_j Y_{ij} \cos(\delta_i - \delta_j - \gamma_{ij}) \tag{4}$$

$$Q_i = d_i^q + \sum_{j=1}^{N_e} V_i V_j Y_{ij} \sin(\delta_i - \delta_j - \gamma_{ij}) \tag{5}$$

where P_i and Q_i are the injected active and reactive power into the network in ith bus. d_i^p and d_i^q account for ith bus active and reactive demand. N_e is the number of electrical buses, V_i and δ_i show the ith bus voltage amplitude and phase. Finally, Y_{ij} and γ_{ij} are the power system admittance matrix ijth element amplitude and phase, respectively.

5.2.2 Electrical Network: Voltage Constraints

Violation of the electrical buses voltage from predetermined range is not acceptable. So, the following relation is considered to satisfy the mentioned constraint:

$$V^{\min} \le V_i \le V^{\max} \tag{6}$$

where $V^{\min}$ and $V^{\max}$ are the minimum and maximum margins of the voltage amplitude.

5.2.3 Electrical Network: Generation Units Limits

The operation of power plants in electrical power network is permitted only in their technical range.

$$P_{i_e}^{\min} \le P_{i_e} \le P_{i_e}^{\max} \tag{7}$$

$$Q_{i_e}^{\min} \le Q_{i_e} \le Q_{i_e}^{\max} \tag{8}$$

where P_{i_e} and Q_{i_e} are the i_eth power plant generated active and reactive power, while $P_{i_e}^{\min}, P_{i_e}^{\max}$ and $Q_{i_e}^{\min}, Q_{i_e}^{\max}$ present the generated active and reactive power boundaries, respectively.

5.2.4 Electrical Network: Line Power Flow

The passing power through the electrical lines should not be more than the line's capacity. The relation (9) illustrates this limitation:

$$-\mathrm{EPF}_i^{\max} \leq \mathrm{EPF}_i \leq \mathrm{EPF}_i^{\max} \tag{9}$$

where the EPF_i is the passing power through the ith line and $\mathrm{EPF}_i^{\max}$ shows its capacity.

5.2.5 Natural Gas Network: Pipe Lines Flow

To model the natural gas network, we start from pipes' flow modeling. The gas flow relation through the pipe which is connected between nodes i and j was identified:

$$f_{ij}^{\text{line}} = \xi_{ij}^{\text{line}} si(\pi_i, \pi_j)\sqrt{si(\pi_i, \pi_j)(\pi_i^2 - \pi_j^2)} \tag{10}$$

where f_{ij}^{line} is the mass flow of pipe and ξ_{ij}^{line} shows the pipe's constant which depends on the pipe's characteristics like interior diameter, length, gas compressibility, and gas temperature. π_i represents the ith node pressure. In addition, $si(\pi_i, \pi_j)$ is a typical sign function which its value is equal to 1 if $\pi_i > \pi_j$ and is equal to -1 for $\pi_i < \pi_j$.

5.2.6 Natural Gas Network: Compressors

The pressure drop in the natural gas network forces the use of the compressors to compensate. Compressors consume energy for pressure compensation. Nevertheless, the compressors are not able to increase the pressure beyond their increase rate. The following equations present these constraints for gas-powered compressor.

$$f_{ij}^{\text{comp}} = \xi_{ij}^{\text{comp}} f_{jk}^{\text{line}} (\pi_j - \pi_i) \tag{11}$$

$$\mathrm{Rate}_{\text{comp}}^{\min} \leq \frac{\pi_j}{\pi_i} \leq \mathrm{Rate}_{\text{comp}}^{\max} \tag{12}$$

where f_{ij}^{comp} shows the mass flow of compressor consumption and ξ_{ij}^{comp} is compressor constant. It is worth mentioning that the discussed compressor is placed between nodes i and j which its discharge flange is connected between nodes j and k. In addition, $\mathrm{Rate}_{\text{comp}}^{\min}$ and $\mathrm{Rate}_{\text{comp}}^{\max}$ are the upper and lower limits of the compressor increase rate.

5.2.7 Natural Gas Network: Pressure Limits

Nodes pressure in the natural gas network should be within acceptable range to satisfy the following technical circumstance.

$$\pi^{\min} \leq \pi_i \leq \pi^{\max} \tag{13}$$

where $\pi^{\min}$ and $\pi^{\max}$ are the minimum and maximum operating pressure of the nodes, respectively.

5.2.8 Natural Gas Network: Gas Supply Units Constraints

Natural gas suppliers should comply with their production limitation during the optimization as demonstrated in Eq. (14):

$$S_{i_g}^{\min} \le S_{i_g} \le S_{i_g}^{\max} \tag{14}$$

where S_{i_g} is the production amount of i_gth supplying unit, and $S_{i_g}^{\min}, S_{i_g}^{\max}$ are the unit's upper and lower boundaries. In addition, the conversion of suppliers' mass production to power can be expressed by Eq. (15).

$$P_{i_g} = S_{i_g} \times \mathrm{GHV} \tag{15}$$

where GHV is gas heat value of the natural gas.

5.2.9 Natural Gas Network: Gas System Energy Balance

To satisfy the gas power balance in each node, the mass flows are converted to the powers by Eqs. (16), (17):

$$P_{ij}^{\mathrm{line}} = \mathrm{GHV} \times f_{ij}^{\mathrm{line}} \tag{16}$$

$$P_{ij}^{\mathrm{comp}} = \mathrm{GHV} \times f_{ij}^{\mathrm{comp}} \tag{17}$$

Then, the gas power balance for each node can be indicated as follows:

$$S_i \times \mathrm{GHV} = (d_i^{\mathrm{gas}} \times \mathrm{GHV}) + \sum_{j=1}^{N_g} \beta_{ij}^{\mathrm{line}} P_{ij}^{\mathrm{line}} + \sum_{j=1}^{N_g} \beta_{ij}^{\mathrm{comp}} P_{ij}^{\mathrm{comp}} \tag{18}$$

were S_i and d_i^{gas} are the injected gas and gas demand in ith node. $\beta_{ij}^{\mathrm{line}}$ and $\beta_{ij}^{\mathrm{comp}}$ are known binary matrices, which elements show the presences of pipe and compressor between nodes i and j, respectively.

5.2.10 Energy Hubs: Energy Balance

The energy hubs models for the coupling locations of electrical and natural gas networks should be added to the formulation. The hub's energy balance relations have been considered in Eqs. (19)–(21). Eqs. (19), (20) calculate the overall active and reactive power demands of ith electrical bus which is provided by imported power from grid through the kth hub. Eq. (21) computes the hub's internal electrical power balance.

$$d_i^p = d_k^{p,\mathrm{Tran}} + d_k^{p,e_b} \tag{19}$$

$$d_i^q = \frac{d_k^q}{\eta_{\mathrm{Tran}}} \tag{20}$$

$$d_k^p = \eta_{\mathrm{Tran}} d_k^{p,\mathrm{Tran}} + \eta_{\mathrm{CHP}}^{ge} P_k^{\mathrm{CHP}} \tag{21}$$

where $d_k^{p,\mathrm{Tran}}$ and d_k^{p,e_b} are the electrical transformer and boiler input active powers, respectively. d_k^q shows the input reactive power of transformer, η_{Tran} and

η^{ge}_{CHP} represents the transformer and CHP's gas to electricity efficiencies. Finally, the P_k^{CHP} demonstrates the kth hub CHP input natural gas power.

After satisfaction of the electrical energy balance, the hub should have the ability to meet its heat energy demand. Then, the produced heat by the hub's devices should not be lower than hubs heating demand as explained in Eq. (22).

$$P_k^{h_b} \leq \eta_{G_{\text{boil}}} P_k^{G_{\text{boil}}} + \eta^{gh}_{\text{CHP}} P_k^{\text{CHP}} + \eta_{E_{\text{boil}}} d_k^{p,e_b} \tag{22}$$

where $P_k^{h_b}$ is the heat demand of kth hub, $\eta_{G_{\text{boil}}}$, $\eta_{E_{\text{boil}}}$, and η^{gh}_{CHP} are gas-powered boiler, electrical boiler, and CHP's gas to heat efficiencies, respectively. The input power of the gas-powered boiler has been shown with $P_k^{G_{\text{boil}}}$. In addition, the natural gas network ith node mass demand can be calculated from Eq. (23) relation.

$$d_i^{\text{gas}} \times \text{GHV} = P_{k,t}^{G_{\text{boil}}} + P_{k,t}^{\text{CHP}} \tag{23}$$

The assumption of this formulation is that the demands of electrical buses and natural gas node, which are not included in the energy hubs, are supplied directly from the grid. The next assumption is that the hub's input gas power is the only demand of the natural gas network.

5.2.11 Energy Hubs: Devices Capacity

Any optimization in the energy hub should consider the devices' technical limitations.

$$0 \leq p_i \leq capa_i^{\max} \quad \forall i = \text{Transformer, E} - \text{boiler, G} - \text{boiler, CHP} \tag{24}$$

where the $capa_i^{\max}$ is the ith device maximum capacity and p_i shows the ith device input power.

6 MULTICARRIER ENERGY SYSTEM OPTIMAL ENERGY FLOW: SIMULATION AND RESULT DISCUSSION

The proposed formulation in Section 5 is simulated by GAMS software in this section for different case studies. Modified IEEE 14 bus power system, which is linked with a 17-node natural gas network by distributed energy hubs, has been adopted as test system. Fig. 2 presents the test system and Table 2 shows the details of distributed energy hubs. The network information is available in [19]. The electrical networks bus demands are as same as the IEEE 14 bus standard test system which its demand is 259 MW and 73.5 MVAr. The hub's heating demands have been presented in Fig. 3. The proposed model in this chapter describes the fundamental mathematical model of problem, but using some techniques such as decomposition methods or evolutionary algorithms can make these formulation more suitable for large-scale optimizations.

The studied cases are for analyzing the impact of electricity and gas networks coupling on different components operation and whole system cost. For that, first, the networks operation have been considered independently

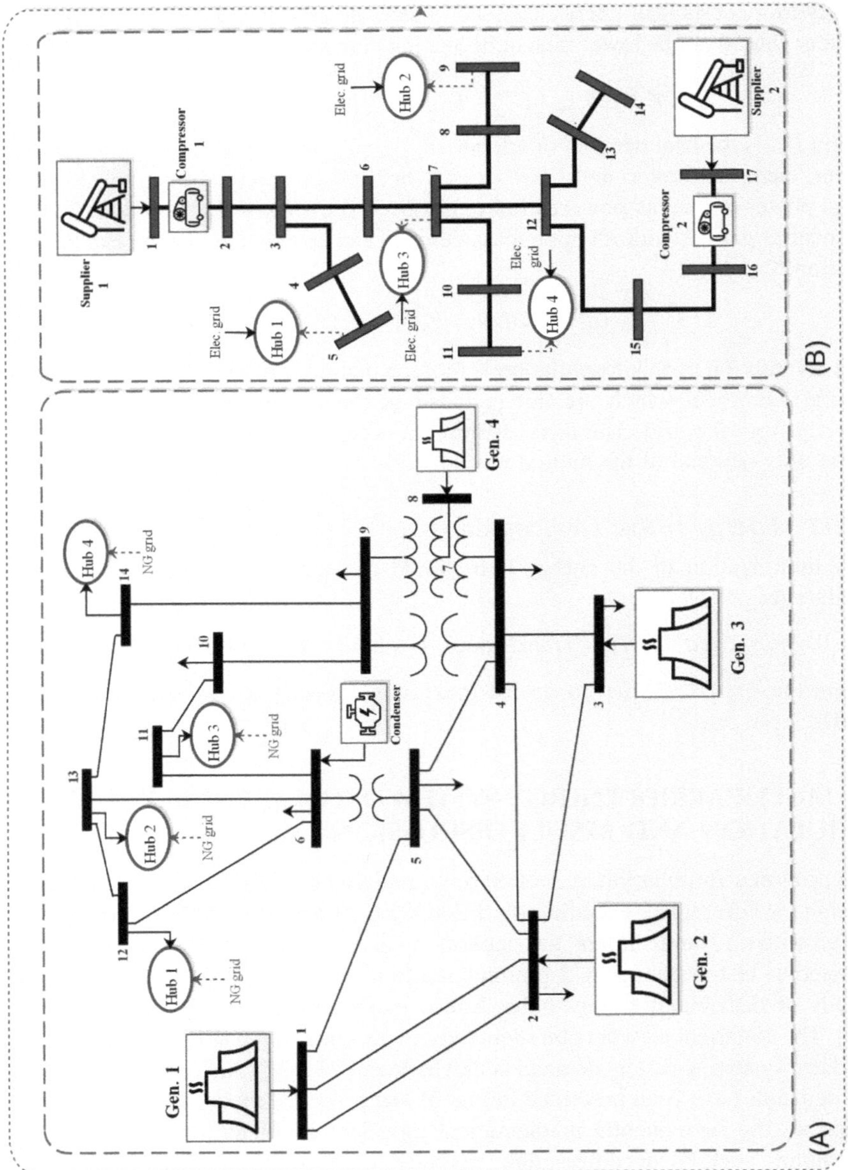

FIG. 2 Test system: (A) 14-bus IEEE test system, (B) 17-node natural gas network.

TABLE 2 Energy Hubs Information

Hub Number	Electrical Bus	NG Node	Devices			
			Electrical Boiler	*Gas-Powered Boiler*	*CHP*	*Transformer*
1	12	5	✓	✓	✓	✓
2	13	9	✓	✓	✓	✓
3	11	7	✓	✓		✓
4	14	11	✓	✓	✓	✓

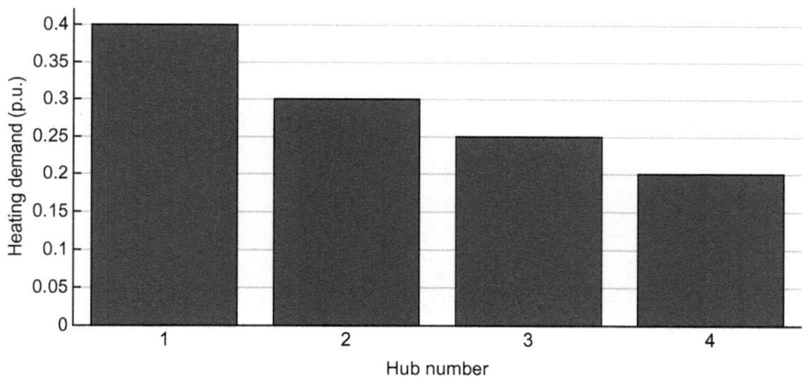

FIG. 3 Hubs heating demand.

and, second, the coupled operation has been investigated after integration. Coupled condition is provided by implementation of four distributed energy hubs into the networks. In the independent operation or the conditions when there is no energy hub in heat demand locations, the assumption is that the heat demand will be supported by gas-powered boiler with efficiency of 0.9 in hub's connected natural gas node.

Table 3 presents the energy networks operation costs for both independent and coupled condition. As shown in the table, the presence of energy hubs result in both electricity and natural gas networks operation costs reduction. Accordingly, whole system operation cost has been decreased about 6%. In other words, the provided flexibility by energy hubs integration reduces the total cost.

Linking the electrical network and natural gas network by energy hubs changes the optimal operation points of both networks components.

TABLE 3 Operation Cost ($)

Condition	Electricity Network Cost	Natural Gas Network Cost	Whole System Cost
Independent	8079	1308	9387
Coupled	7835	1011	8847

TABLE 4 Electrical Machines Output (p.u.)

	Independent		Coupled	
Machine	*P*	*Q*	*P*	*Q*
Generator 1	1.945	0	1.937	0
Generator 2	0.367	0.184	0.366	0.175
Generator 3	0.289	0.227	0.270	0.227
Generator 4	0.081	0.158	0.048	0.155
Condenser	–	0.240	–	0.240

TABLE 5 Natural Gas Components Operation (p.u.)

Component	Independent	Coupled
Supplier 1	0.764	1.411
Supplier 2	0.514	0
Compressor 1	0	0.067
Compressor 2	0	0

The electrical machines and natural gas suppliers are readjusted in new operation point. Tables 4 and 5 show the networks components operation points for both conditions.

Finally, the optimal operation settings of the energy hubs' devices have been reported in Table 6. As it is observable in the table, the CHPs have been operating with full capacity and electrical boilers have been shut down.

TABLE 6 Hubs Devices Input Power (p.u.)-Coupled Condition

Device	Hub 1	Hub 2	Hub 3	Hub 4
CHP	0.05	0.05	–	0.05
Transformer	0.041	0.115	0.0350	0.129
E-boiler	0	0	0	0
G-boiler	0.417	0.366	0.378	0.194

7 THE IMPACT OF INTEGRATED ASPECT ON POWER SYSTEM OPTIMIZATION PROBLEM

Consideration of energy systems optimization simultaneously affects the power systems typical optimization problems. Challenges, such as unit commitment or security constrained optimal power flow, become more critical when the natural gas network or other carriers are considered. However, these considerations make the optimizations more complicated, but, enhance their comprehensiveness, as well. Reference [20] proposes the security constrained optimal power flow of the power system beside natural gas flow problem. The unit commitment topic considering the power system security criterion has been discussed in [21]. The networks interdependency impact on transmission expansion of the networks can be found in [22]. Serving the flexible ramp condition for the power system in multienergy systems planning has been done in [23].

In recent literature, this integrated point of view has expanded the technical issues to include the market subjects or networks cyber security topics. These studies show the wide research area of multicarrier energy systems. Reference [24] develops a the methodology for identification and protection of the coupled electricity-natural gas networks, while and coupling approaches for the electricity and gas market has been proposed in [25].

8 CONCLUSION

The increasing number of gas-fired power plants in the electrical network and the augmentation of their share in the energy provision pattern have coupled two wide spread energy systems, electricity and natural gas, to each other. Furthermore, the entrance of the new technologies such as power to gas units or micro turbines among the energy hubs concept have reinforced the stated linkage. This coupling brings a valuable optimization potential despite increasing complexity. In addition, considering the natural gas constraints gives us a comprehensive and realistic view of the power system optimization problems.

The multicarrier energy systems are not limited to explained electricity and natural gas combination. Other case studies such as electricity and district heating combination has also been fully investigated in literature, as well. Expanding these studies with new carrier systems implementation enriches the multienergy systems concept. These studies focus on the opportunities and limitations, which are implied by the implementation of above concept, especially on the electrical power systems.

The electrical power system optimizations are more challengeable and interesting in the presence of other carriers. New solutions can be found for power systems technical problems by help of multicarrier energy systems provided flexibility. Finally, the world's energy crisis issue can face new conditions and solutions with the expansion of the concept of multicarrier energy systems.

REFERENCES

[1] M. Geidl, G. Koeppel, P. Favre-Perrod, B. Klockl, G. Andersson, K. Frohlich, Energy hubs for the future, IEEE Power Energy Mag. 5 (1) (2007) 24.

[2] M. Geidl, G. Andersson, Optimal power flow of multiple energy carriers, IEEE Trans. Power Syst. 22 (1) (2007) 145–155.

[3] M. Mohammadi, Y. Noorollahi, B. Mohammadi-Ivatloo, H. Yousefi, Energy hub: from a model to a concept—a review, Renew. Sust. Energ. Rev. 80 (2017) 1512–1527.

[4] M. Vahid-Pakdel, S. Nojavan, B. Mohammadi-Ivatloo, K. Zare, Stochastic optimization of energy hub operation with consideration of thermal energy market and demand response, Energy Convers. Manag. 145 (2017) 117–128.

[5] A. Soroudi, B. Mohammadi-Ivatloo, A. Rabiee, Energy hub management with intermittent wind power, in: Large Scale Renewable Power Generation, Springer, New York, NY, 2014, pp. 413–438.

[6] A. Dolatabadi, B. Mohammadi-Ivatloo, Stochastic risk-constrained scheduling of smart energy hub in the presence of wind power and demand response, Appl. Therm. Eng. 123 (2017) 40–49.

[7] M. Moeini-Aghtaie, A. Abbaspour, M. Fotuhi-Firuzabad, E. Hajipour, A decomposed solution to multiple-energy carriers optimal power flow, IEEE Trans. Power Syst. 29 (2) (2014) 707–716.

[8] A. Dolatabadi, B. Mohammadi-Ivatloo, M. Abapour, S. Tohidi, Optimal stochastic design of wind integrated energy hub, IEEE Trans. Ind. Inf. 13 (5) (2017) 2379–2388.

[9] C. Shao, X. Wang, M. Shahidehpour, X. Wang, B. Wang, An MILP-based optimal power flow in multicarrier energy systems, IEEE Trans. Sustain. Energy 8 (1) (2017) 239–248.

[10] S. Nojavan, M. Majidi, A. Najafi-Ghalelou, M. Ghahramani, K. Zare, A cost-emission model for fuel cell/PV/battery hybrid energy system in the presence of demand response program: ε-constraint method and fuzzy satisfying approach, Energy Convers. Manag. 138 (2017) 383–392.

[11] S. Pazouki, A. Mohsenzadeh, S. Ardalan, M.-R. Haghifam, Optimal place, size, and operation of combined heat and power in multi carrier energy networks considering network reliability, power loss, and voltage profile, IET Gener. Transm. Distrib. 10 (7) (2016) 1615–1621.

[12] M. Majidi, S. Nojavan, K. Zare, Optimal stochastic short-term thermal and electrical operation of fuel cell/photovoltaic/battery/grid hybrid energy system in the presence of demand response program, Energy Convers. Manag. 144 (2017) 132–142.

[13] M.C. Bozchalui, S.A. Hashmi, H. Hassen, C.A. Canizares, K. Bhattacharya, Optimal operation of residential energy hubs in smart grids, IEEE Trans. Smart Grid 3 (4) (2012) 1755–1766.
[14] S. Nojavan, M. Majidi, K. Zare, Performance improvement of a battery/PV/fuel cell/grid hybrid energy system considering load uncertainty modeling using IGDT, Energy Convers. Manag. 147 (2017) 29–39.
[15] A. Maroufmashat, M. Fowler, S.S. Khavas, A. Elkamel, R. Roshandel, A. Hajimiragha, Mixed integer linear programming based approach for optimal planning and operation of a smart urban energy network to support the hydrogen economy, Int. J. Hydrog. Energy 41 (19) (2016) 7700–7716.
[16] A. Shabanpour-Haghighi, A.R. Seifi, An integrated steady-state operation assessment of electrical, natural gas, and district heating networks, IEEE Trans. Power Syst. 31 (5) (2016) 3636–3647.
[17] A. Shabanpour-Haghighi, A.R. Seifi, Simultaneous integrated optimal energy flow of electricity, gas, and heat, Energy Convers. Manag. 101 (2015) 579–591.
[18] S.D. Beigvand, H. Abdi, M. La Scala, Optimal operation of multicarrier energy systems using time varying acceleration coefficient gravitational search algorithm, Energy 114 (2016) 253–265.
[19] Smart Energy Systems Lab, Simulation data parameters, 2018, Available from: http://smart.tabrizu.ac.ir/en/page/6129/resources-links/.
[20] C.M. Correa-Posada, P. Sanchez-Martin, Security-constrained optimal power and natural-gas flow, IEEE Trans. Power Syst. 29 (4) (2014) 1780–1787.
[21] C. Liu, M. Shahidehpour, Y. Fu, Z. Li, Security-constrained unit commitment with natural gas transmission constraints, IEEE Trans. Power Syst. 24 (3) (2009) 1523–1536.
[22] X. Zhang, M. Shahidehpour, A.S. Alabdulwahab, A. Abusorrah, Security-constrained co-optimization planning of electricity and natural gas transportation infrastructures, IEEE Trans. Power Syst. 30 (6) (2015) 2984–2993.
[23] X. Zhang, L. Che, M. Shahidehpour, A. Alabdulwahab, A. Abusorrah, Electricity-natural gas operation planning with hourly demand response for deployment of flexible ramp, IEEE Trans. Sustain. Energy 7 (3) (2016) 996–1004.
[24] C. Wang, W. Wei, J. Wang, F. Liu, F. Qiu, C.M. Correa-Posada, S. Mei, Robust defense strategy for gas-electric systems against malicious attacks, IEEE Trans. Power Syst. 32 (4) (2017) 2953–2965.
[25] M. Gil, P. Dueñas, J. Reneses, Electricity and natural gas interdependency: comparison of two methodologies for coupling large market models within the European Regulatory Framework, IEEE Trans. Power Syst. 31 (1) (2016) 361–369.

Chapter 15

Optimal Management of Hydrothermal-Based Micro-Grids Employing Robust Optimization Method

Morteza Nazari-Heris, Sajad Madadi and Behnam Mohammadi-Ivatloo
Faculty of Electrical and Computer Engineering, University of Tabriz, Tabriz, Iran

1 INTRODUCTION

Micro-grids (MGs) are defined as practical solutions for integrating several distributed generation units, energy storage systems (ESS), and local electrical load demands [1]. A combination of distributed generation units, hydrothermal systems, and energy storage technologies, MGs take advantage of low emission of pollutant gases, low cost of supplying load demand, and attaining reliable energy [2,3]. The optimal energy management of MGs can be defined as obtaining optimal set points of the system components, which shows a considerable sensitivity to the uncertainties associated with wind power generation, price of power market, and load demand [3]. Energy storage units are introduced in different technologies including fuel cell [4], electrical energy storage units [5], pumped storage units [6], compressed air energy storage (CAES) [7], lead acid battery [8], and sodium sulfur (NaS) battery [9].

Hydrothermal systems are an important hydro-based power production system, which includes a number of hydro plants and thermal generation units. The investment cost of hydro units is high; however, the operational cost of such generation units is ignorable. On the other hand, the investment cost of thermal plants is lower with respect to hydro units; however, the operation cost of thermal plants is high. Accordingly, the integration of hydro plants and thermal units are effective in terms of economic viewpoints and environmental aspects [10]. The optimal generation scheduling of hydrothermal system aims to obtain optimal power generation of hydro and thermal plants to meet the power load demand. Several equality and inequality constrains should be considered in optimal generation scheduling of hydrothermal systems, including water

Classical and Recent Aspects of Power System Optimization. https://doi.org/10.1016/B978-0-12-812441-3.00015-X
© 2018 Elsevier Inc. All rights reserved.

balance of the hydro units, power balance of the hydrothermal system, limitations of water discharge of hydro units, water storage limits of hydro reservoirs, and limitations of power generation of hydro and thermal pants [11]. Several methods have been applied for solving the scheduling problem of hydrothermal systems including dynamic programming [12], Lagrange relaxation [13], decomposition methods [14], mixed integer programming modeling [15], and linear programming [16], genetic algorithm [17], particle swarm optimization [18], teaching learning based optimization [19], and differential evolution [20].

The uncertainties associated with power system parameters have been handled using a variety of uncertainty modeling methods, which include scenario-based modeling [21], a risk-based stochastic optimization model [22], information gap decision theory [23], Monte Carlo simulation [24], point estimate method [25], possibilistic [26], robust optimization (RO) method [27], and interval based analysis [28]. Recently, RO method is implemented on power system problems for dealing with uncertainties of price uncertainty, wind power generation uncertainty, and load uncertainty, which include optimal bidding and offering strategies of energy storage units [29], unit commitment problem [30], transmission network expansion planning [28], economic dispatch of MGs [31], and hybrid energy systems [32].

In this chapter, optimal short-term scheduling of a MG including hydrothermal system, WT, and a pumped storage unit is investigated. The minimum operational cost of the studied MG is provided considering uncertainties of power market price, which is handled using robust optimization method. The obtained optimal solution is reported and analyzed to evaluate the performance, which proves the effectiveness of the proposed model for providing optimal solution of hydrothermal system, WT, and pumped storage unit.

The remainder of this chapter is as follows: Section 2 studies the problem formulation of robust scheduling of MG containing hydrothermal system, WT, and pumped storage unit. Case study and simulation results are provided in Sections 3 and 4, respectively. Finally, the paper is concluded in Section 5.

2 PROBLEM FORMULATION

2.1 Conventional Thermal Units

The operation cost of thermal plants is commonly modeled as a quadratic function, which can be stated as:

$$F_i^t\left(P_{si}^t\right) = a_i\left(P_{si}^t\right)^2 + b_i P_{si}^t + c_i + C_i^{RU} R_i^U(t) + C_i^{RD} R_i^D(t), \tag{1}$$

where, power production and operation cost of ith thermal plant at tth time are indicated by P_{si}^t, and F_i^t, respectively. The cost coefficients of ith thermal plant are demonstrated by a_i, b_i and c_i. Minimum and maximum limitations of power production of thermal plants should be considered as follows:

$$P_{sj,\min} \le P_{sj}^t \le P_{sj,\max};\ j = 1, 2, \ldots, N_s. \tag{2}$$

The ramp-up and ramp-down limitations of the thermal units should be considered in the optimal short-term scheduling of the MG as follows:

$$P_i(t) - P_i(t-1) \le S_i^U v_i(t-1) + P_i^{SU} y_i(t) \quad \forall i, \forall t, \tag{3}$$

$$P_i(t-1) - P_i(t) \le S_i^D v_i(t) + P_i^{SD} z_i(t) \quad \forall i, \forall t. \tag{4}$$

2.2 Hydro Units

The hydro power production can be stated as follows:

$$\sum_{t=1}^{T} P_h(t) \le E_h^{\max} \quad \forall h, \tag{5}$$

where, $P_h(t)$ is the generated hydropower at tth time interval. $E_h^{\max}$ is the maximum total power can be generated by the hth hydro unit.

Lower and upper limits of hydro units can be stated as:

$$P_{hj,\min} \le P_{hj}^t \le P_{hj,\max}; \quad j = 1, 2, \ldots, N_h. \tag{6}$$

2.3 Pumped Storage Unit

Pumped storage plant is the most practical storage technology implemented in the world. The operation of pumped storage can be stated utilizing the following equations:

Power generation and consumption at tth time interval can be stated as:

$$\begin{aligned} P_p^T(t) &= \sigma_p^T \times q_p^T(t) \quad \forall p, \forall t \\ P_p^P(t) &= \sigma_p^P \times q_p^P(t) \quad \forall p, \forall t, \end{aligned} \tag{7}$$

where, $P_p^T(t)$ and $P_p^P(t)$ are the respective indicators of power generation and consumption of pth storage unit at tth time interval. In addition, conversion factor water flow to produced and consumed power are indicated by σ_p^T and σ_p^P, respectively. Water storage of upper and lower reservoir of the pth storage unit at the beginning of tth time interval can be stated as:

$$\begin{aligned} v_p^U(t+1) &= v_p^U(t) + q_p^P(t) - q_p^T(t) \quad \forall p, \forall t \\ v_p^L(t+1) &= v_p^L(t) - q_p^P(t) + q_p^T(t) \quad \forall p, \forall t \end{aligned} \tag{8}$$

where, water storage of upper and lower reservoir of the pth storage unit at tth time interval are indicated by $v_p^U(t)$ and $v_p^L(t)$. The minimum and maximum values of upper and lower reservoir of the pth storage unit should be taken into account in optimal scheduling of MG as follows:

$$\begin{aligned} v_p^{U,\min} \le v_p^U \le v_p^{U,\max} \quad \forall p, \forall t \\ v_p^{L,\min} \le v_p^L \le v_p^{L,\max} \quad \forall p, \forall t, \end{aligned} \tag{9}$$

where, the minimum and maximum water storage values of upper reservoir are defined by $v_p^{U,\min}$ and $v_p^{U,\max}$, respectively. Moreover, the respective minimum and maximum water storage values of lower reservoir are indicated by $v_p^{L,\min}$ and $v_p^{L,\max}$. The limitation of water flow turbined and pumped can be written as:

$$\begin{aligned} 0 \leq q_p^T(t) \leq Q_p^{\max} \quad \forall p, \forall t \\ 0 \leq q_p^P(t) \leq Q_p^{\max} \quad \forall p, \forall t, \end{aligned} \tag{10}$$

where, water flow turbined and pumped at tth time interval are indicated by $q_p^T(t)$ and $q_p^P(t)$. Maximum water flow in turbined/pumped modes is defined by $Q_p^{\max}$.

2.4 Network Constraints

The limitations of reserve-up and reserve-down related to thermal units should be considered as follows:

$$\begin{aligned} r_i^U(t) \leq R_i^U(t) \\ r_i^D(t) \leq R_i^D(t). \end{aligned} \tag{11}$$

The load-shedding should be limited to its maximum amount as follows:

$$L_j^{shed}(t) \leq L_j(t). \tag{12}$$

The limitations of reserve-up and reserve-down related to hydro plants should be considered as follows:

$$\begin{aligned} r_h^U(t) \leq R_h^U(t) \\ r_h^D(t) \leq R_h^D(t). \end{aligned} \tag{13}$$

The limitations of reserve-up and reserve-down related to pumped storage units should be considered as follows:

$$\begin{aligned} r_p^U \leq v_p^L(t) \times \sigma_p^T - P_p^T(t) \\ r_p^D \leq v_p^U(t) \times \sigma_p^P - P_p^P(t). \end{aligned} \tag{14}$$

2.5 Objective Function

The main objective of the proposed model in minimizing operation cost of thermal plants, reserve-up and reserve-down, and power purchased from power market, which can be stated as:

$$\begin{aligned} \min \sum_{i=1}^{T} \sum_{i \in I} \left(a_i (P_{si}^t)^2 + b_i P_{si}^t + c_i + C_i^{RU} R_i^U(t) + C_i^{RD} R_i^D(t) \right) \\ + \sum_{j \in I} (C_i^u r_i^U(t) - C_i^D r_i^D(t)) + \sum_{q \in Q} \left(\Delta W_q(t) - \Delta W_q^{spill}(t) \right) \\ + \sum_{j \in J} \left(V_j^{LOL} L_j^{shed}(t) \right) + \sum_{h \in H} (C_h^u r_h^U(t) - C_h^D r_h^D(t)) \\ + \max \sum_{i=1}^{T} \lambda^R(t) P_g(t), \end{aligned} \tag{15}$$

where, $\sum_{i=1}^{T} \lambda^R(t) P_g(t)$ determines the worst case of the MG operation, which is handled using robust optimization method in this chapter. The inner optimization problem can be written as follows, which can be handled using dual theory:

$$\begin{aligned} &\max\ \left(\lambda_{Buy,R}(t) - z_{Buy}(t)\hat{\lambda}_h\right) \times P_g(t) \\ &s.t.\ z_{Buy}(t) \leq 1\ :\ \xi_1^t,\ \forall t, \\ &\sum_{t=1}^{T} \left(z_{Buy}(t)\right) \leq \Gamma :\ \beta, \\ &z_{Buy}(t) \geq 0. \end{aligned} \tag{16}$$

The objective function of the problem can be revised as follows:

$$\begin{aligned} \min &\sum_{i=1}^{T} \sum_{i \in I} \left(C_i P_i(t) + C_i^{RU} R_i^U(t) + C_i^{RD} R_i^D(t)\right) \\ &+ \sum_{j \in I} \left(C_i^u r_i^U(t) - C_i^D r_i^D(t)\right) + \sum_{q \in Q} \left(\Delta W_q(t) - \Delta W_q^{spill}(t)\right) \\ &+ \sum_{j \in J} \left(V_j^{LOL} L_j^{shed}(t)\right) + \sum_{h \in H} \left(C_h^u r_h^U(t) - C_h^D r_h^D(t)\right) \\ &+ \xi_1^t + \Gamma\beta. \end{aligned} \tag{17}$$

3 CASE STUDY AND SIMULATION RESULTS

The studied test system, which is a six-bus system, is demonstrated in Fig. 1.

According to this figure, the test system includes two thermal units. The thermal unit characteristics are reported in Table 1. Also, the transmission line data is provided in Table 2. The specification of the wind turbines that is used in the wind farms is represented in Table 3. In addition, the load demands, which connected into bus 3, 4 and 6, are illustrated in Fig. 2. The power which, wind

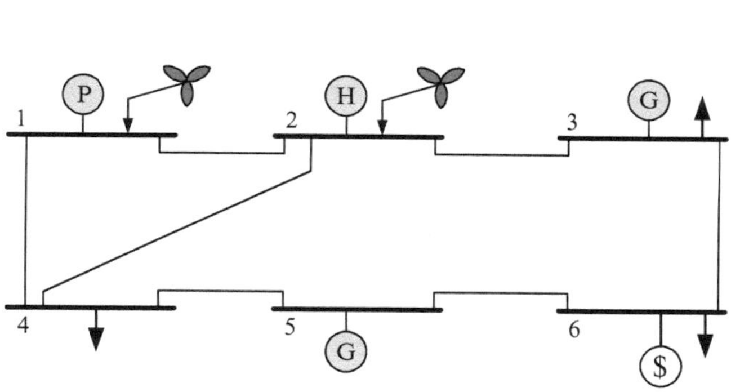

FIG. 1 The studied test system.

TABLE 1 Thermal Unit Characteristics

Unit	P_{Min}	P_{Max}	a	b	c	SUC	RU	RD	C_i^{RD}	C_i^{RU}
1	10	50	0.0697	26.2438	31.67	31.67	40	15	13	16
2	10	100	0.0697	26.2438	6.78	21.67	50	40	12	13

TABLE 2 Line Data

Line Number	From	To	Resistance	Reactance	Capacity
1	1	2	0	0.13	100
2	2	3	0	0.15	85
3	2	4	0	0.23	75
4	4	5	0	0.17	100
5	4	1	0	0.13	65
6	5	6	0	0.16	95
7	6	3	0	0.14	100

TABLE 3 Specification of the Wind Turbines

V_{cin}	V_r	V_{co}	P_r
5.5	13.91	26.67	2

farms can produce, is shown in Fig. 3. The data related to pump storage unit are provided in Table 4.

4 SIMULATION RESULTS

This section provides the simulation results of implementation of the proposed robust optimization model in optimal scheduling of micro-grids. The deviation of power market price from the forecasted values is considered equal to 25%.

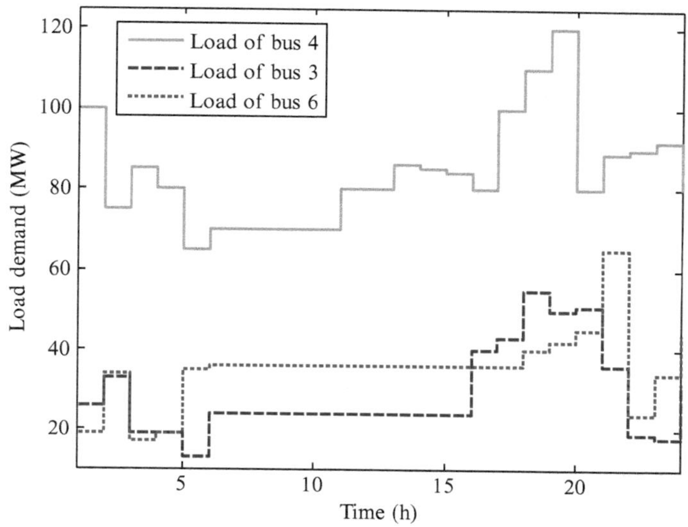

FIG. 2 Load demands of bus 3, 4, and 6.

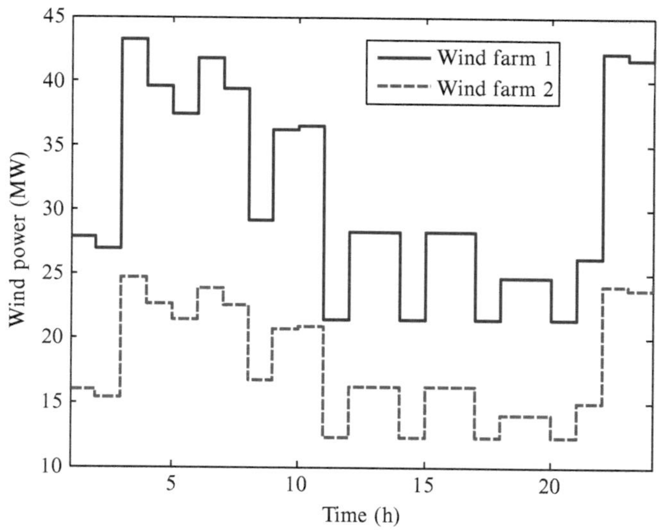

FIG. 3 Power generated by wind farms.

The robust budget is equal to 6. The power market price and its upper bound are shown in Fig. 4. As seen in this figure, the maximum power market price is related to $t = 14$ h, where the price is 35.3483 $/MWh.

The optimal generation scheduling of thermal plants and the power supply form the power market price. As it is obvious from this figure, the power supply

TABLE 4 Pump Storage Data

Parameter	Value
Upper bound of pumping/turbining flow (Hm^3/h)	20
Water-flow to power factor ($MW/Hm^3/h$)	0.8
Power to water-flow factor ($MW/Hm^3/h$)	1.2
Initial volume of upper reservoir (Hm^3)	40
Upper bound of volume of upper reservoir (Hm^3)	80
Lower bound of volume of upper reservoir (Hm^3)	0
Initial volume of lower reservoir (Hm^3)	40
Upper bound of volume of lower reservoir (Hm^3)	80
Lower bound of volume of lower reservoir (Hm^3)	0

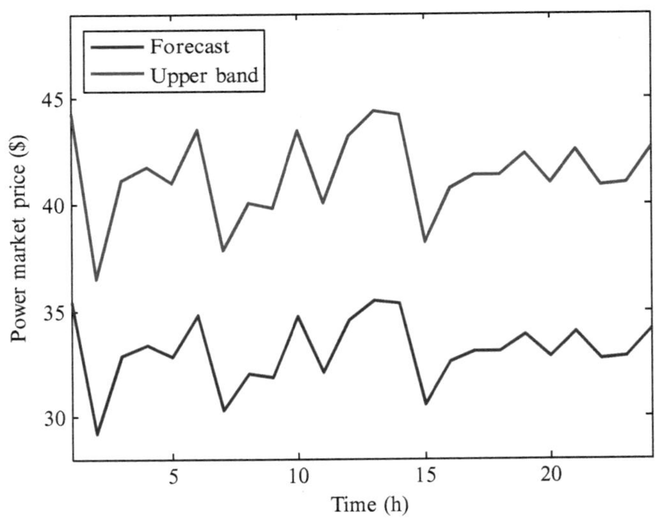

FIG. 4 Upper bound and forecasted power market price.

from the main grid continues during this time interval. In addition, thermal unit 2 had more cooperation in supplying the power load demand of the system (Fig. 5).

The wind power dispatch for two wind farms are shown in Fig. 6. It should be mentioned that the obtained results are the variable power of wind farms.

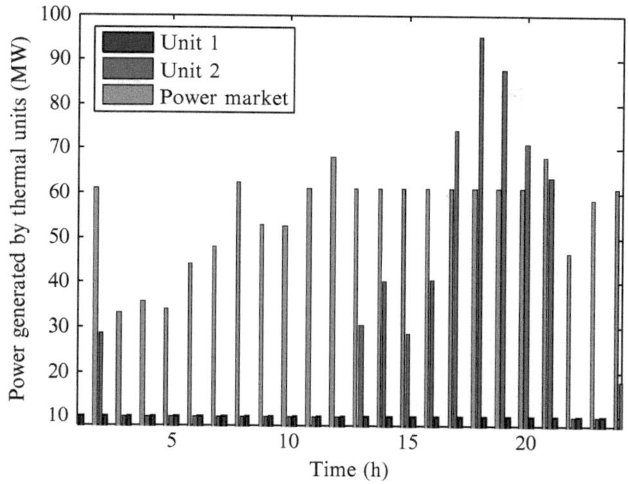

FIG. 5 Power generated by thermal units.

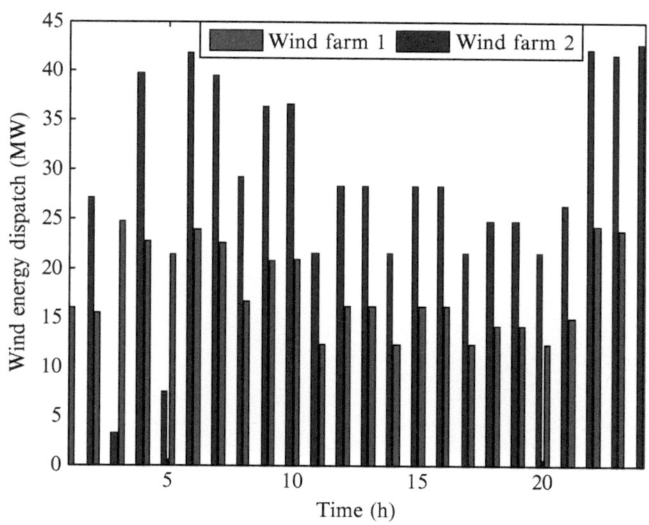

FIG. 6 Wind energy dispatch (MW).

The power dispatch of hydro unit during the scheduling time horizon is provided in Table 5. As it is obvious from this table, the hydro unit has cooperated in power generation at $t = 1$, 11, and 19 h.

Table 6 provides the pumped storage dispatch during the 24 h' time interval.

Upregulation of the thermal unit 1 is shown in Fig. 7. Thermal unit 2 does not participate in upregulation.

TABLE 5 Hydro Unit Dispatch

Time	Power (MW)
1	20.165
2	0
3	0
4	0
5	0
6	0
7	0
8	0
9	0
10	0
11	25.285
12	0
13	0
14	0
15	0
16	0
17	0
18	0
19	14.550
20	0
21	0
22	0
23	0
24	0

TABLE 6 Pumped Storage Dispatch

Time	Charging Mode	Discharging Mode
1	6.365	0
2	0	0
3	4.213	0
4	4.325	0
5	0	0
6	0	0
7	0	0
8	0	1.565
9	0	0
10	0	0
11	0	0
12	0	7.82
13	0	0
14	0	0
15	0	0
16	0	0
17	0	0
18	0	0
19	0	0
20	0	0
21	0	7.155
22	0	0
23	0	0
24	0	0

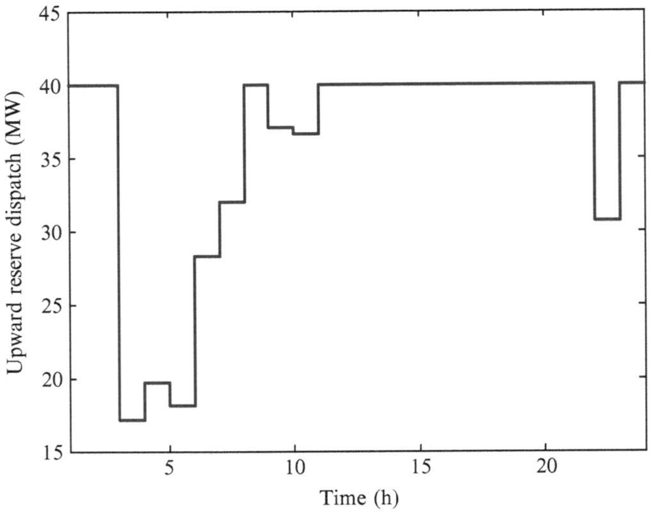

FIG. 7 Upregulation of the thermal unit 1.

5 CONCLUSION

This chapter studied the optimal robust scheduling of MGs in day-ahead market, where the studied MG contains hydrothermal systems, wind farms, and pump storage unit. In addition, the studied MG is interconnected with the main grid, which makes it capable of buying power during scheduling time horizon. The uncertainty associated with power market price has been considered and studied using robust optimization method. An effective and strong optimization technique is employed for providing the minimum operating cost of the studied MG. The obtained solutions prove the practicality and high performance of the proposed model for obtaining optimal robust scheduling of hydrothermal system, WT and pumped storage unit. The operation cost of the studied MG is obtained equal to $76222.125 for the scheduling time interval.

REFERENCES

[1] M. Nazari-Heris, S. Abapour, B. Mohammadi-Ivatloo, Optimal economic dispatch of FC-CHP based heat and power micro-grids, Appl. Therm. Eng. 114 (2017) 756–769.

[2] F. Katiraei, M.R. Iravani, P.W. Lehn, Micro-grid autonomous operation during and subsequent to islanding process, IEEE Trans. Power Deliv. 20 (1) (2005) 248–257.

[3] S. Nojavan, M. Majidi, N.N. Esfetanaj, An efficient cost-reliability optimization model for optimal siting and sizing of energy storage system in a microgrid in the presence of responsible load management, Energy 139 (2017) 89–97.

[4] M. Ehsani, Y. Gao, A. Emadi, Modern Electric, Hybrid Electric, and Fuel Cell Vehicles: Fundamentals, Theory, and Design, CRC Press, Florida, 2017.

[5] M. Nazari-Heris, B. Mohammadi-Ivatloo, G.B. Gharehpetian, Optimal short-term scheduling of combined heat and power systems considering electrical energy storage, in: 3rd International Conference of Iran Energy Association (IEA), Tehran, Iran, 2017.
[6] A.E. Nezhad, M.S. Javadi, E. Rahimi, Applying augmented ε-constraint approach and lexicographic optimization to solve multi-objective hydrothermal generation scheduling considering the impacts of pumped-storage units, Int. J. Electr. Power Energy Syst. 55 (2014) 195–204.
[7] E.A. Bouman, M.M. Øberg, E.G. Hertwich, Environmental impacts of balancing offshore wind power with compressed air energy storage (CAES), Energy 95 (2016) 91–98.
[8] R. Dufo-López, J.M. Lujano-Rojas, J.L. Bernal-Agustín, Comparison of different lead–acid battery lifetime prediction models for use in simulation of stand-alone photovoltaic systems, Appl. Energy 115 (2014) 242–253.
[9] S.I. Kim, W.I. Park, K. Jung, C.-S. Kim, An innovative electronically-conducting matrix of the cathode for sodium sulfur battery, J. Power Sources 320 (2016) 37–42.
[10] M. Nazari-Heris, B. Mohammadi-Ivatloo, G. Gharehpetian, Short-term scheduling of hydro-based power plants considering application of heuristic algorithms: a comprehensive review, Renew. Sust. Energ. Rev. 74 (2017) 116–129.
[11] M. Nazari-Heris, B. Mohammadi-Ivatloo, A. Haghrah, Optimal short-term generation scheduling of hydrothermal systems by implementation of real-coded genetic algorithm based on improved Mühlenbein mutation, Energy 128 (2017) 77–85.
[12] T. Homem-de-Mello, V.L. de Matos, E.C. Finardi, Sampling strategies and stopping criteria for stochastic dual dynamic programming: a case study in long-term hydrothermal scheduling, Energy Syst. 2 (1) (2011) 1–31.
[13] V.N. Dieu, W. Ongsakul, Improved merit order and augmented Lagrange Hopfield network for short term hydrothermal scheduling, Energy Convers. Manag. 50 (12) (2009) 3015–3023.
[14] T.N.D. Santos, A.L. Diniz, A new multiperiod stage definition for the multistage benders decomposition approach applied to hydrothermal scheduling, IEEE Trans. Power Syst. 24 (3) (2009) 1383–1392.
[15] A. Ahmadi, J. Aghaei, H.A. Shayanfar, A. Rabiee, Mixed integer programming of multiobjective hydro-thermal self scheduling, Appl. Soft Comput. 12 (8) (2012) 2137–2146.
[16] G.W. Chang, et al., Experiences with mixed integer linear programming based approaches on short-term hydro scheduling, IEEE Trans. Power Syst. 16 (4) (2001) 743–749.
[17] A. Haghrah, B. Mohammadi-Ivatloo, S. Seyedmonir, Real coded genetic algorithm approach with random transfer vectors-based mutation for short-term hydro-thermal scheduling, IET Gener. Transm. Distrib. 9 (1) (2014) 75–89.
[18] A. Mahor, S. Rangnekar, Short term generation scheduling of cascaded hydro electric system using novel self adaptive inertia weight PSO, Int. J. Electr. Power Energy Syst. 34 (1) (2012) 1–9.
[19] P.K. Roy, Teaching learning based optimization for short-term hydrothermal scheduling problem considering valve point effect and prohibited discharge constraint, Int. J. Electr. Power Energy Syst. 53 (2013) 10–19.
[20] J. Zhang, S. Lin, W. Qiu, A modified chaotic differential evolution algorithm for short-term optimal hydrothermal scheduling, Int. J. Electr. Power Energy Syst. 65 (2015) 159–168.
[21] A. Zakariazadeh, S. Jadid, P. Siano, Stochastic operational scheduling of smart distribution system considering wind generation and demand response programs, Int. J. Electr. Power Energy Syst. 63 (2014) 218–225.
[22] N. Zhang, et al., A convex model of risk-based unit commitment for day-ahead market clearing considering wind power uncertainty, IEEE Trans. Power Syst. 30 (3) (2015) 1582–1592.

[23] S. Nojavan, M. Majidi, K. Zare, Risk-based optimal performance of a PV/fuel cell/battery/grid hybrid energy system using information gap decision theory in the presence of demand response program, Int. J. Hydrog. Energy 42 (16) (2017) 11857–11867.

[24] G. Baudry, C. Macharis, T. Vallee, Range-based multi-actor multi-criteria analysis: A combined method of multi-actor multi-criteria analysis and Monte Carlo simulation to support participatory decision making under uncertainty, Eur. J. Oper. Res. 264 (1) (2018) 257–269.

[25] M.J. Bordbari, A.R. Seifi, M. Rastegar, Probabilistic energy consumption analysis in buildings using point estimate method, Energy 142 (2018) 716–722.

[26] M. Mousazadeh, S. Torabi, M. Pishvaee, F. Abolhassani, Health service network design: a robust possibilistic approach, Int. Trans. Oper. Res. 25 (1) (2018) 337–373.

[27] A. Najafi-Ghalelou, S. Nojavan, K. Zare, Heating and power hub models for robust performance of smart building using information gap decision theory, Int. J. Electr. Power Energy Syst. 98 (2018) 23–35.

[28] A. Kavousi-Fard, A. Khosravi, S. Nahavandi, A new fuzzy-based combined prediction interval for wind power forecasting, IEEE Trans. Power Syst. 31 (1) (2016) 18–26.

[29] S. Nojavan, A. Najafi-Ghalelou, M. Majidi, K. Zare, Optimal bidding and offering strategies of merchant compressed air energy storage in deregulated electricity market using robust optimization approach, Energy 142 (2018) 250–257.

[30] V.K. Tumuluru, D.H. Tsang, A two-stage approach for network constrained unit commitment problem with demand response, IEEE Trans. Smart Grid 9 (2) (2016) 1175–1183.

[31] S.A. Alavi, A. Ahmadian, M. Aliakbar-Golkar, Optimal probabilistic energy management in a typical micro-grid based-on robust optimization and point estimate method, Energy Convers. Manag. 95 (2015) 314–325.

[32] A. Martinez-Mares, C.R. Fuerte-Esquivel, A robust optimization approach for the interdependency analysis of integrated energy systems considering wind power uncertainty, IEEE Trans. Power Syst. 28 (4) (2013) 3964–3976.

Chapter 16

Geomagnetically Induced Currents: A Threat to Modern Power Systems

David T.O. Oyedokun* and Pierre J. Cilliers*,†

**University of Cape Town, Cape Town, South Africa,* †*South African National Space Agency, Hermanus, South Africa*

GLOSSARY

***Ap*-index**	the *Ap*-index provides a daily average level for geomagnetic activity on a linear scale
BGS	British Geological Survey
Co-rotating interaction regions (CIR)	transition zones between slow-moving solar wind streams (~300 km/s) and fast-moving streams from coronal holes interacting with the Earth's magnetic field
Coronal mass ejection (CME)	ejection of solar mass in the form of a cloud of plasma from the surface of the Sun following an eruptive event on the Sun
Coronagraph	a camera for taken photos of the Sun with a telescopic attachment designed to block out the direct light from the Sun so that nearby objects such as the erupting solar wind—which otherwise would be hidden in the Sun's bright glare—can be resolved
Disturbance storm time (DST)	index which expresses the strength of the ring current around Earth caused by solar protons and electrons
Geomagnetically induced current (GIC)	current induced in power lines due to fluctuations in the geomagnetic field

Classical and Recent Aspects of Power System Optimization. https://doi.org/10.1016/B978-0-12-812441-3.00016-1
© 2018 Elsevier Inc. All rights reserved.

Geomagnetic disturbances (GMDs)	fluctuations in the geomagnetic field on the surface of the Earth due to currents in the magnetosphere and ionosphere which interact with background magnetic field generated in the Earth's core
Interplanetary magnetic field (IMF)	a part of the Sun's magnetic field that is carried into interplanetary space by the solar wind
Lagrangian point (L1)	a point 220 Earth Radii (1.5 million km) upstream from Earth toward the Sun at which the gravitational force of the Earth and Sun are equal and opposing
LT	local time as determined by the position of the Sun at the location of interest
Magnetopause	the boundary between the Earth's magnetosphere and interplanetary space, about 65,000 km above the Earth
Solar flare	a sudden, rapid, and intense variation in brightness observed on the surface of the Sun, which occurs when magnetic energy that has built up in the solar atmosphere is suddenly released
Solar wind (SW)	the continuous flow of charged particles from the Sun which permeates the solar system
Space weather	conditions on the Sun and in the solar wind, magnetosphere, ionosphere, and thermosphere that may influence the performance and reliability of space-borne and ground-based technological systems, and may endanger human life or health
Sudden storm commencement (SSC)	an abrupt increase in the northward component of the geomagnetic field, which marks the beginning of a geomagnetic storm
Sunspots	spots or patches that appear from time to time on the Sun's surface, appearing dark by contrast with their surroundings due their being at a lower temperature than their surroundings

1 INTRODUCTION

The motivation for including a chapter on geomagnetically induced currents (GIC) in the context of power system optimization is to familiarize the reader with a potential threat to power system stability and availability, which requires the inclusion of particular elements in the system to optimize the resilience of the system to this threat. The flow of GIC through transmission lines and transformers across a power network could have negative consequences. These include an increase in the reactive power demanded by GIC-laden transformers [1], transformers operating within the region of nonlinearity due to half-wave saturation [2], excessive heating [3] in transformers leading to thermal damage, malfunction of transmission line protection schemes [4] and voltage problems in affected sections of the network [5,6].

The literature on GICs originally dealt with GICs in power lines in high-latitude countries where GICs are most likely to occur. This includes research on GICs in Finland [7] and research referenced in the article. Boteler et al. researched GICs in Canada [8,9], and referenced similar work. Several papers have appeared on GICs generated during landmark geomagnetic disturbances, including the Quebec storm of March 1989 [10] and the Halloween storm of October 2003 [11,12]. The research reported by Gaunt and Coetzee [13] established the awareness that GIC-related damage to power systems can occur in mid- and low-latitude regions. Subsequently, a number of papers appeared on GICs in mid- and low-latitude regions such as Scotland [11], Brazil [14], Spain [15], China [16], Japan [17], New Zealand [18] and Australia [19]. A recent special collection issue of the journal Space Weather [20] presented an overview of the literature that appeared in AGU journals on the current understanding of several aspects of GICs.

This chapter contains contributions by researchers in Space Weather and Electrical Engineering. The contributions by space weather researchers put the generation of geomagnetic disturbances (GMD) which impact power system within the context of space weather. The contributions by researchers in Electrical Engineering provide the methodology to estimate GICs in a power network during a GMD and an overview of the possible technological mitigation measures which should be considered in the optimization of power systems.

The first part of the chapter (Sections 1–5) focuses on the origins and propagation of space weather events from the Sun to the Earth and provides an overview of the indices that are used to characterize the intensity of GMDs. Some models are presented which are used for prediction of the arrival time and intensity of GMDs from solar wind parameters. The first section concludes with the means available to Power System operators to be notified about space weather events that may lead to GMDs. The section gives an overview of the coupling between space weather and the electrical power system. Once the coupling has been established, the resulting effect of the flow on GIC in the power system can be discussed.

The second part of the chapter (Sections 6–8) provides a methodology for calculating GICs in electrical power systems, and discusses aspects of modeling, measurement, and mitigation of GICs in power systems.

GICs are the natural result of changing geomagnetic fields. During geomagnetic disturbances caused by a solar storm, the changes in geomagnetic fields are more prominent when compared to geomagnetically quiet times, and give rise to larger GICs. During solar storms, enormous explosions of plasma are ejected from the Sun's surface into interplanetary space. These ejections are called Coronal Mass Ejections (CMEs) [21], which are known to disrupt the solar wind [22] throughout interplanetary space. The subsequent interaction of the solar wind with the Earth's magnetic field is known as a geomagnetic storm [23].

During geomagnetic storms, solar wind pressure and speed can suddenly increase on average from 2 to 30 nPa and from 400 to 2000 km/s respectively [24]. Geo-effective CMEs lead to large fluctuations of the Earth's magnetic fields, which induce an electric field according to Faraday's law of induction [25]. This gives rise to quasi-DC currents (<0.1 mHz) in electric power systems via the grounded neutrals of power transformers [26].

The various consequences of the flow of GIC in power systems pose a threat in terms of power system stability, security, and the reliability of electrical power supply. Modern society is heavily dependent on a stable electrical power supply. There is a close relationship between advancing the economy and the availability of reliable electricity supply. Industries such telecommunication, medicine, transportation, banking, food processing, manufacture, mining, and education, to mention a few, thrive on a stable supply of electrical power. The increasing interconnectedness of the power transmission network allows for increased redundancy which provides a means by which alternative sources of power can be made available to different load centers. These networks, especially their high voltage sections are prone to the adverse effects of GICs. This necessitates that network planners and operators consider the threat posed by GICs to the power system earnestly and develop strategies for operating and protecting the power system during intense geomagnetic disturbances.

In this chapter, space weather as the root cause of GICs will be discussed in some detail. This discussion includes mechanisms of GIC flow in the power system, how GICs are modeled, calculated, and measured, and ways by which GICs can be mitigated or blocked.

2 DEFINITION OF SPACE WEATHER

Space weather refers to physical conditions in the solar interior, in the solar atmosphere, in the solar wind, and in the Earth's magnetosphere, ionosphere, and thermosphere that can influence the performance and reliability of space-borne and ground-based technological systems, including power distribution systems.

Geomagnetically induced currents in power systems are induced by the transient currents in the Earth's magnetosphere and ionosphere which induce perturbations in the Earth's magnetic field known as geomagnetic storms. Geomagnetic storms can be considered as the end link of the space weather chain extending from the Sun to the surface of the Earth and below.

The space weather chain which leads to geomagnetic disturbances start on the Sun (see Fig. 1). A combination of complex phenomena occurs on the Sun

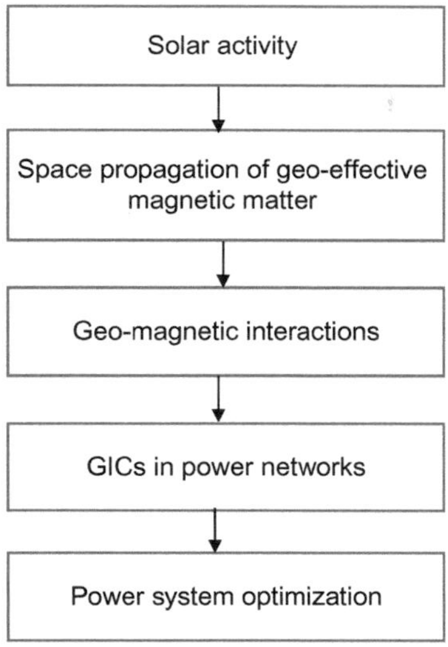

FIG. 1 Flowchart to demonstrate the link between power system optimization and solar activity, as discussed in the following sections.

which give rise to eruptions of the solar atmosphere into interplanetary space. These phenomena include the dynamo mechanism in the interior of the Sun which causes its magnetic field, convection of plasma along magnetic field lines which extend through the solar corona, and differential rotation of the Sun which leads to local turbulence in the solar magnetic field. The solar magnetic field extends through the outer layers of the Sun into the interplanetary space [27].

2.1 Sunspots

Sunspots provide the first indications of the possibility of solar eruptions that may precede geomagnetic storms on the Earth. Sunspots are regions on the Sun's photosphere that appear darker than the surrounding areas on the visible solar disk due to reduced surface temperature associated with concentrations of magnetic field flux and intense magnetic activity on the Sun. They usually appear in pairs of opposite magnetic polarity. The number of sunspots has a cyclical increase and decrease over an approximately 11-year period known as the solar cycle. Individual sunspots may persist from a few days to months. Sunspots are accompanied by secondary phenomena including coronal loops, prominences and reconnection events. Most solar flares and coronal mass ejections originate in the magnetically active regions associated with sunspot groups.

The solar rotation period as determined from the observation of sunspots is 27.27 days as seen from the Earth. Differential solar rotation refers to an increase in the solar rotation period from 24.47 sidereal days at the solar equator to approximately 38 sidereal days at the poles. The 11-year solar cycle and the approximately 27-day solar rotation period of the Sun are useful for the prediction of geomagnetic disturbances on the Earth, which tend to repeat with the same periodicities. Sunspot numbers are not to be confused with eruptive solar activity, since solar flares and other activity on the solar surface may occur at any time, and not all sunspots give rise to eruptive solar activity.

2.2 Solar Flares, Coronal Holes, High Speed Streams, and CIRs

The eruptive events on the Sun which ultimately lead to GICs are ongoing and transient emissions of high energy particles from the surface of the Sun. Ongoing emissions occur in the form of high speed streams of particles known as the solar wind from active regions known as coronal holes, which are regions in the Sun's corona where magnetic field lines "open out" into space, allowing solar atmospheric plasma to escape more freely from the solar gravitational field. The solar wind comprises mostly protons and electrons. Coronal holes which emit high speed streams can occur even when there are no sunspots, as during the waning period of solar cycle 24 when a spotless Sun produced intense auroras and minor geomagnetic storms with activity levels of $Kp = 5$ occurring every day for the period 11–15 October 2017 due to the Earth passing through a high speed solar stream [28].

Flares can also occur when solar filaments on the surface of the Sun become unstable and cause an explosive collapse on the surface of the Sun, producing a "Hyder Flare"—a type of solar flare with no underlying sunspot. Solar filaments are tube-shaped clouds of dense plasma held above the surface of the Sun by magnetic forces. They can stretch out over a distance of 300,000 km from end to end.

Geomagnetic storms may also occur during the absence of sunspots due to co-rotating interaction regions (CIRs) interacting with the Earth's magnetic field. The CIRs are transition zones between slow-moving solar wind streams (~300 km/s) and fast-moving streams from coronal holes (~700 km/s) and which produce density gradients and shock waves [29]. The CIRs exhibit a recurrence with the solar rotation period. CIRs are characterized by a density enhancement on a leading edge of solar wind speed rise.

2.3 Coronal Mass Ejections

Eruptive transient emissions that result in large clouds of plasma leaving the surface of the Sun at supersonic speeds are known as Coronal Mass Ejections (CMEs). The CMEs erupt from magnetically active regions on the Sun known as sunspots. The number of sunspots have a cyclical variation of 11 years, corresponding to the reversal of the solar magnetic field every 11 years. The cyclic magnetic behavior can be explained by the Sun's differential rotation, meridional circulation, and large-scale convective motions [30]. CMEs are most frequent near or soon after the peak of the solar cycle when the number of sunspots reach a maximum.

CMEs are the main cause of interplanetary-magnetospheric disturbances leading to the phenomena known as geomagnetic storms or geomagnetic disturbances (GMDs). GMDs are the result of complex and nonlinear chaotic activities involving the Sun, the interplanetary medium, and the Earth's magnetosphere and ionosphere.

Not all active regions emit CMEs. Only a small percentage of the CME's are geo-effective, i.e. emitted towards the Earth. Energetic and fast CMEs originate mostly from the solar active regions, and slower-moving CMEs are associated with erupting prominences [31]. The CME productivity increases with the size of the sunspots [32]. The underlying chaotic mechanisms make the prediction of these phenomena challenging. The science of predicting CMEs is still relatively immature in terms of practical applications that could be used for GIC mitigation purposes. Despite an increased understanding of solar physics, it is not yet possible to satisfactorily predict the timing or size of solar eruptions. Large CMEs can happen any time and are not limited to the peak of the solar cycle. The approximately linear relation between the number of sunspots [33] and the likelihood of CMEs and the saturation relation between the sunspot area and the number of CMEs per day are valuable for the prediction of CMEs [34] from sunspot observations.

Coronal holes are most likely to cause geomagnetic disturbances on the Earth during the descending phase of the sunspot cycle when they form nearest to the ecliptic plane, often in the absence of sunspots.

Unlike flares, geomagnetic storms do not track well with the sunspot cycle. Geomagnetic storms may occur at any point in the solar cycle, even though they are more likely to occur during the peak of the solar cycle due to a larger number of active regions. In general, there is a time lag from the sunspot maximum to the time when the geomagnetic activity reaches a maximum. The orientation of the interplanetary magnetic field (IMF), which seems to be independent of the sunspot cycle shown in Fig. 2, is one of the key factors that give rise to the dissociation between the sunspot peak and the peak of geomagnetic activity.

2.4 The Observation of CMEs

Because the most intense GICs are known to originate in CMEs, it is important to understand how they are observed. CMEs are observed from several ground-based observatories around the Earth and from space-based observatories.

Space-based observations are done from the Solar and Heliospheric Observatory (SOHO) and Advanced Composition Explorer (ACE) and DSCOVR

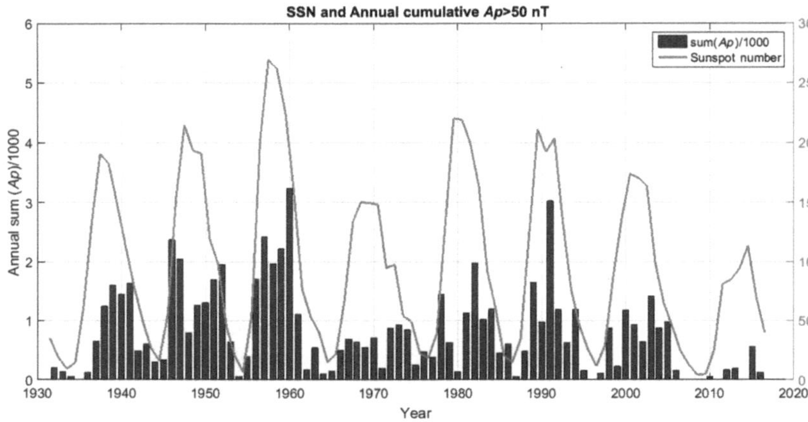

FIG. 2 Correlation between Sunspot cycle distribution and geomagnetic storms in terms of annual smoothed (13-month) sunspot number (SSN) and *Ap*-index from 1932 to 2017. $Ap > 50\,\text{nT}$ is taken as the threshold for a major geomagnetic storm. The bars show the sum of *Ap* for all days of the year for which $Ap > 50\,\text{nT}$. The yearly mean sunspot numbers are courtesy of WDC-SILSO, Royal Observatory of Belgium, Brussels http://sidc.be/silso/datafiles#total. The *Ap*-index values are from the WDC Kyoto http://swdcwww.kugi.kyoto-u.ac.jp/kp/index.html.

satellites at the L1 Lagrangian point, 220 Earth Radii (1.5 million km) upstream from Earth, the Geostationary Operational Environmental Satellites (GOES), and more recently from the STEREO pair of identical satellites co-orbiting with the earth around the Sun, and the Solar Dynamic Observatory (SDO), a geo-synchronous satellite. These satellites are equipped with X-ray sensors in various frequency bands, as well as particle and magnetic sensors.

CMEs and CIRs can currently be observed with coronagraphs on board SOHO, STEREO, and other satellites. A coronagraph is a CCD camera aimed at the Sun. It has an occulting disk that shields the light from the brightest part of the solar disk so that the flares associated with CMEs can be observed. Halo CMEs as observed from Earth or satellites on the Sun-Earth line have flares that seemingly emanate in all directions surrounding the disk and are thus taken to be directed toward or directly away from Earth. SOHO has various instruments to enhance its observation of the solar disk, including the LASCO instrument, with a wide enough field of view to permit an estimation of the initial velocity of a CME. The two satellites of the STEREO mission look at the Sun from different angles and allow the side-view of CMEs which permits the estimation of the propagation direction and velocity of CMEs with much better accuracy than SOHO. The initial velocity of CMEs is one of the most significant parameters that determines the transit times of CMEs. Real-time observations of CMEs from satellites are available via the internet. The SDO mission provides very high-resolution UV and extreme UV imaging of the solar surface and allows for the exact determination of the geo-effectivity of CMEs.

CMEs can be observed from ground-based coronagraphs such as the MK4 white coronagraph at Mona Lua Observatory in Hawaii and others around the world. Photospheric magnetic field observations carried out by the National Solar Observatory and the Mt Wilson observatory in California provide critical inputs for estimating the complexity of magnetic active areas on the Sun, and the properties of the solar wind prior to its arrival at the L1 Lagrangian point.

CMEs are also observed with interplanetary scintillation (IPS) receivers that are terrestrial instruments able to track some CMEs, via the principle of scintillation of radio sources due to electron density inhomogeneities in the solar wind. The flares accompanying CMEs also emit radio bursts over a wide band of frequencies 10–800 MHz. The solar radio spectrum is monitored by the e-Callisto network comprising 136 solar spectrographs in 133 countries. In the case of internet outage during an intense space weather storm, the data from the e-Callisto solar spectrographs can be used to infer the occurrence of a CME and estimate the velocity of the CME from the rate of change of the frequency of the radio emissions [35].

Magnetic field lines connect the local magnetic North and South pole regions on the Sun, with both regions typically occurring within a single sunspot group. The solar plasma spirals back and forth along the magnetic field lines. From time to time, the field lines flare when they open up into interplanetary space and carry the plasma into space. CMEs occur as a result of such flares. During solar flares, radio waves and X-rays are emitted which provide early warning indicators of potential space weather impacts on the Earth. The velocity of the CMEs can be determined from satellite images of the solar corona. Only about 2% of CMEs are geo-effective. The CME velocity as determined from the leading edge of eruptions recorded on coronagraphs is supersonic and ranges from 100 to 2000 km/s, with the average being close to 470 km/s [31]. The speed of a CME is correlated with the type of radio bursts it produces. The fastest CMEs tend to produce type II–IV radio burst pairs, while single type II or type IV bursts tend to be associated with events of intermediate speed [31]. Geo-effective CMEs arrive at the Earth within 2–4 days after leaving the Sun. The majority of intense GMDs (83% during solar cycle 23) are associated with CMEs with the remainder produced by co-rotating interactive regions (CRIs) that result from the interaction between the fast and slow solar wind emanating from coronal holes [36].

2.5 Solar Wind: Propagation of CMEs Through Interplanetary Space and IMF

The Sun interacts with the Earth's environment via the solar wind (SW) which propagates from the surface of the Sun through interplanetary space. The solar wind is modulated by CMEs and CIRs and when it reaches the Earth, it transfers energy into the Earth's magnetosphere and ionosphere. This energy transfer leads to enhancement of the current systems in these regions, which in turn cause geomagnetic field fluctuations known as geomagnetic storms.

2.5.1 Interplanetary Magnetic Field

Whether or not the arrival of the solar wind will cause a geomagnetic storm is determined by the direction of the IMF at the Earth. The solar wind carries the Sun's magnetic field with it as it propagates in interplanetary space. When the component of the interplanetary magnetic field vector which is perpendicular to the ecliptic plane, B_z, is Southward directed and hence antiparallel to that of the Earth's magnetic field, a condition occurs that allows a physical process called "magnetic reconnection" to take place on the dayside of the magnetopause, during which energy is transferred from the solar wind into the Earth's magnetosphere. Reconnection can, however, also take place at higher latitudes when the solar wind impacts the Earth with a Northward directed IMF [37].

2.5.2 Halo CME's

Solar observations show that CMEs that appear to surround the occulting disk of the observing coronagraphs, known as halo CMEs, have the highest probability of impacting the Earth's magnetosphere [38] if they are Earth-directed or geoeffective. Not all halo CMEs are associated with GMDs, and nonhalo CMEs can also cause intense GMDs, if they arrive at Earth with high speed and an enhanced southward component of the IMF [33].

2.5.3 Modeling of the Solar Wind

Numerical models are important tools in simulating and characterizing the time evolution of solar wind, since there are no in situ observations of the solar wind during its progression from the solar corona to the L1 Lagrangian point. Many solar wind models have been developed for the prediction of arrival time, duration, and intensity of storms. The Wang-Sheeley-Arge (WSA)-Enlil model [39,40] is one of the current advanced models used by NOAA and other facilities in forecasting the arrival time of CMEs.

The WSA model is a coronal model that combines observations and plasma physics theory to represent the background solar wind flow. The WSA model can be used to determine the solar wind speed and interplanetary magnetic field from 5Rs (Rs is radius of the Sun) up to the location of the Earth. Its inputs are the estimates of the solar wind velocity leaving the corona. The Wang-Sheeley-Arge model therefore generates steady co-rotating ambient background flow. A map of the solar corona is used as input to another part of the model, which then propagates the solar wind flow out to 21.5Rs. The empirically derived solar wind speed and magnetic field polarity at 21.5Rs are then fed into three-dimensional magnetohydrodynamic (MHD) numerical model (Enlil) that propagates the resulting flow out to Earth. The Enlil simulation can be requested online at the Community Coordinated Modeling Centre (CCMC) website of NASA (http://ccmc.gsfc.nasa.gov/).

2.6 Phases of the Geomagnetic Storm

2.6.1 SSC

The first indication of a geo-effective GMD is the sudden increase in the intensity of the geomagnetic field. The arrival of the solar wind at the bow of the magnetosphere leads to a compression of the magnetopause and a resulting sudden increase in the magnetic field on the day-side of the Earth. This event is known as the sudden storm commencement (SSC) and is typically of short duration (minutes). When SSC events occur in low and mid-latitude regions it results in geomagnetic field changes of $<200\,\text{nT}$, which is significantly less than the variations associated with electrojet-driven disturbances at high latitudes [41]. However, the rapid change in the magnetic field during SSCs has been linked to the occurrence of large GICs observed in low and mid-latitude regions where no auroral electrojet increases are present. During a SSC event on 23 March 1991 LT in North America, some of the largest GICs ever observed in the United States occurred with GICs exceeding 100 A in one of the transformers of a nuclear power plant near Philadelphia. At the Meadowbrook 500 kV substation in western Virginia, GIC levels of 47 and 66 A were observed in two transformers that were connected in parallel. Large GICs in several other mid- and low-latitude regions have been observed coincident with the SSC onset of geomagnetic storms [42].

2.6.2 Main Phase

The sharp decrease in the geomagnetic field caused by the increase in the electric current in the magnetosphere and ionosphere is known as the main phase of the geomagnetic storm. During the main phase of the geomagnetic storm, electric current in the magnetosphere creates a magnetic force that pushes out the boundary between the magnetosphere and the solar wind. The direction of the storm time ring current in the magnetosphere during the main phase of the storm is from West to East, thus creating a magnetic field on the ground which is in the opposite direction as the Earth's magnetic field lines. This results in a decrease in the geomagnetic field which can last for several hours or even 2 days in the case of a severe storm.

3 INDICES RELATED TO THE INTENSITY OF GMDS

The intensity of GMDs are measured in terms of the deviation of the geomagnetic field from its daily variation during geomagnetically quiet times. Both the rate of change of the geomagnetic field, (in nT/min) and the direction change of the field (in degrees/min) express the instantaneous deviation of the geomagnetic field. Various indices are derived from the instantaneous variation of the horizontal component of the geomagnetic field by subtracting the mean (ΔH) or combining the measurements from several observatories and taking the maximum deviation of the horizontal component of the magnetic field from the quiet-time values over a period of time (SYM-H, *Kp*).

The intensity of GMDs are correlated with the intensity of the X-ray flares emitted before the CMEs leave the surface of the Sun. The largest GMDs for which the corresponding X-ray flare intensity is available, are all associated with X-ray flare intensities above the X1 level (X-ray flux of 10^{-4} W/m^2). This correlation is useful for prediction of the impact of GMDs.

The impact of GMDs are not only related to their intensity, but also to the duration and re-occurrence of storms within a short period, that is, before the power systems have had time for sufficient recovery. The largest GMDs occur on the Sun-facing side of the Earth during the main phase of the storm. During the main phase there is a sharp decrease in the geomagnetic field, caused by the increase in the westward flowing ring current. The main phase can last for hours. Significant GMDs may also occur on the night-side of the Earth due to short-duration fluctuations of the geomagnetic field (in the order of minutes) associated with substorms arising from the reconnection of the magnetic field lines on the antisunward side of the Earth.

The DST (Disturbance Storm Time) index is related to the decrease of the geomagnetic field due to east-to-west currents which occur in the magnetosphere when the solar wind interacts with the magnetosphere. The ionospheric current systems which increase during GMDs are the equatorial electrojet and the auroral electrojet. The auroral electrojet causes the strongest GMDs, which typically occur in high-latitude regions of the Earth (60–70 degrees north and south).

The planetary three-hourly K-index, *Kp*, is a logarithmic index which is derived from the maximum deviation of the horizontal component of the geomagnetic field from its quiet-time values over a 3-hour period at four equatorial magnetometer stations. The *Kp*-index ranges from 0 to 9 with 1 being calm and 5 or more indicating a geomagnetic storm. Likewise, local K-indices are derived for magnetic observatories. The *ap*-index is a linearized version of the *Kp*-index and the *Ap* index is the daily average of the eight three-hourly *ap*-index values.

A correlation between the number of transformer failures and a weighted storm severity with *Kp* values between 7 and 9 provides evidence for storm-induced transformer failures, even in mid-latitude regions considered to have low GIC-risk [13].

4 MODELING GEOMAGNETIC STORM INTENSITY FROM SOLAR WIND PARAMETERS

A combination of solar and interplanetary parameters has been used as inputs in Neural Network (NN) models to predict the occurrence and strength of GMDs with an occurrence probability of 87% [36].

Lotz and Cilliers [43] developed a NN model to predict a geomagnetic field-derived proxy for GIC, namely the sum of the variation in the northward and eastward components of the geomagnetic field at Hermanus over a ten-minute period, using as inputs the IMF magnitude, solar wind speed, and fluctuations in solar wind particle density.

4.1 Extreme Event Analysis

Power utilities are interested in the levels and likelihood of extreme GIC events, to plan for the mitigation of extreme geomagnetic storms. Measured GIC data is generally proprietary and location-dependent. Hence, extreme events are usually characterized in terms of the expected extreme values of the geo-electric field (E-field).

4.2 Carrington Event

The most extreme geomagnetic storm on record, known as the 1859 Carrington event, was so intense that ground-based magnetometers were saturated at Greenwich and Kew in the United Kingdom. It lasted 5 days. Winter et al. [44] used a spectral analysis technique and a 1D conductivity profile in the United Kingdom and estimated the maximum electric field during the Carrington event to be on the order of 9 V/km [44]. The scaling was based on the only nonsaturated magnetic record of the storm in the form of 1-hour resolution data from the Colaba Magnetic Observatory in India.

Using Extreme Value Theory Thomson et al. extrapolated 30 years of digital data for Europe and estimated the maximum rate of change of the geomagnetic field in Europe to be in the range of 1500–4000 nT/min for 1 in 100 year returns and 2000–6000 nT/min for 1 in 200 year returns [45].

The Planning Guide of the Geomagnetic Disturbance Task Force of the National Energy Regulator Commission (NERC) in the United Kingdom proposed an extreme E-field value of 8 V/km with a scaling factor for latitude and ground conductivity to be used as the extreme value for planning for mitigation of such events.

4.3 Estimation of 1 in 100-year E-field Levels

Extrapolation of measurement and modeling of the E-field can be used to estimate the 1 in 100-year E-field values (Fig. 3).

Ngwira et al. used the Quebec 1D conductivity profile in a study of the global behavior of the ground horizontal geomagnetic and the maximum induced geoelectric field variations with latitude by considering twelve severe/extreme geomagnetic storm events [46]. They found that there is a latitude threshold for maximum E-field values exceeding 500 mV/km at about 50–55 degrees of magnetic latitude for most major or extreme

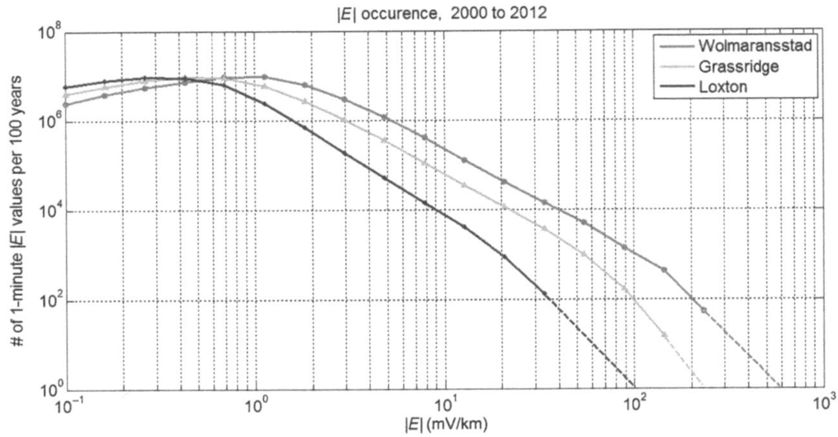

FIG. 3 Estimation of 1 in 100-year values of the magnitude of the E-field $|E|$ based on magnetic field measurements from the INTERMAGNET station at Hermanus over the period 2000 to 2012 and three different sets of surface impedance values derived from 1D conductivity models for Wolmaranstad (high resistivity), Grassridge (intermediate), and Loxton (low resistivity) in South Africa. The range of 1 in 100-year $|E|$ values is 190–590 mV/km for these mid-latitude stations, but the value of the extreme E-field for locations with a low conductivity Earth model differ by an order of magnitude from that of locations with a high conductivity model.

geomagnetic storms, with stronger E-field values occurring polewards from the threshold. The latitude threshold expands equatorward during extreme storms.

5 DISSEMINATION OF RELIABLE SPACE WEATHER DATA/INFORMATION

Constant monitoring of the space environment from a range of vantage points is needed, together with timely dissemination of reliable data to the service users, to maintain the integrity of power systems and other technologies.

The 19 regional space weather warning centers (RWCs) of the International Space Environment Service (ISES) are the primary providers of high-quality scientific observations, results, and models as well as a number of space weather products to local customers, including space weather warnings and now-casting of geomagnetic conditions. The RWCs integrate data from satellites (GOES, SOHO, ACE, SDO, STEREO) with local measurements of geomagnetic fields and ionospheric conditions, and simulations from space weather models, to provide regular space weather updates and predictions.

The raw measurements are fed into advanced computational models and tools to give both near-real-time information and advanced warning of upcoming space weather conditions that may affect a diverse range of user systems to help protect regional critical infrastructure from damage by space weather events. Space weather dissemination will enable end-users in a wide range of affected sectors to mitigate the effects of space weather on their systems, reducing costs and improving reliability.

Of particular interest for the impact on power systems is the current geomagnetic activity, expressed through both the real-time data and through various indices derived from the real-time data.

Forecasting of space weather indices given by space weather centers range from a few hours to a few years and are done by means of both linear and nonlinear prediction methods such as Neural Networks. The geomagnetic and solar activity forecast service of the British Geological Survey makes the following forecasts available [47]:

- Daily local geomagnetic activity (*DRX* index), global geomagnetic activity (*Ap* index), and solar activity (*F10.7* index) for up to 27 days ahead, using a linear prediction method.
- Daily global geomagnetic activity (*Ap* index) for up to 3 days ahead, using a nonlinear prediction method.
- Smoothed monthly solar (*F10.7* and *SSN* indices) and geomagnetic (*Aa* index) activity for the next solar cycle, using a linear prediction method.

BGS [47] in the United Kingdom provides GIC related parameters to the UK National Grid control room via a web based system called MAGIC ('Monitoring and Analysis of GIC'). Through MAGIC, BGS also provides forecasts of geomagnetic activity, information about disturbances in the solar wind that cause geomagnetic storms, and provides an up-to-the-minute UK-wide simulation of GIC flow in the transmission system.

6 GICs IN ELECTRICAL POWER SYSTEMS

The physical principles which lead to GICs are based on Faraday's and Ohm's laws: Rapidly changing currents in the Earth's magnetosphere and ionosphere during a space weather storm create temporal variations in the geomagnetic field (B-field) that induce a geo-electric field, based on the Faraday's law.

$$dB/dt = -\mathrm{curl}\,(E). \quad (1)$$

The fluctuating B-field then drives currents in conducting materials including power lines, pipelines, and the Earth itself, of which the lithosphere is the conducting solid outer layer. Currents induced in the ground also contribute to the geomagnetic disturbance field and to the geoelectric field occurring at the surface of the Earth during geomagnetic storms. In power systems, the closed conducting path for GICs, satisfying Ohm's law, is formed when the neutrals of wye-connected transformers at the ends of a transmission line are grounded. This path includes the impedance of the transmission line, transformers, and series reactors. Due to the low frequency of GICs, typically $<10\,\mathrm{mHz}$, the

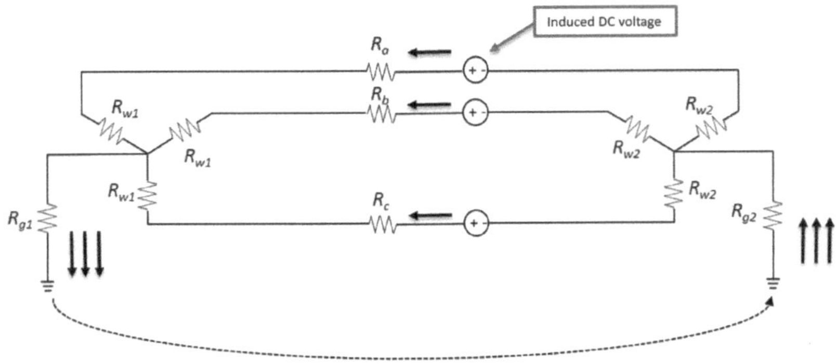

FIG. 4 Illustration of GIC flow in transformers and transmission lines. *(Based on GIC flow in power systems, 8th January, 2014, Available from: http://www.powerworld.com/files/GICGroundLoop.jpg.)*

reactance of the conducting path is neglected, and only the DC resistance is represented in transmission line models.

Fig. 4 illustrates the flow of GIC in the transmission lines, grounded neutrals, and windings or wye-connected transformers.

In Fig. 4, R_w represents the transformer winding resistance, R_g represents the transformer grounding resistance: R_a, R_b, and R_c represent the resistance of each phase of the transmission line.

The storm-time induction of electrical fields in the Earth's lithosphere can drive low frequency currents in electric power grids which interfere with the operation of the grids through the generation of increased VAR demand, and saturation of transformers, with concomitant increase in vibration and power line harmonics, and can cause rapid aging of transformers, and blackouts [48]. Reactive power is needed to support system voltage. Variation in additional reactive power demand which exceeds available network capacity may be lead to fluctuations in the voltage level at certain nodes in the network. If this persists (either over or under voltage scenarios), power system protection will act to island parts of the network. This tripping in itself may constitute a large disturbance which can affect voltage stability of the network.

This action may further lead to over or under voltage scenarios based on the composition of the islands in respect of reactive power availability. A widely reported case of blackout due to GIC is the Hydro-Quebec blackout on 13 March 1989 [49]. Millions people were left in the dark for 12 hours towards the end of winter [49, 50]. In Sweden, the flow of GIC in the Swedish power grid during the Halloween storm on 30 October 2003 led to a large-scale blackout, affecting about 50,000 customers for about one hour [12]. The economic cost of such blackouts easily run in hundreds of millions of dollars.

7 MODELING GICs IN ELECTRICAL POWER SYSTEMS

The estimation of GICs in power systems comprise a geophysical part and an engineering part. The key problem in the geophysical part is to determine the geo-electric field that drives GICs in power systems. The engineering part finds the currents induced in each line and in each grounded transformer by means of circuit analysis. The electric field can be calculated from the measured magnetic field using measured surface impedance values or surface impedance values derived from models of the Earth's conductivity.

7.1 Geophysical Part

The E-field values are determined in the frequency domain from

$$E(\omega) = \frac{1}{\mu} Z(\omega) B(\omega), \tag{2}$$

where $E(\omega)$ is the electric field intensity, $B(\omega)$ is the magnetic field flux density and $Z(\omega)$ is the surface impedance at frequency ω and μ is the permeability of the Earth's surface, taken to be the same as that of free space.

The process for deriving the time-domain representation of the E-field via the Fourier Transform is shown in Fig. 5.

The surface impedance can be calculated using the plane wave model for the B-field and various Earth conductivity models including homogeneous models, 1D Conductivity profiles, 2D and 3D models [51]. It is best to use local geomagnetic field measurements, together with local surface impedance values

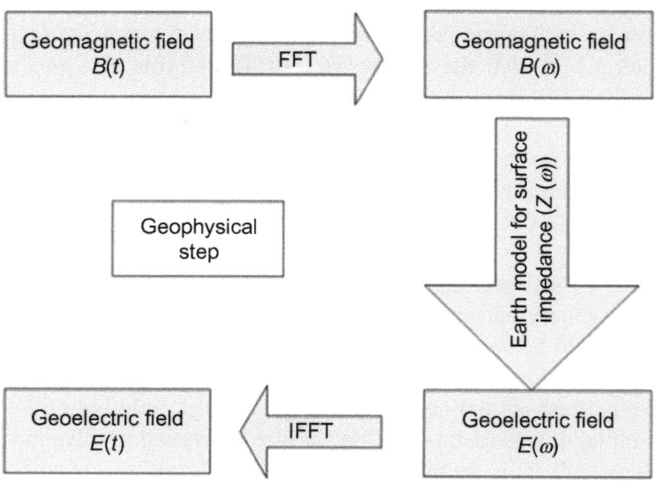

FIG. 5 Flow diagram for the geophysical step in the modeling GIC from geomagnetic data. *(Based on K. Zheng, et al., Effects of geophysical parameters on GIC illustrated by benchmark network modeling, IEEE Trans. Power Deliv., 28 (2) 2013 1183–1191.)*

in Eq. (2) when calculating the E-field induced in the power lines. However, due to the cost and requirements of a clean magnetic environment, magnetic field measurements are mostly done at magnetic observatories that are not located near high voltage power lines, and only at a few locations in a country. Likewise, surface impedance measurements are not generally done countrywide. Surface impedance is often inferred from the geology of the region and the conductivities of the various geological layers of the Earth under the power lines.

As a first approximation, the magnetic field measured at an observatory in the region of interest is assumed to be valid for the whole region in the form of a vertically propagating uniform plane wave over the whole region. This method, which was first introduced by Viljanen [52] is based on Faraday's law of induction. The simplest model for the surface impedance is to assume that the Earth can modeled as a homogeneous half-space with a constant and uniform conductivity, for which case there is an analytic expression for the surface impedance, that is,

$$Z(\omega) = \sqrt{\frac{\mu\omega}{\sigma}} e^{i\pi/4}. \tag{3}$$

The horizontal components of the E-field have an orthogonal relation to the horizontal components of the B-field [53], that is,

$$\begin{aligned} E_y(\omega) &= -\frac{1}{\mu} Z(\omega) B_x(\omega) , \\ E_x(\omega) &= \frac{1}{\mu} Z(\omega) B_y(\omega). \end{aligned} \tag{4}$$

The time domain representation of the horizontal components of the E-field can be determined from the inverse Fourier Transforms of $E_x(\omega)$ and $E_y(\omega)$ obtained by Eq. (4), or by a time series approximation of the inverse Fourier transform [52].

The plane wave model for the magnetic field together with a homogeneous Earth conductivity model has been used to give estimates of the E-field which result in fairly good estimates of the GICs, compared to measured values [54].

To derive a local geomagnetic field under the power lines, the method of spherical elementary current systems (SECS) has been used to interpolate the geomagnetic field between measurement locations [55]. The results obtained using SECS was closer to the measured GIC when compared to the uniform plane wave method for the calculation of the electric field [24,56].

The complex image method (CIM) has also been used to solve the geophysical problem [57]. This method adopts a layered conductivity model of the Earth. In 2003, Pulkkinen et al. [58] introduced a combination of SECS and CIM. Due to its sophisticated complex method, CIM is not a preferred option for operational purposes.

The local surface impedance and ground conductivity structure can be determined from GIC measurements.

Where magnetotelluric measurements of the full tensor components of the surface impedance are available the horizontal components of the E-field can be determined from the following equation

$$\begin{bmatrix} E_x(\omega) \\ E_y(\omega) \end{bmatrix} = \frac{1}{\mu} \begin{bmatrix} Z_{xx}(\omega) & Z_{xy}(\omega) \\ Z_{yx}(\omega) & Z_{yy}(\omega) \end{bmatrix} \begin{bmatrix} B_x(\omega) \\ B_y(\omega) \end{bmatrix}. \tag{5}$$

7.2 The Engineering Part: Network Calculation

Two widely used methods to calculate the GICs flowing in the network from induced electrical field are the Lehtinen-Pirjola and Nodal Admittance Matrix methods.

7.2.1 Lehtinen-Pirjola Method

This method [59,60] is illustrated using a three-bus network, as shown in Fig. 6. The induced electric fields are shown on the lines as E_{ij} and E_{jk}. The elements R_{ij} and R_{jk} represent the resistance of the transmission line while R_i, R_j and R_k represent the resistances at each substation. The latter includes the substation grounding resistance and transformer winding resistance. For simplicity, the resistances of each phase in each component in the conducting path are assumed to be equal. This allows for single phase equivalent treatment of all the components in the conducting path.

Using Norton's theorem [61,62], the electric field induced in line L_{ij} with resistance R_{ij} is converted to an equivalent current source h_{ij} in parallel with the admittance of the line y_{ij}. These are defined in Eqs. (6), (7).

The parameters h_{ij} and y_{ij} in Fig. 7 are defined as:

$$h_{ij} = \frac{E_{ij}}{R_{ij}}, \tag{6}$$

$$y_{ij} = \frac{1}{R_{ij}}. \tag{7}$$

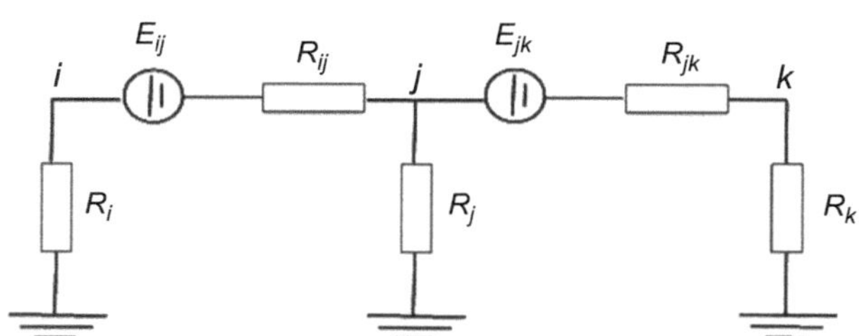

FIG. 6 Three-bus network with induced electric field.

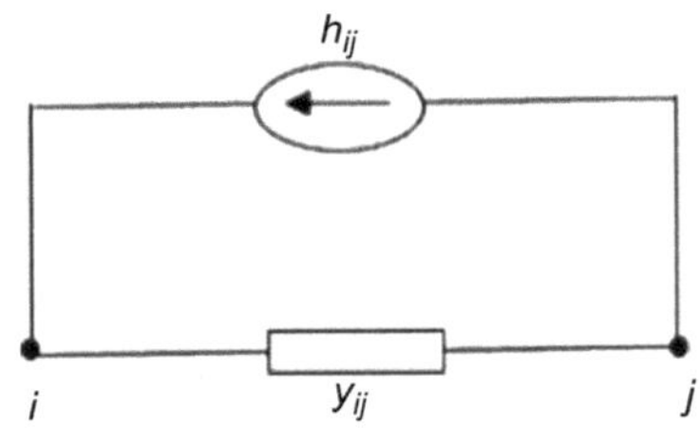

FIG. 7 Norton's current equivalent.

Substituting Eq. (7) in Eq. (6), the current in line i_{ij} due to the induced electric field E_{ij} can be written as:

$$h_{ij} = E_{ij}y_{ij}. \tag{8}$$

The sum of all the currents at node i is derived using Kirchhoff's law for currents as:

$$i_i = \sum_{j=1}^{N} i_{ji} = -\sum_{j=1}^{N} i_{ij}, \tag{9}$$

where N is the total number of nodes. The currents in line ij is:

$$i_{ij} = E_{ij}y_{ij} + y_{ij}(v_i - v_j). \tag{10}$$

Factoring the common term y_{ij} in Eq. (10) will yield:

$$i_{ij} = y_{ij}\left[E_{ij} + (v_i - v_j)\right], \tag{11}$$

and substituting Eq. (11) into Eq. (9) will yield:

$$i_i = -\sum_{j=1}^{N} y_{ij}\left[E_{ij} + (v_i - v_j)\right]. \tag{12}$$

Assuming the path to ground from each node has zero resistance, the node voltages will be zero. Therefore, the current in the branches will be exactly equal to current sources. Hence, the sum of the current sources is equal to the current that flows to ground. This is given in Eq. (13) using Kirchhoff's' current law:

$$J_i^e = -\sum_{j=1}^{N} E_{ij}y_{ij} \quad i \neq j. \tag{13}$$

Substituting Eq. (13) in Eq. (12) then gives:

$$i_i = J_i^e - \sum_{j=1}^{N} (v_i - v_j)y_{ij} \quad i \neq j. \tag{14}$$

And expanding Eq. (14) yields:

$$i_i = J_i^e - \sum_{j=1}^{N} v_i y_{ij} + \sum_{j=1}^{N} v_j y_{ij} \quad i \neq j. \tag{15}$$

The first summation in Eq. (15) represents the diagonal elements of a network admittance matrix:

$$Y_{ii}^j = \sum_{j=1}^{N} y_{ij} \quad i \neq j. \tag{16}$$

The second summation represents the dependence of the current i_i on all the other voltages and gives the off-diagonal elements of the network admittance matrix:

$$Y_{ij}^j = -y_{ij} \quad i \neq j. \tag{17}$$

Substituting Eqs. (16), (17) in Eq. (15) gives:

$$i_i = J_i^e - \sum_{j=1}^{N} v_j Y_{ij}^j. \tag{18}$$

In matrix form, Eq. (18) can be written for all the nodes as:

$$[I^e] = [J^e] - [Y^j][V^j], \tag{19}$$

where the elements of the column matrix $[I^e]$ are the nodal currents, elements of the column matrix $[V^j]$ are the nodal voltages and $[J^e]$ is the column matrix of the current sources at each node.

The nodal voltage at each node can be derived as the product of the earthing impedance and the nodal current given in Eq. (19):

$$[V^j] = [Z^e][I^e], \tag{20}$$

where $[Z^e]$ is the earthing impedance matrix.

Substituting Eq. (20) in Eq. (19) gives:

$$[I^e] = [J^e] - [Y^j][Z^e][I^e]. \tag{21}$$

Re-arranging Eq. (21) in terms of $[I^e]$ gives

$$[I^e]\left([1] + [Y^j][Z^e]\right) = [J^e], \tag{22}$$

where $[1]$ is a unit matrix.

Matrix $[I^e]$ can be calculated by taking the inverse of $([1] + [Y^j][Z^e])$ and multiplying the result by $[J^e]$ given in Eq. (22):

$$[I^e] = \left([1] + [Y^j][Z^e]\right)^{-1}[J^e]. \tag{23}$$

The current in each node in $[I^e]$ is the GIC flowing through the node to ground.

By applying the principle of superposition, when the electric field induced in the network is decomposed into the eastern component and the northern component, $[J^e]$ in Eq. (22) will be calculated twice:

$$J^e_{i(x)} = -\sum_{j=1}^{N} E(x)_{ij} y_{ij} \quad i \neq j, \tag{24}$$

where $J^e_{i(x)}$ is the sum of the current source at a node due to the eastern component of the electric field, and

$$J^e_{i(y)} = -\sum_{j=1}^{N} E(y)_{ij} y_{ij} \quad i \neq j, \tag{25}$$

where $J^e_{i(y)}$ is the sum of the current source at a node due to the northern component of the induced electric field.

Therefore, the nodal current as a result of the induced electric field will be:

$$[I^e] = \left([1] + \left[Y^j\right][Z^e]\right)^{-1}\left[J^e_x\right] + \left([1] + \left[Y^j\right][Z^e]\right)^{-1}\left[J^e_y\right]. \tag{26}$$

For an electric field of 1 V/km in both eastern and northern components for node i:

$$a_i = \left([1] + \left[Y^j\right][Z^e]\right)^{-1}\left[J^e_x\right], \tag{27}$$

where a is nodal current corresponding to the eastern electric field of 1 V/km, and:

$$b_i = \left([1] + \left[Y^j\right][Z^e]\right)^{-1}\left[J^e_y\right], \tag{28}$$

where b is the nodal current corresponding to the northern electric field of 1 V/km.

This allows for Eq. (26) to re-written as:

$$[I^e] = [a_i]E_x + [b_i]E_y, \tag{29}$$

where E_x represents the induced electric field in the northern direction and E_y represents the electric field in the eastern direction.

Eqs. (27), (28) are fixed network parameters for a specific network. Therefore, once the a and b network parameters are calculated, the nodal currents can be calculated as shown in Eq. (29). It should be noted that the status of the transmission network is dynamic due to faults that may trip off transmission lines, or maintenance related switching. As a result, the a and b parameters may not be accurate when the network configuration has changed. Research is ongoing into statistical approaches to determining and applying the a and b in estimating GICs in the transmission network.

7.2.2 Nodal Admittance Matrix (NAM) Method

In the NAM method, the electric field input, which is in series with the transmission line, is converted to its Norton equivalent circuit. The nodal voltages are calculated by matrix inversion and multiplied by the nodal admittances [63]. The NAM method was compared with the Lehtinen-Pirjola method by Boteler and Pirjola and they were found to be mathematically equivalent [60].

From Figs. 16 and 17, h_{ij} and y_{ij} in Fig. 7 are defined as:

$$h_{ij} = \frac{E_{ij}}{R_{ij}}, \tag{30}$$

$$y_{ij} = \frac{1}{R_{ij}}. \tag{31}$$

The summation of the current equivalents at node j is represented by Eq. (32).

$$H_j = \sum_{i \neq j}^{n} h_{ji}, \tag{32}$$

From the equation above, summing up the equivalent currents at each node in Fig. 6 are:

$$H_i = h_{ij}, \tag{33}$$

$$H_j = h_{ji} + h_{jk}, \tag{34}$$

$$H_k = h_{kj}. \tag{35}$$

The total current flowing through line i, j is:

$$i_{ij} = h_{ij} + \left(v_i - v_j\right) y_{ij}, \tag{36}$$

where v_i and v_j are the voltage at node i and node j respectively.

The total current flowing through node i is:

$$i_i = v_i . y_i. \tag{37}$$

Since there is only one line that is connected to bus i, the nodal current at bus i is the same as the line current i_{ij} in Eq. (38).

$$i_i = i_{ij}. \tag{38}$$

Substitution of the R.H.S of Eq. (36) into Eq. (38) gives:

$$i_i = h_{ij} + \left(v_j - v_i\right) y_{ij}. \tag{39}$$

Similarly, at bus j, the sum of the line currents connected to it, is given in Eq. (38).

$$i_j = i_{ij} + i_{jk}. \tag{40}$$

Substitution of the R.H.S of Eq. (36) and i_{jk} into Eq. (40) gives:

$$i_j = h_{ij} + (v_j - v_i)y_{ij} + h_{jk} - (v_j - v_k)y_{jk}. \tag{41}$$

At bus k, the only current flowing through this node is the line current i_{jk}:

$$i_k = i_{jk}. \tag{42}$$

Substituting for i_{jk}, gives:

$$i_k = h_{jk} - (v_k - v_j)y_{jk}. \tag{43}$$

Replacement of the L.H.S of Eqs. (39), (41), and (43) with the R.H.S of Eq. (37) gives:

$$v_i.y_i = h_{ij} + (v_j - v_i)y_{ij}, \tag{44}$$

$$v_j.y_j = h_{ij} + (v_j - v_i)y_{ij} + h_{jk} - (v_j - v_k)y_{jk}, \tag{45}$$

$$v_k.y_k = h_{jk} - (v_k - v_j)y_{jk}. \tag{46}$$

Re-arranging the above three equations

$$h_{ij} = v_i.y_i - (v_j - v_i)y_{ij}, \tag{47}$$

$$h_{ij} + h_{jk} = v_j.y_j - (v_j - v_i)y_{ij} + (v_j - v_k)y_{jk}, \tag{48}$$

$$h_{jk} = v_k.y_k + (v_k - v_j)y_{jk}. \tag{49}$$

The three equations above are written in matrix form as shown in Eq. (50).

$$\begin{bmatrix} h_i \\ h_{ij} + h_{jk} \\ h_{jk} \end{bmatrix} = \begin{bmatrix} y_i + y_{ij} & -y_{ij} & 0 \\ -y_{ij} & y_j + y_{ij} + y_{jk} & -y_{jk} \\ 0 & -y_{jk} & y_{jk} + y_k \end{bmatrix} * \begin{bmatrix} v_i \\ v_j \\ v_k \end{bmatrix}. \tag{50}$$

Eq. (50) is summarized as shown in Eq. (51).

$$[H] = [Y][V], \tag{51}$$

where $[Y]$ is the network admittance matrix, $[V]$ is the nodal voltage matrix and $[H]$ is the matrix of Norton's currents. From Eq. (51), the nodal voltage is determined as shown below,

$$[V] = [Y]^{-1}[H]. \tag{52}$$

In Eq. (52), $[Y]^{-1}$ is the inverse of [Y] as is calculated as shown in Eq. (53).

$$[Y]^{-1} = \frac{1}{\det Y}(\text{adjoint of } Y). \tag{53}$$

When substituting Eq. (52) in Eq. (37), the GIC flowing through the grounded neutral of the transformer(s) at a node in the network is derived as,

$$[i] = [y_0][Y]^{-1}[H]. \tag{54}$$

If there are multiple transformers connected in parallel in a substation in the network, the GIC flowing through each winding of each transformers can be calculated as follows:

Let n = the number of three-phase transformers connected in parallel, i = the GIC flowing through a substation in the network, the GIC flowing through the winding of one of the transformers, i_w, will be,

$$i_w = \frac{i}{3 \times n}$$

It is assumed here that the DC winding resistance of each of the transformers are the same. In practice the windings may not have the same resistance and hence the GIC flowing through each winding may differ slightly from i_w.

i. Consideration for special transformers:

The primary and secondary winding in an autotransformer are linked, implying that both magnetic and electrical circuits are coupled through the common winding. Where an autotransformer connects two buses, its winding resistance is divided between the coupling resistance and a resistance to the neutral terminal (if grounded) [25,64]. In cases where an autotransformer connects more than one bus, the autotransformer is treated as a transmission line [64]. Fig. 8 shows the windings of a single-phase auto transformer. The voltage across the common winding is V_2, the voltage across the full winding is V_1. I_1 and I_2 are the currents flowing in and out of the node A and B respectively. I_n is the current flowing through the common winding. If the common winding is grounded, I_n is the GIC that will flow through ground.

7.3 Example of GIC Calculation

Consider the geomagnetic disturbance of March 1989, the measured horizontal and vertical components of the geomagnetic field at Hermanus Magnetic Observatory during a 24-h period as shown in Figs. 9 and 10. Calculate the GIC that

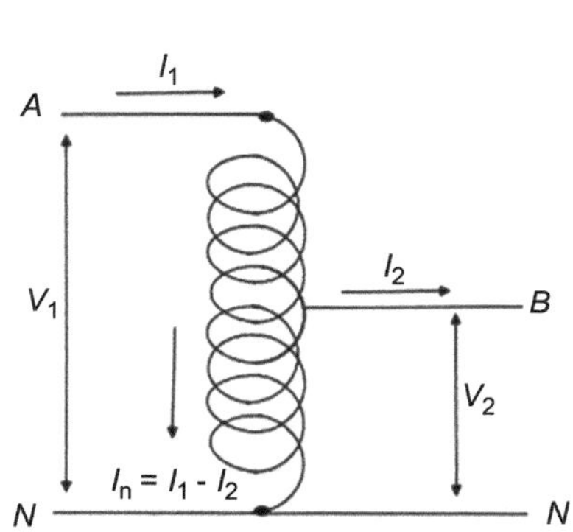

FIG. 8 Winding representation of a single-phase auto transformer.

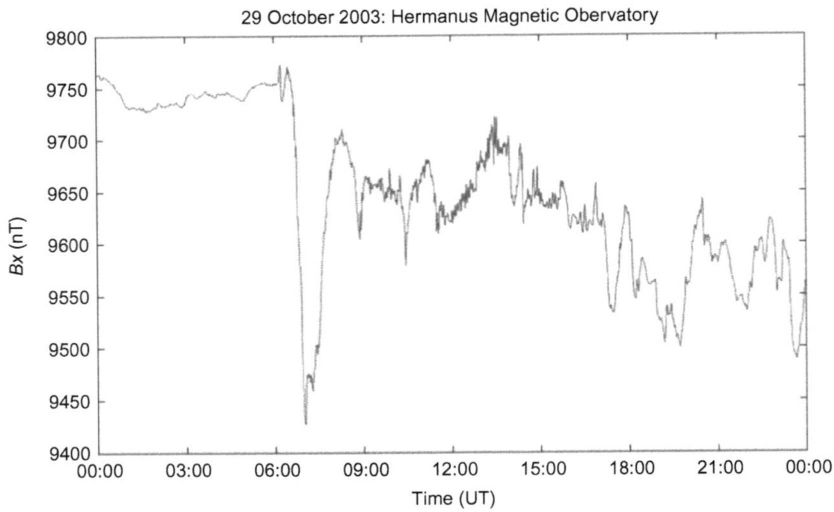

FIG. 9 Northward directed horizontal component of the magnetic field.

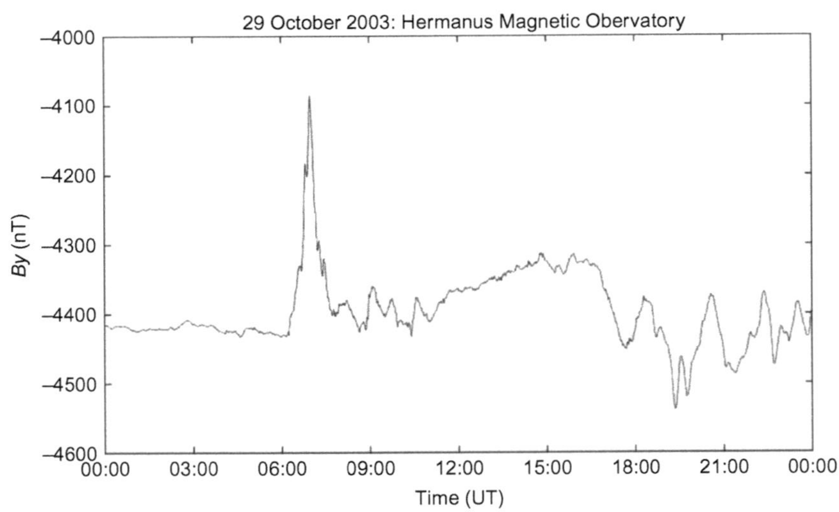

FIG. 10 Eastward directed horizontal component of the magnetic field.

will flow through the substations of the 400-kV power transmission network shown in Fig. 11.

The rate of change of the geomagnetic field in accordance with Faraday's law is the factor that drives the geoelectric field which in turn drives GICs in the transmission lines and transformers. The rate of change of the northward and eastward directed horizontal components of the geomagnetic field are shown in Figs. 12 and 13, respectively.

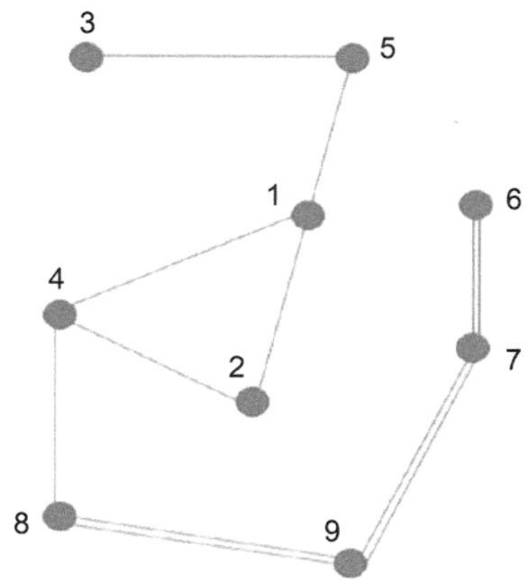

FIG. 11 Transmission network with nine substations and twelve transmission lines not drawn to scale.

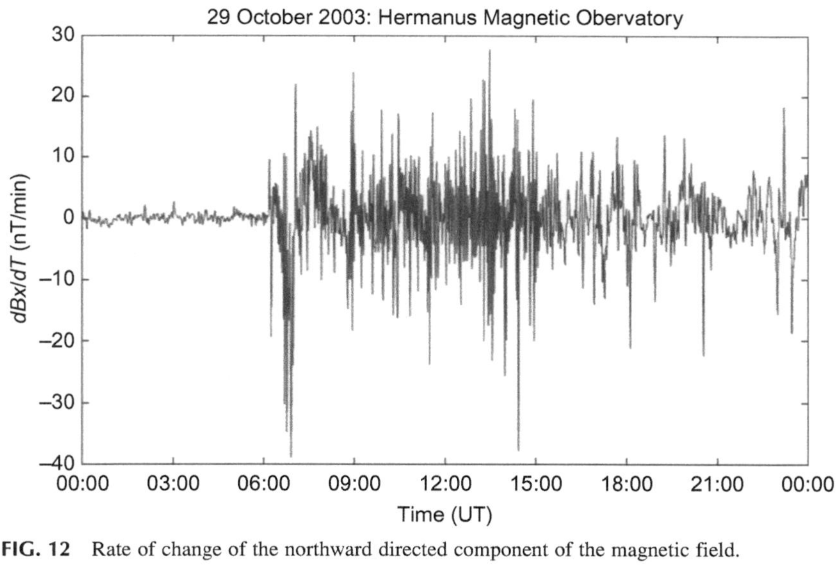

FIG. 12 Rate of change of the northward directed component of the magnetic field.

Using a homogenous Earth model with a conductivity of 2.7 mS/m, the northward and eastward directed horizontal components of the electric field are calculated shown in Figs. 14 and 15.

The network is made up of twelve 400 kV transmission lines and nine substations. The ground resistance in each substation are assumed to be the same. The relevant network data are given in Tables 1 and 2.

The substation resistance per phase in Table 2 was calculated using Eq. (55). A more detailed explanation on how to calculate the substation resistances can be found in the research work published by Boteler and Pirjola [60].

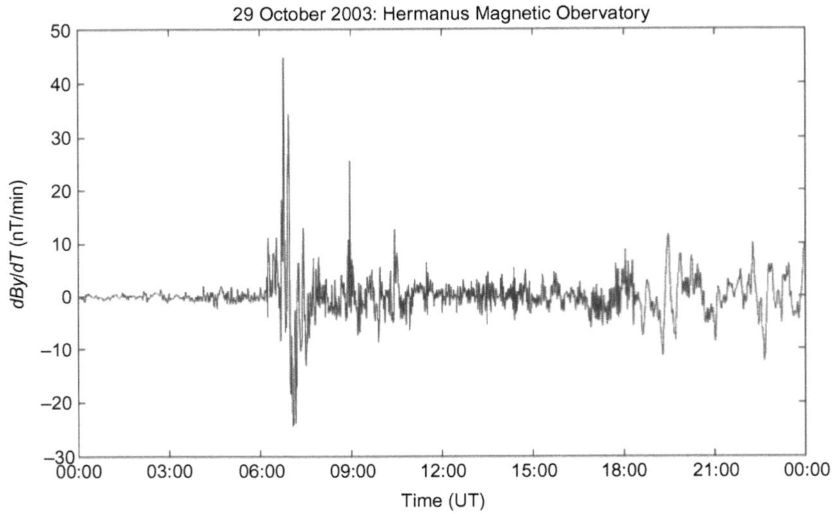

FIG. 13 Rate of change of the eastward directed component of the magnetic field.

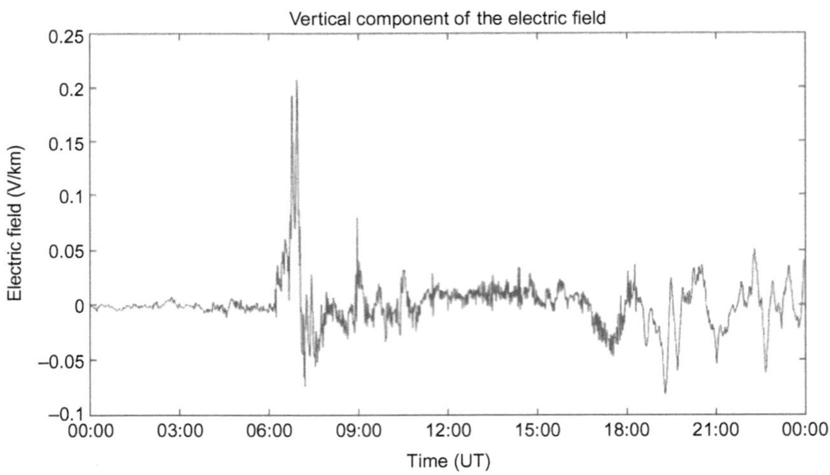

FIG. 14 Northward horizontal component of the electric field.

$$R_j = {}^1/_3\left[{}^{r_t}/_{n_t} + {}^{r_r}/_{n_r}\right] + R_e, \tag{55}$$

where,

R_j is the substation resistance per phase.
R_e is the substation grounding resistance
r_t is the transformer resistance per phase.

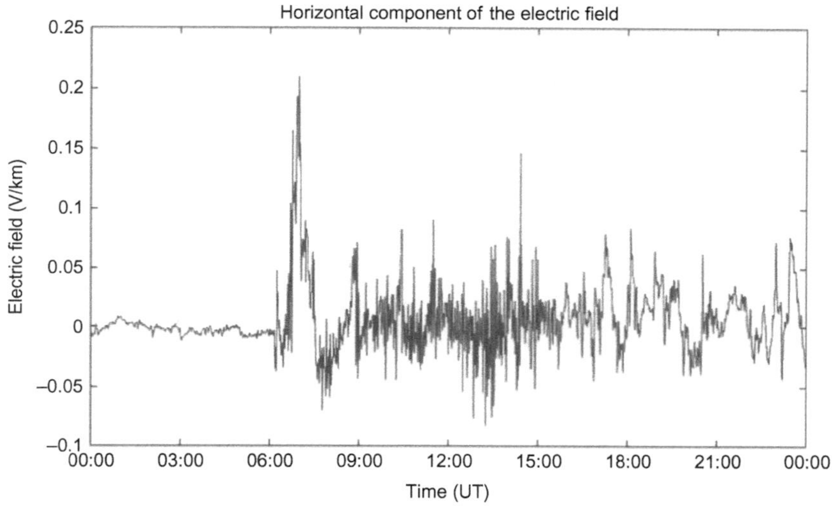

FIG. 15 Eastward horizontal component of the electric field.

r_r is the reactor resistance per phase.
n_t is the number of transformers connected in parallel.
n_r is the number of reactors connected in parallel.

Using the nodal admittance in Section 7.2.2, the GICs flowing in the network is calculated. It should be noted that the electric field was multiplied by a scalar factor of 5 for ease of illustration. Figs. 16–18 show the GIC flowing through substation 5, 6, and 9.

Two of the factors that determine the magnitude of the GIC induced in the network, assuming the ground parameters and network structure are unchanged, are the orientation of the transmission lines and the orientation of the induced geoelectric field. The maximum induced electric field in the network will be realized when most of the transmission lines are aligned with the electric field. Fig. 19 shows the transmission lines and substations in the Cartesian plane with corresponding longitude and latitude coordinates. In this example, the maximum GIC flowing in the network was observed at 06:57 UT when the angle of the induced electric field was −41.95 degrees. The magnitude of the GIC flowing through the nine substations at this time is shown in Fig. 20, where the diameter of the circle is proportional to the magnitude of the GIC.

The changing magnitude and direction of the induced electric field during a geomagnetic disturbance drives the GIC flowing through the substations to vary in real time, which is corroborated by the GIC profiles shown in Figs. 16–18. The results shown in Figs. 16–20 show that the magnitude of the GIC flowing through the nodes in the network varies in real time. So also, the vulnerability of the transformers to GICs vary.

TABLE 1 Transmission Line Data

Line Number	Substation Number		Line Resistance (Ω/phase)	Line Length (km)	From Substation		To Substation	
	From	*To*			*Latitude (degrees)*	*Longitude (degrees)*	*Latitude (degrees)*	*Longitude (degrees)*
1	3	5	1.75	76	−29.6	17.18	−29.63	17.88
2	5	1	2.29	99	−29.63	17.88	−29.3	18.8
3	1	2	4.52	196	−29.3	18.8	−29.49	20.79
4	1	4	7.78	337	−29.3	18.8	−26.54	18.12
5	2	4	9.70	420	−29.49	20.79	−26.54	18.12
6	4	8	10.30	446	−26.54	18.12	−22.59	17.37
7	4	8	10.30	446	−26.54	18.12	−22.59	17.37
8	8	9	0.67	29	−22.59	17.37	−22.52	17.08
9	8	9	0.67	29	−22.59	17.37	−22.52	17.08
10	9	7	3.76	163	−22.52	17.08	−21.5	16.03
11	9	7	3.81	165	−22.52	17.08	−21.5	16.03
12	7	6	12.10	524	−21.5	16.03	−17.4	14.22

TABLE 2 Substation Data

Substation Number	Grounding Resistance, R_e (Ω)	Number of Transformers at Substation (n_t)	Transformer Resistance per Phase, t_r (Ω)	Number of Reactors at Substation (n_r)	Reactor Resistance per Phase, r_r (Ω)	Substation Resistance per Phase (R_j) (Ω)
1	0.12	1	0.15	0	0.1	0.17
2	0.12	1	0.15	0	0	0.17
3	0.12	2	0.15	0	0	0.15
4	0.12	1	0.15	1	0.1	0.20
5	0.12	1	0.15	0	0	0.17
6	0.12	2	0.15	0	0	0.15
7	0.12	2	0.15	0	0	0.15
8	0.12	2	0.15	1	0.1	0.18
9	0.12	2	0.15	1	0.1	0.18

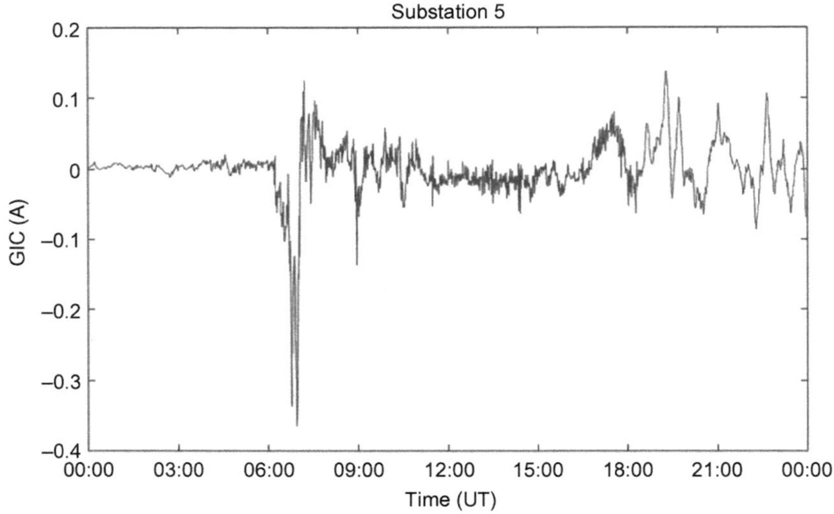

FIG. 16 GIC flowing through the neutral grounding point in substation 5.

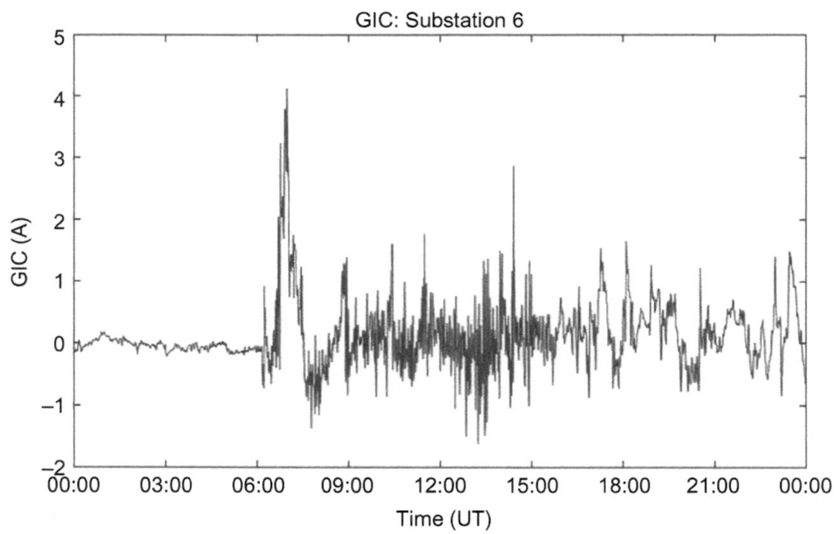

FIG. 17 GIC flowing through the neutral grounding point in substation 6.

8 GIC MEASUREMENTS

Setting up instrumentation to measure GICs in the transmission network is one of the credible ways through which GIC calculation models can be verified and the flow of GICs confirmed and monitored. Measurements can be taken using Hall-effect sensors around the neutrals of the transformer in the substation which can

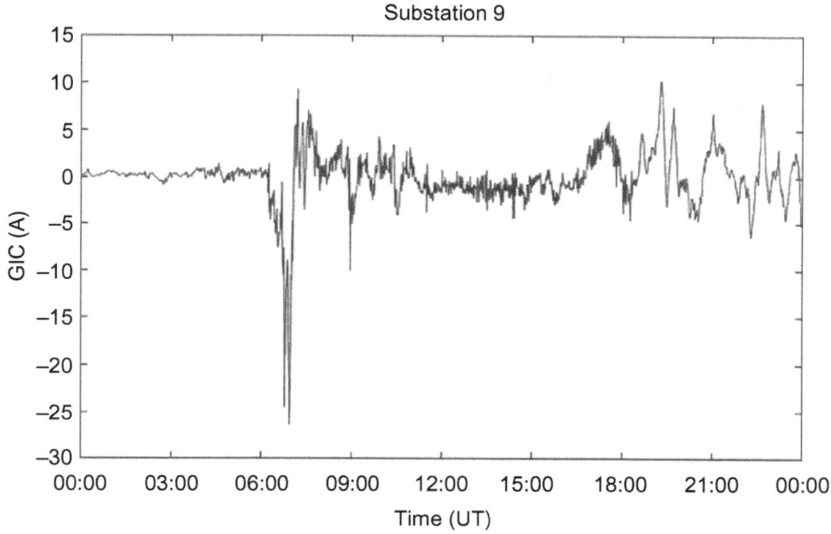

FIG. 18 GIC flowing through the neutral grounding point in substation 9.

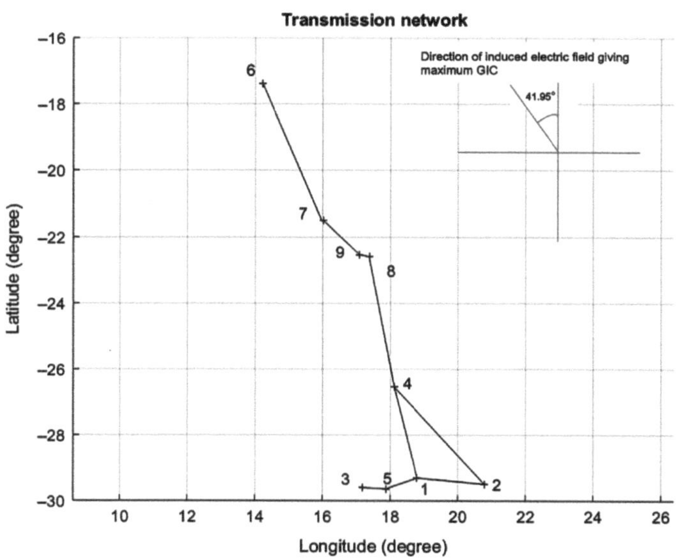

FIG. 19 Geographic locations of the nine nodes of the modelled transmission network and direction of the electric field inducing maximum GIC in the transmission network.

be intrusive. The Differential Magnetometer Method (DMM) is a nonintrusive method which has been used in Namibia and South Africa [65]. In this method, the magnetic field caused by the GIC flowing in a transmission line is measured using a magnetometer under the line and compared with the reference magnetic field measured by a magnetometer away from the line.

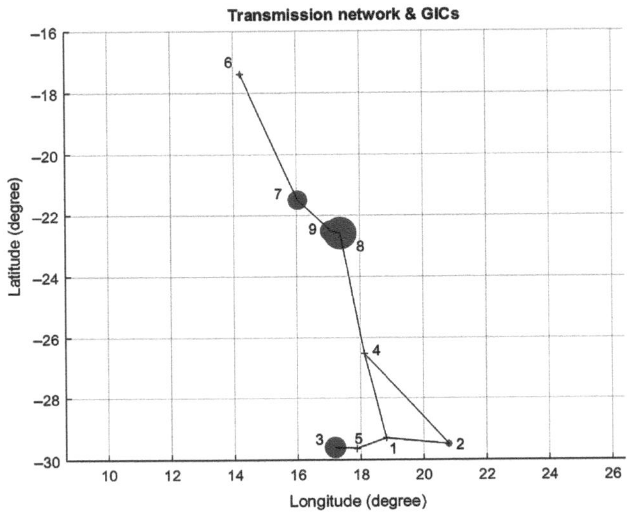

FIG. 20 The magnitude of the GIC flowing through substations is shown proportional to the radius of the blue dot. This occurred at 06:57 UT When the sum of the absolute values of the GIC in each node was the highest.

9 POWER NETWORK OPTIMIZATION FOR GIC

The power grid supplies critical loads, noncritical loads, and loads of varying levels of priority in-between. If it were possible, shutting off the power network just before an impending geomagnetic storm would be a quick solution to avert damage to network equipment such as power transformers. Not only is this an expensive exercise, black-starting an entire grid could take weeks depending on the availability and spread of black-start facilities.

Though the electrical and magnetic circuits of a transformer are interlinked, it is possible to consider electrical based and magnetic based optimization strategies to alleviate the potential adverse effect of GIC on power transformer and the power system.

9.1 Optimization Strategies

9.1.1 DC Blocking Capacitors

To mitigate the flow of GIC, DC current blocking devices have been designed and installed in high voltage transmission grids [66]. In Finland, neutral-point reactors have been used decrease the GIC magnitudes in its high-voltage system [67]. In England, series capacitors have been used to block GIC [68] in transmission lines. However, studies by Erinmez et al. [69] showed that it would be advantageous to strategically place these devices to prevent the risk of adversely increasing the flow of GIC in other parts of the network. Extensive off-line modeling of GIC flow in the transmission network is needed to fully understand

TABLE 3 Showing the GIC in Each Substation at 06:57 UT When the Sum of the Absolute Values of the GIC in Each Node is the Highest

Bus 1	Bus 2	Bus 3	Bus 4	Bus 5	Bus 6	Bus 7	Bus 8	Bus 9	Sum
3.3 A	7.2 A	23.9 A	0.7 A	0.3 A	4.6 A	22.1 A	37.1 A	24.2 A	123.5 A

the exposure of the network to damaging levels of GICs, hotspots in the network that are highly vulnerable, and what the impact of DC capacitors would be.

Consider the installation of DC blocking capacitors at selected substations based on the following criteria.

(a) The top two substations with the highest amount of GIC when the sum of the GICs flowing through the network is at its peak.
(b) The substation with the third largest GIC

For criteria (a), the absolute values of the GIC flowing through each substation is shown in Table 3.

The top two substations in Table 3 are 8 and 9. After the installation of DC blocking capacitors in these substations, the flow of GIC in the network was recalculated. It was found that no GIC flowed through substation 8 and 9 as expected. Blocking substation 8 and 9 created a long GIC path into substation 4, which had the most significant change in GIC. The GIC flowing through substation 4 increased by a factor of 26 during the peak of the storm. The rest of the substations showed comparatively insignificant changes. The GIC before and after this network change is shown in Fig. 21.

For criteria (b), the absolute values of the GIC flowing through each substation is shown in Table 4.

The effect of the blocking capacitor in substation 3 can be seen across the network with varying impact. In substation 1, the GIC reduced by >33-fold while there was an increase in bus 5 by 66-fold.

Figs. 22 and 23 shows the GIC profile for substation 1 and 5, respectively.

This example illustrates the challenge of network optimization to mitigate the adverse effect of GIC flowing in the network.

An ideal optimization scheme would start with setting the desired maximum GIC that will be allowed to flow through various substations in the network when exposed to a known severe geomagnetic such as the March 1989 GMD. Parameters to consider when setting such limits should include the transformer core type, size, and transformer health status. The latter of which can be deduced from maintenance records. The core type is of critical importance as the magnetic circuit for each transformer core has unique characteristics. The next step considers the economic constraint placed on implementing such a solution. For instance, in a network on 490 nodes, if it is only financially viable

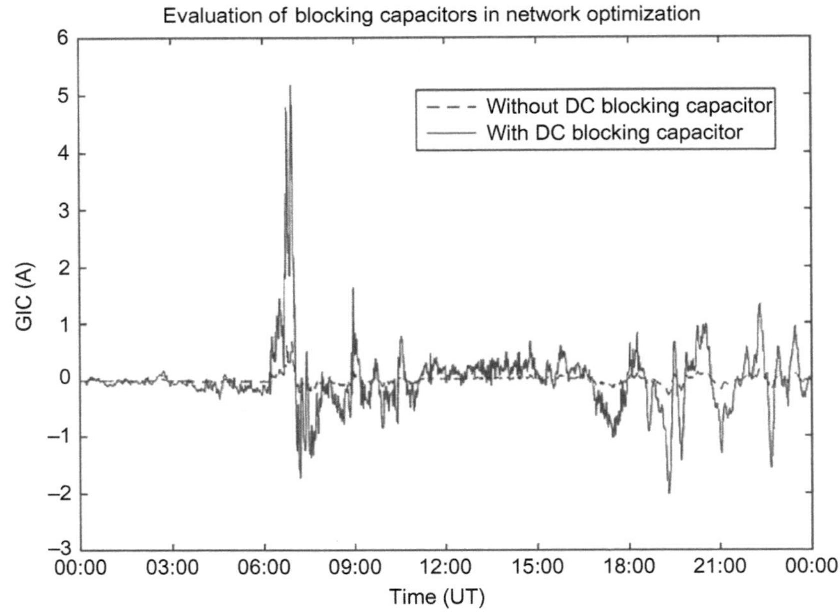

FIG. 21 The flow of GIC through substation 4 with or without blocking capacitors in bus 8 and 9.

to install 50 DC blocking capacitors, then the number of DC blocking capacitors is taken as one of several inputs to the objective function. Other criterion such as ground conductivity may be taken into consideration. Since the capacitor placement is a fixed installation, factors such as the orientation of the lines and location of substations would be analyzed in depth to understand the impact of the GMD. These parameters, in desired order, can then be used to optimally select the substations where the capacitors can be installed.

9.1.2 Magnetically Unloading Power Transformers

The root course of high reactive power consumption by transformers during a GMD is the increase in harmonics associated with the transformer operating in the region of nonlinearity, which is called saturation. In this region, the magnetic field density does not increase linearly with the increase in magnetic field intensity. The knee-point between the region of linearity and saturation is normally seen as the point below which the transformer is operated. On the other hand, the grid code of several countries allows the nodal voltage in the transmission network to fall within an acceptable range such as 0.9–1.1 pu. A number of methods that could be used to magnetically unload transformers are been tested using power system simulation software. It is important to study the impact of such methods, to determine its feasibility for specific network configurations and operating condition.

TABLE 4 Showing the GIC (A) in Each Substation at 06:57 UT With and Without DC Blocking Capacitor at Substation 3

	Bus 1	Bus 2	Bus 3	Bus 4	Bus 5	Bus 6	Bus 7	Bus 8	Bus 9	Sum
No DC capacitor	3.3	7.2	23.9	0.7	0.3	4.6	22.1	37.1	24.2	123.5
With DC capacitor	0.1	6.9	0.0	0.9	19.8	4.6	22.1	37.2	24.2	115.7

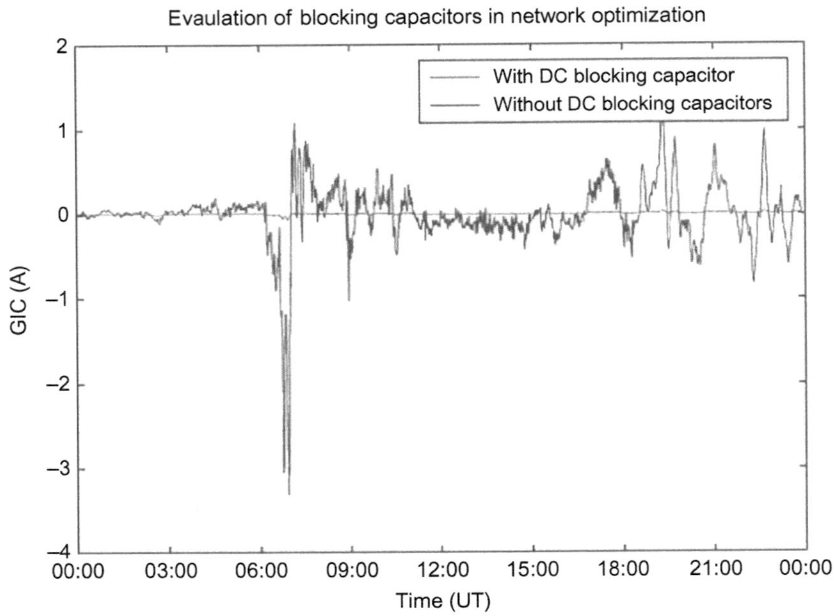

FIG. 22 GIC flowing through substation 1 with and without DC blocking capacitor in substation 3.

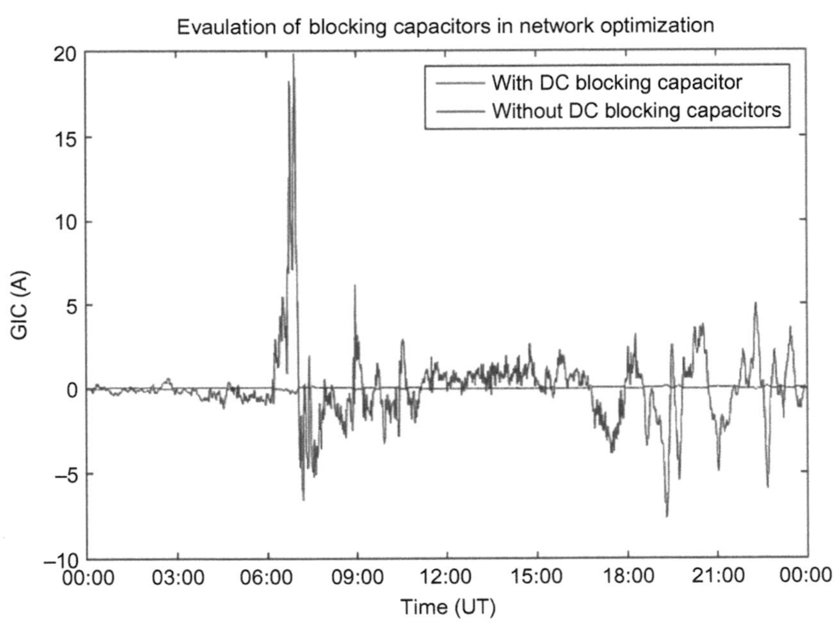

FIG. 23 GIC flowing through substation 5 with and without DC blocking capacitor in substation 3.

REFERENCES

[1] W. Radasky, J. Kappenman, Impacts of geomagnetic storms on EHV and UHV power grids, in: 2010 Asia-Pacific Symposium on Electromagnetic Compatibility (APEMC), 2010, pp. 695–698.

[2] K. Zheng, et al., Effects of geophysical parameters on GIC illustrated by benchmark network modeling, IEEE Trans. Power Del. 28 (2) (2013) 1183–1191.

[3] J.G. Kappenman, Geomagnetic storms and their impact on power systems, IEEE Power Eng. Rev. 16 (5) (1996) 5–8.

[4] D. Boteler, The impact of space weather on the electric power grid (Chapter 4), in: C.J. Schrijver, F. Bagenal, J.J. Sojka (Eds.), Heliophysics Active Stars, their Astrospheres, and Impacts on Planetary Environments, first ed., Cambridge University Press, Cambridge, UK, March 2016 (Online).

[5] L. Marti, J. Berge, R.K. Varma, Determination of geomagnetically induced current flow in a transformer from reactive power absorption, IEEE Trans. Power Del. 28 (3) (2013) 1280–1288.

[6] T.S. Molinski, Why utilities respect geomagnetically induced currents, J. Atmos. Solar Terr. Phys. 64 (16) (2002) 1765–1778.

[7] A. Viljanen, R. Pirjola, Geomagnetically induced currents in the Finnish high-voltage power system, Surv. Geophys. 15 (4) (1994) 383–408.

[8] D.H. Boteler, Assessment of geomagnetic hazard to power systems in Canada. Nat. Hazards 23 (2–3) (2001) 101–120.

[9] D.H. Boteler, R.M. Shier, T. Watanabe, R.E. Horita, Effects of geomagnetically induced currents in the BC Hydro 500 kV system, IEEE Trans. Power Del. 4 (1) (1989) 818–823.

[10] R. Pirjola, Geomagnetically induced currents during magnetic storms, IEEE Trans. Plasma Sci. 28 (6) (2000) 1867–1873.

[11] N.B. Trivedi, et al., Geomagnetically induced currents in an electric power transmission system at low latitudes in Brazil: a case study, Space Weather 5 (4) (2007).

[12] A. Pulkkinen, S. Lindahl, A. Viljanen, R. Pirjola, Geomagnetic storm of 29–31 October 2003: geomagnetically induced currents and their relation to problems in the Swedish high-voltage-power transmission system, Space Weather 3 (2005).

[13] C. Gaunt, G. Coetzee, Transformer failures in regions incorrectly considered to have low GIC-risk, in: 2007 IEEE Lausanne Power Tech, 2007, pp. 807–812.

[14] J.M. Torta, et al., Geomagnetically induced currents in a power grid of northeastern Spain, Space Weather 10 (6) (2012).

[15] L. Liu, et al., Strong magnetic storm's influence on China's Guangdong power grid, Chin. J. Geophys. 51 (4) (2008) 694–699.

[16] S. Watari, et al., Measurements of geomagnetically induced current in a power grid in Hokkaido, Japan, Space Weather 7 (3) (2009).

[17] R. Marshall, et al., Geomagnetically induced currents in the New Zealand power network, Space Weather 10 (8) (2012).

[18] R. Marshall, et al., Observations of geomagnetically induced currents in the Australian power network, Space Weather 11 (1) (2013) 6–16.

[19] T. Forbes, A review on the genesis of coronal mass ejections, J. Geophys. Res. Space Phys. (1978–2012) 105 (A10) (2000) 23153–23165.

[20] C.T. Gaunt, Why space weather is relevant to electrical power systems, Space Weather 14 (2016) 2–9. https://doi.org/10.1002/2015SW001306.

[21] J.A. Jacobs (Ed.), Geomagnetism, Vol. 4, Elsevier, 2016.

[22] A.J. Dessler, E.N. Parker, Hydromagnetic theory of geomagnetic storms, J. Geophys. Res. 64 (12) (1959) 2239–2252.
[23] E. Bernhardi, P. Cilliers, C. Gaunt, Improvement in the modelling of geomagnetically induced currents in southern Africa, S. Afr. J. Sci. 104 (7–8) (2008) 265–272.
[24] V. Albertson, et al., Geomagnetic disturbance effects on power systems, IEEE Trans. Power Del. 8 (3) (1993) 1206–1216.
[25] C.T. Gaunt, Reducing uncertainty—responses for electricity utilities to severe solar storms, J. Space Weather Space Climate 4 (2014) A01.
[26] Y. Kamide, A.C. Chian, Handbook of the Solar-Terrestrial Environment, Springer Science & Business Media, New York, NY, 2007.
[27] Spaceweather.com, What's up in space, 2017, October 18, Available from: http://spaceweather.com/archive.php?view=1&day=18&month=10&year=2017.
[28] A.P. Rouillard, et al., First imaging of corotating interaction regions using the STEREO spacecraft, Geophys. Res. Lett. 35 (10) (2008).
[29] D.H. Hathaway, R.M. Wilson, E.J. Reichmann, A synthesis of solar cycle prediction techniques, J. Geophys. Res. Space Phys. 104 (A10) (1999) 22375–22388.
[30] J. Gosling, et al., The speeds of coronal mass ejection events, Sol. Phys. 48 (2) (1976) 389–397.
[31] R.C. Canfield, H.S. Hudson, D.E. McKenzie, Sigmoidal morphology and eruptive solar activity, Geophys. Res. Lett. 26 (6) (1999) 627–630.
[32] A.W. Thomson, et al., Surface electric fields and geomagnetically induced currents in the Scottish Power grid during the 30 October 2003 geomagnetic storm, Space Weather 3 (11) (2005).
[33] N. Gopalswamy, S. Yashiro, S. Akiyama, Geoeffectiveness of halo coronal mass ejections, J. Geophys. Res. Space Phys. 112 (A6) (2007).
[34] K. Ramesh, Coronal mass ejections and sunspots—solar cycle perspective, Astrophys. J. Lett. 712 (1) (2010) L77.
[35] International Network of Solar Radio Spectrometers, Available from: http://www.e-callisto.org/.
[36] J. Uwamahoro, L. McKinnell, J. Habarulema, Estimating the geoeffectiveness of halo CMEs from associated solar and IP parameters using neural networks, in: AnnalesGeophysicae, 2012, p. 963.
[37] W. Gonzalez, F. Mozer, A quantitative model for the potential resulting from reconnection with an arbitrary interplanetary magnetic field, J. Geophys. Res. 79 (28) (1974) 4186–4194.
[38] D. Webb, et al., Relationship of halo coronal mass ejections, magnetic clouds, and magnetic storms, J. Geophys. Res. Space Phys. 105 (A4) (2000) 7491–7508.
[39] A. Parsons, et al., Wang-Sheeley-Arge–Enlil cone model transitions to operations, Space Weather 9 (3) (2011).
[40] D. Odstrcil, et al., Initial coupling of coronal and heliospheric numerical magnetohydrodynamic codes, J. Atmos. Sol. Terr. Phys. 66 (15) (2004) 1311–1320.
[41] S. Tsunomura, Characteristics of geomagnetic sudden commencement observed in middle and low latitudes, Earth Planets Space 50 (9) (1998) 755–772.
[42] J.G. Kappenman, Storm sudden commencement events and the associated geomagnetically induced current risks to ground-based systems at low-latitude and midlatitude locations, Space Weather 1 (3) (2003).
[43] S. Lotz, P. Cilliers, A solar wind-based model of geomagnetic field fluctuations at a mid-latitude station, Adv. Space Res. 55 (1) (2015) 220–230.
[44] L.M. Winter, et al., Spectral scaling technique to determine extreme Carrington-level geomagnetically induced currents effects, Space Weather (2017).

[45] A.W. Thomson, E.B. Dawson, S.J. Reay, Quantifying extreme behavior in geomagnetic activity, Space Weather 9 (10) (2011).
[46] C.M. Ngwira, et al., Extended study of extreme geoelectric field event scenarios for geomagnetically induced current applications, Space Weather 11 (3) (2013) 121–131.
[47] British Geological Survey, Geomagnetic and Solar Activity Forecast Service, 2018. Available from: http://www.geomag.bgs.ac.uk/data_service/space_weather/forecast.html (accessed 29.05.18).
[48] J. Kappenman, A perfect storm of planetary proportions, IEEE Spectr. 49 (2) (2012).
[49] L. Bolduc, GIC observations and studies in the Hydro-Qu_ebec power system, J. Atmos. Sol. Terr. Phys. 64 (2002) 1793–1802.
[50] D.H. Boteler, R.J. Pirjola, H. Nevanlinna, The effects of geomagnetic disturbances on electricalsystems at the Earth's surface, Adv. Space Res. 22 (1998) 17–27.
[51] D. Boteler, R. Pirjola, Modeling geomagnetically induced currents, Space Weather 15 (1) (2017) 258–276.
[52] A. Viljanen, R. Pirjola, Statistics on geomagnetically-induced currents in the Finnish 400 kV power system based on recordings of geomagnetic variations, J. Geomagnetism Geoelectricity 41 (1989) 411–420.
[53] L. Cagniard, Basic theory of the magneto-telluric method of geophysical prospecting, Geophysics 18 (3) (1953) 605–635.
[54] J. Koen, C.T. Gaunt, Geomagnetically induced currents in the southern African electricity transmission network, in: Power Tech Conference Proceedings, 2003 IEEE Bologna, vol. 1, 2003, 7–14 pp.
[55] O. Amm, Ionospheric elementary current systems in spherical coordinates and their application, J. Geomagnetism Geoelectricity 49 (7) (1997) 947–955.
[56] C.M. Ngwira, et al., Limitations of the modeling of geomagnetically induced currents in the South African power network, Space Weather 7 (10) (2009).
[57] D. Boteler, R. Pirjola, Modelling geomagnetically induced currents produced by realistic and uniform electric fields, IEEE Trans. Power Del. 13 (4) (1998) 1303–1308.
[58] A. Pulkkinen, O. Amm, A. Viljanen, Ionospheric equivalent current distributions determined with the method of spherical elementary current systems, J. Geophys. Res. 108 (A2) (2003) 1053.
[59] M. Lehtinen, R. Pirjola, Currents produced in earthed conductor networks by geomagnetically-induced electric fields, in: AnnalesGeophysicae, 1985, pp. 479–484.
[60] D. Boteler, R. Pirjola, Comparison of methods for modelling geomagnetically induced currents, in: AnnalesGeophysicae, 2014, pp. 1177–1187.
[61] P.A. Tipler, G. Mosca, Physics for Scientists and Engineers, W.H. Freeman and Co., New York, 2008.
[62] P.E. Gray, C.L. Searle, Electronic Principles: Physics, Models, and Circuits, Wiley, New York, 1969.
[63] A.A. Trichtchenko, D.H. Boteler, A. Foss, GIC modelling for an overdetermined system, in: Canadian Conference on Electrical and Computer Engineering, 2006, CCECE'06, 2006, pp. 394–397.
[64] J. Berge, R.K. Varma, A software simulator for geomagnetically induced currents in electrical power systems, in: Canadian Conference on Electrical and Computer Engineering, 2009, CCECE'09, 2009, pp. 695–700.
[65] E. Matandirotya, et al., Differential magnetometer method applied to measurement of geomagnetically induced currents in Southern African power networks, Space Weather 14 (3) (2016) 221–232.

[66] J. Kappenman, et al., GIC mitigation: a neutral blocking/bypass device to prevent the flow of GIC in power systems, IEEE Trans. Power Del. 6 (3) (1991) 1271–1281.
[67] R.J. Pirjola, A. Viljanen, Research of geomagnetically induced currents (GIC) in Finland, in: 2007 7th International Symposium on Electromagnetic Compatibility and Electromagnetic Ecology, 2007, pp. 269–272.
[68] A.W. Thomson, et al., Present day challenges in understanding the geomagnetic hazard to national power grids, Adv. Space Res. 45 (9) (2010) 1182–1190.
[69] I.A. Erinmez, J.G. Kappenman, W.A. Radasky, Management of the geomagnetically induced current risks on the national grid company's electric power transmission system, J. Atmos. Sol. Terr. Phys. 64 (5) (2002) 743–756.

Chapter 17

Application of Evolutionary Algorithm for Multiobjective Transformer Design Optimization

S. Tamil Selvi*, S. Baskar† and S. Rajasekar‡
*SSN College of Engineering, Kalavakkam, India, †Thiagarajar College of Engineering, Madurai, India, ‡NEC Asia Pacific Pte. Ltd, Singapore

1 INTRODUCTION

1.1 Need for Energy Efficient Transformer Design

1.1.1 High Aggregate Total Losses

Electric power typically passes through six to eight transformer units before being utilized. Although, all Distributed Transformers (DTs) have high efficiencies, a large total loss of energy results, due to the large quantity of DTs. Energy losses throughout the world's electrical distribution networks amount to 1279 TWh [1]. Transmission lines contribute most of these losses. After transmission lines, transformers are the loss-making components and there is a cumulative effect of losses in the transformers, which adds up to a significant amount of electricity.

1.1.2 High Economic Cost

Transformer losses in the form of heat not only reduce the transformer life by causing damage to the insulation, but also constitute a significant amount of economic cost. Many countries have realized this fact and a study conducted in United States shows that transformer losses are estimated to be 2%–3% of the total electric energy, accounting for approximately to 25 billion dollars annually [2]. So, finding ways to decrease the transformer losses is one of the important factors toward reducing the failure rate and costs.

In deregulated electricity markets, as the price of electrical energy varies every hour, so does the cost of transformer losses. The seasonal load variations

Classical and Recent Aspects of Power System Optimization. https://doi.org/10.1016/B978-0-12-812441-3.00017-3
© 2018 Elsevier Inc. All rights reserved.

also increase the benefits associated with efficient transformers, particularly if the season of maximum load is coincident with the maximum energy prices. As the system investment and energy costs continue to increase, electric utilities are more and more interested in installing energy-efficient transformers at their distribution networks.

1.1.3 High Total Life Time Cost

Energy efficient transformers are more expensive but use less energy, resulting in lower transformer Total life time cost (TLTC) than less energy efficient transformers. TLTC is the total life cycle cost, which considers the future operating costs of a unit over its lifetime, brought back into present day cost and then added to its total purchase price. Transformers with low TLTC possess low losses and are expected to have longer life. Selecting Energy Efficient Distribution Transformers "SEEDT" project [3] also concluded that electricity distribution companies and commercial and industrial users should use the TLTC method for making transformer purchasing decisions.

1.1.4 More Greenhouse Gas Emissions

Energy-efficient transformers have reduced total losses, i.e., reduced load and no-load losses. Energy-efficient transformers reduce energy consumption and consequently reduce the generation of electrical energy to accommodate these losses and thereby the greenhouse gas emissions. European Copper Institute studies indicated that improving energy efficiency of existing European stock of transformers by 40% would result in about 22 TWh annual energy savings, which is equivalent to annual reduction in green house gas (GHG) emissions of about 9 million tonnes of CO_2. As per the statistics taken in September 2012 [4], India alone contributes 6% of the total CO_2 emissions in world. One-third of the CO_2 emissions in India are due to electricity generation and heat. Reduction of GHG emission is becoming topical issue due to the growing concern for global warming and climate change. Actions that can immediately reduce GHG emissions and cost for an electric utility are the use of energy efficient transformers and renewable energy sources. But disadvantages with renewable energy are (i) difficult to generate large quantities of required electricity; (ii) reliability of supply, as they often rely on the weather (which is unpredictable and inconsistent) for its source of power; and (iii) the high cost of installation for renewable energy technology. All the above problems drive the use of fossil fuel and the only alternative to reduce the CO_2 emissions is the use of energy efficient transformers, which can reduce the amount of power needed to accommodate the losses, i.e., energy consumption, and thereby reduce the need to operate the generators that dump heat and carbon dioxide in to the atmosphere.

It is obvious from the discussions that a transformer with low losses has longer life and is cost effective as well. However, it has been a challenge for the transformer manufacturers to develop an efficient manufacturing technique, efficient manufacturable transformer design techniques, and efficient types of core materials, with the aim of energy efficient transformers for the utilities and other transformer users. Major efforts have been devoted in the recent years to develop new design in the field of transformer. In this direction, energy efficient transformer design (TD) becomes inevitable and vital rather than simply dimensioning various parts of the transformer based on experience or thumb rule, and it deserves extensive research.

1.2 Transformer Design

The aim of the transformer design is to obtain dimension for all parts of the transformer to furnish the data for the transformer manufacturer.

1.2.1 Transformer Design Process

Preliminary Design Details

The process of transformer design is implemented with hundreds of input parameters including description parameters, design parameters, special parameters, performance parameters, cost parameters, etc. [5]. The performance variables include name plate details, allowable no-load and load losses, short circuit impedance, full load minimum allowed efficiency, maximum allowed temperature rise, voltage regulation, etc., based on current standards. Core variables include mainly the core stacking factor, mass density, magnetization curve, and variations of specific core loss for different values of maximum magnetic flux densities at certain frequencies for various magnetic materials. Conductor variables include the resistivity of copper at maximum specified temperature, type of internal and external winding, typical practical values for insulation of the conductor, distance and insulation between windings and core, mass density of the LV and HV conductor, distance between two adjacent cores, maximum ambient and winding temperature, direction space factor for turns and layer, HV taps, etc. Cost variables required are unit price of various main materials such as iron, copper, sheet steel, mineral oil, insulating paper, duct strips, and corrugated panel, etc. Other variables to be considered are insulation outside HV winding, insulation between layers of HV winding, insulation between LV and HV winding, insulation between LV winding and core, insulation between layers of LV winding, number of ducts, width of the LV and HV duct strips, thickness of LV and HV duct strips, mass density of sheet steel, mineral oil, insulating paper, and duct strips, width, height, and weight per unit area of the corrugated panel, etc.

Conventional Transformer Design

A generalized conventional design procedure is outlined below [6] with the flowchart as shown in Fig. 1.

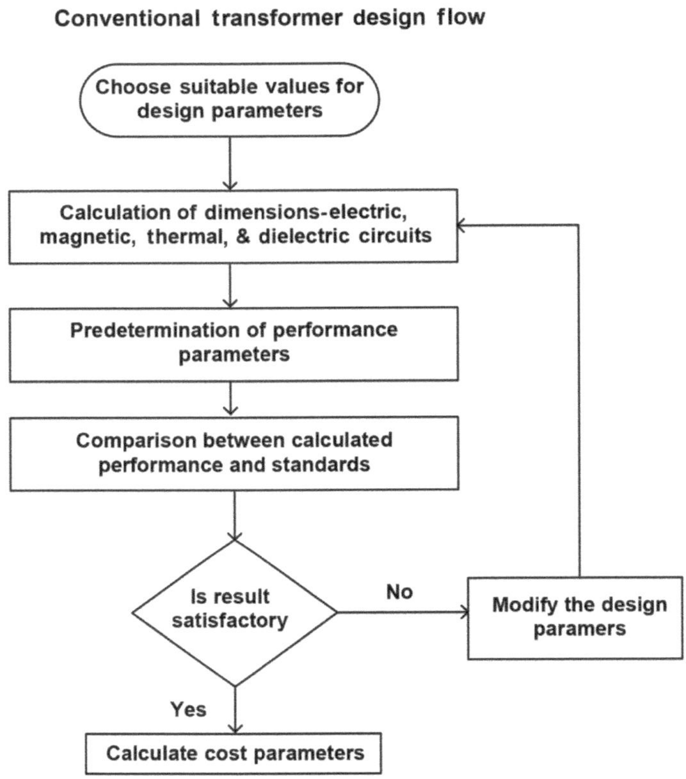

FIG. 1 Conventional transformer design flow.

Step 1: Initially, ***assumption of basic design parameters*** such as, specific magnetic loading (flux density), specific electric loading, space factor, stacking factor etc., are made.

Step 2: Next, the ***design process*** is completed involving analytical calculations to determine various dimensions of electric circuit, magnetic circuit, dielectric design, thermal design, etc.

Step 3: ***Predetermination of performance*** of the transformer is then done based on the calculated dimensions. This includes calculation of transformer performance parameters (characteristics), such as performance under load, and no-load conditions, efficiency, magnetizing current, temperature rise, short circuit impedance, etc., from the dimensions obtained in the previous design process.

Step 4: Finally, ***comparison between calculated performance and desired performance*** (defined by customer specification and Indian Standard Specification) is done. If no satisfactory result is obtained, designer must modify the basic assumed quantities of step 1, and redesign, so as to bring the final design closer to the need.

In a conventional design procedure, modification is not generally a simple task, it requires the skill and intuition of a designer, as there are many input quantities that can be changed. The design process is thus iterative, whether done hand by hand or by digital computers.

Design Quality

Quality of a design is a measure of how well a transformer is designed to achieve its stated purpose. A major feature of the design is the specifications. If the transformer does not meet the specifications, then the quality of design is low. The designer is required to know basic data of primary importance, international technical Standards, and customer specifications, based on which design is developed.

Specifications of the transformer, provided by the customer are Volt-Ampere (KVA or MVA), voltage ratio (primary volt/secondary volt on no load), vector group reference, currents in primary and secondary windings, number of phases, load loss at 75°C, connections-high voltage and low voltage windings, maximum temperature rise, and percent tappings.

Published International Standards

Standards play an important part in the choice, design, manufacture, and operation of a transformer. In an effort to standardize, all countries have established national rules, which are revised as required. Such work is conducted by organizations, which include representatives of manufacturers and users: in India by the Indian Standards Institution; in Britain by the British Standards Institution; in the United States, by the American Standards Association. The international electro technical commission (IEC), created jointly by various countries, publishes internationally accepted recommendations, so products can be sold globally.

The transformers are required to comply in all respects to the relevant Indian Standard Specifications with latest amendments, IS 2026-1977 or International & Internationally recognized standard, with latest amendments, IEC 60076-1993. A few important Indian and IEC Standards related to design of transformers are given in Table 1.

1.2.2 Transformer Design Considerations

The overall design is dictated by two main basic quantities such as "Specific Magnetic Loading" (flux density in the core) and "Specific Electric Loading" (current density in the winding). Thus, in general, flux density and current density are the basic design variables.

Choice of Flux Density

The choice of operating flux density in the core has a very significant impact on the overall size, material cost, and performance of a transformer, due to the following reasons.

TABLE 1 Standards

International and Internationally Recognized Standard	Indian Standard Specifications	Description
IEC 60076-1	IS 2026-1	Transformers: General
IEC 60076-2	IS 2026-2	Temperature rise
IEC 60076-3	IS 2026-3	Insulation levels and dielectric tests
IEC 60076-4	IS 2026-4	Terminal markings, tappings, and connections
IEC 60076-5	IS 2026-5	Ability to withstand short-circuit

Effect of Higher Flux Density

The value of flux density (B_m) in the core determines the core area (A_{net}), because

$$A_{net} = \phi_m / B_m \tag{1}$$

where ϕ_m is flux.

When flux density is slightly increased, (with flux kept constant), the iron core area is reduced and therefore there is a reduced iron weight and saving of iron cost. Moreover, mean turn length of HV and LV windings are also reduced, leading to lower copper losses. But on the other hand, due to the increase in flux density, no load losses, i.e., eddy current loss (P_e) and hysteresis loss (P_h) are increased, because.

$$P_e \, \alpha \, B_m^2 f^2 t^2 \text{ Watts} \tag{2}$$

$$P_h \, \alpha \, B_m^{1.6} f \text{ Watts} \tag{3}$$

where t is the thickness of insulation; f is frequency.

But increase in no load losses result in reduced efficiency, and increased temperature rise.

Effect of Over Excitation

Over excitation is the cause for rise in flux density. It is a known fact that flux depends on voltage/frequency ratio, because

$$\phi_m \, \alpha \, V/f \tag{4}$$

where V is the emf induced.

This is the system condition of simultaneous over-voltage and under-frequency condition. When the transformer is subjected to over excitation, flux and thereby flux density in the core is increased. When the operating flux density is even slightly more than the saturation flux density (B_{sat}) of the core material, relatively higher magnetizing current is drawn and eventually causes overheating and damage to the laminations. Over excitation conditions need to be considered in the design. In this way, for a continuous over excitation condition of $a\%$, general guideline to use operating peak flux density is given below,

$$B_m = \frac{B_{sat}}{\left(1 + \frac{a}{100}\right)} \tag{5}$$

It is therefore necessary to give considerations in choosing value for flux density while designing and it is a good practice to keep the flux density in silicon steel core ranges from 1.6 to 1.8 Tesla to avoid high no-load losses and magnetizing current.

Choice of Current Density

The operating magnetic flux density is the parameter that determines the loss in the magnetic core. Similarly, current density in the windings determines the loss in the windings. When the current density is increased, cross-sectional area of the windings is reduced and hence, the volume and in turn copper weight are reduced. On the other hand, copper loss, which varies as a square of current density, is increased causing efficiency to reduce. Moreover, temperature rise will increase and injure the insulation [7].

The choice of the current density must be done in such a way that the maximum temperature of the transformer due to losses is below the insulation class temperature. Current density chosen should guarantee the level of losses and cooling conditions required. However, a designer must compare the increased cost due to the improved cooling method required with the economy in material due to the choice of increased value of current density. In short, current density is governed by load losses, temperature class of insulation, and short circuit current withstanding ability. Maximum limit for current density is calculated as

$$\text{Maximum current density} = \frac{j}{Z_{sc}} \tag{6}$$

where Z_{sc} is the short circuit impedance in %; j is the short circuit current density in A/mm^2, which can be calculated using,

$$\theta_2 = \theta_0 + j^2.y.t1\,.10^{-3}\ {}^{\circ}\mathrm{C} \tag{7}$$

where

θ_2 = Maximum permissible average winding temperature, which is 250°C for copper conductor and 200°C for aluminium conductor;

θ_0 = Initial temperature of winding, which is 105°C;

$t1$ = Duration of short circuit. It is 2 s;

$y = \text{Function of } \frac{1}{2}(\theta_2 + \theta_0)$, in accordance with (Clause 9.15: Table 6, IS2026 Part I).

It is therefore necessary to give considerations in choosing value for current density while designing.

1.2.3 Effect of Various Factors on Transformer Performance Variables

Generally, the cost of a transformer decreases when increasing the no-load loss and load loss of the transformer. However, the performance and thereby life expectancy of the transformer deteriorates for higher transformer losses. Hence, there is a need to highlight the various factors that affect performance parameters such as no-load losses, load losses, and short circuit impedance before discussing the design demands.

No Load Loss

The losses which practically occur in an unloaded transformer are the iron losses, which are the sum of the hysteresis and eddy current losses. It is the power dissipated in the magnetic core subjected to a time varying magnetizing force. They are constant for a given applied voltage and unaffected by the load on the transformer. It is highly desirable that the no-load loss of the transformer should be reduced to minimum because the loss is constant and continuous during the whole of the time the transformer is connected to the supply. The no-load loss can be decreased by one of the following ways:

- Raising the number of turns of LV winding
- Reducing the magnetic flux density
- Increasing the core cross-sectional area
- Reducing the core window height

Load Loss

It includes I^2R loss in the current carrying parts, stray loss due to eddy currents induced in other structural parts of the transformer. The load loss can be decreased by one of the following ways:

- Reducing the number of turns
- Increasing the magnetic flux density
- Increasing the cross-sectional area of HV winding
- Increasing the cross-sectional area of LV coil
- Increasing the core window height
- Reducing current density

Short Circuit Impedance

The short circuit impedance is the percentage of the rated voltage to be supplied at the primary coil of the transformer, when the secondary coil is shorted, for having rated current in the primary coil. It represents the impedance of the transformer. The higher the impedance voltage, the higher will be the regulation. The lower the impedance voltage, higher will be the short-circuit current during short-circuit condition. Voltage regulation, load distribution during parallel operation of transformers, and short-circuit current are determined only based on short-circuit impedance. The impedance voltage can be decreased by reducing the turns of LV winding and increasing the core window height.

It is clear from the effect of various parameters on performance that there is a cross interaction between the design variable and performance variables. For example, no-load loss is decreased with the decrease of magnetic induction but unfortunately, the load loss is increased. Similarly, by raising the number of turns of LV winding, no-load loss can be decreased, whereas short circuit impedance gets increased.

While a low iron loss is undoubtedly a great advantage, it should not be obtained at the expense of a very high copper loss and thereby the probable sacrifice of reliability. A high copper loss represents a distinct disadvantage to the coil insulation because of the possibility of excessive undetermined local heating occurring in the interior of the coils. Therefore, it is far more practical to design for a closer agreement between copper and iron loss than to look for a very low iron loss at the expense of the copper loss or the electrical and thermal design of the transformer. Thus, it is a challenging task to select a suitable value for every design variable to avoid unnecessary serious effects on the performance variables.

1.2.4 Transformer Design Demands

TD is carried out based on the design constraints given in the international standards, and user specifications. There are many aspects of constraint conditions in transformer design procedure [5]. The first aspect is performance index constraint, for example, the value of impedance voltage, no-load loss, load loss, no-load current, and temperature rise cannot cross the limit value. The second aspect is constructional, i.e., structure and technology conditions constraints. The third aspect is thermal constraint. The fourth aspect is dielectric constraint. The last aspect is the size and weight of the transformer, which must be limited in some range, considering transport and setting room.

(i) *Induced emf constraint*:

The relation between induced voltage in the secondary winding and the maximum flux density should be

$$V_2 = 4.44f\,N_2\,B_m\,A_{net} \tag{8}$$

(ii) *Turns ratio constraint:*

The voltage ratio is equal to the ratio of primary winding turns (N_1) to secondary winding turns (N_2)

$$V_1/V_2 = N_1/N_2 \tag{9}$$

(iii) *No load loss constraint:*

The designed no-load loss (P_{NLL}) must be smaller than a guaranteed maximum no-load loss (P_{GNLL}).

$$P_{NLL} < P_{GNLL} \tag{10}$$

(iv) *Load loss constraint:*

The designed load loss (P_{LL}) must be smaller than guaranteed maximum load loss (P_{GLL}).

$$P_{LL} < P_{GLL} \tag{11}$$

(v) *Total loss constraint:*

The total loss (*TL*) of the transformer is equal to the sum of its no-load loss and load loss, the total loss must be smaller than a maximum total loss (TL_{max}).

$$TL < TL_{max} \tag{12}$$

(vi) *Impedance voltage constraint:*

If the impedance is too low, short circuit currents are high. This increases the danger that the windings will be damaged by mechanical forces developing under short-circuit emergency condition. But, the higher the short-circuit impedance, the higher the voltage regulation, which can cause extra eddy current and stray losses, which reduce transformer efficiency. Moreover, it results in higher winding and oil temperature rise, which again increases the copper content and requires additional cooling.

Thus, when designing a transformer, the magnitude of impedance voltage is so selected that it makes the transformer as strong mechanically and thermally as possible, on one hand, and provides the maximum possible efficiency on the other hand. To strike a balance between these two conflicting requirements, impedance voltage (U_k) is required to be in between a minimum guaranteed impedance voltage ($U_{ks,\ min}$) and maximum guaranteed impedance voltage ($U_{ks,\ max}$).

$$U_{ks,\ min} < U_k < U_{ks,\ max} \tag{13}$$

(vii) *Flux density constraint:*

The maximum flux density, B_m is required to be smaller than a saturation flux density, B_{sat}

$$B_m < B_{sat} \tag{14}$$

(viii) *Heat transfer constraint:*

A transformer should be designed in such a way that the total heat produced by the total loss is lower than the total heat (TH_{ccr}) resulting from (i) heat that can be transferred due to transformer tank by convection, (ii) heat that can be transferred due to transformer tank by radiation, and (iii) heat that can be transferred through radiators.

$$TL < TH_{\text{ccr}} \tag{15}$$

(ix) *Temperature rise constraint:*

The temperature rise/gradient is the difference between the temperature of the part under consideration (usually the average winding rise or the hottest-spot winding rise) and the ambient temperature. The average winding temperature rise of a transformer is the arithmetic difference between the average winding temperature of the hottest winding and the ambient temperature. The rise in temperature of winding and oil should not exceed the predetermined values of 50°C rise for winding and 40°C rise for oil. Hence, it is necessary to ensure that the temperature rise (ΔT) caused by the losses remains within acceptable limits (ΔT_{max}).

$$\Delta T < \Delta T_{max} \tag{16}$$

(x) *No-load current constraint:*

The no-load current is expressed as a percentage of the value of the rated primary current. The transformer no-load current (i_o) is required to be smaller than maximum no-load current, ($i_{o,max}$).

$$i_0 < i_{max} \tag{17}$$

(xi) *Efficiency constraint:*

The transformer efficiency is sometimes required to be greater than a minimum efficiency.

$$\eta > \eta_{\min} \tag{18}$$

(xii) *Insulation thickness constraint:*

Maximum voltage appearing in the winding is between two layers at any point in the winding. During induced over voltage test, double the voltage at double frequency is applied between the layers. The thickness of insulation required between the layer (inter layer insulation) must withstand the induced voltage test. Specifically, the induced over voltage ($Induced_{LV}$) must be smaller than the maximum induced voltage ($Induced_{LV,\,max}$) that the insulation can withstand. Similarly, the constraint should satisfy for HV winding also.

$$Induced_{Lv} < Induced_{LV,\,max} \tag{19}$$

$$Induced_{Hv} < Induced_{HV,\,max} \tag{20}$$

(xiii) *Voltage regulation constraint:*

The voltage regulation must be less than the maximum voltage regulation (ΔV_{max}).

$$\Delta V < \Delta V_{max} \tag{21}$$

Thus, the design should be satisfactory with respect to performance criteria, dielectric strength, mechanical endurance, and thermal stresses in the event of short-circuit.

1.3 Transfomer Design Optimization

The transformer design flowchart, shown below in Fig. 2, depicts the complex design process, which involves magnetic circuit design, electric circuit design, dielectric design, tank design, and thermal design. Steps in the design process include computation of core section dimensions, window dimensions, yoke dimensions, frame dimensions, clearances, insulating paper thickness, number of ducts, duct size, LV winding section dimensions, HV winding section dimensions, performance parameters, tank dimensions, cooling required (corrugated panel/radiator), oil capacity, weight of various parts, and finally total transformer cost. Each step is based on a set of analytical equations for obtaining several electrical and geometrical parameters of the design process.

1.3.1 Nature of Transformer Design Problem

Iron and copper are the two active materials of a transformer. The objective functions such as transformer cost, volume, weight or loss, etc., are directly related to these active materials and vary with certain basic quantities, called design variables. The objective and constraint functions are inter-related in complex ways and composite functions of these design variables. No explicit analytical equation linking the design variables to the objective functions and constraint functions can be provided, because the final value is obtained after several intermediate calculations (relevant equations can be found in the references, [5]). The design variables are mixed in type, i.e., integer as well as discrete. TD calculations require accessing data from several look-up tables, graph, chart for the evaluation of specific core loss at various flux densities, magnetizing current, winding gradient, oil gradient, and heat transmission, etc. Moreover, the TD constraints are very hard in nature (strictly meeting the Standards and Specifications) and any amount of violation in each constraint requires adjustment (increase or decrease) of the design variables and redesigning.

Thus, TD is a complex problem in terms of both modeling as well as solving. Therefore, the design of transformer is characterized as constrained, multiobjective, multidimensional, mixed variable, nonconvex, nonlinear programming problem (multimodal problem). An optimal transformer design cannot be not simply obtained by solving a set of equations, instead, it requires an iterative process. Some of the difficulties in the design of transformers and a suitable technique to solve the TD problem are discussed as follows.

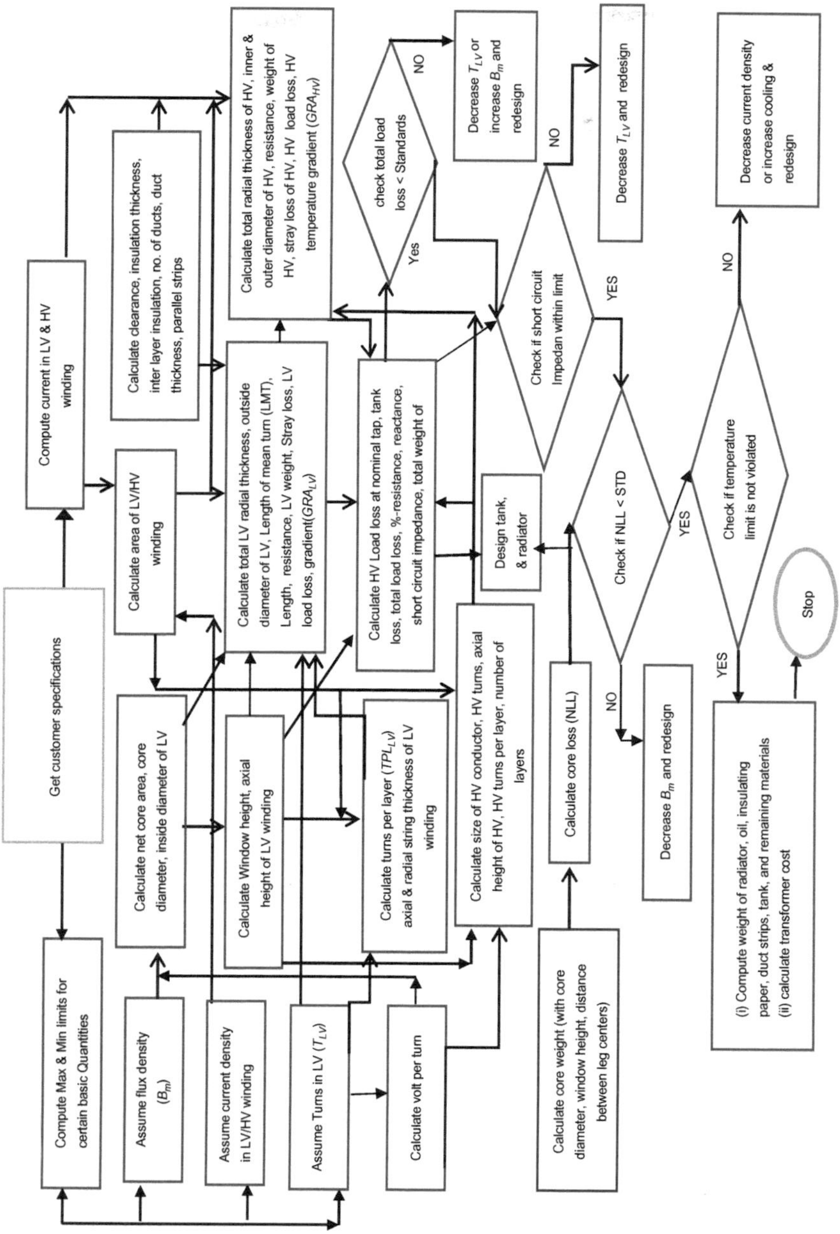

FIG. 2 Transformer design flowchart.

1.3.2 Difficulties in Transformer Design Problem

Choice of Design Variables and Their Values

The transformer design procedure involves selection of significant design variables among several variables of the transformer and requires an assumption of suitable values for these variables. Variables that have large influence on objective functions, constraint functions, and performance variables are the best ones to select as design variables. Generally, the greater the number of design variables selected, the better the optimal results are. But the complexity of the problem will be high. From the design considerations and design constraints, it is found that variables such as current densities of HV winding, current densities of LV winding, flux density, LV turns, and window height can be considered as design variables in transformer design process. Choosing the right values for the design variables while taking in to account various design considerations is a difficult and challenging task requiring balancing energy efficiency with low material costs. For example, high flux density saves iron material and conductor material, whereas it increases core loss.

International Standards

The transformer design is carried out based on international standards given. The design is worked out by using various methods based on accumulated experience realized in different formula, equations, tables, and charts. The design should be satisfactory with respect to various electrical constraints, performance constraints, manufacturer constraints, thermal withstand of windings in the event of short circuit, dielectric strength, and mechanical endurance, etc. Transformer design remains a challenge due to these conflicting constraints/requirements that must be satisfied.

Customer Specification Request

The transformer design is also based on the user specifications. On one hand, electric utilities and some industrial transformer users, at least those who realize that losses amount to a significant cost, use TLTC analysis and choose an energy efficient transformer, despite the fact it is expensive initially. But on the other hand, some commercial customers and contractors who have not realized the consequence of energy losses choose transformers with lowest possible purchase cost, leading to high TLTC, which is not actually economical. These two customers have contradictory ideas and it falls within the duties of the manufacturers to meet all contradictory needs of customers.

1.3.3 Need for Transformer Design Optimization

The nature of the TD problem and the above stated difficulties prove that the transformer design is a challenging task requiring finding an optimum balance between transformer cost and performance, while satisfying various design constraints. Customer specifications request, market globalization, and international standards have taken the manufacturers to face competitive environment for maintaining the sales figures. Ever-increasing competition in the global market has put tremendous responsibilities on the transformer designers to increase transformer life expectancy, reliability, energy efficiency, while reducing cost, because high quality, low cost transformers have become the key to survival. So, the transformer manufacturers are expected to develop a manufacturable transformer design, with the aim of energy efficient transformers for the utilities and other transformer users.

Subsequently, techniques that are employed for the solution of transformer design must be able to deal with the design considerations, to provide a design objective optimum, while remaining cost-effective. A design tool capable of producing manufacturable transformer design in a very short time has become essential for the manufacturers and designers to compete successfully in a global economy.

Several studies have been carried out on application of various techniques for solving the TD problem. However, it is difficult to meet today's transformer design demands via conventional design techniques. In this scenario, empirical models become ineffective. From the overview of research papers about transformers, efforts are focused on prediction of specific transformer characteristics, techniques adopted for transformer design optimization, transformer standards, post design transformer performance, modeling, and recent trends on transformer technology [8]. But, overall Transformer Design Optimization (TDO) remains an active area, focusing on the optimization of multiple parameters of transformer performance including no-load losses, short-circuit impedance, and several other approaches for TDO implementation that have not yet been investigated. So, the modern and efficient approach is to use optimization techniques to find the optimum combination of design variables to satisfy various design requirements. This chapter has therefore adopted optimization of transformer design, by minimizing TD objective while simultaneously satisfying all the constraints imposed by international standards, and customer specifications.

1.4 Need for Multiobjective Evolutionary Algorithm for TDO Problem

Objective function of transformer design optimization problem can be minimization of no-load loss [9,10], minimization of load loss [11], maximization of efficiency [12–15], maximization of rated power [16], minimization of active part mass (sum of weight of core and winding) [16], minimization of active part cost (sum of costs of winding and core), minimization of main material cost (sum of costs of core, winding, insulating paper, duct strip, corrugated panel, insulating paper, sheet steel), minimization of manufacturing cost (sum of costs

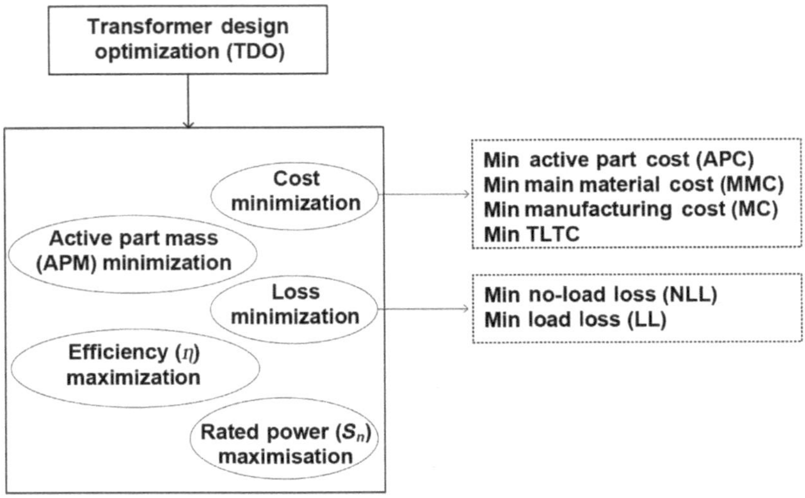

FIG. 3 Various objectives of TDO problem.

of active materials, remaining materials, labor), and minimization of TLTC [12,13,17–26], as in Fig. 3.

After the 1970s, several Mathematical Programming Techniques (MPT)-based TD procedures such as calculus of variation [27], geometric programming [16], integer programming [18], branch and bound technique [21], mathematics appeared in the literature. After 2005, research papers show a combination of numerical technique and MPT such as finite element-boundary element technique [28], knowledge-based system [14,15], A parallel mixed integer programming in combination with finite element method (FEM) technique [17], mixed integer, nonlinear programming tailored to branch bound algorithm [21] for TD.

After 1999, TD research activities were geared up, due to the various applications of artificial intelligence techniques. Neural network for iron loss reduction [9], Genetic Algorithm (GA) for transformer construction cost minimization [25], combined neural network-GA approach for iron loss reduction [10], active part cost minimization using simulated annealing technique [20], heuristic solution for minimization of manufacturing cost [19], TLTC minimization using covariance matrix adaptation evolution strategy [29], and bacterial foraging algorithm for minimizing active part cost [13] have been applied for solving Single Objective TDO (SOTDO) problem.

In the way of Multiobjective TDO (MOTDO), bacterial foraging algorithm with sum of weighted objectives for purchase cost minimization and efficiency maximization [12], multiobjective differential evolution for minimization of mass and losses [30,31], and multiobjective particle swarm optimization minimizing core and copper volume [32] have been advanced.

1.4.1 Literature Survey Summary

TDO research literatures have dealt mostly minimization of material cost. Overall, TLTC minimization, which is the best investment for long term, is scarcely addressed in the literatures. Designing and buying a transformer based on material cost/purchase cost minimization, without considering the future losses is not a right decision. It is important to incorporate the cost of losses into economic evaluation of DT to produce energy efficient TD. Hence, there is a prospect in designing transformer with minimum TLTC.

(i) In literatures, MPT are predominantly used in solving the TDO. But the nature of TD problem is multimodal and complex calculations interacting with graphical data cannot be handled accurately by the conventional derivative based MPT. Optimal solution is not guaranteed by MPT.
(ii) Research efforts are focused mainly in solving SOTDO problem rather than formulating the TD problem as MOTDO problem. But TD problem is a Multiobjective optimization problem (MOP), as it has many conflicting design objectives, such as purchase cost and total loss, purchase cost, and TLTC, etc. so, it necessitates expressing the TD problem as MOP and such formulation can only aid finding a trade-off between contradictory objectives.

But, very few research publications are available for MOTDO and most of them have applied the conventional approach, wherein MOPs are solved by converting them into SOTDO problem, resulting in single solution using either weighted sum method [12], or lexicographic method, or goal programming method, or ϵ-constraint method, etc. But for getting many trade-off solutions, these approaches need to be executed many times with appropriate weight coefficients. Unlike the classical methods, Multiobjective evolutionary algorithms (MOEAs) are naturally suitable for solving MOPs as they use population-based strategy for achieving an approximation of Pareto set in a single run and can handle discrete, multimodal, discontinuous problems with mixed variables. In literatures where MOEAs are applied for solving MOTDO problem [30–32], considerations are given only to MOEAs rather than overall transformer design aspects. Hence, this chapter has made an attempt to solve the MOTDO problem using recent MOEA, without compromising the aspect of energy efficient transformer design.

2 MULTIOBJECTIVE EVOLUTIONARY ALGORITHM: AN OVERVIEW

2.1 Multiobjective Optimization Problem

Consider a decision-maker who wishes to optimize M objectives such that the objectives are noncommensurable, and the decision-maker has no clear preference of the objectives relative to each other. Mathematically, a minimization of MOP is defined as follows:

Find an n-dimensional design vector, $x = \{x_t : \forall t = 1, \ldots n\}$ in the solution space X, that minimizes a given set of M objectives, $F(x) = \{f_i(x) : \forall i = 1, \ldots M\}$. The solution space is restricted by a series of constraints and bounds on the design variables [33].

If all the M objective functions are for minimization, a feasible solution y is said to dominate another feasible solution ' ($y > z$), if and only if, $f_i(y) \leq f_i(z)$, for $i = 1, 2, \ldots M$ and $f_j(y) < f_j(z)$ for least one objective function j. A solution is said to be Pareto optimal if it is not dominated by any other solution in the solution space. The nondominated solutions are depicted in Fig. 4. A Pareto optimal solution cannot be improved with respect to any objective without worsening at least one other objective. The set of all feasible nondominated solutions in the solution space is referred to as the Pareto optimal set, and for a given Pareto optimal set, the corresponding objective function values in the objective space are called the Pareto front [33].

MOP requires an efficient multiobjective optimization method. Traditionally, such MOPs are treated as single objective optimization problems and solved using single objective mathematical optimization techniques. Conventional mathematical methods available to solve MOPs are time consuming and require several steps of processing to obtain the Pareto optimal solution set.

Moreover, objectives under consideration conflict with each other and optimizing x with respect to a single objective often results in unacceptable results with respect to the other objectives. A perfect multiobjective solution that simultaneously optimizes each objective function is almost impossible. A reasonable solution to a multiobjective problem is to investigate a set of solutions, each of which satisfies the objectives at an acceptable level without being dominated by any other solution.

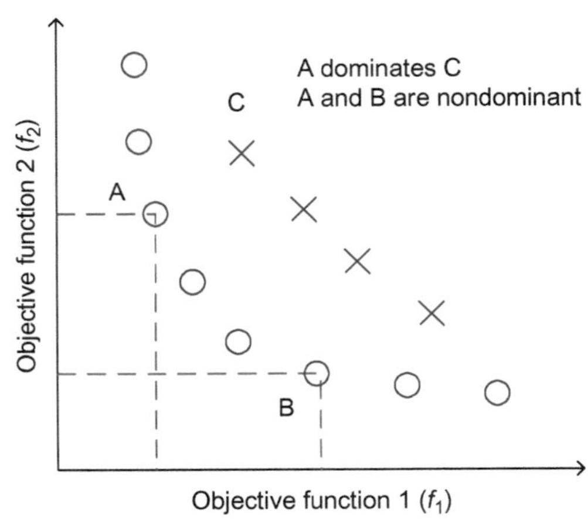

FIG. 4 Nondominated solutions.

2.2 Multiobjective Evolutionary Algorithm

The concept of solving MOP using EA was first suggested by Rosenberg in the late 1960s. Development of MOEAs has cushioned researchers and engineers by solving MOPs effectively producing Pareto optimal solution set, in a single optimization run. With this concern in mind, a MOEA should achieve the following three conflicting goals [33], as shown in Fig. 5.

The ultimate goal of a multiobjective evolutionary algorithm is to identify solutions in the Pareto optimal set. However, identifying the entire Pareto optimal set, for many multiobjective problems, is practically impossible due to its size. In addition, for many problems, especially for combinatorial optimization problems, proof of solution optimality is computationally infeasible. Therefore, a practical approach to multiobjective optimization is to investigate a set of solutions (the best-known Pareto set) that represent the Pareto optimal set. With these concerns in mind, a MOEA should achieve the following three conflicting goals.

- The best-known Pareto front should be as close as possible to the true Pareto front. The first goal is best served by focusing the search on a particular region of the Pareto front.
- Solutions in the best-known Pareto set should be uniformly distributed and diverse over the Pareto front to provide the decision-maker a true picture of trade-offs. Second goal demands the search effort to maintain a diverse population to obtain solutions uniformly distributed over the Pareto front.
- The best-known Pareto front should capture the whole spectrum of the Pareto front. This requires investigating solutions at the extreme ends of the objective function space. The third goal aims at extending the Pareto front at both ends, exploring new extreme solutions.

MOEAs are classified into Pareto ranking based algorithms and decomposition-based algorithms. Pareto ranking based algorithms introduced in 1990s are

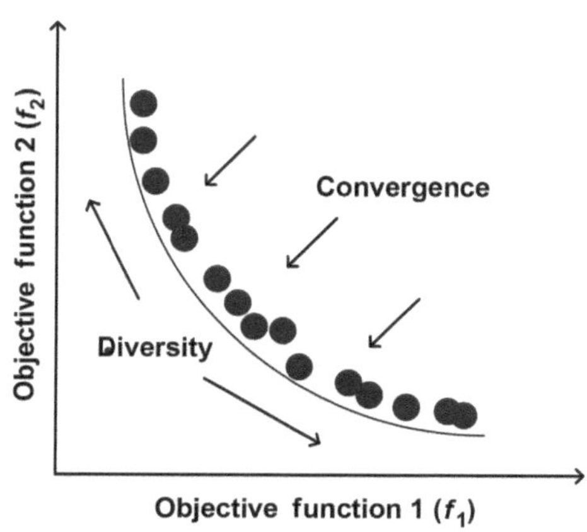

FIG. 5 Conflicting goals of Pareto front.

nondominated sorting algorithms. They were Nondominated sorting GA (NSGA 1994), Strength Pareto Evolutionary Algorithm (SPEA 1999), Pareto Archived Evolutionary Strategy (PAES 2000), and NSGA-II 2002, that use nondominated sorting technique with suitable diversity operator to determine the Pareto front. Among all these algorithms, except for problems with three or more objectives, nondominated sorting and the use of diversity maintenance operators (crowding distance, clustering operator, external archive, etc.) are not applicable. Decomposition based MOEA, (MOEA/D) proposed by Zhang in 2007 [34] has overcome this difficulty, by decomposing the MOP into several subproblems to obtain the Pareto front.

2.2.1 Nondominated Sorting Genetic Algorithm

Nowadays several multiobjective EAs (MOEAs) have gained popularity with the advantage of generating the whole Pareto-optimal solutions in a single simulation. The main principles of MOEA are to push the entire population toward Pareto-optimal front (POF) and to maintain diversity among the solutions in the POF. The Pareto-based approach of NSGA-II, which is popular and offers excellent multiobjective evolutionary algorithms, has been used in many engineering multiobjective problems, due to its efficient nondominant ranking procedure in giving different levels of Pareto frontiers [35]. In this chapter, we have applied an elitist nondominated sorting genetic algorithm, NSGA-II, one of the state-of-the art multiobjective optimization algorithms for solving the MOTDO problem, because Deb [35] suggested that NSGA- II outperforms two other contemporary MOEAs: PAES, SPEA, in terms of finding a diverse set of solutions and in converging near the true Pareto-optimal set [29].

A MATLAB code is developed for NSGA-II algorithm and published on-line at MATLAB CENTRAL by the authors. It is available in the below link.

http://in.mathworks.com/matlabcentral/fileexchange/49806-matlab-code-for-constrained-nsga-ii-drs-baskar–s-tamilselvi-and-p-r-varshini.

The steps involved in the working of NSGA-II are detailed below.

1: [Initial Population]: Initially generate a random parent population, P_t of size Np at generation t.

2: [Fitness Evaluation] Calculate all the objective function values separately for the entire population.

3: [Rank Assignment] Sort the population based on nondomination levels and assign a rank for each solution based on the following procedure:

- Nondomination sorting
- Crowding distance (CD) calculation

4: [Offspring Population] Apply the following genetic operators of real GA, on P_t to create an offspring population Q_t of same size Np [35,36]

- Binary tournament parent selection
- SBX crossover
- Polynomial mutation

5: [Intermediate population] Combine parent and offspring populations to implement elitism and form an intermediate population, R_t of size $2Np$.

6: [Fitness Evaluation] Evaluate the fitness for $2Np$ population solutions using multiple objective functions.

7: [Nondominated Sorting] Perform nondominated sorting over $2Np$ population to rank and divide the individuals into different nondominated POFs. Assign rank for the POFs in such a way that the first nondominating POF with rank one and so on.

8: [CD Calculation] Calculate crowding distance of all the solutions [35]. Higher values of crowding distance increase the probability of the solution to be selected for the next generation.

9: [Survival Selection] Select Np member for the new population P_{t+1} from the combined population $2Np$, on the basis of ranking and crowding distance as below:

- Choose individuals of the best ranked POFs first, followed by the next best POF and so on till Np individuals are obtained.
- Because the intermediate population R_t has a size of $2Np$, discard those fronts which could not be accommodated.
- In case there is space only for a portion of a POF to accommodate in the new population, choose individuals among those in the front that are from the least crowded regions (determined by a crowding distance operator) and fill up the required number, Np for the new population, P_{t+1}.

10: [Replacement] Form the parent population for the next generation P_{t+1}.

11: [Termination Criteria] Check if maximum number of generations is met. If yes, get the Pareto optimal solution. Otherwise, increment the generation count and go to step 4.

Nondominated Sorting Ranking Procedure

The aim of rank assignment in NSGA-II algorithm is summarized as follows. The solution that is not dominated by any other solutions in the population is inserted in to the first front. They are given the highest fitness value and rank 1. These solutions are then removed from the parent solution set and nondominated sorting is repeated for the second time. Second front from the remaining chromosomes is created and assigned the second rank, after identifying the next best nondominated Pareto set. This methodology is iterated until all the solutions are classified and assigned a rank. The algorithm for nondominated sorting is explained later.

a. Initialize the front counter, $r=0$.
b. Increment the front index, $r=r+1$.
c. Find the individuals that are not dominated by any other solution, from the population [35].
d. Assign front index (rank), r to these set of nondominated individuals.
e. Exclude these individuals from the population
f. If population is empty then stop, otherwise repeat the steps from 2.

2.2.2 Crowding Distance Operation

The crowding distance helps to measure the density of nondominated solutions. This distance shows an estimate of relative density of nondominated solutions surrounding a particular solution in the front pool. Let the number of nondominated solutions be Q; and dis_j be the value of distance for the jth nondominated solution. Distance calculation steps are given as follows.

a. Let $dis_j = 0$ for all solutions in Q.
b. Let dis_1 and dis_Q be the max distance value (say, infinity).
c. For solutions between 1 and Q, calculate the distances. That is,

for $j = 2$ to $Q - 1$,

$$dis_j = dis_j + \sum_{M=1}^{2} \frac{f_M(j+1) - f_M(j-1)}{f_M(Q) - f_M(1)} \tag{22}$$

3 MOTDO PROBLEM FORMULATION

MOTDO of a three-phase, oil-immersed, shell-type transformer is carried out in this chapter. The proposed methodology optimizes two (M) conflicting objectives simultaneously, subject to various constraints using seven (n) design variables. Mathematically, MOTDO problem can be defined as,

Find, $x = \{x_t : \forall t = 1, \ldots n\}$ which.
Minimize, $F(x) = \{f_i : \forall i = 1, \ldots M\}$ simultaneously.
Subject to $h_j(x) \le d_j$, for $j = 1, 2, \ldots, q$

$$x_t \ge 0, \quad \text{for } t = 1, 2, \ldots, n$$

$$x_{t,min} \le x_t \le x_{t,max}, \quad \text{for } t = 1, 2, \ldots, n \tag{23}$$

where, f_i is the ith objective function; M is the number of objectives; h_j is the jth inequality constraint; x_t is the tth design variable; n is the number of design variables; d_j are constants of constraints; q is the number of constraints; $x_{t,min}$ and $x_{t,max}$ are lower and upper limit on x_t, respectively.

3.1 Transformer Design Objectives

The two contradictory transformer design objectives considered for simultaneous optimization in this chapter are minimization of purchase cost—(f_1) and minimization of TLTC—(f_2) [29].

(a) *Minimum Purchase Cost Design—Objective function* f_1*:*
This objective minimizes the total cost of transformer main materials (C_{TM}) including core, LV conductor, HV conductor, insulating paper, duct strips, oil, corrugated panel, and sheet steel [5] with,

$$\text{Min}, f_1 = C_{TM} = \{C_{FE,i}.G_{FE,i} + C_{COND}.G_{LV} + C_{COND}.G_{HV} + C_{INS}.G_{INS} + C_{DS}.G_{DS} + C_{OIL}.G_{OIL} + C_{CORR}.G_{CORR} + C_{SS}.G_{SS}\} \quad (24)$$

where G_{FE}, G_{LV}, G_{HV}, G_{INS}, G_{DS}, G_{OIL}, G_{CORR}, G_{FE}, G_{CU}, G_{INS}, G_{DS}, G_{OIL}, G_{CORR}, G_{SS}, and G_{SS} are weight of magnetic material, LV conductor, HV conductor, insulating paper, duct strips, oil, corrugated panel, and sheet steel in kg, respectively; $C_{FE,\ i=1,\ 2,\ \ldots 10}$, C_{COND}, C_{SS}, C_{OIL}, C_{INS}, C_{DS}, and C_{CORR} are the unit cost of magnetic material, LV/HV conductor, sheet steel, oil, insulating paper, duct strips, and corrugated panel, respectively.

(b) *Minimum TLTC Design—Objective function f_2:*

The optimal design of transformer must minimize the sum of transformer cost and running cost. In this objective, a transformer's selling price and cost due to losses are considered. So, material saving and energy saving are the two important aspects in the minimization of TLTC. This objective minimizes TLTC of transformer (C_{TLTC}),[5] using,

$$\text{Min}, f_2 = C_{TLTC} = \left\{ \frac{(C_{TM} + C_{rem} + C_{lab})}{(1 - S_m)} + AA.P_{NLL} + BB.P_{LL}\,LTC \right\} \quad (25)$$

where, C_{TLTC}, AA, BB, C_{rem}, C_{lab}, S_m, P_{NLL}, and P_{LL} are total life time cost in euro, no-load loss cost rate in euro per watts, load loss cost rate in euro per watts, remaining materials cost in euro, labour cost in euro, sales margin, designed no-load loss in watts, and designed load loss in watts, respectively.

3.2 Transformer Design Variables

Design variables are described in the following vector (dv) as

$$v = \{x_1, x_2, x_3, \ldots x_6\} \quad (26)$$

where

x_1 is width of the core leg in mm-(integer variable);
x_2 is height of the core window in mm-(integer variable);
x_3 is maximum magnetic flux density in Tesla-(discrete variable);
x_4 is current density in LV winding in A/mm^2-(discrete variable);
x_5 is current density in HV winding in A/mm^2-(discrete variable);
x_6 is voltage per turn in volts-(discrete variable);
x_7 is Type of magnetic material-(integer variable).

The chromosomes are generated in the population with the following representation:

x_1	x_2	x_3	x_4	x_5	x_6	x_7

For the integer design variables, x_1, x_2, and x_7, real values generated within their lower and upper bounds are rounded off to the nearest integer in the fitness function evaluation code. The discrete design variables, $x_3 - x_6$, are handled in the MATLAB code as follows.

- x_3 is maximum flux density takes five values 1.6, 1.65, 1.7, 1.75, 1.8 Tesla.

  x_3 = round(4*rand)*0.05 + 1.6

- x_4 is LV current density, takes 4 values 3.0, 3.2, 3.4, 3.6 A/mm^2

  x_4 = round(3*rand)*0.2 + 3

- x_5 is LV current density, takes 5 values 3.0, 3.15, 3.3, 3.45, 3.6 A/mm^2

  x_5 = round(4*rand)*0.15 + 3

- x_6 is the volt per turn, takes 1501 values between 0 and 15 in step of 0.01, as

  [0, 0.01, 0.02, 0.03, ... 1.0, 1.01, 1.02, ... 11.0, 11.01, 11.02, ... 15] volts.
  x_6 = round(1500*rand)*0.01

3.3 Transformer Design Constraints

Some of the major constraints considered for solving MOTDO problem are detailed later [29].

3.3.1 Performance Constraints

Transformer performance must satisfy international standards, specified in IEC 60076-1.

- No-load loss constraint:

$$P_{NLL} \leq P_{GNLL}(1 + e_1) \quad (27)$$

- Load loss constraint:

$$P_{LL} \leq P_{GLL}(1 + e_1) \quad (28)$$

- Total loss constraint:

$$(P_{NLL} + P_{LL}) \leq (P_{GNLL} + P_{GLL})(1 + e_2) \quad (29)$$

- Impedance voltage constraint:

$$U_{ks,min}(1 - e_2) \leq U_k \leq U_{ks,max}(1 + e_2) \quad (30)$$

3.3.2 Manufacturer Constraints

These constraints are based on the transformer manufacturer specifications.

- Ratio of width of the core leg to the height of the core window should be less than or equal to unity;

$$x_1 \leq x_2 \tag{31}$$

- 2-multiplied core leg thickness (T_{cl}) must be between a minimum of half core leg width and a maximum of 90% core leg width;

$$0.5x_1 \leq 2T_{cl} \leq 0.9x_1 \tag{32}$$

3.3.3 Constraints of Material Performance

- Heat transfer constraint: Total heat produced by total loss of transformer must be smaller than the total heat dissipated by convection and radiation (H_{diss}) through cooling arrangement;

$$P_{NLL} + P_{LL} \leq H_{diss} \tag{33}$$

- Magnetic flux density constraint:

$$x_{4,min} \leq x_4 \leq x_{4,max} \tag{34}$$

where

e_1 is the IEC standard tolerance for no-load and load loss constraints in % and e_2 is IEC standard tolerance for total loss and impedance voltage constraints in percentage.

4 RESULTS AND DISCUSSION

The design procedure is presented for three-phase, oil-immersed, shell-type, wound-core, distribution transformer. A design example of 400 kVA, 50 Hz, 20/0.4 KV, three-phase, shell type, wound core transformer with vector group, Dyn11 has been considered.

4.1 Objective Function: Sample TLTC Calculation

Design data shows the transformer of 750 kVA, 50 Hz, three-phase oil immersed DT. Electricity price (*EP*) is \$0.075 per kWh. Discount rate (*d*) relating the future costs to present day dollars is 7%. The transformer lifetime is 15 years (*N*). Transformer will be loaded to 75% of nameplate rating for 6000 h per year (HPY), and zero load for 2760 h per year. Compute TLTC.

Design example: Purchase price (*BP*)—10,194\$
No load losses (P_{NLL})—1647 W
Load losses (P_{LL}) at 100%—9507 W
Load losses at 75%—5348 W

Total life time cost (TLTC) is calculated using

$$TLTC = BP + PV_{NLL} + PV_{LL} \tag{35}$$

where

BP is the bid price (purchase cost) in \$.
PV_{NLL} is the present value of no load losses in \$.
PV_{LL} is the present value of load losses in \$

- To calculate present value of no load losses—PV_{NLL}

No load losses per year = 1.647 kW × 8760 hpy = 14,427 kwhr.

The annual cost (\$/year) of transformer no load losses, C_{NLL} can be calculated using

$$C_{NLL} = P_{NLL}EP.HPY.10^{-3} \tag{36}$$

$$C_{NLL} = 14427 \times 0.075 = 1,082.08\,\$$$

Present value of no load losses for 15 years (PV_{NLL}) is calculated with.

$$PV_{NLL} = C_{NLL}.PV_m \tag{37}$$

Where, PV_m present value multiplier.

$$PV_m = \sum_{i=1}^{N} \frac{1}{(1+d)^N} = \frac{(1+d)^N - 1}{d.(1+d)^N} \tag{38}$$

$$PV_{NLL} = 1082.08 \times \left((1+0.1)^{15} - 1\right)/(0.1) \times (1+0.1)^{15}$$

$$= 1,082.08 \times (4.1772 - 1)/0.41772$$

$$PV_{NLL} = 8,229.74\$$$

- To calculate present value of load losses—PV_{LL}

Load losses per year = 5.348 kW × 6000 hpy = 32088 kwhr

Annual cost of load losses, C_{LL} = 5.348 kW × 6000 h/y × \$0.075/kWh = 2406.60\$

Present value of load losses for 15 years:

$$PV_{LL} = 2406.60 \times (4.1772 - 1)/0.41772$$

$$= 18,304.72\$$$

- TLTC

$$TLTC = 10,194.00 + 8229.74 + 18,304.72 = 36,728.46\$$$

The steps for calculating the other objective function, purchase cost, can be obtained from the source [5], where it is explained clearly by Georgilakis.

4.2 Fitness Function Evaluation

Fitness function, ($FF(x)$) is determined using penalty parameter-less constraint handling method [35]. The fitness of a feasible solution is always fixed and is equal to its objective function value ($f_{M=1,2}$), which is purchase cost and TLTC. However, infeasible solutions are compared based on the constraint violation. Here, the fitness of the infeasible solutions not only depends on the overall constraint violation, but also on the fitness of the worst feasible solution. If z^{max} is the objective function value of the worst feasible solution in the current population, then the fitness value is the sum of z^{max} and the overall constraint violation $\sum_{j=1}^{q}\langle err_j(x)\rangle$. If there is no feasible solution in the current population, z^{max} is zero.

$$FF(x)=\left\{\begin{array}{ll} f_M(x), & for\, M=1,2 \text{ if feasible} \\ z^{max}+\sum_{j=1}^{q}\langle err_j(x)\rangle, & \text{otherwise} \end{array}\right\} \tag{39}$$

where

$q=9$ (no. of constraints) excluding the boundary constraints;
$\sum_{j=1}^{q}\langle err_j(x)\rangle$ is the overall constraint violation;
$err_j(x)$ is the amount of violation of jth constraint;
$< >$ indicates the absolute value of the operand.

4.2.1 Overall Constraint Violation

The overall constraint violation sum ($\langle err_j(x)\rangle$) is the sum of absolute values of the amount of violation of every jth constraint. The violation amount is calculated for every constraint as follows for the TDO problem considered.

Referring Eq. (23),

$$err_j(x)=h_j(x)-d_j \text{ for the constraints } h_j(x)\le d_j$$

$$err_j(x)=d_j-h_j(x) \text{ for the constraints } h_j(x)\ge d_j$$

The following normalized error calculation of all the constraints are given below.

$$err_1(x)=[(P_{NLL}+P_{LL})/((P_{GNLL}+P_{GLL})*(1+e_2))]-1;$$
$$err_2(x)=[P_{NLL}/(P_{GNLL}*(1+e_1))]-1;$$
$$err_3(x)=[P_{LL}/(P_{GLL}*(1+e_1))]-1;$$
$$err_4(x)=[U_k/(U_{ks}*(1+e_2))]-1;$$
$$err_5(x)=[(P_{NLL}+P_{LL})/H_{diss}]-1;$$
$$err_6(x)=1-[(2*T_{cl})/(0.5*x_2)];$$
$$err_7(x)=(x_1/x_2)-1;$$
$$err_8(x)=1-[U_k/(U_{ks}*(1-e_2))];$$
$$err_9(x)=((2*T_{cl})/(0.9*x_1))-1;$$

The overall constraint violation, ($\sum_{j=1}^{q}\langle err_j(x)\rangle$) which is the sum of absolute values of the amount of violation of q constraints, is calculated finally. The remaining constraints not included in this portion are the boundary constraints, handled during population generation, and fitness evaluation code.

4.3 MOTDO Result by NSGA-II

The upper and lower limit for seven decision variables, no-load loss rate, load loss rate (*AA*, *BB*), unit cost of various materials, C_{rem}, C_{lab}, S_m, $e_1=15\%$, $e_2=10\%$ have been taken from [21]. Transformer with loss category AB' according to CENELEC standards of $P_{GNLL}=750\,\text{W}$, $P_{GLL}=4600\,\text{W}$, and $U_{ks}=4\%$ has been selected. Specific core loss data at 50 Hz frequency for 10 magnetic materials (M3-0.27, M4-0.27, MOH-0.23, MOH-0.27, 23ZDKH90, 23ZH90, 23ZH95, 27ZDKH95, 23ZDMH85, and 27ZDMH) are derived from [37] web link. Copper sheet is used for LV winding and wire is used for HV coil. MATLAB coding for MOTDO is developed using MATLAB 7.4 on Intel core, i3 processor Laptop, operating at 3.2 GHZ, with 3 GB RAM. The code is attached in the chapter in the appendix. NSGA-II has been applied to solve the MOTDO problem. Two TDO objectives such as purchase cost (f_1), and TLTC (f_2) are minimized simultaneously by NSGA-II. The control parameters are set as follows.

- Population size—*Np* is set as 100.
- Stopping criterion—Maximum number of function evaluations is chosen as 20,000.
- Crossover probability and mutation probability rate are selected as 0.9 and 0.14, respectively.
- η_c—Distribution index for crossover, which is set as 5.
- η_m—Mutation distribution index, which is set as 20.

4.3.1 Updating New Pareto Front

Due to the difficulty of the MOTDO problem, 20 independent runs have been executed and from all the optimization runs, nondominated solutions of the first POF are filtered and combined to form a new Pareto set. After 20 runs, nondominated sorting [35] is performed on the Pareto pool for extracting the exact true new Pareto optimal front, as shown in Fig. 6. The tradeoff curve thus extracted has better diversity characteristics than the POF obtained from single run of NSGA-II for TDO problem. A trade-off is achieved between these competing objectives using fuzzy mechanism.

4.3.2 Post Optimality Analysis

For MOP, multiple-criteria decision-making is necessary. This work utilizes fuzzy based set theory [38,39], to extract the best compromise TD over the new Pareto optimal front. This method uses membership function and extracts a compromise solution that has maximum membership value. The membership function for the *i*th objective function of a nondominated solution *k* in the new Pareto optimal front is defined as,

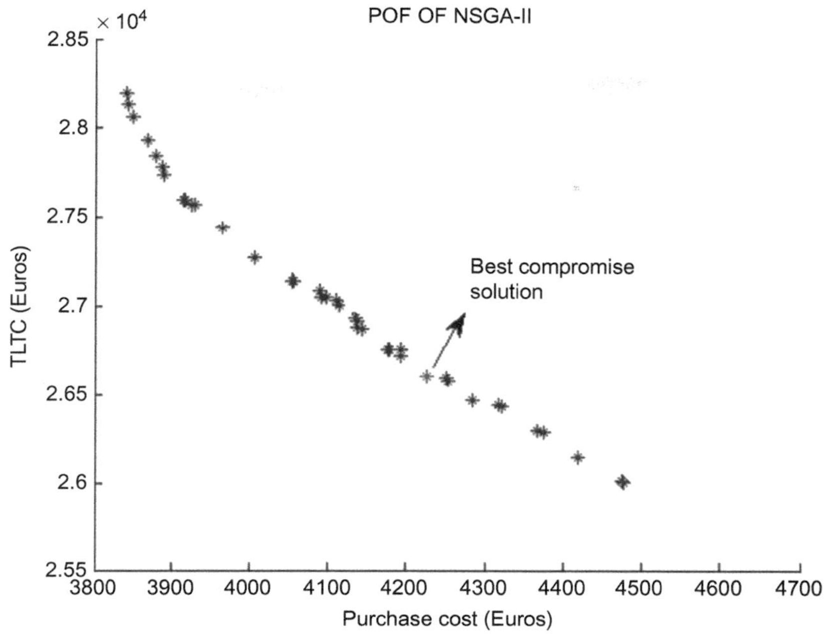

FIG. 6 Pareto optimal front of NSGA-II.

$$MF_i^k = \begin{cases} 1, & f_i \le f_i^{min} \\ \dfrac{f_i^{max} - f_i}{f_i^{max} - f_i^{min}}, & f_i^{min} < f_i < f_i^{max} \\ 0, & f_i \ge f_i^{max} \end{cases} \tag{40}$$

For each nondominated solution k, the normalized membership function is calculated as

$$MF^k = \frac{\sum_{i=1}^{M} MF_i^k}{\sum_{k=1}^{y} \sum_{i=1}^{M} MF_i^k} \tag{41}$$

where y is the number of nondominated solutions.

This procedure thus derives a candidate solution, negotiating the conflicting two objectives of several solutions of new Pareto optimal front to help the decision makers. The optimized values for the design variables and objective variables are tabulated in Table 2 and Fig. 7, respectively, for the compromise solution thus obtained from fuzzy mechanism. The complete design data for the compromise solution is also provided in the Table 3.

TABLE 2 Comparison of MOTDO and SOTDO Results

Minimization of purchase cost (SOTDO)—f_1 [40]				
MINLP	**HA**	**HS**	**GA**	**DE**
4402	4383	4449	4494	4539

Best compromise solution of MOTDO			
Design variables/ performance variables/ TD objectives	**UTOPIA results**		**MOTDO compromise solution**
	f_1	f_2	
x_1 mm	224	245	245
x_2 mm	227	245	245
x_3 Tesla	1.8	1.7	1.75
x_4 A/mm^2	3.6	3.0	3.2
x_5 A/mm^2	3.6	3.0	3.0
x_6 (Volts)	13.22	13.20	13.2
x_7	4	5	3
C_{TM} Euro	**3817**	4629	**4227**
C_{TLT} Euro	28,394	**25,882**	**26,609**

Bold represents the objective considered for single objective optimization.

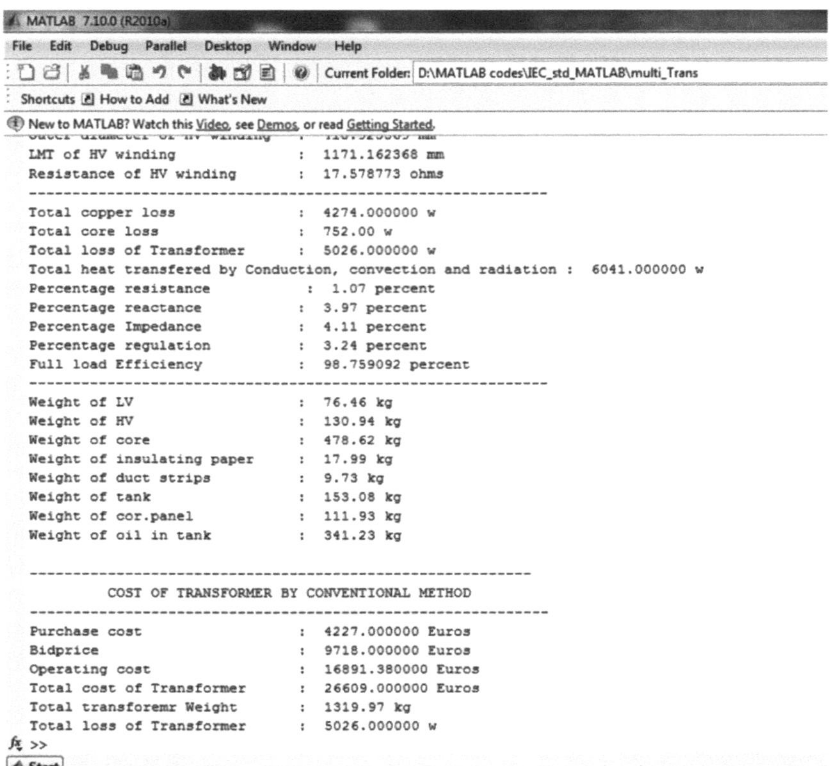

FIG. 7 Design results of MOTDO problem in MATLAB.

TABLE 3 Design Data for the Compromise Solution of MOTDO Problem

Rating of Transformer	
400 KVA, 3phase, 20,000/400 V, delta/star Transformer	
Design variables	
Voltage per turn	: 13.20 V
Current density in HV winding	: 3.00 A/sq.mm
Current density in LV winding	: 3.20 A/sq.mm
Maximum flux density	: 1.75 wb/sq.m
Width of core leg	: 245.000000 mm
Height of core window	: 245.000000 mm
Type of magnetic material	: 3
Core details	
Thickness of core leg	: 71.834876 mm
Window width of small core	: 70.473736 mm
Window width of large core	: 140.947473 mm
LV design data	
Current in LV winding per phase	: 577.350269 Amps
Area of LV conductor	: 180.421959 sq.mm
Total Number of LV Turns	: 17
Number of layers of LV winding	: 17
Width/height of LV conductor	: 226.000000 mm
Thickness of LV winding	: 20.166736 mm
Total radial thickness of LV winding	: 28.586736 mm
Inner diameter of LV winding	: 255.976912 mm
Outer diameter of LV winding	: 313.199722 mm
Length of mean turn (LMT) of LV	: 893.607315 mm
Resistance of LV winding	: 0.001765 Ω
HV design data	
Total number of HV turns (20 KV)	: 1515
Max number of HV turns (+5% tap)	: 1591
Min number of HV turns (−5% tap)	: 1439

Continued

TABLE 3 Design Data for the Compromise Solution of MOTDO Problem—cont'd

Area of HV conductor	: 2.116402 sq.mm
Diameter of HV conductor	: 1.600000 mm
Current in HV winding	: 6.349206 Amps
Number of layers in HV	: 17
Number of HV turns per layer	: 91
Thickness of HV winding	: 35.247000 mm
Total radial thickness of HV winding	: 41.887000 mm
Total height of HV winding	: 161.000000 mm
Inner diameter of HV winding	: 327.039722 mm
Outer diameter of HV winding	: 418.923569 mm
LMT of HV winding	: 1171.162368 mm
Resistance of HV winding	: 17.578773 Ω
Performance parameters	
Total copper loss	: 4274.000000 w
Total core loss	: 752.00 w
Total loss of transformer	: 5026.000000 w
Total heat transferred by Conduction, convection and radiation	: 6041.000000 w
Percentage resistance	: 1.07%
Percentage reactance	: 3.97%
Percentage impedance	: 4.11%
Percentage regulation	: 3.24%
Full load Efficiency	: 98.759092%
Tank dimensions	
Total width of transformer	: 583.000000 mm
Total height of transformer	: 753.000000 mm
Total length of transformer	: 1033.000000 mm
Weight details	
Weight of LV	: 76.46 kg
Weight of HV	: 130.94 kg

TABLE 3 Design Data for the Compromise Solution of MOTDO Problem—cont'd

Weight of core	: 478.62 kg
Weight of insulating paper	: 17.99 kg
Weight of duct strips	: 9.73 kg
Weight of tank	: 153.08 kg
Weight of corrugated panel	: 111.93 kg
Weight of oil in tank	: 341.23 kg
Transformer objectives	
Purchase cost	: 4227.000000 Euros
Bid price	: 9718.000000 Euros
Operating cost	: 16891.380000 Euros
Total cost of transformer	: 26609.000000 Euros
Total transformer weight	: 1319.97 kg
Total loss of transformer	: 5026.000000 w

4.3.3 Efficient Design Result

The new Pareto optimal front obtained is displayed in Fig. 6, after 20 independent runs gives the set of nondominated solutions of the MOTDO problem. The extreme points in the Pareto front are called as Utopia points. They are the SOTDO results, i.e., the extreme utopia points given the optimal values of minimization of single objective purchase cost and TLTC, individually. Minimum transformer purchase cost value (f_1) can be taken from the extreme point in the top of the new Pareto optimal front. Similarly, minimum TLTC value (f_2) for the minimization of TLTC objective alone can be fetched from the bottom extreme point of the front.

The minimum purchase cost value taken from the top of the NSGA-II Pareto has been compared with those of SOTDO results [40], i.e., minimization of purchase cost objective alone using deterministic method: mixed integer nonlinear programming (MINLP), and several stochastic methods such as Heuristic Algorithm (HA), Harmonic Search Algorithm (HSA), Genetic Algorithm (GA), and Differential Evolution (DE) in Table 2. It is found that NSGA-II has found the minimum purchase cost of 3817 Euros, which is 15.32%, 14.82%, 16.55%, 17.7%, and 18.91% lower than MINLP, HA, HAS, GA, and DE methods.

The best compromise design solution extracted from Pareto after post optimality analysis gives a transformer purchase cost of 4227 Euros, as seen in

Tables 2 and 3. It is 410 Euros higher than the Utopia/SOTDO result of minimum purchase cost, which is only 3817 Euros, as in Table 2. But on the other hand, this best compromise solution saves 1785 Euros (28394–26,609 Euros), compared to the Utopia result of minimum TLTC.

Minimum TLTC is cost-effective in long term, although it requires a higher initial cost. Designing/buying a transformer based on material cost/purchase cost minimization, without considering the future losses is not efficient. It is important to incorporate the cost of losses into economic evaluation of DT to produce energy efficient transformer design. Transformers with minimum TLTC are the most energy efficient ones among competing transformers, as minimum TLTC is responsible for higher life expectancy, lower losses, higher efficiency, and more reliability. MOTDO results render low TLTC rather than low first cost of investment/purchase cost that is offered by SOTDO. Hence, simultaneous minimization of purchase cost and TLTC with MOEAs is necessary instead of optimization of purchase cost objective alone for TDO problem.

5 CONCLUSION

This multiobjective problem formulation is a powerful tool for decision making as this approach analyses all the trade-offs between conflicting TD criteria. Multi-objective formulation for TDO problem presented in this chapter aims to improve reliability, design, and efficiency of the transformer and, in turn, its life expectancy. To satisfy all the needs of various types of customers, manufacturers, utilities, etc., minimization of competing objectives, purchases cost and TLTC, subjected to account performance and thermal and practical manufacturing constraints has been considered. The validity of the methodology is illustrated by its investigation on a sample 400kVA transformer. The compromise solution obtained from MOTDO is found superior to the SOTDO results, with respect to minimum total life time cost, which is the way for obtaining energy efficient transformer design. Thus, the MOP formulation for design optimization is essential and the solution for this complex MOTDO problem must be found using MOEAs.

Finite element-based electromagnetic analysis can assess the design obtained by the proposed methodology employing analytical equations, because it gives accuracy closer to experimental results. Such comparison indicates the improvement of accuracy and eliminates the manufacturing problems that arise due to the deviation of core loss estimation from the actual values. Hence two main aspects can be identified for future research work in this area: (a) Development of FEM package to estimate transformer core loss and short circuit impedance. (b) Application of evolutionary algorithm in conjunction with FEM to carry out transformer design optimization.

A APPENDIX

A.1 MATLAB Code for Three-Phase Shell Type Transformer Design

```
clear all
KVA=400;f=50;
voltperturn=  13.2;
D=245;                                  % width of core leg in mm
G=245;                                  %Height of the window in mm
Bm=1.75;                                % max flux density in core tesla
HVcurrentdensity = 3.0                  %A/mm^2
LVcurrentdensity = 3.2;                 %A/mm^2
magmaterial=3;                          % M3-0.27,M4-0.27,MOH-0.23,MOH-
                                        0.27,23ZDKH90,23ZH90,23ZH95,27ZDKH95,ZDMH
                                        -0.23,ZDMH-0.27]

%perphase voltage
HVPh1 = 20000;
HVPh2 = 20000;

LV = 400;
LVPh = LV/sqrt(3);
LVturns=round(LVPh/voltperturn);

tap=5/100;
HVPhmaxtap=HVPh1*(1+tap);
HVPhmin=HVPh2*(1-tap);

TurnsHVmax=round(HVPhmaxtap/voltperturn);
TurnsMainHV=round(TurnsHVmax*HVPh1/HVPhmaxtap);
TurnsHVmin=round(TurnsMainHV*HVPhmin/HVPh1);


%perphase current in conductor
HVcurrent1 = KVA*1000/(HVPhmaxtap*3);
HVcurrent2= KVA*1000/(HVPhmin*3);
HVcurrent = KVA*1000/(HVPh1*3);
LVcurrent = KVA*1000/(LVPh*3);

%Dimension of HV conductor          (% mm square)
HVarea = HVcurrent1/(HVcurrentdensity);
LVarea = LVcurrent/(LVcurrentdensity);
dhv = sqrt(4*HVarea/3.14);           % bare conductor dia(mm)
dhv=roundn(dhv,-1);

% Thickness of core leg
Eu=(voltperturn*34.9*10^5*25.4^2/(2*10^4*0.965*D*Bm*f));

%parameters
Ilvl=0.28;Ihvhv=6.64;Ihvl=0.28;       %mm
Dlvc=6.5; Dhvc=39;                    %mm
CCEE=3;                               %mm
LDSPlv=0.909;LDSPhv=1;
TDSPhv=0.98;Ilvc=1.5;Ihvlv=6.92;

% Design of LV winding
LayersLV=LVturns;
TDlv=G-2*CCEE-2*Dlvc;                 % width of LV layer in mm
WidthLV=TDlv;                         % Width of LV conductor with insulation
tlv=(LVarea/WidthLV);
BLDlv=(tlv+Ilvl)*LayersLV/LDSPlv;     %thickness of LV winding in mm
SPLDlv=BLDlv+Ilvc+Ihvlv;              % total thickness of LV winding in mm
```

```
% Design of HV winding
TDhv=G-2*CCEE-2*Dhvc;                    % width of HV layer in mm
WidthHV=dhv+0.111;                       % width of HV conductor
TurnsperlayerHV=((TDhv/WidthHV)-1)*TDSPhv;
nosc=floor(HVPh1/HVPh2);
TurnsperlayerHVsc=floor(TurnsperlayerHV/nosc);
TurnsperlayerHV=nosc*TurnsperlayerHVsc;
LayersHV=round(TurnsHVmax/TurnsperlayerHV);
BLDhv=(WidthHV+Ihvl)*(LayersHV/LDSPhv)+1.4;     % thickness of HV winding
SPLDhv=BLDhv+Ihvhv;                             %total thickness of HV
winding mm

% window width
F1=SPLDlv+SPLDhv;                          % small individual core
F2=F1*2;                                   % big core

% Weight of core (kg)
CMT1=2*(F1+G)+2*3.14*((Eu/2)+3.5)-28;      %mm
SCW=CMT1*D*Eu*0.965*7650*10^-9;            %weight of small core in kg
CMT2=2*(F2+G)+2*3.14*((Eu/2)+3.5)-28;      %mm
LCW=CMT2*D*Eu*0.965*7650*10^-9;
wtcore=2*(SCW+LCW);

% Total iron loss (KW)
Bmvec =   [1.6 1.65 1.7 1.75 1.8];
spFelossvectypes1 =1.54*[0.9 1.05 1.17 1.3 1.45; 1 1.1 1.22 1.37 1.56;0.76
0.83 0.92 1.02 1.15;0.8 0.9 1 1.1 1.22;0.68 0.74 0.8 0.9 1;0.72 0.76 0.84 0.9
1.08;0.74 0.8 0.9 0.98 1.1;0.77 0.8 0.9 1 1.1;0.61 0.65 0.72 0.78 0.86;0.63
0.68 0.74 0.8 0.88];

% 10 mag.materials[M3-0.27,M4-0.27,M0H-0.23,M0H
%0.27,23ZDKH90,23ZH90,23ZH95,27ZDKH95,ZDMH-0.23,ZDMH-0.27]

spFelossvec=spFelossvectypes1(magmaterial,:);
Bcnew = Bm*(ones(1,length(Bmvec)));
pos=Bmvec-Bcnew;
posnew=find(pos==0);
if isempty(posnew)==1
    pos1=find(pos>0);
    vecpos=pos1(1)-1;
else
    vecpos=posnew;
end
spFelosscore=spFelossvec(vecpos);
Totalcoreloss = round(wtcore * spFelosscore);
corecost=[3.6 3.5 4 3.9 4.5 4.4 4.3 4.2 6.1 6];

% cu loss
k=8;DuctLV=8;DuctHV=10;
MW=2*Eu+k;                              % width of coil former
ML=D+2*CCEE;                            % length of coil former
DMC=(2*(ML+MW)-10.992)/3.14;
D3=DMC+2*Ilvc;
A3=3.14*(D3^2)/4;
D5=D3+2*BLDlv;                          % external dia of LV
A5=3.14*(D5^2)/4;
LGlv=MW;
A6=DuctLV*3.33*LGlv*2;
A7=A5+A6;
D7=sqrt(4*A7/3.14);
MTlv=(D3+D7)*3.14/2;                    %LMT of LV in mm
CLlv=MTlv*LayersLV*3/1000;              %m
Rlv=0.020968*CLlv/(3*LVarea);
LLlv=round(3*Rlv*(LVcurrent^2)*1.04);
D9=D7+2*Ihvlv;
```

```
A9=3.14*(D9^2)/4;
D11=D9+2*BLDhv;
A11=3.14*(D11^2)/4;
LGhv=MW+2*BLDlv+2*Ihvlv;
A12=DuctHV*3.33*LGhv*2;
A13=A11+A12;
D13=sqrt(4*A13/3.14);
MThv=(D9+D13)*3.14/2;                 % LMT of HV
CLhv1=MThv*TurnsMainHV*3/1000;        % length of HV
Rhv=0.020968*CLhv1/(3*HVarea);
LLhv1=3*Rhv*(HVcurrent^2)*1.04;
eddyloss=0;
LL1=LLlv+LLhv1+eddyloss;
CLhv2=MThv*TurnsHVmin*3/1000;
HVcurrentsc=HVcurrent2/nosc;
Rhv2=0.020968*CLhv2/(3*HVarea);
LLhv2sc=3*Rhv2*(HVcurrentsc^2)*1.04;
LLhv2=nosc*LLhv2sc;
LL2=LLlv+LLhv2+eddyloss;
Totalculoss=round(LL1);

%Total loss
Totalloss = Totalculoss+Totalcoreloss;

% weight of cu
wtLV=CLlv*LVarea*8856*1.05*10^-6;    % for 3 phases in kg
wtHV=CLhv1*1.05*HVarea*8856*1.25*10^-6;
wtcu = wtLV+wtHV;

% Efficiency % at full load UPF
Eff_full = KVA*1000/(KVA*1000+Totalloss)*100;

% Resistance in percentage
Rt = Rhv+Rlv*(HVPh1/LVPh)^2;
Perresis = Totalculoss*100/(KVA*1000);

% reactance in percentage
Lmt = (MTlv+MThv)/(2*1000);           %m
AvgHtLVHV = (TDlv+TDhv)/2;
Perreac = ((2*3.14^2*f*4*10^-
7*Lmt*round(LVturns)*LVcurrent*(Ihvlv+(BLDlv+BLDhv)/3))/(AvgHtLVHV*voltpertur
n))*100; %

% Regulation
pf=0.8;
cosphy=pf;sinphy=sin(acos(pf));
Perreg = Perresis*cosphy+Perreac*sinphy;

% Impedance
Perimp = sqrt(Perreac^2+Perresis^2);

%Design of tank
TLTlv=12.48;TLThv=14.2;TE=38.1;
TCD=2*(DuctLV+DuctHV)*3.33;          % Total thickness of cooling duct
OLIlv=1.5+Ilvl;
OLIhv=LayersHV*Ihvl;
TLC=ML+2*F1+TCD+TLTlv+TLThv+OLIlv+OLIhv+TE;
TW=floor(TLC);                        % tank width
TL=floor(2*(4*Eu+3*F1+k)+k+12);       % tank length in mm
TH=floor(G+2*Eu+365);                 % tank height
```

```
% calculation of winding and oil gradient(temp rise)
DuctareaLV=DuctLV*4*MW*TDlv/25.4^2; %inch square
GapductareaLV=D7*3.14*TDlv/25.4^2;
totductareaLV=DuctareaLV+GapductareaLV;
LPSlv=LLlv/(3*totductareaLV);
ACR=60;
GraLV=5.8/0.49*LPSlv;
AvgraLV=2.09*GraLV;                  % temp b/w oil and LV
DuctareaHV=DuctHV*4*LGhv*TDhv/25.4^2;
GapductareaHV=D7*3.14*TDhv/25.4^2;
totductareaHV=DuctareaHV+GapductareaHV;
LPShv=LLhv2/(3*totductareaHV);
ACR=60;                              % average cu temprise
GraHV=5.8/0.49*LPShv;
AvgraHV=2.09*GraHV;                  % temp b/w oil and HV
AOR=ACR-max(AvgraLV,AvgraHV);        % average oil temp rise in deg celcius

% weight of corrugated panel & safe heat transfer in watts
AORvec = [30 32.5 35 37.5 40 42.5 45 47.5 50 52.5 55 57.5 60];
TRCvec = [220 240 260 280 300 315 350 360 390 400 440 480 480];
TCCvec = [140 150 160 170 190 210 220 240 250 280 280 310 320];

AORnew = AOR*(ones(1,13));
pos1=AORvec-AORnew;
posnew1=find(pos1==0);
if isempty(posnew1)==1
    pos2=find(pos1>0);
    vecpos1=pos2(1);
else
    vecpos1=posnew1;
end
TRC=TRCvec(vecpos1(1));
TCC=TCCvec(vecpos1(1));
TCA=(2*(TL+TW)*TH+2*TL*TW)/1000000;        %Tank convection area
TCL=TCA*TCC;

% heat transfer due to transformer tank convection`
Dpanel=180;HCP=450;
TRA=(2*(TL+TW)+(4*Dpanel*sqrt(2)))*TH/1000000; % in sq.m
TRL=TRA*TRC;                          % heat transfer due to tank radiation
NCPTL=floor((TL-60)/44)+1;
NCPTW=floor((TW-60)/44)+1;
NCP=2*(NCPTL+NCPTW);
CPA=2*Dpanel*HCP*NCP/1000000;
CPC=330;
CPL=CPA*CPC;                                % heat transfer due to cor panel
TLRTT=round(TCL+TRL+CPL);

% heat transfer due to transformer tank radiation,convection & cor panel
Wcorpanel=9.87*CPA;

% weight of insulator
Slvtp=(MW+(D3*3.14))*(G-2*CCEE)/1000000; % met square
Slvli=MTlv*(G-2*CCEE)*LayersLV/1000000;
Slvedi=MW*(G-2*CCEE)*DuctLV*2/1000000;
Slvpdi=D7*3.14*(G-2*CCEE)/1000000;
NPGI=(Ihvlv-3.3-2*0.41)/Ihvl;
Slvepi=4*3.14*D13*((G-2*CCEE)+1.8*BLDhv)/1000000;
Slvgi=D7*3.14*(G-2*CCEE)*NPGI/1000000;
Shvli=MThv*((G-2*CCEE)+4*19)*LayersHV*1.25/1000000 ;
Shvedi=LGhv*(G-2*CCEE)*2*DuctHV/1000000;
Shvpdi=D13*3.14*(G-2*CCEE)/1000000;
Shvsub=MThv*(G-2*CCEE)*5/1000000;
NPEGI=(Ihvhv-3.3-2*0.41)/Ihvl;
Shvgi=D13*3.14*(G-2*CCEE)*NPEGI/1000000;
Shvibp=MThv*25*LayersHV/1000000 ;
WIM15=3*1.1*1.5*(Slvtp+Shvibp);
WIM41=3*1.1*0.3*Slvepi;
```

```
WIM28=3*1.1*0.2*(Slvli+Slvgi+Shvli+Shvgi+Shvsub);
WIM015=3*1.1*0.13*(Slvedi+Slvpdi++Shvedi+Shvpdi);
WIM=(WIM15+WIM41+WIM28+WIM015);

% Weight of duct strips
Dw=11;
N1=floor(MW*DuctLV*2/25);
N2=floor(3.14*D7/25);
N3=floor((MW+2*BLDhv)*DuctHV*2/35);
N4=floor(3.14*D13/35);
Wduct=3*1.1*1.25*(N1+N2+N3+N4)*(G-2*CCEE)*Dw*3/1000000;

% Weight of mineral oil
volcorpanel=8*HCP*Dpanel*NCP/1000000;     % litres
volcu=wtcu/8856*1000;
volcore=wtcore/7650*1000;
voltank=TL*TW*TH/1000000;                 % m.cube
volmineraloil= (voltank+volcorpanel-volcore-volcu)/1000; % cubic m
wtminoil=0.95*870*volmineraloil;          % kg

%Weight of tank
tankthick = 5.5*10^-3;                    %5mm
wtoftank=tankthick*7.65*10^3*TCA;         %kg

% cost
purchasecost2=round(0.6215*(wtminoil*1.72+wtoftank*1.03 +wtcu*12.01
+wtcore*6.01+WIM*7.72+Wduct*8.58+Wcorpanel*1.2));
purchasecost1=round(wtminoil*1.1+wtoftank*0.7 +wtcu*7.7
+wtcore*corecost(magmaterial)+WIM*5+Wduct*5.5+Wcorpanel*0.8);
operatingcost=8.31*Totalcoreloss+2.49*Totalculoss; %  euro
Clabcrm=2564;

Bidprice=round((purchasecost1+Clabcrm)/(1-0.3012));
if (Totalcoreloss<(750*1.15) && Totalculoss<(4600*1.15) &&
Totalloss<(1.1*5350) && AOR<60 && Perimp<(1.1*4.5) && Perimp>(0.9*4.5) &&
D<=G && Perreg<4 && TLRTT>(LL2+Totalcoreloss) && (2*Eu)<(0.9*D) &&
(0.5*D)<2*Eu)
    TOC = round(Bidprice+operatingcost);
else
    penalty1=100000;
    TOC = round(Bidprice+operatingcost+penalty1);
end

% Total weight of transformer
Tweight=wtminoil+wtoftank+wtcu+wtcore+WIM+Wduct+Wcorpanel;

%Total loss
Totalloss = Totalculoss+Totalcoreloss;
```

REFERENCES

[1] Kefalas, T, Kladas, A, Tsili, M, Amoiralis, EI, & Georgilakis, PS 2010, 'Energy conservation by low loss distribution transformers adapted to load characteristics', Schneider Electric. Available from: <http://www.oas.gr/wp-content/uploads/2010/09/Energy-Conservation-by-Low-Loss-Distribution-Transformers-Adapted-to-Load-Characteristics.pdf>.

[2] M. Carlen, D. Xu, J. Clausen, T. Nunn, V.R. Ramanan, D.M. Getson, Ultra high efficiency of transformers, Contribution to IEEE-PES, New Orleans, 2010.

[3] SEEDT, Selecting Energy Efficient Distribution Transformers, 2008. A Guide for Achieving Least-Cost Solutions, Intelligent Energy Europe, by Polish Copper Promotion Centre and European Copper Institute.

[4] D. Maheswaran, K.K. Kailas, V. Rangaraj, W. Kumar, Energy efficiency in electrical systems, Proceedings of IEEE Intenational conference on power electronics, drives, and energy systems, pp. 16–19, 2010.

[5] P.S. Georgilakis, Spotlight on Modern Transformer Design, Springer-Verlag London Limited, 2009.

[6] S.K. Sen, Principles of electrical machine design with computer programmes, Oxford and IBH publishing & co Pvt. ltd, 1987.

[7] X. Liu, Y. Yang, F. Yang, A. Jadoon, Numerical research on the losses characteristic and hot-spot temperature of laminated core joints in transformer, Appl. Therm. Eng. 110 (2017) 49–61. ISSN 1359-4311.

[8] E.I. Amoiralis, M. Tsili, A. Kladas, Transformer design and optimization: a literature survey, IEEE Trans. Power Delive. 24 (4) (1999) 1999–2024.

[9] D.L.P. Feil, P.R. Silva, D.P. Bernardon, T.B. Marchesan, M. Sperandio, L.H. Medeiros, Development of an efficient distribution transformer using amorphous core and vegetable insulating oil, Electr. Power Syst. Res. 144 (2017) 268–279. ISSN 0378-7796.

[10] P.S. Georgilakis, N.D. Doulamis, A.D. Doulamis, N.D. Hatziargyriou, S.D. Kollias, A novel iron loss reduction technique for distribution transformers based on a combined genetic algorithm-neural network approach, IEEE Trans. Syst. Man Cybern. 31 (2001) 16–34.

[11] J. Tsivgouli Anastassia, A. Tsili Marina, G. Kladas Antonios, S. Georgilakis Pavlos, T. Souflaris Athanassios, A.D. Skarlatini, Geometry optimization of electric shielding in power transformers based on finite element method, J. Mater. Process. Technol. 181 (2007) 159–164.

[12] S. Subramanian, S. Padma, Bacterial foraging algorithm based multi objective optimal design of single phase transformer, J. Comput. Sci. Eng. 6 (2) (2011) 1–6.

[13] S. Subramanian, S. Padma, Optimization of transformer design using bacterial foraging algorithm, Int. J. Comput. Appl. 19 (3) (2011) 52–57.

[14] M.A. Arjona, C. Hernandez, M. Cisneros-Gonzalez, Hybrid optimum design of a distribution transformer based on 2-D FE and a manufacturer design methodology, IEEE Trans. Magn. 46 (8) (2010) 2864–2867.

[15] C. Hernandez, M.A. Arjona, D. Shi-Hai, Object-oriented knowledge based system for distribution transformer design, IEEE Trans. Magn. 44 (10) (2008) 2332–2337.

[16] T. Orosz, P. Sőrés, D. Raisz, A.Z. Tamus, Analysis of the green power transition on optimal power transformer designs, Period. Polytech. Elect. Eng.Comput. Sci. 59 (3) (2015) 125–131.

[17] I. Amoiralis Eleftherios, A. Tsili Marina, S. Georgilakis Pavlos, G. Kladas Antonios, T. Souflaris Athanassios, A parallel mixed integer programming-finite element method technique for global design optimization of power transformers, IEEE Trans. Magn. 44 (6) (2008) 1022–1025.

[18] I. Amoiralis Eleftherios, S. Georgilakis Pavlos, M.A. Tsilia, Design optimization of distribution transformers based on mixed integer programming methodology, J. Optoelectron. Adv. Mater. 10 (5) (2008) 1178–1183.

[19] S. Georgilakis Pavlos, A. Tsili Marina, T. Souflaris Athanassios, A heuristic solution to the transformer manufacturing cost optimization problem, J. Mater. Process Technol. 181 (2007) 260–266.

[20] S.V. Daneshmand, H. Heydari, A diversified multi objective simulated annealing and genetic algorithm for optimizing a three phase HTS transformer, IEEE Trans. Appl. Supercond. 26 (2) (2016) 1–10.

[21] I. Eleftherios, S. Georgilakis Pavlos, A. Tsili Marina, G. Kladas Antonios, Global transformer optimization method using evolutionary design and numerical field computation, IEEE Trans. Magn. 45 (3) (2009) 1720–1723.

[22] P.S. Georgilakis, Recursive genetic algorithm-finite element method technique for the solution of transformer manufacturing cost minimization problem, IET Electr. Power Appl. 3 (6) (2009) 514–519.

[23] T.D. Kefalas, S. Magdaleno-Adame, Techno-economic comparative evaluation of mixed and conventional magnetic wound cores for three-phase distribution transformers, Electr. Power Syst. Res. 155 (2018) 331–339.

[24] S. Zhang, Q. Hu, X. Wang, Z. Zhu, Application of chaos genetic algorithm to transformer optimal design. in: International Workshop on Chaos-Fractals Theories and Applications, IEEE Computer Society, 2009.

[25] L. Hui, H. Li, H. Bei, Y. Shunchang, Application research based on improved genetic algorithm for optimum design of power transformers, Electrical Machines and Systems, ICEMS, Proceedings of the fifth international conference, 1 2001, pp. 242–245.

[26] M. Ramezani, H. Falaghi, M.P. Moghaddam, M.R. Haghifam, Genetic based algorithm for optimal placement of distribution transformers, IEEE Power Eng. Soc. (2006) 1–5.

[27] C.J. Wu, F.C. Lee, Minimum weight EI core and pot core inductor and transformer designs, IEEE Trans. Magn. 16 (5) (1980) 755–757.

[28] M. Tsili, A. Kladas, P. Georgilakis, A. Souflaris, D. Paparigas, Numerical techniques for design and modeling of distribution transformers, J. Mater. Process. Technol. 161 (2005) 320–326.

[29] S. Tamil Selvi, S. Baskar, Modified parameter optimization of distribution transformer design using covariance matrix adaptation evolution strategy, Int. J. Electr. Power Energy Syst. 61 (2014) 208–218.

[30] L.D.S. Coelho, V.C.M. Mariani, Combining of chaotic differential evolution and quadratic programming for economic dispatch optimization with valve-point effect, IEEE Trans. Power Syst. 21 (2) (2006) 989–996.

[31] L.D.S. Coelho, V.C. Mariani, M.V. Ferreira da Luz, J.V. Leite, Novel gamma differential evolution approach for multiobjective transformer design optimization, IEEE Trans. Magn. 49 (5) (2013) 2121–2124.

[32] A.A. Adly, S.K. Abd-El-Hafiz, A performance-oriented power transformer design methodology using multi-objective evolutionary optimization, J. Adv. Res. 6 (3) (2015) 417–425.

[33] A. Konak, D.W. Coit, A.E. Smith, Multi-objective optimization using genetic algorithms: a tutorial, Reliab. Eng. Syst. Saf. 91 (9) (2006) 992–1007.

[34] Q. Zhang, H. Li, MOEA/D: a multi-objective evolutionary algorithm based on decomposition, IEEE Trans. Evol. Comput. 11 (6) (2007) 712–731.

[35] K. Deb, A. Pratap, S. Agarwal, T. Meyarivan, A fast and elitist multiobjective genetic algorithm: NSGA-II, IEEE Trans. Evol. Comput. 6 (2) (2002) 182–197.

[36] Y. Cui, Z. Geng, Q. Zhu, Y. Han, Review: multi-objective optimization methods and application in energy saving, Energy 125 (2017) 681–704.
[37] http://www.nsc.co.jp/en/product/kind/sheet/product-list.html.
[38] M.A. Abido, Multi objective evolutionary algorithms for electric power dispatch problem, IEEE Trans. Evol. Comput. 10 (3) (2006) 315–329.
[39] S. Tamilselvi, S. Baskar, L. Anandapadmanaban, K. Mohaideen Abdul Kadhar, P.R. Varshini, Chaos-Assisted Multi Objective Evolutionary Algorithm to the Design of Transformer, in: Soft Computing, Springer, 2016, pp. 1–18.
[40] E.I. Amoiralis, M.A. Tsili, D.G. Paparigas, A.G. Kladas, Global transformer design optimization using deterministic and nondeterministic algorithms, IEEE Trans. Ind. Appl. 50 (1) (2014) 383–394.

Chapter 18

A Multiobjective Teaching-Learning Algorithm for Power Losses Reduction in Power Systems

Juan M. Ramirez*, Miguel A. Medina† and Carlos A. Coello Coello*,‡
**Centro de Investigacion y de Estudios Avanzados del Instituto Politecnico Nacional, Zapopan, Mexico, †Centro Nacional de Control de Energia, Mexico, ‡Av. Instituto Politécnico Nacional 2508, Madero, Mexico*

NOMENCLATURE

ABC	artificial bee colony algorithm
DE	differential evolution
GA	genetic algorithm
HS	harmony search
MOEA/D	multiobjective evolutionary algorithm based on decomposition
MOP	multi objective problem
MOTLA/D	multiobjective teaching learning algorithm based on decomposition
NSGA-II	nondominated sorting genetic algorithm II
OPF	optimal power flow
ORPD	optimal reactive power dispatch
PSO	particle swarm optimization
TLBO	teaching learning-based optimization
Ω	feasible region

1 INTRODUCTION

Over the past half-century, optimal power flow (OPF) has become one of the most important and widely studied problems in power system operation, analysis, and planning. The OPF was proposed in the early 1960s based on the

Classical and Recent Aspects of Power System Optimization. https://doi.org/10.1016/B978-0-12-812441-3.00018-5
© 2018 Elsevier Inc. All rights reserved.

economic dispatch problem. In general, the OPF problem is a large-scale, highly constrained, nonlinear, nonconvex optimization problem which may contain both continuous and discrete control variables. The main purpose of the optimal power flow is to optimize a selected objective function, such as cost, planning, or reliability by controlling power flow within an electrical network without violating network power flow constraints or system and equipment operating limits. Many different classes of the OPF problem have been developed to address specific instances of the problem, using varying assumptions and selecting different objective functions to be optimized, different sets of control variables, and different system constraints. The resulting optimization problems have different names depending on the particular objective function being addressed and the constraints under consideration. The optimal reactive power dispatch (ORPD) is a special case of the optimal power flow problem and is one of the major issues in terms of secure and economic operation. Generally, the control variables of the ORPD consist of voltage magnitudes of generator buses, output of static reactive power compensators, transformer tap-settings, and shunt capacitors/reactors. The aim of the reactive power dispatch problem is to minimize real power transmission losses of the network, while maintaining the system voltage profile in an acceptable range by regulating generator bus voltages, transformer tap-settings, and reactive power generation of reactive power sources. Furthermore, physical and operation constraints must be maintained within the allowable limits. Because output shunt capacitors/reactors and tap-settings of transformers are discrete variables, but some other variables are continuous, the reactive power dispatch problem can be modeled as a large-scale, mixed-integer, nondifferentiable, and nonlinear problem. Additionally, the difficulty of solving the ORPD problem significantly increases with the increment of network's size and complexity.

Thus, optimal reactive power (ORP) plays a significant role in the secure operation of power systems. One of the main tasks of a power system operator is to manage the system in such a way that its operation is safe and reliable. Its main aim is to determine the optimal operating capacity and the physical distribution of the compensation devices such as voltage rating of generators, reactive power injection of shunt capacitors/reactors, and tap ratios of the tap setting transformers, to ensure a satisfactory voltage profile, while minimizing the transmission losses. Active power line losses are small while reactive power line losses are large. Reducing the reactive power losses enables more active power to be transferred over a single line. Due to the continuous growth in the demand for electricity with unmatched generation and transmission capacity expansion, voltage stability has emerged as a challenge to power system planning and operation. Therefore, a voltage stability index should also be considered as an objective of the ORP problem.

It is important that each system and control area handles capacitive and inductive reactive resources at proper levels to maintain the voltages within established high and low limits. Reactive generation scheduling, transmission and switching, and load shedding, if necessary, should be implemented to maintain these levels. Likewise, each control area should provide its reactive power requirements, including appropriate reserves to protect the voltage levels for contingency conditions.

The optimal reactive power problem is a nonlinear, nonconvex, overdetermined system, a large-scale optimization problem with both continuous and discrete variables; additionally, its high dimensionality represents a major difficulty. This problem is quite important for power system security. In this chapter, the basic objective is to estimate proper adjustments on the control variables, such as generator bus voltages and tap setting transformers that help to maintain an acceptable voltage profile and minimize the reactive power losses; one voltage stability metric (L_{index}) is also used. Thus, an optimal formulation that contributes to attain these purposes becomes appropriate. In general, it may include several objective functions, possibly in conflict among them.

This type of optimization problem has a set of possible solutions (named Pareto-optimal set), which represents the best commitment (feasible) among the objectives [1]. Several optimization techniques have been proposed to solve such optimal reactive power problems. From them, two major approaches may be identified [2]:

(1) The first approach is based on the use of evolutionary algorithms such as differential evolution (DE) [3], nondominated sorting genetic algorithm II (NSGA-II) [4], particle swarm optimization (PSO) [5], an improved hybrid evolutionary programming technique [6], and artificial bee colony algorithm (ABC) [7].

(2) The second approach is based on conventional methods. They include gradient-based methods, nonlinear programming (NLP), quadratic programming (QP), linear programming (LP), interior point methods [8–12], the weighting method [13], and the ε-constraint method [14].

These conventional methods are based on an estimation of the global minimum. However, due to difficulties of differentiability, nonlinearity, and nonconvexity, these methods do not guarantee reaching the global optimum [15]. Thus, these methods present limitations when dealing with certain types of problems. For instance, they cannot be used when the objective function is not available in an algebraic form. This has motivated the development of alternative methods, such as meta-heuristics. Over the years, meta-heuristics (from which evolutionary algorithms is a particular subclass) have become a popular choice for solving complex optimization problems, due to their flexibility, generality (they are less sensitive to the actual shape or continuity of the Pareto front than conventional methods), and ease of use. Additionally, most meta-heuristics require little or no specific domain knowledge.

Multiobjective optimization is a design methodology that optimizes a set of objective functions systematically and simultaneously. This type of optimization problem has a set of possible solutions (named Pareto-optimal set), which represent the best commitment (or trade-off) among the objectives. In the open research, several methods have been suggested and successfully applied for solving multiobjective power systems optimization problems. These approaches are generally divided into two categories: classical mathematical optimization algorithms and Pareto-based optimization algorithms.

The classical methods transform the multiobjective optimization problem into a single objective optimization problem usually by either aggregating the objective functions into a single weighted function, or simply by optimizing one objective function and treating the remaining as constraints. Therefore, the resulting single optimization problem can be solved by deterministic or nondeterministic (heuristic) algorithms. However, these classical single-objective approaches have several limitations to solve multiobjective optimization problems: (1) it requires a priori knowledge about the relative importance of the objective functions; (2) the aggregated function leads to only one solution, and therefore, it requires multiple simulation runs to obtain the Pareto-optimal set; (3) the trade-offs among objective functions cannot be easily evaluated; and (4) the solution may not be attainable unless the optimal Pareto set is convex. Thus, classical optimization methods are not suitable to solve multiobjective optimization problems. Furthermore, because real life problems involve several objectives and system engineers may desire to know all possible optimization solutions of all objectives simultaneously, Pareto-based approaches have been proposed as an alternative to address the shortcomings of classical single-objective methods. These algorithms handle the objective functions simultaneously as competing objectives.

In this chapter, a multiobjective teaching learning algorithm based on decomposition (MOTLA/D) is proposed for solving the reactive power handling problem. To minimize the reactive power losses and a voltage stability index [16], the proposed algorithm estimates the following optimal values: (i) generator bus voltages; (ii) tap setting transformers. The effectiveness of the proposed approach is demonstrated and compared with respect to the multiobjective evolutionary algorithm based on decomposition (MOEA/D) [17], which is representative of the state-of-the-art on the subject; both methods are applied on three test systems: 9-, 26-, and 118-buses.

The rest of the chapter is organized as follows. Section 2 exposes a brief review of previous work. Section 3 presents the problem approached in this chapter. Section 4 describes the formulation. Section 5 summarizes the decomposition strategy. In Sections 6–8, the general framework of the proposed approach is summarized. Results of a comparative study are presented in Section 9. Finally, conclusions are provided.

2 A BRIEF REVIEW OF PREVIOUS WORK

Several review articles about evolutionary algorithms in power systems applications have been published since the 1990s: [18–21]. This section reviews the technical literature related to applications of multiobjective evolutionary algorithms to voltage stability enhancement, reactive power dispatch, and economic dispatch optimization problems in power systems, which are the topics dealt with in this chapter. The review in this section has been drawn from refereed journal articles from the following journals:

- IEEE Transactions on Power Systems
- International Journal of Electrical Power & Energy Systems
- Electric Power Systems Research
- Energy Conversion and Management
- Engineering Applications of Artificial Intelligence

These journals have published a large quantity of heuristic algorithms in the context of power system applications. Books and conference proceedings have not been included.

Novel modern heuristics algorithms have been reported in Power systems applications. Chaohua Dai et al. [22] proposed a seeker optimization algorithm (SOA) for the optimal reactive power dispatch considering three objective functions: active power losses, voltage deviation, and voltage stability index. SOA is a novel population-based heuristic search algorithm, which attempts to simulate the act of human searching for real-parameter optimization. The algorithm's performance is studied on the IEEE 57 and 118-bus power systems.

Liao et al. [23] presented a new method called multiobjective optimization by reinforcement learning (MORL) to solve the optimal power system dispatch and voltage stability problem by achieving two objectives: reduction of fuel cost and enhancement of voltage stability. Reinforcement Learning stems from the idea of trial and-error learning and focuses on goal-directed learning from interactions between learner and environment.

Panigrahi et al. [24] extend the bacteria foraging meta-heuristic into the domain of multiobjective optimization to solve the multiobjective environmental/economic dispatch (EED) problem. The idea of bacteria foraging principle is based on the fact that natural selection tends to eliminate animals with poor foraging strategies through methods for locating, handling, and ingesting food, and to favor the propagation of genes of those animals that have successful foraging strategies. In this work, the multiobjective bacteria foraging (MOBF) optimization technique was compared in respect to other multiobjective evolutionary algorithms on the standard IEEE 30-bus test system.

Sivasubramani and Swarup [25] model the optimal power flow problem as a nonlinear constrained multiobjective optimization problem where power losses, fuel cost, and voltage stability objectives are considered into the formulation. The problem is solved by a multiobjective harmony search (MOHS) algorithm.

The Harmony search algorithm has been recently developed in an analogy with improvisation process where musicians always try to polish their pitches to obtain a better harmony.

Niknam et al. [26] present a multiobjective teaching-learning-based optimization algorithm (θ-MTLBO) to solve the dynamic economic/emission dispatch problem considering ramp-rate limits and valve-point effects. The teaching-learning-based optimization algorithm is a new efficient optimization algorithm which has been inspired by learning mechanism in a class. The proposed algorithm is applied on three test systems and compared with respect to several heuristic optimization techniques and with a popular optimization software named general algebraic modeling system (GAMS), which uses linear programming to optimize the problems. In Ref. [27] a multiobjective optimization algorithm based on modified teaching-learning-based optimization (MTLBO) algorithm is proposed to solve the optimal location of automatic voltage regulators (AVRs) in distribution systems in presence of distributed generators (DGs).

Khorsandi et al. [28] propose a fuzzy based modified artificial bee colony (MABC) algorithm to solve the optimal power flow problem. The problem was formulated as a multiobjective mixed-integer nonlinear problem where the simultaneous minimization of the total fuel cost of thermal units, total emission, total real power losses, and voltage deviation were considered. The artificial bee colony algorithm is a relatively new optimization technique which simulates the intelligent foraging behavior of a bee swarm. The proposed approach is applied on the IEEE 30-bus and IEEE 118-bus test systems.

Additional new modern meta-heuristic algorithms for multiobjective optimization can be found in the literature. This dissertation proposes the multiobjective variant for two popular meta-heuristics of current interest, namely, the artificial bee colony algorithm (ABC), and the teaching-learning-based optimization algorithm (TLBO). These algorithms are the two with the highest efficiencies in their class. The effectiveness of these algorithms has been tested in several benchmark functions and they have been compared with those of state-of-the-art algorithms such as: differential evolution (DE), particle swarm optimization (PSO), and their variants [29,30]. Moreover, the performance and effectiveness of these algorithms to solve large scale optimization problems and real-world optimization problems, including problems in power systems, have been demonstrated [26–28,31,32].

There are multiobjective variants of the TLBO and ABC in the literature such as those presented in Refs. [26–28,33]. These variants are Pareto-based approach. Thus, these methods require some other techniques for ranking and distribution of the solutions (e.g., crowding distance, fitness sharing, niching). However, as above mentioned, these methods cannot always provide good results, especially when dealing with complex multiobjective optimization problems. Therefore, in this dissertation the proposed multiobjective variants of TLBO and ABC algorithm are based on the MOEA/D framework. This framework decomposes a multiobjective problem into several single-objective optimization subproblems with neighborhood relations. In this way, a set of approximate solutions to the Pareto front is achieved by minimizing each subproblem instead of using Pareto ranking.

The performance of the multiobjective evolutionary algorithms based on decomposition in power system problems, has not been fully investigated. A publication related with the application of MOEA/D in reactive power dispatch problem is presented on Innovative Smart Grid Technologies Conference Europe, 2010 *IEEE PES* [23]. In this work, Liao, H. L et al., compared a novel multiobjective optimization by reinforcement learning with the first version of the MOEA/D to solve the economic power dispatch with voltage stability enhancement.

This chapter proposes a multiobjective teaching-learning algorithm for managing reactive power. The proposition is applied on several power test systems with different dimensionality.

3 MULTIOBJECTIVE POWER FLOW FORMULATION

Over the past half-century, the optimal power flow (OPF) has become one of the most important and widely studied problems in power system operation, analysis, and planning. In general, the OPF problem is a large-scale, highly constrained, nonlinear, nonconvex optimization problem which may contain both continuous and discrete control variables. The main purpose of the OPF is to optimize a selected objective function, cost for instance, with planning or reliability purposes by controlling power flow within an electrical network without violating network constraints or system and equipment operating limits [34].

Many different OPF classes have been developed to address specific instances of the problem, among them the reactive power dispatch and economic dispatch problems have always been the major issues in terms of secure and economic power systems operation. These problems can be considered into the analysis of multi objective optimization. The objective functions in these problems have many variants, which include transmission network losses, transmission capacity, and investment of compensation devices. In addition, some technical indices, such as voltage stability indices to prevent security margins, may be used as objectives. This research takes the voltage stability into account within the short-term operation planning context, where an optimal preventive action must be found to enhance voltage stability including operational limits.

In this chapter, the minimization of the total power transmission losses has been selected as an objective function in the optimal reactive power dispatch and economic dispatch problems. An optimal power flow problem is formulated as a multiobjective optimization problem, where three objective functions are considered for minimization, while satisfying a number of equality and inequality constraints. The problem is formulated in the sequel.

3.1 Problem Objectives

The multiobjective optimal power flow addressed in this chapter uses the following objective functions: (i) minimization of the total fuel cost, (ii) minimization of the total power transmission losses, and (iii) voltage stability enhancement through the minimization of the L-index.

3.1.1 Fuel Cost Minimization

Fig. 1 depicts a conventional cost curve for the ith generator with active power output P_{gi}.

The objective is to minimize the total fuel cost F_{cost}. Conventionally, the fuel cost for ith generator with active power output P_{gi} is modeled as a quadratic function of the active power generation, and the total fuel cost F_{cost} in ($/h) may be expressed as,

$$F_{\text{cost}} = \sum_{i=1}^{N_g} a_i + b_i P_{gi} + c_i P_{gi}^2 \tag{1}$$

where N_g is the number of generators; a_i, b_i, and c_i are the cost coefficients of the ith generator; and P_{gi} is the corresponding active power output.

3.1.2 Active Power Losses Minimization

Transmission losses constitute economic loss providing no benefits. Therefore, the objective is to minimize the active power losses (P_{loss}) through the transmission lines. If we express these losses in terms of bus voltages and associated angles, then the losses may be calculated by,

$$P_{loss} = \sum_{k=1}^{nl} g_k \left[V_i^2 + V_j^2 - 2V_i V_j \cos\left(\theta_i - \theta_j\right)\right] \tag{2}$$

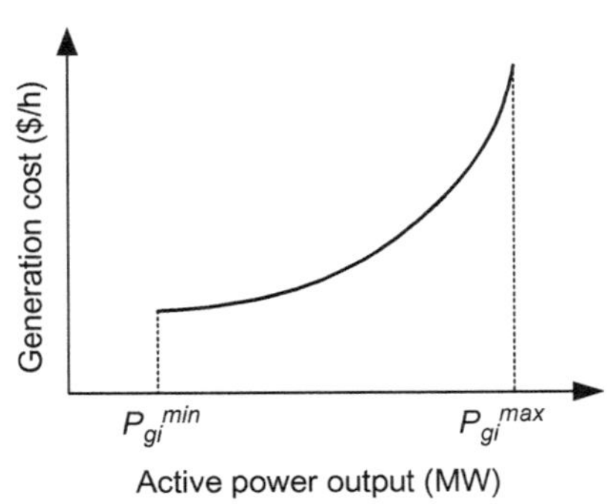

FIG. 1 Operating cost curve for the ith generator.

where nl is the number of transmission lines, g_k is the conductance of the kth transmission line connecting the ith and jth bus; V_i, V_j, θ_i, and θ_j are the voltages magnitudes and phase angles of ith and jth bus, respectively.

3.1.3 Voltage Stability Enhancement

A conventional way looking for voltage stability assessment is the use of indices, which estimate the proximity to voltage instability and determine those buses exhibiting weak stability. Nowadays, there are a variety of indexes that help to assess the steady-state voltage stability [35–37].

In this research, voltage stability enhancement is achieved through minimizing the voltage stability index L_{index} [16], which can evaluate the steady state voltage stability margin of each bus. The L-index value lies between 0 (no load) and 1 (voltage collapse). This value implicitly includes the load effect. The bus with the highest L_{index} value will be the most vulnerable, and therefore, this method helps to identify weak areas that require reactive power critical support. The L_{index} is calculated in the following way [16].

The network equations in terms of bus admittance matrix may be written as,

$$I_{bus} = Y_{bus} V_{bus} \tag{3}$$

The buses are broken down into two categories: (i) the set of load buses (α_L) and (ii) the set of generator buses (α_G). Thus, Eq. (3) becomes,

$$\begin{bmatrix} I^L \\ I^G \end{bmatrix} = \begin{bmatrix} Y_1 & Y_2 \\ Y_3 & Y_4 \end{bmatrix} \cdot \begin{bmatrix} V^L \\ V^G \end{bmatrix} \tag{4}$$

It is assumed that the transmission system is linear and allows a representation in terms of a hybrid matrix H. Therefore, rearranging the above equation,

$$\begin{bmatrix} V^L \\ I^G \end{bmatrix} = H \cdot \begin{bmatrix} I^L \\ V^G \end{bmatrix} = \begin{bmatrix} Z^{LL} & F^{LG} \\ K^{GL} & Y^{GG} \end{bmatrix} \cdot \begin{bmatrix} I^L \\ V^G \end{bmatrix} \tag{5}$$

where V^L and I^L are load voltage and current vectors; V^G and I^G are generator voltage and current vectors; Z^{LL}, F^{LG}, K^{GL}, and Y^{GG} are submatrices of the hybrid matrix H.

Matrix H is generated from the admittance matrix (Y_{bus}) by a partial inversion, where the load buses' voltage vector is exchanged for the current vector. This representation may then be utilized to define a voltage stability index for load buses, namely L_j which is defined by,

$$L_j = \left| 1 - \frac{\sum_{i \in \alpha_G} F_{ji} V_i}{V_j} \right| \tag{6}$$

For stable conditions, $0 \le L_j \le 1$, must not be violated for any load bus j. A global indicator L_{index} describing the whole system's stability is defined by,

$$L_{index} = \max_{j \in \alpha_L} (L_j) \tag{7}$$

L_{index} in Eq. (7) is associated with the worst bus in the sense of voltage stability. The L_{index} minimization implies to take such bus toward a less stressed condition.

3.2 Problem Constraints

The minimization of the above functions must satisfy a number of equality and inequality constraints. The OPF equality constraints reflect the physics of the power system by imposing Kirchhoff's Laws to the electrical network, and the inequality constraints of the OPF reflect the limits on physical devices, as well as the limits created to ensure system security. These constraints are described in the following sections.

3.2.1 *Equality Constraints*

The equality constraints are the balance of the active and reactive power described by the set of power flow equations. They may be expressed as follows,

$$P_{gi} - P_{di} - \sum_{j=1}^{N_b} |V_i||V_j||Y_{ij}| \cos(\delta_i - \delta_j - \theta_{ij}) \tag{8}$$

$$Q_{gi} - Q_{di} - \sum_{j=1}^{N_b} |V_i||V_j||Y_{ij}| \sin(\delta_i - \delta_j - \theta_{ij}) \tag{9}$$

where, N_b is the number of buses, P_{gi} is the ith active power generation, Q_{gi} is the ith reactive power generation, P_{di} is the ith active power load, Q_{di} is the ith reactive power load, and $|Y_{ij}|$ is the ijth element of the bus admittance matrix.

3.2.2 *Inequality Constraints*

These constraints represent the system operating limits as follows,

(A) *Generators*: These constraints are associated to the generator voltages (V_g), active power output (P_g), and reactive power output (Q_g) by lower and upper limits as follows,

$$V_{gi}^{\min} \le V_{gi} \le V_{gi}^{\max}, \quad i = 1, \ldots, N_g \tag{10}$$

$$P_{gi}^{\min} \le P_{gi} \le P_{gi}^{\max}, \quad i = 1, \ldots, N_g \tag{11}$$

$$Q_{gi}^{\min} \le Q_{gi} \le Q_{gi}^{\max}, \quad i = 1, \ldots, N_g \tag{12}$$

where N_g is the number of generators.

(B) *Transformers*: Transformers tap settings are restricted by their minimum and maximum limits as follows:

$$T_i^{min} \leq T_i \leq T_i^{max}, \quad i = 1, \ldots, N_t \tag{13}$$

where N_t is the number of transformers.

(C) *Shunt VAR*: Reactive power injections at buses are restricted by their minimum and maximum limits as follows:

$$Q_{ci}^{min} \leq Q_{ci} \leq Q_{ci}^{max}, \quad i = 1, \ldots, N_c \tag{14}$$

(D) *Load bus voltage*: Each load bus is restricted by its limits as follows:

$$V_i^{min} \leq V_i \leq V_i^{max}, \quad i = 1, \ldots, N_{PQ} \tag{15}$$

where N_{PQ} is the number of load buses.

4 MULTIOBJECTIVE OPTIMAL POWER FLOW FORMULATION

Considering the above-mentioned objective functions, equality, $g(x,u)$, and inequality constraints, $h(x,u)$, the multiobjective optimal power flow is defined as the optimization of the following study case.

Minimization of fuel cost and power losses

$$\begin{aligned} &\text{minimize } [F_{cost}, P_{loss}] \\ &\text{Subject to: } \; g(x, u) = 0 \\ &\qquad\qquad\quad\; h(x, u) = 0 \end{aligned}$$

where g is the equality constraints representing the nonlinear power flow Eqs. (8) and (9); h represents the inequality constraints through the system operating limits (10)–(15); and x is the vector of dependent variables including:

1. Active power P_{g1} generated by the slack bus
2. Load bus voltage V_L
3. Generator reactive power output Q_G

Hence x can be expressed as:

$$x = \left[P_{g1}, V_{L1}, \ldots, V_{LN_{PQ}}, Q_{G1}, \ldots, Q_{GN_g}\right] \tag{16}$$

where N_{PQ}, N_g, are the number of load buses and the number of generators, respectively.

Likewise, u is the vector of control or independent variables including:

1. Generator active power outputs P_g except the slack bus P_{g1}
2. Generator bus voltage V_g
3. Transformer tap setting T
4. Shunt VAR compensation Q_C

Hence, the vector of control variables (u) is expressed as,

$$u = \left[P_{g2}, \ldots, P_{gN_g}, V_{g1}, \ldots, V_{gN_g}, T_1, \ldots, T_{Nt}, Q_{c1}, \ldots, Q_{cN_c}\right] \tag{17}$$

where N_t and N_c denote the number of regulating transformers and number of shunt compensations, respectively. It is worth noting that the decision variables are self-constrained by the optimization algorithm.

4.1 Handling of Equality and Inequality Constraints

The equality constraints of power balance equations shown in (8), (9) are handled by the Newton-Raphson based power flow calculations; therefore, there is no need to integrate them into the objective function.

Due to the optimal power flow results may violate the inequality constraints, it is common to use penalization to handle the inequality violation. The violations are multiplied by its corresponding penalty coefficient and added to the objective function to constitute a fitness function. It is simple to implement and easy to understand. However, it introduces new parameters into the algorithm and the value of such coefficients significantly influences the solution reached by algorithm. There is not a specific rule to choose the suitable penalty coefficient values and the trial-error strategy is applied in most situations. The electricity quality offered by power system is vastly important and a large variation could result in a huge economic loss. For this consideration, a non-parameter strategy is adopted to handle the inequality constraints.

In this chapter Deb's heuristic, constrained handling technique [38] is used in the proposed MOTLA/D and MOABC/D methods. This technique uses a tournament selection operator in which two solutions are selected and each other compared. Therefore, this approach does not require the use of the penalty function method.

The following heuristic rules are implemented for the selection:

i. If one solution is feasible and the other infeasible, then the feasible solution is preferred.
ii. If both are feasible solutions, then the solution with better objective function value is preferred.
iii. If both solutions are infeasible, then the solution with least constraint violation is preferred.
iv. If both solutions are infeasible and have the same number of constraint's violations, then the solution with a smaller objective function value is preferred.

These rules are implemented at the end of the main phases of the proposed algorithms, i.e., at the end of the teacher and the learner phases for MOTLA/D, and at the end of the employed and onlooker phases for the MOABC/D. Instead of accepting the new solution x_{new}, if it gives better function value at the end of these phases, Deb's constraint handling rules are used to select x_{new} based on the above heuristic rules.

5 METAHEURISTIC ALGORITHMS BASED ON DECOMPOSITION

"*A metaheuristic is formally defined as an iterative generation process which guides a subordinate heuristic by combining intelligently different concepts for exploring and exploiting the search space, learning strategies are used to structure information in order to find efficiently near-optimal solutions* [39]". Some well-known algorithms falling within this category are: Genetic algorithms (GAs) [40], Differential Evolution (DE) [41], Particle Swarm Optimization (PSO) [42,43], Artificial Bee Colony (ABC) [44,45], and the Harmony Search (HS) [46].

Problems with multiple objectives arise in a natural manner in many disciplines and their solutions have been a challenge to researchers for a long time [47]. These problems are known as multiobjective optimization, multicriteria optimization, or vector optimization problem and can be defined (in words) as the problem of finding [48]: "a vector of decision variables which satisfies constraints and optimizes a vector function whose elements represent the objective functions. These functions form a mathematical description of performance criteria which are usually in conflict with each other. Hence, the term "optimize" means finding such a solution which would give the values of all the objective functions acceptable to the decision maker."

The mathematical definition of a multiobjective problem (MOP) is important in providing a basis of understanding between the interdisciplinary nature of every possible solution technique (deterministic, nondeterministic), i.e., search algorithms [47]. Without loss of generality, a multiobjective problem may be stated as:

Finding a vector $\vec{x}^* = \left[x_1^*, x_2^*, \ldots, x_n^*\right]^T$ which optimizes the vector function:

$$\text{Minimize } \boldsymbol{F}\left(\vec{x}\right) = \left[f_1\left(\vec{x}\right), f_2\left(\vec{x}\right), \ldots, f_k\left(\vec{x}\right)\right]^T \tag{18}$$

Subject to the constraints:

$$g_i\left(\vec{x}\right) \leq 0, \quad i = 1, \ldots, m \tag{19}$$

$$h_j\left(\vec{x}\right) = 0, \quad j = 1, \ldots, p \tag{20}$$

where a solution $\vec{x} \in \Re^n$ is a vector of n decision variables: $\vec{x} = (x_1, x_2, \ldots, x_n)^T$. Notice that T indicates *transposition*. The inequality and equality constraints in

(19) and (20), respectively, define the feasible region Ω and any solution $\vec{x} \in \Omega$ defines a feasible solution. One of the notable differences between single-objective and multiobjective optimization is that in the later one the objective functions constitute a multidimensional space, in addition to the usual decision variable space ($\Re^n$). This additional K-dimensional space is called *objective space*, $\Re^k$. Therefore, for each solution $\vec{x}$ in the decision variable space, there exists a point $\vec{z}$ in the objective space, $\vec{z} \in \Re^k$. The vector function $\boldsymbol{F}\left(\vec{x}\right)$ is a function which maps the set of solutions from the decision variable space to the objective space: $\boldsymbol{F}: \vec{x} \in \Re^n \rightarrow \Re^k$, Fig. 2. The k-elements of $\boldsymbol{F}\left(\vec{x}\right)$ represents the values of the objective functions.

Vector $\vec{x}^*$ denotes the optimal solution. However, because the objective functions of a multiobjective optimization problem usually conflict with each other, they cannot be optimized simultaneously. Therefore, there no exists a unique optimal solution, instead a satisfactory trade-off between the conflicting objectives must be found. This leads to a set of optimal solutions known as *Pareto-optimal* solutions or *nondominated* solutions. In multiobjective optimization, the goal is to find the best possible compromise (or trade-off) among the objectives because, frequently, one objective can be improved only at the expense of worsening another. Therefore, it is necessary to establish some criteria to define what would be considered as an "optimal" solution.

To describe the concept of "optimality" for a multiobjective problem, which was generalized by Vilfredo Pareto [49] on the concept "*Pareto optimum,*" the following definitions are provided [47]:

Definition 1 (Pareto Dominance)

A solution $\vec{x} = (x_1, x_2, \ldots, x_n)$ is said to dominate another solution $\vec{y} = (y_1, y_2, \ldots, y_n)$ (denoted by $\vec{x} \preceq \vec{y}$) if and only if $\vec{x}$ is partially less than $\vec{y}$ on the objective space, i.e., $\forall i \in \{1, \ldots, k\}$, $f_i\left(\vec{x}\right) \leq f_i\left(\vec{y}\right) \wedge \exists i \in \{1, \ldots, k\}: f_i\left(\vec{x}\right) < f_i\left(\vec{y}\right)$.

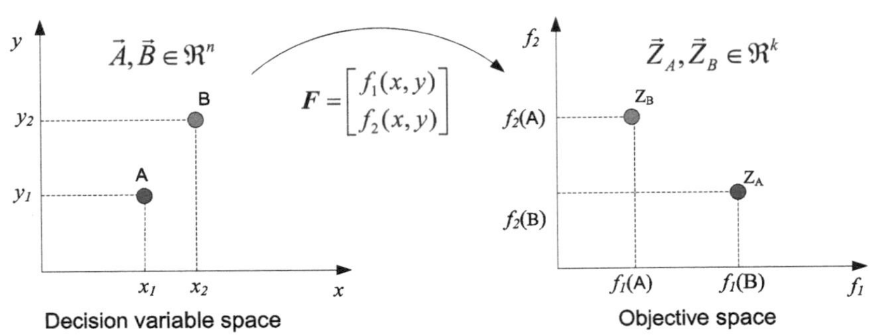

FIG. 2 Example of the multidimensional space in a multiobjective problem.

Definition 2 (Pareto Optimality)
A solution $\vec{x} \in \Omega$ is said to be a *Pareto-optimal* solution $\vec{x}^*$, if and only if there is no other solution $\vec{y} \in \Omega$ such that $\vec{y} \preceq \vec{x}$.

Definition 3 (Pareto-Optimal Set)
For a given MOP, the Pareto-optimal set ($\overline{PS}$) is defined by $\overline{PS} = \left\{ \vec{x} \in \Omega \,\middle|\, \vec{x} \text{ is Pareto} - \text{optimal solution} \right\}$.

Definition 4 (Pareto Front)
For a given MOP and a Pareto-optimal set $\overline{PS}$, the Pareto front $\overline{PF}$ is defined as the image of $\overline{PS}$ in the objective space, i.e., $\overline{PF} = \left\{ F\left(\vec{x}\right) \,\middle|\, \vec{x} \in \overline{PS} \right\}$.

Based on the concept of "*Pareto dominance,*" the comparison between two decision vectors $\vec{x}$ and $\vec{y}$ can lead to the following three possibilities:

(1) $\left(\vec{x}\ \textbf{strictly dominates}\ \vec{y}\right): \vec{x} \prec \vec{y} \quad \textit{iff} \quad f_i\left(\vec{x}\right) < f_i\left(\vec{y}\right) \quad \text{for } i = 1,..,k$

(2) $\left(\vec{x}\ \textbf{weakly dominates}\ \vec{y}\right): \ \vec{x} \preceq \vec{y} \quad \textit{iff} \quad f_i\left(\vec{x}\right) \leq f_i\left(\vec{y}\right) \quad \text{for } i = 1,..,k$

(3) $(\vec{x}$ is **incomparable** to $\vec{y}): \vec{x} \sim \vec{y} \quad \textit{iff} \quad f_i(\vec{x}) \nleq f_i(\vec{y}) \wedge f_i(\vec{y}) \nleq f_i(\vec{x}) \quad \text{for } i = 1,..,k$

A graphical illustration of the previous concepts about *Pareto dominance*, is provided in Fig. 3.

When the Pareto-optimal solutions (or nondominated solutions) are plotted in the objective space, they are collectively known as the *Pareto front*. The Pareto front dominates all other possible solutions, and in many cases, is located on the boundary of the objective space (i.e., feasible solution space). In Fig. 4 a **bold** dash line is used to mark this boundary for a bi-objective problem which is known as the *Pareto front*.

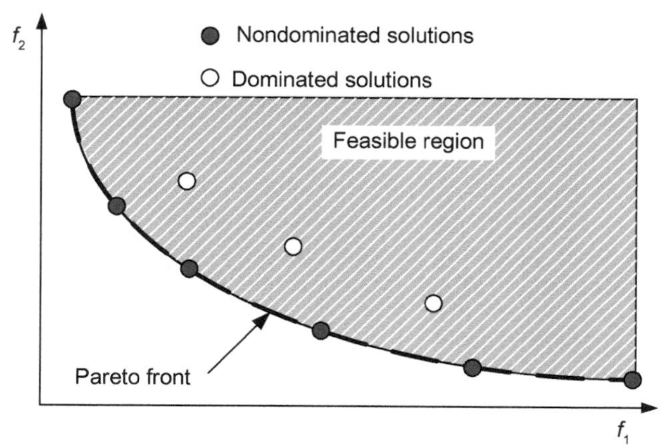

FIG. 3 Illustrative example of the concepts about Pareto dominance.

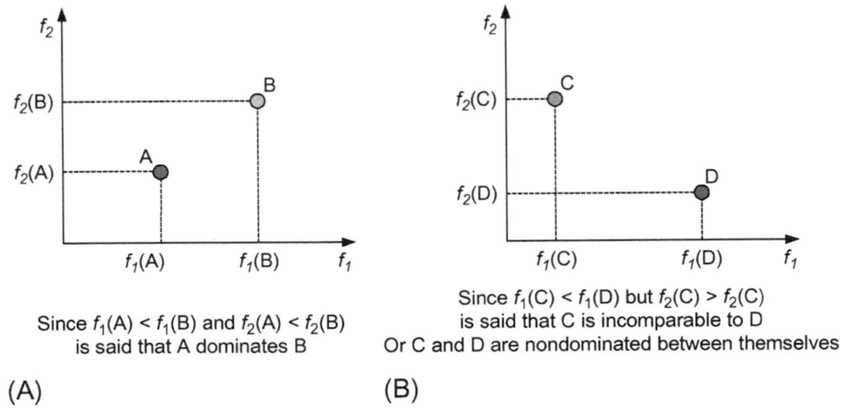

FIG. 4 Graphical illustration of the *Pareto front* of a biobjective minimization problem.

In general, it is not easy to find an analytical expression of the line or surface that contains the Pareto-optimal solutions, and in most cases, it turns out to be impossible. The normal procedure to generate the Pareto front is to compute many solutions in the feasible region Ω and their corresponding $f(\Omega)$. When there is a sufficient number of these, it is then possible to determine the nondominated points and to generate a Pareto front [47].

5.1 Decomposition of a Multiobjective Problem

It is well-known that a Pareto-optimal solution to a multiobjective optimization problem, under certain conditions, could be an optimal solution of a single-objective optimization problem in which the objective is an aggregation function of all the individual objectives. Therefore, the approximation of the Pareto front can be decomposed into several single-objective optimization subproblems. This is the basic idea behind many traditional mathematical programing methods for approximating the Pareto front. Several methods for constructing aggregation functions can be found in the literature (see, e.g., [50]). These methods use a weighted vector to define a scalar function. In this way, and under certain assumptions (e.g., the minimum is unique, the weighting coefficients are positive, etc.), a Pareto-optimal solution is achieved by minimizing such function. Among these methods, probably the two most widely used are the *Weighted Sum* and the *Tchebycheff* approaches.

- *Weighted Sum Method*

The classical aggregating approach to solve a multiobjective optimization problem is the *weighted sum method*. This method consists of assigning a weight w_i to each objective function so that the multiobjective problem is converted to a single objective problem with a scalar objective function of the form:

$$\begin{aligned}&\text{Minimize } g(\vec{x}\,|w) = \sum_{i=1}^{k} w_i f_i(\vec{x}) \\ &\text{subject to } x \in \Omega\end{aligned} \tag{21}$$

where w_i is a vector of weights usually set by the decision maker such that, $\sum_{i=1}^{k} w_i = 1$ and $w_i \geq 0$ for all $i = 1\ldots\ k$. If all of the weights are positive, then minimizing (21) provides a sufficient condition for the Pareto optimality, which means that the minimum of (21) is always a Pareto-optimal solution of the multiobjective problem [51]. As with most methods involving scalar weights w_i, setting one or more of the weights to zero may result in weakly Pareto-optimal points [52]. The relative value of the weights generally reflects the relative importance of the objectives. This is another common characteristic of weighted methods. This method is easy to use, however, there are a few recognized difficulties with the weighted sum method [52–55]:

(1) This method does not provide a well-distributed set of solutions along the Pareto front, with a consistent change in weights, even with a consistent Euclidean distance between consecutive solutions.
(2) This method is unable to obtain points on nonconvex portions of the Pareto front.

- *Weighted Tchebycheff Approach*

Under this scheme, the scalar optimization problem can be stated as [56]:

$$\begin{aligned}&\text{Minimize } g(\vec{x}\,|w, z^*) = \max_{i\in\{1,\ldots,k\}} \left\{ w_i \left| f_i(\vec{x}) - z_i^* \right| \right\} \\ &\text{subject to } x \in \Omega\end{aligned} \tag{22}$$

where w_i is a weighting vector such that, $w_i \geq 0$ for all $i = 1, \ldots, k$, and $\sum_{i=1}^{k} w_i = 1$. The vector $z^* = (z_1^*, \ldots, z_k^*)$ represents the reference point, i.e., $z_i^* = \min\left\{ f_i(\vec{x})\,|x \in \Omega \right\}$, $i = 1, \ldots, k$.

Whereas the *weighted sum method* discussed previously always yields Pareto-optimal solutions, this method may provide the complete Pareto-optimal set, through weights variations; it provides a necessary condition for Pareto optimality [50]. In addition, the solution using the *Tchebycheff* approach is always weakly Pareto optimal, and if the solution is unique, then it is Pareto optimal. This means that for each Pareto-optimal solution x^* there exists a weight vector "w" such that x^* is an optimal solution of (22) and each optimal solution of (22) is a Pareto-optimal solution of (18). Thus, it is able to obtain different Pareto-optimal solutions by modifying the weighting vector.

The advantages of the *Tchebycheff* approach are as follows:

(1) It provides a clear interpretation of minimizing the largest difference between f_i and the reference point z_i.
(2) It can provide the complete Pareto-optimal set.
(3) It always provides a weakly Pareto-optimal solution.
(4) It is relatively well suited for generating the complete Pareto-optimal set (with weight variation)

The main disadvantage is that it requires the minimization of each objective when using the referent point, which can be computationally expensive.

6 TEACHING LEARNING-BASED OPTIMIZATION

Teaching learning-based optimization (TLBO) algorithm was proposed by Rao et al. [30] in 2011 as a metaheuristic algorithm inspired on the philosophy of teaching-learning process in a class between the teacher and learners (students). TLBO has emerged as a simple and efficient technique for solving single-objective complex benchmark problems and real-world problems, including power system applications. It has been well tested on many optimization problems [30,32,57–59]. Thus, it is attractive to extend the TLBO to solve multiobjective optimization problems [26,27,33,60,61].

Metaheuristics, in general, require parameters that affect their performance. For example, DE depends on the mutation strategy adopted, and on its intrinsic control parameters such as its scaling factor and the crossover rate. PSO requires learning factors, the variation of the inertia weight and the maximum value of velocity. ABC requires the number of employed bees, onlooker bees, and a value of limit. HS, requires harmony memory consideration rate, pitch adjusting rate and the number of improvisations. In contrast, TLBO does not require any specific parameter to be tuned, which facilitates its implementation and use [58]. This represents another remarkable characteristic of TLBO.

The main idea of TLBO is explained in the sequel [30]. Fig. 5 shows the distribution of marks obtained by the learners of two different classes. A normal distribution is assumed for the marks, but in actual practice it can have skewness. The normal distribution is defined as,

$$f(x) = \frac{1}{\sigma\sqrt{2\pi}} e^{\frac{-(x-\mu)^2}{2\sigma^2}} \tag{23}$$

where σ is the variance, μ is the mean, and x is the independent variable.

It is evident from Fig. 5 that the teacher in class 2 is better than the teacher in class 1 because the mean of the grades, M_2, obtained by the learners in class 2 attain better results than M_1. Therefore, it can be stated that a good teacher produces a better average in learners. Students also learn from interaction between themselves, which also helps in their results. However, in practice a teacher can only move the mean of a class toward some extent depending on the capability

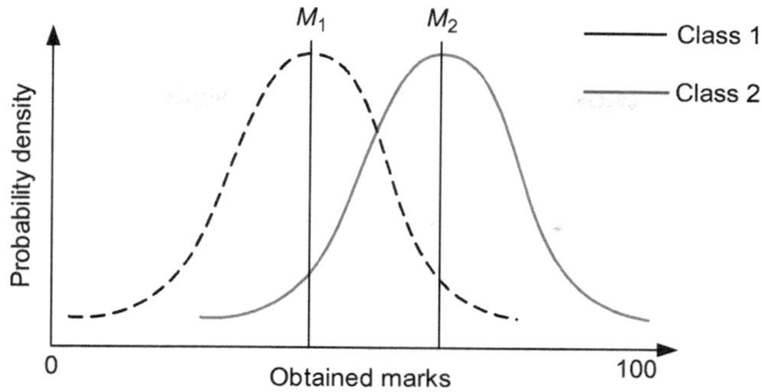

FIG. 5 Distribution of marks obtained by learners in two different classes.

of the class [30]. Consequently, because a teacher, who is the most experienced person on a subject in the society, influences the learner's behavior to attain good marks or grades, and it is expected that the teacher increases the knowledge level of the whole class depending on their skills. Therefore, the teacher will put maximum effort into teaching their students, but students will gain knowledge according to the quality of teaching delivered by a teacher and the quality of students attending the class [30]. In addition, the students can also gain knowledge by discussing, communicating, discovering, and interacting each other.

Based on the above teaching process, a mathematical model of a novel optimization technique called teaching-learning-based optimization (TLBO) was developed by Rao et al. [30]. Like any other nature-inspired algorithm, TLBO is a population-based method that uses a population of solutions to proceed to the global solution. For TLBO, the population is described as a class of learners. In optimization algorithms, the population consists of different design variables. In TLBO, the different design variables represent the number of subjects offered to learners, and the grade attained by learner is analogous to the "fitness" as in any other population-based optimization technique. The teacher is considered as the best solution obtained so far. The process of TLBO is based on two main phases. The first phase is the "teacher phase", which involves learning from the teacher, and the second phase is the "learner phase," which involves learning through the interaction among learners. It should be noted that the output solutions of the "teacher phase" are the input solutions for the "learner phase."

The pseudo-code of the TLBO algorithm may be summarized in the sequel.

1. Initialization
2. Evaluation
3. Iteration = 1
4. **Repeat**
5. Teacher Phase
6. Keep the best solutions
7. Learner Phase

8. Keep the best solutions
9. iteration = iteration + 1
10. Until Iteration = Maximum Number of Iterations

A brief description of the main phases is provided in the following subsections.

6.1 Teacher Phase

A good teacher is one who brings their learners up to their level in terms of knowledge. But in practice, this is not possible, and a teacher can only move the mean of a class up to some extent depending on the capability of the class. This follows a random process which is interfered by many factors [30].

Let M_i be the mean of the class and $T_{best,i}$ be the teacher of that class (best solution so far) at the ith iteration. Hence, $T_{best,i}$, will try to move the mean of the class (M_i) toward its own level. Thus, the new mean will be $T_{best,i}$, designated as $M_{new,i}$. The difference between the mean of the class (M_i) and the new mean ($M_{new,i}$) is expressed by [30]:

$$\Delta_i = r_i(M_{new,i} - T_F M_i) \tag{24}$$

where T_F is a teaching factor that weights the current mean value. The value of T_F can be either 1 or 2, which is determined randomly with equal probability as $T_F = \text{round}[1 + rand(0,1)]$, and r_i is a random number within [0, 1]. TLBO uses the current best solution to improve the existing solution, thereby increasing the convergence rate. The difference (24) updates the current solution according to the following expression [30]:

$$x_{new,i} = x_{old,i} + \Delta_i \tag{25}$$

x_{new} is accepted if it improves the function value.

6.2 Learning Phase

A learner interacts randomly with other learners through group discussions, presentations, formal communications, etc. Thus, each learner may acquire new knowledge if the others have more knowledge than them. The modification of the learners is expressed as [62]:

$$\begin{array}{l}
\textit{for } i = 1 \text{ to number of learners} \\
\quad \text{Randomly select two learners } x_i \text{ and } x_j \text{ such that } x_i \neq x_j \\
\quad \text{if } f(x_i) < f(x_j) \\
\quad\quad x_{new,i} = x_i + r_i(x_i - x_j) \\
\quad \text{else} \\
\quad\quad x_{new,i} = x_i + r_i(x_j - x_i) \\
\quad \text{end if} \\
\text{end for}
\end{array} \tag{26}$$

x_{new} is accepted if it accomplishes a better objective function value. The new solutions update the initial learners and the teaching-learning process continues until the stopping criterion is achieved.

7 MULTIOBJECTIVE TEACHING-LEARNING ALGORITHM BASED ON DECOMPOSITION

This is a proposed metaheuristic algorithm [63]: the multiobjective teaching-learning algorithm based on decomposition (MOTLA/D). This approach is an extension of the teaching-learning-based optimization algorithm previously described, and its main aim is to deal with multiobjective optimization problems employing a decomposition framework. This framework decomposes a multi-objective problem into several single-objective optimization subproblems. In this way, a set of approximate solutions to the *Pareto front* is achieved by minimizing each subproblem through neighborhood relationships, instead of using a Pareto ranking method. This characteristic makes MOTLA/D different from the multiobjective variants of the published TLBO algorithm because the proposed approach does not need a density estimator that distributes the solutions along the *Pareto front*.

In this chapter, MOTLA/D uses the *Tchebycheff* approach to decompose a multiobjective optimization problem into single-objective subproblems by choosing N uniformly distributed weighting vectors. The goal of the ith subproblem is to minimize

$$g\left(x|w^i,z^*\right)=\max_{j\in\{1,..,m\}}\left\{w_j^i|f_j(x)-z_j{}^*|\right\} \tag{27}$$

where $w^i=\{w_1^i,\dots,w_m^i\}$ is the weighting vector with $i=1, 2, \dots, N$, and $w_j^i\geq 0$ for all objective functions$j=1, 2, \dots, m$. $\sum_j w_j^i=1$ and vector $z^*=(z_1^*,\dots,z_m^*)$ represents the reference point, i.e., $z_j{}^*=\min\left\{f_j(x)\middle|x\in\Omega\right\}$. If N is large enough and these weights are uniformly distributed, then under a mild condition, the N optimal solutions to these subproblems will be a good approximation to the *Pareto front*. In case of two objective functions, i.e., $m=2$ the weighting vector, $w^i=(w_1^i,w_2^i)$ may be set as,

$$\begin{array}{c} w_1^i=j/N \\ w_2^i=(N-j)/N \end{array},\quad j=1,2,\dots,N \tag{28}$$

This method for generating weighting vectors works well for the formulation in this proposition. Nevertheless, other methods may be used as well.

The ith subproblem is associated with weighting vector w^i and its scalar function is denoted as $g(x|w^i)$. MOTLA/D solves simultaneously these subproblems based on the philosophy of teaching-learning process in a class among teacher and learners, in a similar way as that described in the TLBO [30]. Due to $g(x|w)$ is a continuous function of w, two subproblems are likely to have similar solutions if their weighting vectors are close to each other [64]. Therefore, any information about the weighting vectors close to w^i should be helpful for optimizing $g(x|w^i)$. Based on this observation, the neighborhood of subproblem i, defined as $B(i)$, contains a predetermined number (T_n) of subproblems with the closest weighting vectors with respect to w^i. The Euclidean distance is used to measure the closeness between any two weighting vectors, while it is assumed that $i \in B(i)$. That is, the ith subproblem is its own neighbor. The neighborhood for each subproblem represents a group of learners or a class and is responsible for solving such subproblem. The neighborhood size T_n should be much smaller than the population size N [64]. The main steps of the proposed MOTLA/D may be summarized as follows.

(1) *Initial learners*

At the first step, the algorithm generates a randomly distributed initial population of learners ($x_{i,j}$) within the range of the parameters' boundaries by,

$$x_{i,j} = x_j^{\min} + rand(0,1) \cdot \left(x_j^{\max} - x_j^{\min}\right) \tag{29}$$

where $i = 1, \ldots, N$ and $j = 1, \ldots, D$. The population size is N, i.e., the potential solutions, and D is the number of decision variables. $x_j^{\min}$ and $x_j^{\max}$ are the lower and upper limits for the jth decision variable, respectively.

(2) *Selection of the class*

The class for the i-th sub-problem is selected between the neighborhood $B(i)$ and the entire population N according to,

$$C_{i\text{th}} = \begin{cases} B(i) & \text{if } rand < \delta \\ \{1, \ldots N\} & \text{otherwise} \end{cases} \tag{30}$$

where $rand$ is a random number within [0, 1] and δ is the probability to select the neighborhood $B(i)$ as the colony.

(3) *Teacher phase*

In this phase, the class of the ith subproblem may be expressed as,

$$C_{i\text{th}} = \begin{bmatrix} x_{1,1} & x_{1,2} & \cdots & x_{1,D} \\ x_{2,1} & x_{2,2} & \cdots & x_{2,D} \\ \vdots & \vdots & \ddots & \vdots \\ x_{\Omega_T,1} & x_{\Omega_T,2} & \cdots & x_{\Omega_T,D} \end{bmatrix} \tag{31}$$

where Ω_T is the size of the class and D is the number of design variables. Within the teacher phase, the mean of the class (M_{class}) for each particular design variable is calculated column-wise,

$$M_{class,i} = [mean_1, mean_2, \ldots, mean_D] \tag{32}$$

The teacher (M_{new}) for the ith subproblem represents the best learner of the class $C_{i\text{th}}$. Thus, the teacher is determined by,

$$M_{new,i} = \left\{ x_i \,\middle|\, \min_{x_i \in \Omega_T} g\left(x_i \middle| w^i, z^*\right) \right\} \tag{33}$$

The current solutions are updated according to the difference between the mean of the class (M_{class}) and the new mean (M_{new}),

$$x_{new} = x_{i,j} + rand_{i,j} \cdot (M_{new,i} - T_F M_{class,i}) \tag{34}$$

where index i corresponds to the current ith subproblem, $rand_{i,j}$ is a random number within the interval [0, 1]. T_F is the teaching factor, which value can be either 1 or 2; this is decided randomly with equal probability as $T_F = \text{round}[1 + rand(0,1)]$. If one or more decision variables of the new solution exceeds its predetermined boundaries, the jth variable is set to an acceptable value through,

$$x_{new} = \begin{cases} x_j^{\min}, & if\ x_{new,j} \le x_j^{\min} \\ x_j^{\max}, & if\ x_{new,j} \ge x_j^{\max} \end{cases} \tag{35}$$

The new solution (x_{new}) is accepted if it improves the function value and it replaces the old one (x_i).

(4) *Learner phase*

In this phase for the ith subproblem, two learners x_j and x_k are selected randomly such that $i \neq j \neq k$. A new solution (x_{new}) is generated as follows,

$$\begin{array}{l} \text{if } f(x_j) < f(x_k) \\ \quad x_{new} = x_i + rand \cdot (x_j - x_k) \\ \text{else} \\ \quad x_{new} = x_i + rand \cdot (x_k - x_j) \\ \text{end} \end{array} \tag{36}$$

Additionally, a polynomial mutation operator is applied to maintain solutions' diversity. This operator uses the polynomial distribution,

$$\delta_j = \begin{cases} (2 \cdot r_j)^{\frac{1}{\mu+1}} - 1, & \text{if } r_j < 0.5 \\ 1 - [2 \cdot (1 - r_j)]^{\frac{1}{\mu-1}}, & \text{if } r_j \ge 0.5 \end{cases} \tag{37}$$

where r_j is a random number in the interval [0, 1], and μ is a mutation distribution index. The mutated element is given by,

$$x_{new} = x_{new,j} + \left[x_j^{\max} - x_j^{\min}\right] \delta_j \tag{38}$$

where x_j^{max} and x_j^{min} are the upper and lower limits for the jth decision variable, respectively. For the new solution, if one or more decision variables exceed their predetermined boundaries, the jth variable is set to an acceptable value, (38). The new solution (x_{new}) is accepted if it improves the function value and replaces the old one (x_i).

(5) *Updating the class*

At the end of each phase (*teacher* and *learner*) the next strategy is applied to update the class of the ith subproblem [17]. Set $n = 0$ and then do the following:

$$\begin{array}{ll} \text{(a)} & \text{if } n = s_r \text{ or } C \text{ is empty} \\ & \textbf{break}\text{. Otherwise randomly picks an index } i \text{ from } C \\ \text{(b)} & \text{if } g(x_{new}|w^i) \le g(x_i|w^i), \text{ Then set } x_i = x_{new} \\ & \text{and } n = n+1 \\ \text{(c)} & \text{remove } i \text{ form } C \text{ and go to } (a) \end{array} \tag{39}$$

where (s_r) is the maximum number of solutions replaced by the new solution.

Summarizing, the proposed MOTLA/D algorithm may be described as in the next pseudo-code.

Step 1) Initialization

- Generate a well-distributed set of N weighting vectors by (28)
- Find the neighborhood of each sub-problem $B(i)$
- Generate the initial population by using Eq. (29) and evaluate its fitness.
- Initialize the reference point z^*

Step 2)

For $i = 1$ to N

- Determine the class (C_i) according to Eq. (30)
- ***Teacher phase***: Create a new solution (x_{new}) by Eq. (34)
- Update the reference point z^*
- Update solutions by Eq. (39)

End For i

Step 3)

For $i = 1$ to N

- Determine the class (C_i) according to Eq. (30)
- ***Learner Phase***: Create a new solution (x_{new}) by Eq. (36) and Eq. (38)
- Update the reference point z^*
- Update solutions by Eq. (39)

End For i

Step 4) Stop Criterion

If the stop condition is satisfied, (such as getting the maximum number of iterations or the maximum number of function evaluations), then stop MOTLA/D. Otherwise, go to **Step 2)**.

7.1 Remarks About MOTLA/D

The flow chart of the proposed MOTLA/D algorithm is presented in Fig. 6. Notice that algorithm MOABC/D (multiobjective artificial bee colony based on decomposition) is included as well. Although the proposed approaches are based on the MOEA/D (multiobjective evolutionary algorithm based on decomposition) framework [64], an important difference between the original scheme of decomposition in MOEA/D and the proposed one is the number of stages. While MOEA/D uses only one stage to apply a search strategy, in the proposed scheme two individual stages are implemented, see Fig. 6. The main phases of each algorithm are included within each of these stages. The applied phase depends on the used algorithm. This scheme allows that each phase explores individually the searching space, and also helps to compose hybrid algorithms easily by including the desired strategy within the stages.

It is noteworthy, that the *onlooker phase* and the *learner phase* of the proposed MOABC/D and MOLTA/D, respectively, have been modified from the original ABC and TLBO. These two phases in the original ABC and TLBO create a new solution from the random selection of two parents. This strategy may increase the probability that algorithms remain trapped in local minima. Therefore, to prevent premature convergence and to avoid getting trapped in local minima, a new strategy was needed. In the proposed onlooker and learner phase, in order to create a new solution for the ith subproblem, three parents (x_i, x_j, and x_k) are selected such that $x_i \neq x_j \neq x_k$. Additionally, in the learner phase, a polynomial mutation operator is applied to maintain the solutions' diversity.

8 PERFORMANCE ASSESSMENT

The ultimate goal of a multiobjective optimization algorithm is to identify solutions in the Pareto-optimal set. However, identifying the entire Pareto-optimal set is practically impossible. In addition, for many problems, especially for combinatorial optimization problems, the solution optimality proof is computationally infeasible [65]. Therefore, a practical approach to multiobjective optimization is to investigate a set of solutions (the best-known Pareto set) that represents the Pareto-optimal set as well as possible. With these concerns in mind, a multiobjective optimization approach should achieve the following three conflicting goals [66]:

(1) The estimated *Pareto front* should be as close as possible to the true *Pareto front*. Ideally, it should be a subset of the Pareto-optimal set.
(2) Solutions in the estimated Pareto set should be uniformly distributed and diverse over the true *Pareto front* in order to provide the decision-maker a true picture of tradeoffs.
(3) The estimated *Pareto front* should capture the whole spectrum of the true *Pareto front*. This requires investigating solutions at the extreme ends of the objective function space.

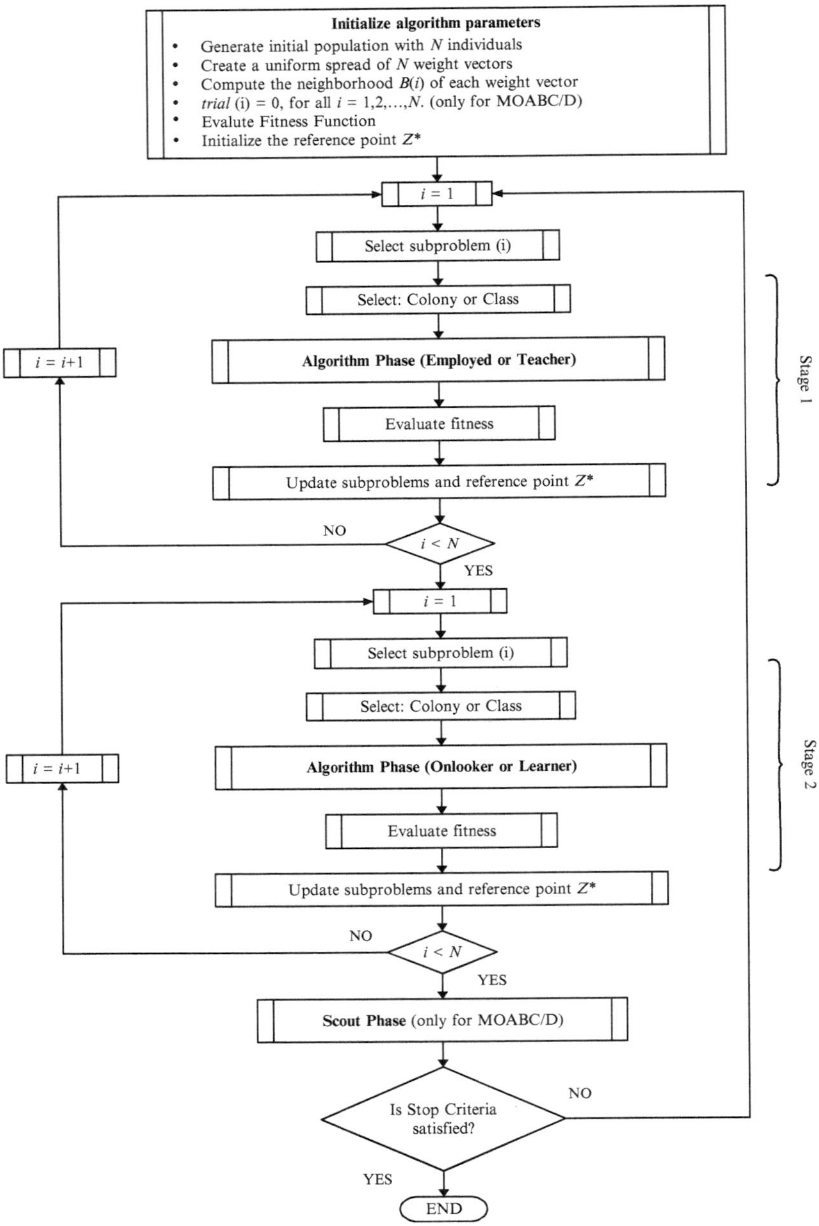

FIG. 6 Flowchart of MOABC/D and MOTLA/D.

These three tasks cannot be measured adequately by a performance measure. Therefore, in order to assess the performance of the proposed algorithms the following performance measures are adopted.

8.1 Inverted Generational Distance

The concept of the generational distance (GD) was introduced by David Veldhuizen and Gary Lamont as a way of estimating how far the Pareto-optimal solutions obtained by an algorithm are from those in the *Pareto front* of the problem [67]. The inverted generational distance (IGD), as its name suggests, it is an inverted variation of the widely used GD performance metric. There are different ways of computing and averaging the distances in the GD (e.g., [67,68]). The version of the IGD used in this chapter inverts the γ version in [68]. This means, the *Pareto front* is used as the reference point and all its elements are compared with respect to the approximation generated by the algorithms. This measure is described as follows.

Let P^* be a set of points uniformly distributed on the *Pareto front* and A be the approximation obtained by an algorithm. IGD represents the average distance from P^* to A defined as,

$$IGD(A,P^*) = \frac{\sum_{v \in P^*} d(v,A)}{|P^*|} \tag{40}$$

where $d(v,A)$ is the minimum Euclidean distance between v and the points in A. If the points in the set P^* can appropriately represent the Pareto front, IGD may measure both the spread and convergence of set A. The smaller the value of this metric is, the better is the performance of the algorithm. A value of IGD equal to zero implies that all obtained solutions lie on the true *Pareto front* and have the best possible spreads.

8.2 Spacing (S_p)

This performance measure was proposed by Schott [69], and it quantifies the spread of solutions (i.e., how uniformly distributed the solutions are) along a Pareto front approximation. This is defined by,

$$S_p = \sqrt{\frac{1}{|n-1|}\sum_{i=1}^{n}\left(\overline{d}-d_i\right)^2} \tag{41}$$

where n is the number of nondominated solutions, $d_i = \min_{i,i\neq j}\sum_{m=1}^{k}\left|f_m^i - f_m^j\right|$, $i,j = 1,2, \ldots, n$, and k denotes the number of objectives, while $\overline{d} = \sum_{i=1}^{n} d_i/|n|$. A value of zero implies that all solutions are uniformly spread (i.e., the best possible performance).

8.3 Coverage of Two Sets *C(A,B)*

This performance measure was proposed by Zitzler et al. [66]. This metric compares two sets of nondominated solutions (A, B) and evaluates the percentage of individuals in one set dominated by the individuals on the other set. This performance measure is defined by,

$$C(A,B) = \frac{|\{b \in B | \exists\ a \in A : a \preceq B\}|}{|B|} \tag{42}$$

The value $C(A,B) = 1$ means that all points in B are dominated by or equal to all points in A. The opposite, $C(A,B) = 0$ represents the situation when none of the solutions in B are covered by the set A. It is worth noting that both $C(A,B)$ and $C(B,A)$ have to be considered, because $C(A,B)$ is not necessarily equal to $1 - C(A,B)$. When $C(A,B) = 1$ and $C(B,A) = 0$ then, the solutions in A completely dominate the solutions in B (i.e., this is the best possible performance for A).

9 CASE STUDIES

This chapter compares the effectiveness and performance of the proposed algorithm with respect to that of the MOEA/D. Both MOTLA/D and MOEA/D have been applied to three test systems. In the first case study, we consider the 9-bus test system; this system consists of 9 transmission lines and 3 generating units. The system model and data can be found in Ref. [70]. The second case study is related to the IEEE-26 bus test system, which has 26 buses, 46 branches, 6 generators, 7 transformers, and 9 shunt capacitors. The detailed data of this problem can be found in [71]. Finally, in the third case study, the IEEE 118-bus test system is used. The system has 54 generator buses, 64 load buses, and 186 transmission lines with 9 tap setting transformers. The complete system data are taken from Ref. [72]. For each case study, 20 independent runs are performed. The number of subproblems considered by each algorithm is 100 for cases 1 and 2, and 200 for case 3. It is worth mentioning that the stopping criterion of each algorithm is the number of generations N_{gen} (120, 180, and 200 generations for cases 1, 2, and 3, respectively). For all test instances, the control parameter settings utilized by the MOTLA/D and MOEA/D are summarized in the following. The neighborhood size (T_{size}) is 30. The distribution index (μ), used in the polynomial mutation, is 20. The parameter of scale factor (F_s) associated with MOEA/D, which represents the amount of perturbation added to the main parent, is 0.5. The Crossover rate (P_{cr}) associated with MOEA/D, which determines the quantity of elements to be exchanged by the crossover operator, is 1. Finally, a mutation rate $P_m = 1/n$ is taken into account, where n is the number of decision variables. This parameter indicates the probability that each decision variable has of being changed.

9.1 Performance Measures

There are two goals in multiobjective optimization: (a) to achieve convergence to the Pareto-optimal set and (b) to obtain a well-distributed set of solutions along the Pareto front. These two tasks cannot be measured adequately by one performance measure each. Therefore, in order to assess the algorithms' performance two performance measures are adopted: (i) spacing (41) and (ii) coverage of two sets (42).

9.2 Experimental Results and Comparison

The advantage of evolutionary algorithms is that they have minimum requirements regarding the problem formulation; objectives can be easily added, removed, or modified. Likewise, in this application, they are well-suited to tackle highly complex problems such as those existing in power systems.

All algorithms compared were implemented in MATLAB 7.3 and run on a PC with a Pentium core duo processor operating at 2 GHz with 2 GB RAM. Three test systems were used: the IEEE 9-bus, IEEE 26-bus, and the IEEE 118-bus systems, operating under their corresponding base case. For each test power system and each algorithm, 20 runs were executed. The following results correspond to the best solution attained by each algorithm, with respect to the coverage of two set performance measure.

9.2.1 Case Study 1: 9-Buses Test System

The decision variables are related to the generator voltage V_{gi}, and range in the interval [1.0, 1.05] p.u. Table 1 summarizes the best solution for minimum reactive losses calculated through MOTLA/D and MOEA/D. Notice that for the optimized case, a reduction of the reactive losses and voltage stability index is attained. Both algorithms reduce the losses in 38.54%, which represents an important proportion of the losses with respect to the base case. Regarding

TABLE 1 Case Study 1: Best Solutions Calculated by Both MOTLA/D and MOEA/D

Decision Variables	Base Case	f_{Rloss}		f_{Lindex}	
		MOTLA/D	*MOEA/D*	*MOTLA/D*	*MOEA/D*
V_{g1} (p.u.)	1.04	1.05	1.05	1.05	1.05
V_{g2} (p.u.)	1.02533	1.0376	1.0377	1.05	1.05
V_{g3} (p.u.)	1.02536	1.0328	1.0331	1.0499	1.0499

the voltage stability index, L_{index}, this has been decreased in 10.99% relative to the base case. In summary, the objective functions become:

$$f_{Rloss(Base\ Case)} = 0.0755\,\text{p.u.} \quad f_{Lindex(Base\ Case)} = 0.1673$$
$$f_{Rloss(MOTLA/D)} = 0.0464\,\text{p.u.} \quad f_{Lindex(MOTLA/D)} = 0.1489$$
$$f_{Rloss(MOEA/D)} = 0.0464\,\text{p.u.} \quad f_{Lindex(MOEA/D)} = 0.1489$$

9.2.2 Case Study 2: IEEE 26-Buses Test System

The decision variables are related to the generator voltage V_{gi}, and range in the interval [1.0, 1.05] p.u. Likewise, another decision variable is the transformer tap setting T_i, which ranges in the interval [0.95, 1.05].

The best solution for minimum reactive losses (R_{loss}) and voltage stability index (L_{index}) is summarized in Table 2. The minimum R_{loss} and L_{index} for the base case is 0.6302 p.u. and 0.1241, respectively. As can be noticed, MOTLA/D estimates $R_{loss} = 0.1487$ p.u. and $L_{index} = 0.1023$, while MOEA/D attains $R_{loss} = 0.2105$ p.u. and $L_{index} = 0.0995$. This means that MOTLA/D reaches 76.4% reduction in losses and 17.56% reduction in L_{index} with respect to the base case. Meanwhile, MOEA/D reaches 66.59% reduction in losses and 19.82% reduction in L_{index} with respect to the base case. It is assumed that the tap positions vary among 32 positions (16 up, and 16 down), and the closest is selected in Table 2. In summary, the objective functions become:

$$f_{Rloss(Base\ Base)} = 0.6302\,\text{p.u.} \quad f_{Lindex(Base\ Case)} = 0.1241$$
$$f_{Rloss(MOTLA/D)} = 0.1487\,\text{p.u.} \quad f_{Lindex(MOTLA/D)} = 0.1023$$
$$f_{Rloss(MOEA/D)} = 0.2105\,\text{p.u.} \quad f_{Lindex(MOEA/D)} = 0.0995$$

9.2.3 Case Study 3: 118-Buses Tests System

The decision variables are related to the generator voltage V_{gi}, and range in the interval [0.98, 1.05] p.u. Likewise, another decision variable is the transformer tap setting T_i, which ranges in the interval [0.95, 1.05].

Table 3 summarizes the optimal values for the two objective functions (R_{loss}) and (L_{index}) estimated by both algorithms. The minimum R_{loss} and L_{index} for the base case became 7.8223 p.u. and 0.0693, respectively. MOTLA/D reduced reactive losses from 7.8223 p.u. to 6.9097 p.u. (a reduction of approximately 11.66%) and improved the L_{index} from 0.0693 to 0.0630 (a reduction of approximately 9.1%). Meanwhile, MOEA/D reduced reactive losses from 7.8223 p.u. to 6.9116 p.u. (a reduction of approximately 11.64%) and improved the L_{index} from 0.0693 to 0.0630 (a reduction of approximately 9.1%). It is assumed that taps vary among 32 positions (16 up, and 16 down), and the closest is selected in Table 3. In summary, the objective functions become:

$$f_{Rloss(Base\ Case)} = 7.8223\,\text{p.u.} \quad f_{Lindex(Base\ Case)} = 0.0693$$
$$f_{Rloss(MOTLA/D)} = 6.9097\,\text{p.u.} \quad f_{Lindex(MOTLA/D)} = 0.0630$$
$$f_{Rloss(MOEA/D)} = 6.9116\,\text{p.u.} \quad f_{Lindex(MOEA/D)} = 0.0630$$

TABLE 2 Case Study 2: Best Solutions Calculated by Both MOTLA/D and MOEA/D

		f_{Rloss}		f_{Lindex}	
Decision Variables	**Base Case**	*MOTLA/D*	*MOEA/D*	*MOTLA/D*	*MOEA/D*
V_{g1} (p.u.)	1.025	1.0265	1.0466	1.05	1.0498
V_{g2} (p.u.)	1.02	1.0109	1.0112	1.0429	1.0365
V_{g3} (p.u.)	1.03	1.021	1.0137	1.0498	1.0420
V_{g4} (p.u.)	1.045	1.0498	1.0487	1.05	1.0499
V_{g5} (p.u.)	1.045	1.0248	1.0445	1.05	1.0379
V_{g26} (p.u.)	1.015	1.0459	1.0216	1.0499	1.0500
T_3	0.96 (−13)	1.0135 (4)	0.9705 (−9)	0.95 (−16)	0.9850 (−5)
T_6	0.96 (−13)	0.95 (−16)	1.0493 (13)	0.95 (−16)	1.0129 (1)
T_8	1.017 (5)	1.0016 (1)	0.9840 (−6)	0.95 (−16)	0.9579 (−14)
T_9	1.05 (16)	0.9637 (−12)	0.9703 (−10)	0.95 (−16)	0.9500 (−16)
T_{10}	1.05 (16)	0.9735 (−8)	0.9598 (−13)	0.95 (−16)	0.9507 (−16)
T_{15}	0.95 (−16)	0.964 (−12)	0.9507 (−16)	0.95 (−16)	0.9515 (−16)
T_{18}	0.95 (−16)	0.9768 (−7)	0.9730 (−9)	0.95 (−16)	0.9500 (−16)

TABLE 3 Case Study 3: Best Solutions Calculated by Both MOTLA/D and MOEA/D

		f_{Rloss}		f_{Lindex}	
Decision Variables	**Base Case**	*MOTLA/D*	*MOEAD*	*MOTLA/D*	*MOEAD*
V_{g1} (p.u.)	0.955	1.0356	1.0334	1.0285	1.0273
V_{g4} (p.u.)	0.998	1.05	1.05	1.0497	1.05
V_{g6} (p.u.)	0.99	1.0472	1.0451	1.0459	1.0401
V_{g10} (p.u.)	1.05	1.0499	1.05	1.0491	1.05
V_{g19} (p.u.)	0.963	1.0435	1.037	1.0497	1.0397
V_{g24} (p.u.)	0.992	1.0497	1.0436	1.0498	1.047
V_{g27} (p.u.)	0.968	1.037	1.0371	1.049	1.0327
V_{g31} (p.u.)	0.967	1.033	1.031	1.0362	1.035
V_{g36} (p.u.)	0.98	1.0495	1.0493	1.0493	1.0495
V_{g40} (p.u.)	0.97	1.0398	1.0351	1.0415	1.0461
V_{g42} (p.u.)	0.985	1.0499	1.0435	1.0499	1.0498
V_{g54} (p.u.)	0.955	1.0187	1.0122	1.0203	1.0118
T_8	0.985 (−5)	0.9935 (−2)	0.9903 (−3)	0.9871 (−4)	0.9911 (−3)
T_{32}	0.96 (−13)	1.000 (1)	1.0006 (1)	1.0013 (1)	0.9995 (−1)
T_{36}	0.96 (−13)	0.9948 (−2)	0.9965 (−5)	1.0006 (1)	1.0032 (2)
T_{51}	0.955 (−14)	0.983 (−6)	0.9857 (−5)	0.9725 (−9)	0.9814 (−6)
T_{93}	0.96 (−13)	1.0259 (9)	1.0177 (6)	1.05 (16)	1.0223 (8)
T_{95}	0.985 (−5)	1.0179 (6)	1.013 (5)	1.0377 (12)	1.0136 (5)

9.2.4 Comparison of MOTLA/D and MOEA/D

For each case study, MOTLA/D and MOEA/D are evaluated using the two performance measures (41) and (42). The results are summarized in Tables 4 and 5. Each of these tables present the average and the standard deviation (in brackets) of each performance measure for each case study. The best results are displayed in **boldface**.

Notice in Table 4 that the proposed approach (MOTLA/D) outperformed MOEA/D in all cases regarding the *Coverage of two sets (C)*. This indicates that the proposed approach produces more solutions that dominate (according to Pareto optimality) the solutions produced by MOEA/D. The difference among the nondominated solutions produced by MOTLA/D and MOEA/D is more noticeable in cases 2 and 3. According to Table 4, in the case study 2, MOTLA/D produced solutions which dominate 55% of the solutions generated by MOEA/D. In contrast, MOEA/D produced solutions that dominate only 30% of the solutions generated by MOTLA/D. In the case study 3, the solutions obtained by MOTLA/D dominate about 41% of the solutions generated by MOEA/D; in contrast, MOEA/D produced solutions that only dominate 25% of the solutions generated by MOTLA/D.

Regarding Spacing (S), MOEA/D attains relatively better results for cases 1 and 2. However, because coverage (which relates to convergence) has

TABLE 4 Results of *Coverage of Two Set(C)* Performance Measure

	C(MOTLA/D,MOEA/D)	C(MOEA/D,MOTLA/D)
Test	*Average* (**Std. Dev.**)	*Average* (**Std. Dev.**)
Case study 1	**0.014** (*0.009*)	0.011 (*0.011*)
Case study 2	**0.545** (*0.415*)	0.293 (*0.335*)
Case study 3	**0.411** (*0.354*)	0.252 (*0.224*)

TABLE 5 Results of *Spacing (S)* Performance Measure

	MOTLA/D	MOEA/D
Test	*Average* (**Std. Dev.**)	*Average* (**Std. Dev.**)
Case study 1	0.0228 (*0.000*)	**0.0197** (*0.002*)
Case study 2	0.0256 (*0.002*)	**0.0184** (*0.003*)
Case study 3	**0.0254** (*0.013*)	0.0338 (*0.014*)

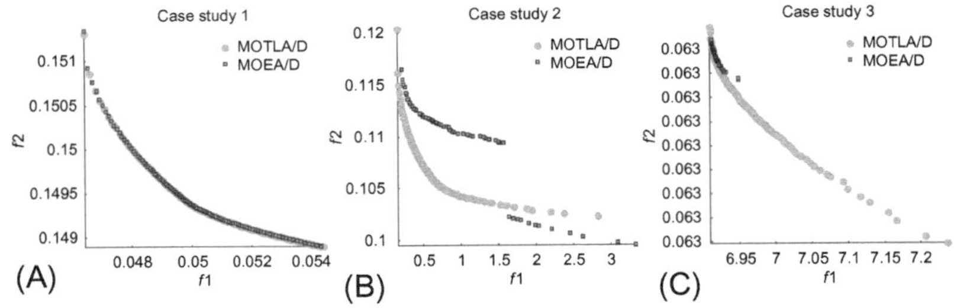

FIG. 7 From left to right: Pareto fronts (*f*1–*f*2) for both the MOTLA/D and MOEA/D (best result): (a) Case study 1; (b) Case study 2; (c) Case study 3.

precedence over spread, we can conclude that our proposed MOTLA/D outperformed MOEA/D in the analyzed cases of study.

The Pareto's fronts obtained by MOTLA/D and MOEA/D for all cases are depicted in Fig. 7. These curves represent the best case, according to the performance measures defined in (41), (42). Notice that both algorithms perform similarly for case study 1. The difference between the approximations obtained by MOTLA/D and MOEA/D is more noticeable in cases 2 and 3. It is noteworthy that MOTLA/D is able to achieve more distributed solutions in the case study 3. A distribution of nondominated solutions as uniform as possible along the Pareto front, ensures that there are not big gaps in the Pareto front and, therefore, all the different types of trade-off solutions are generated. This is relevant, because if big gaps occur, it may happen that the trade-off solution in which we are interested on is not produced (i.e., the solution of concern may be located in the missing portion of the Pareto front).

10 CONCLUSIONS

This chapter proposed a multiobjective teaching learning algorithm based on decomposition (MOTLA/D) for solving a reactive power system problem. The effectiveness and performance of MOTLA/D were compared with respect to those of MOEA/D, which represents a state-of-the-art algorithm, in three cases of study: 9-, 26-, and 118-buses test systems. The results indicate that the proposed algorithm was able to obtain better solutions than MOEA/D in all the analyzed cases. Thus, it may be concluded that the proposed algorithm is a reliable choice for power systems applications. In this chapter, an improvement of both reactive losses and voltage stability were attained. Likewise, some other additional objectives could be taken into account as well.

REFERENCES

[1] W. Stadler, Multicriteria Optimization in Engineering and the Sciences, first ed., Plenum Press, New York, 1988.

[2] M. Belazzoug, M. Boudour, FACTS Placement Multiobjective Optimization for Reactive Power System Compensation, Systems Signals and Devices (SSD), 2010 7th International Multi-Conference, pp. 1–6, June 27–30, 2010.

[3] A.A. Abou El Ela, M.A. Abido, S.R. Spea, Optimal power flow using differential evolution algorithm, Electr. Eng. 91 (2) (2009) 69–78.
[4] S. Jeyadevi, S. Baskar, C.K. Babulal, M. Willjuice Iruthayarajan, Solving multiobjective optimal reactive power dispatch using modified NSGA-II, Int. J. Electr. Power Energy Syst. 33 (2) (2011) 219–228.
[5] M.A. Abido, Optimal power flow using particle swarm optimization, Int. J. Electr. Power Energy Syst. 24 (7) (2002) 563–571.
[6] W. Yan, S. Lu, D.C. Yu, A novel optimal reactive power dispatch method based on an improved hybrid evolutionary programming technique, IEEE Trans. Power Syst. 19 (2) (2004) 913–918.
[7] C. Sumpavakup, I. Srikun, S. Chusabapiputt, A solution to the optimal power flow using artificial bee Colony algorithm. in: International Conference on Power System Technology, vol. 1, International Conference on Power System Technology, vol. 1, 2010, pp. 1–5. ISBN 978-1-4244-5940-7, https://doi.org/10.1109/POWERCON.2010.5666516.
[8] H.W. Dommel, W.F. Tinny, Optimal power flow solutions, IEEE Trans. Power Syst. PAS-87 (1968) 1866–1876.
[9] K.Y. Lee, Y.M. Park, J.L. Ortiz, A united approach to optimal real and reactive power dispatch, IEEE Trans. Power Apparat. Syst. PAS-104 (1985) 1147–1153.
[10] S. Granville, Optimal reactive power dispatch through interior point methods, IEEE Trans. Power Syst. 9 (1) (1994) 136–146.
[11] G.R.M. da Costa, Optimal reactive dispatch through primal-dual method, IEEE Trans. Power Syst. 12 (2) (1997) 669–674.
[12] M.O. Mansour, T.M. Abdel-Rahman, Non-linear VAR optimization using decomposition and coordination, IEEE Trans. Power Syst. PAS-103 (2) (1984) 246–255.
[13] C.C. Kuo, P.H. Chen, C.L. Hsu, Y.T. Chao, The reactive power and voltage control of distribution systems using the normalized weighting method, Transmission and Distribution Conference and Exhibition: Asia and Pacific, 2005, pp. 1–6.
[14] Y.T. Hsiao, H.D. Chiang, C.C. Liu, Y.L. Chen, A computer package for optimal multiobjective VAR planning in large scale power systems, IEEE Trans. Power Syst. 9 (2) (1994) 668–676.
[15] S. Ramesh, S. Kannan, S. Baskar, in: Multi-Objective Evolutionary Algorithm Based Reactive Power Dispatch with Thyristor Controlled Series Compensator, Power Electronics, Drives and Energy Systems (PEDES) & 2010 Power India, 2010 Joint International Conference on, pp. 1–7, Dec. 20–23, 2010.
[16] P. Kessel, H. Glavitsch, Estimating the voltage stability of a power system, IEEE Trans. Power Delivery 1 (3) (1986) 346–354.
[17] H. Li, Q. Zhang, Multiobjective optimization problems with complicated Pareto sets, MOEA/D and NSGA-II, IEEE Trans. Evol. Comput. 13 (2) (2009) 284–302.
[18] S. Madan, K.E. Bollinger, Applications of artificial intelligence in power systems, Electr. Power Syst. Res. 41 (2) (1997) 117–131.
[19] V. Miranda, D. Srinivasan, L. Proença, Evolutionary computation in power systems, Int. J. Electr. Power Energy Syst. 20 (2) (1998) 89–98.
[20] M.R. AlRashidi, M.E. El-Hawary, Applications of computational intelligence techniques for solving the revived optimal power flow problem, Electr. Power Syst. Res. 79 (4) (2009) 694–702.
[21] X. Xia, A.M. Elaiw, Optimal dynamic economic dispatch of generation: a review, Electr. Power Syst. Res. 80 (8) (2010) 975–986.
[22] C. Dai, W. Chen, Y. Zhu, X. Zhang, Reactive power dispatch considering voltage stability with seeker optimization algorithm, Electr. Power Syst. Res. 79 (10) (2009) 1462–1471.

[23] H.L. Liao, Q.H. Wu, L. Jiang, Multi-objective optimization by reinforcement learning for power system dispatch and voltage stability, Innovative Smart Grid Technologies Conference Europe (ISGT Europe) 2010, pp. 1–8. IEEE PES.
[24] B.K. Panigrahi, V.R. Pandi, R. Sharma, S. Das, S. Das, Multiobjective bacteria foraging algorithm for electrical load dispatch problem, Energy Convers. Manag. 52 (2) (2011) 1334–1342.
[25] S. Sivasubramani, K.S. Swarup, Multi-objective harmony search algorithm for optimal power flow problem, Int. J. Electr. Power Energy Syst. 33 (3) (2011) 745–752.
[26] T. Niknam, F. Golestaneh, M.S. Sadeghi, θ-multiobjective teaching-learning-based optimization for dynamic economic emission dispatch, IEEE Syst. J. 6 (2) (2012) 341–352.
[27] T. Niknam, R. Azizipanah-Abarghooee, M. Rasoul Narimani, A new multiobjective optimization approach based on TLBO for location of automatic voltage regulators in distribution systems, Eng. Appl. Artif. Intell. 25 (8) (2012) 1577–1588.
[28] A. Khorsandi, S.H. Hosseinian, A. Ghazanfari, Modified artificial bee colony algorithm based on fuzzy multi-objective technique for optimal power flow problem, Electr. Power Syst. Res. 95 (2013) 206–213.
[29] D. Karaboga, B. Basturk, in: Artificial bee Colony (ABC) optimization algorithm for solving constrained optimization problems, Proceedings of the 12th International Fuzzy Systems Association World Congress on Foundations of Fuzzy Logic and Soft Computing, Cancún, México, vol. 4529, pp. 789–798, 2007.
[30] R.V. Rao, V.J. Savsani, D.P. Vakharia, Teaching–learning-based optimization: a novel method for constrained mechanical design optimization problems, Comput. Aided Des. 43 (3) (2011) 303–315.
[31] B. Akay, D. Karaboga, Artificial bee colony algorithm for large-scale problems and engineering design optimization, J. Intell. Manuf. 23 (4) (2012) 1001–1014.
[32] R.V. Rao, V.J. Savsani, D.P. Vakharia, Teaching–learning-based optimization: an optimization method for continuous non-linear large-scale problems, Inf. Sci. 183 (1) (2012) 1–15.
[33] M.R. Nayak, C.K. Nayak, P.K. Rout, Application of multi-objective teaching learning based optimization algorithm to optimal power flow problem, Proc. Technol. 6 (2012) 255–264.
[34] S. Frank, I. Steponavice, S. Rebennack, Optimal power flow: a bibliographic survey I, Energy Syst. 3 (3) (2012) 221–258.
[35] C.W. Taylor, Power System Voltage Stability, McGraw-Hill, 1993.
[36] T. van Cutsem and C. Vournas, Voltage Stability of Electric Power Systems. Springer, 2007.
[37] C.A. Cañizares (Ed.), Voltage Stability Assessment: Concepts, Practices and Tools, IEEE-PES Power Systems Stability Subcommittee Special Publication, 2003. Product Number SP101PSS (ISBN 0780378695).
[38] K. Deb, An efficient constraint handling method for genetic algorithms, Comput. Methods Appl. Mech. Eng. 186 (2–4) (2000) 311–338.
[39] I. H. Osman and G. Laporte, «Metaheuristics: a bibliography», Ann. Oper. Res., vol. 63, n.o. 5, pp. 511–623, 1996.
[40] D.E. Goldberg, Genetic Algorithms in Search, Optimization, and Machine Learning, first ed., Addison-Wesley, Boston, 1989.
[41] R. Storn, K. Price, Differential evolution—a simple and efficient heuristic for global optimization over continuous spaces, J. Glob. Optim. 11 (4) (1997) 341–359.
[42] J. Kennedy and R. Eberhart, «Particle swarm optimization, 1995. Proc. IEEE Int. Conf. Neural Netw., vol 4, pp. 1942–1948, 1995.
[43] J.F. Kennedy, R.C. Eberhart, Y. Shi, Swarm Intelligence, Morgan Kaufmann Publishers, San Francisco, 2001.

[44] D. Karaboga, An Idea Based on Honey Bee Swarm for Numerical Optimization, Erciyes University, Computer Engineering Department, Technical Report-TR06, 2005.
[45] D. Karaboga, B. Basturk, A powerful and efficient algorithm for numerical function optimization: artificial bee colony (ABC) algorithm, J. Glob. Optim. 39 (3) (2007) 459–471.
[46] Z.W. Geem, J.H. Kim, G.V. Loganathan, A new heuristic optimization algorithm: harmony search, Simulation 76 (2) (2001) 60–68.
[47] C.C. Coello, G.B. Lamont, D.A. van Veldhuizen, Evolutionary Algorithms for Solving Multi-Objective Problems, second ed., Springer, 2007.
[48] A. Osyczka, Multicriteria optimization for engineering design, on Design Optimization J. Gero, Ed. Academic Press, 1985, (pp. 193-227).
[49] V. Pareto, Cours d'économie Politique, Vol. I and II, F. Rouge, Lausanne, 1896-1897.
[50] K. Miettinen, Nonlinear Multiobjective Optimization, Kluwer Academic Publishers, Boston, 1999.
[51] L. Zadeh, «Optimality and non-scalar-valued performance criteria», IEEE Trans. Automat. Cont. , vol. 8, n.o. 1, pp. 59–60, 1963.
[52] R. T. Marler and J. S. Arora, «The weighted sum method for multi-objective optimization: new insights», Struct. Multidiscip. Optim., vol. 41, n.o. 6, pp. 853–862, 2010.
[53] J. Koski, «Defectiveness of weighting method in multicriterion optimization of structures», Commun. Appl. Numer. Methods, vol. 1, n.o 6, pp. 333–337, 1985.
[54] W. Stadler, J. Dauer, Multicriteria optimization in engineering: a tutorial and survey, in: M.-P. Kamat (Ed.), on Structural Optimization: Status and Promise, American Institute of Aeronautics and Astronautics, Washington, DC, 1992, pp. 209–249.
[55] I. Das and J. E. Dennis, A closer look at drawbacks of minimizing weighted sums of objectives for Pareto set generation in multicriteria optimization problems, Struct. Optim., vol. 14, n.o 1, pp. 63–69, 1997.
[56] V. Joseph Bowman Jr., On the relationship of the tchebycheff norm and the efficient frontier of multiple-criteria objectives, in: H. Thiriez, S. Zionts (Eds.), Multiple Criteria Decision Making, Springer, Berlin, Heidelberg, 1976, pp. 76–86.
[57] S. Chandra Satapathy, A. Naik, and K. Parvathi, «High dimensional real parameter optimization with teaching learning based optimization», Int. J. Ind. Eng. Comput., vol. 3, n.o 5, pp. 807–816, 2012.
[58] R. V. Rao and V. Patel, «An elitist teaching-learning-based optimization algorithm for solving complex constrained optimization problems», Int. J. Ind. Eng. Comput., vol. 3, n.o 4, pp. 535–560, 2012.
[59] V. Toğan, Design of planar steel frames using teaching–learning based optimization, Eng. Struct. 34 (2012) 225–232.
[60] R. V. Rao and V. Patel, Multi-objective optimization of heat exchangers using a modified teaching-learning-based optimization algorithm. Appl. Math. Model., vol. 37, n.o. 3, pp. 1147–1162, 2013.
[61] R. Venkata Rao and V. Patel, Multi-objective optimization of two stage thermoelectric cooler using a modified teaching–learning-based optimization algorithm. Eng. Appl. Artif. Intell., vol, 26, n.o. 1, pp. 430–445, 2013.
[62] A. Mohamed, G. B. Jasmon, and S. Yusoff, «A static voltage collapse indicator using line stability factors», J. Ind. Technol., vol. 7, n.o 1, pp. 73–85, 1989.
[63] M.A. de Jesús Medina López, Meta-Heuristic Algorithms Based on Decomposition for Multi-Objective Optimal Power Flow Problems, Dissertation, Cinvestav-Unidad Guadalajara, Julio, 2014.

[64] Q. Zhang and H. Li, «MOEA/D: a multiobjective evolutionary algorithm based on decomposition», IEEE Trans. Evol. Comput., vol. 11, n.o. 6, pp. 712–731, 2007.

[65] A. Konak, D. W. Coit, and A. E. Smith, «Multi-objective optimization using genetic algorithms: a tutorial», Reliab. Eng. Syst. Safety, vol. 91, n.o. 9, pp. 992–1007, 2006.

[66] E. Zitzler, K. Deb, and L. Thiele, «Comparison of multiobjective evolutionary algorithms: empirical results: Evol. Comput., vol. 8, n.o 2, pp. 173–195, 2000.

[67] D.A.V. Veldhuizen, G.B. Lamont, Evolutionary computation and convergence to a Pareto front, Late Breaking Papers at the Genetic Programming 1998 Conference, Stanford University, California, 1998, pp. 221–228.

[68] K. Deb, A. Pratap, S. Agarwal, and T. Meyarivan, «A fast and elitist multiobjective genetic algorithm: NSGA-II», IEEE Trans. Evol. Comput., vol. 6, n.o 2, pp. 182–197, 2002.

[69] J.R. Schott, Fault Tolerant Design Using Single and Multicriteria Genetic Algorithm Optimization, Thesis, Massachusetts Institute of Technology, 1995.

[70] P.W. Sauer, M.A. Pai, Power System Dynamics and Stability, Prentice Hall, 1997.

[71] S. Hadi, Power System Analysis, McGraw-Hill, New York, 2004.

[72] Power System Test Case Archive, Available at:http://www.ee.washington.edu/research/pstca/pf118/pg_tca118bus.htm, 2006.

Index

Note: Page numbers followed by *f* indicate figures, *t* indicate tables, and *b* indicate boxes.

D

E

F

G

H

N

O

P

FSC
www.fsc.org
MIX
Papier | Fördert
gute Waldnutzung
FSC® C083411

Kolibri 360 GmbH
Ferdinand-Jühlke-Straße 7
99095 Erfurt, Deutschland
produktsicherheit@kolibri360.de